ADVANCED ULTRAWIDEBAND RADAR

SIGNALS, TARGETS, AND APPLICATIONS

ADVANCED ULTRAWIDEBAND RADAR

SIGNALS, TARGETS, AND APPLICATIONS

EDITED BY

JAMES D. TAYLOR

CRC Press
Taylor & Francis Group
Boca Raton   London   New York

CRC Press is an imprint of the
Taylor & Francis Group, an **informa** business

Cover Photo

Ultrawideband radar systems can operate a distances from direct contact to orbital ranges.

Left: Photograph of UWB radar measuring animal activity in biological experiments from Chapter 8. (Photo by Lesya Anishchenko, Bauman Remote Sensing Laboratory, Bauman Moscow State Technical University.)

Center: Chapters 6 and 11 show how UWB radar can inspect installed materials such as space vehicle cryogenic tank coatings and aircraft composite structural components. (NASA Photo)

Right: The MIT Lincoln Laboratory Haystack Ultrawideband Satellite Imaging Radar (HUSIR) can image satellites in space. See Chapter 1. (Permission of MIT LL)

CRC Press
Taylor & Francis Group
6000 Broken Sound Parkway NW, Suite 300
Boca Raton, FL 33487-2742

First issued in paperback 2019

© 2017 by Taylor & Francis Group, LLC
CRC Press is an imprint of Taylor & Francis Group, an Informa business

No claim to original U.S. Government works

ISBN-13: 978-0-4665-8657-4 (hbk)
ISBN-13: 978-0-367-86814-7 (pbk)

Library of Congress Cataloging-in-Publication Data

Names: Taylor, James D., 1941- editor.
Title: Advanced ultrawideband radar : signals, targets, and applications / edited by James D. Taylor.
Description: Boca Raton, FL : CRC Press, Taylor & Francis Group, 2016.
Identifiers: LCCN 2016028361| ISBN 9781466586574 (hardback) | ISBN 9781466586604 (ebook)
Subjects: LCSH: Ultra-wideband radar.
Classification: LCC TK6592.U48 A38 2016 | DDC 621.3848/5--dc23
LC record available at https://lccn.loc.gov/2016028361

Visit the Taylor & Francis Web site at
http://www.taylorandfrancis.com

and the CRC Press Web site at
http://www.crcpress.com

Contents

Preface

In late 1990, I had lunch with Dr. Michael Salour, the president of IPITEC Corp, a fiber optics company. When I described ultrawideband radar, he asked "Are there any books about it?" I replied, "No, it's too new of a subject." Michael thought for a minute and said, "Why don't you write one?" Michael's suggestion evolved into the series that started in 1995 with *Introduction to Ultra-Wideband Radar Technology,* followed by *Ultra-Wideband Radar Technology* (2000) and *Ultrawideband Radar Technology Applications and Design* in 2012.

Every time I finish writing and editing a new book, I think it will be the last one. Then I see the progress reported and start another. My special thanks to all of the writers who contributed chapters about their special research areas. Their gracious cooperation and positive responses to my editing and suggested changes greatly eased this otherwise daunting enterprise.

Advanced Ultrawideband Radar: Signals, Targets, and Applications presents the latest progress in ultrawideband (UWB) radar technology and shows new directions for future development and applications. I have edited for the practical engineer or technical manager who wants to learn about what UWB radar can do for them. The writers have shown the applications, technical issues, and solutions supported by the background theory.

For practical purposes, UWB now means all radar systems with a small spatial resolution, generally less than 30 cm (1 ft) or a signal bandwidth greater than 500 MHz. The practical uses of high resolution signals have evolved to provide greater remote sensing capabilities. Advances in integrated circuits and signal processing technology have enabled applications from short-range buried object detection to imaging satellites at geosynchronous orbit altitudes. Crime drama and history channel television enthusiasts often see detectives or archeologists using ground penetrating radar to search for evidence at crime scenes or historical sites. Future air travelers may unknowingly pass through a UWB imaging radar that searches their bodies for weapons or contraband. Highway departments use UWB radars to inspect road beds and bridges.

For the future, imagine a medical radar scanner for use in the field at primitive medical facilities such as a remote clinic or battlefield hospital. The advanced UWB radar technology presented here can expand remote sensing by adding time–frequency analysis of return signals (spectroscopy) for target identification, range measurement, and imaging.

The following chapters will show how a radar could analyze the spectrum of reflected signals. Comparing time–frequency profiles against a data base can provide a way to identify objects. Carried a step further, the radar can then change the transmitted signal spectrum to match the target for an enhanced response using correlation processing. This will open major possibilities for medical applications including internal imaging.

Chapter 1, "Introduction to Advanced Ultrawideband Radar Systems" by James D. Taylor, presents the basic ultrawideband radar concepts and shows how the field has evolved over a quarter century. As an editor, I have seen the term UWB change meaning over the years. Originally it referred to short-range impulse systems with a fractional bandwidth greater than 25% of center frequency. In 2002, the American Federal Communications Commission (FCC) expanded the definition to include systems with a bandwidth greater than 500 MHz. UWB now has a practical definition of any radar with a spatial resolution less than 30 cm (~1 ft) In this book you will see radars operating from 0.5 to 10 GHz for materials penetrating applications, and up to THz frequencies for nondestructive testing.

Applications range from measuring pavement layers to imaging satellites at geosynchronous orbits. Some vehicles now come with UWB radar systems for collision avoidance.

Chapter 2, "Advances in Short-Range Distance and Permittivity Ground-Penetrating Radar Measurements for Road Surface Surveying" by Gennadiy P. Pochanin, Sergey A. Masalov, Vadym P. Ruban, Pavlo V. Kholod, Dmitriy O. Batrakov, Angelika G. Batrakova, Liudmyla A. Varianytsia-Roshchupkina, Sergey N. Urdzik, and Oleksandr G. Pochanin, presents a major advance in precision ranging for ground penetrating radar (GPR). Road surveying presents a major problem for GPR systems because of electrical properties of the layers of concrete, asphalt, and roadbed have little difference and produce weak reflections. These researchers developed a way to measure the thickness of each pavement layer by recording and analyzing the return signal spectrum. This technique of recording and analyzing the reflected waveform has major potential benefits when applied to medical imaging, nondestructive materials testing, and other applications using multiple transmission media.

Chapter 3, "Signals, Targets, and Advanced Ultrawideband Radar Systems" by James D. Taylor, Anatoliy Boryssenko, and Elen Boryssenko, presents the concept of target and UWB signal interactions to identify objects and enhance specific target returns. Every object will reflect/reradiate electromagnetic waves differently. Radar target reflection characteristics depend on the geometry and materials that make the objects act like a set of resonators. UWB signals can act like a delta-Dirac impulse function. Probing a target with a UWB impulse can determine the target characteristics by examining the reflected signal spectrum. Analyzing the return signal with time–frequency analysis techniques can provide a distinct "profile." Comparing target return time–frequency profiles with databases can identify the object class, for example, land mines, aircraft, vehicles, and biological anomaly. Changing the transmitted signal to match the target resonance characteristics can provide a way to search for special classes of objects and suppress other returns. Terence Barrett developed this approach during the Resonance and Aspect Matched Adaptive Radar (RAMAR) demonstrations. This means the advanced UWB radar will require major advances in receiver and transmitter technology. It will need a receiver that can preserve an exact digital file of the UWB waveform with sufficient resolution and signal-to-noise ratio to permit meaningful time–frequency analysis. Fast, close to real time signal capturing of receiver waveforms can be performed digitally or in a mixed analog-digital way with initial analog transformation (time lens, analog-to-information transition) followed by an analog to digital convertor (ADC) with much relaxed requirements compared to the first digital only approach. Generation of target-specific waveforms requires a device that performs inversely to ADC, namely digital to analog converter (DAC) that does not yet exist for practical wide band and UWB. A review of several mixed analog-digital approaches concludes this chapter.

Chapter 4, "Ultrawideband (UWB) Time–Frequency Signal Processing" by Terence W. Barrett, presents the signal analysis methods needed to build a time–frequency profile of an object based on the reflected signal. It explains the signal analysis objectives and examines the methods that can provide a time–frequency profile of targets or classes of targets. Barrett has successfully demonstrated these methods to support his Resonance and Aspect Matched Adaptive Radar (RAMAR) patents.

Chapter 5, "Modeling of Ultrawideband (UWB) Impulse Scattering by Aerial and Subsurface Resonant Objects Based on Integral Equation Solving" by Oleg I. Sukharevsky, Gennady S. Zalevsky, and Vitaly A. Vasilets, presents a way to predict how an object will scatter UWB signals within the 100–300 MHz VHF spectrum to enhance the noncooperative radar identification of objects. The experimental studies mentioned in Chapter 4 describe how the use of radar resonant waveband sounding signals can stimulate second-ary radiation by objects. This chapter describes how to predict those reflected fields.

Chapter 6, "Nondestructive Testing of Aeronautics Composite Structures Using Ultrawideband Radars" by Edison Cristofani, Fabian Friederich, Marijke Vandewal, and Joachim Jonuscheit, examines materials testing with an extremely high frequency (EHF) (10–300 GHz) materials-penetrating radar. Modern aircraft use nonconducting composite materials to reduce structural weight. The inspection of dielectric materials such as ara-mid or fiberglass composites requires methods to examine them in place, which rules out X-Ray and tomographic methods. This chapter presents demonstrations of material pen-etrating radar techniques to locate manufacturing defects and parts that have deteriorated in use. Applying time–frequency analysis could provide greater NDT testing capabilities.

Chapter 7, "Modeling of UWB Radar Signals for Bioradiolocation" by Lanbo Liu, pres-ents a new approach to using UWB radar for human vital sign detection. He describes the approach of using the finite-difference time-domain (FDTD) numerical simulation approach and synthetic computational experiments to investigate the effectiveness of the ultrawideband (UWB) radar for bioradiolocation. The direct use of bioradiolocation can provide contactless monitoring of human vital signs in security surveillance, biomedical engineering applications, searching for victims under collapsed building debris caused, and other applications.

Chapter 8, "Bioradiolocation as a Technique for Remote Monitoring of Vital Signs" by Lesya Anishchenko, Timothy Bechtel, Sergey Ivashov, Maksim Alekhin, Alexander Tataraidze, and Igor Vasiliev, present their investigations into UWB radar medical applications. Bioradars provide a wide range of possibilities for remote and noncontact monitoring of the psycho-emotional state and physiological condition of many macro-organisms. They describe UWB bioradars designed at Bauman Moscow State Technical University (BMSTU), Russia, in an international collaboration with Franklin & Marshall College, Lancaster, PA. The results of their experiments demonstrate that bioradars of BioRASCAN-type may be used for simultaneous remote measurements of respiration and cardiac rhythm parameters. To demonstrate possible applications the writers describe bio-radar-assisted experiments for detection of various sleep disorders. The results show how bioradiolocation can accurately diagnose and estimate the level of obstructive sleep apnea severity. Bioradar could replace polysomnography, the current standard medical evalua-tion method that requires direct patient contact.

Chapter 9, "Noise Radar Techniques and Progress" by Ram M. Narayanan, presents an overview of the development, techniques, and capabilities of noise radar. Noise radar refers to techniques and applications that use random noise (narrowband or broadband) as the transmitted waveform. The receiver performs correlation processing, or dual spec-tral processing of radar returns for target detection and imaging. This chapter reviews the history of noise radar to show the fundamental concepts pertaining to noise radar operation. It describes the developments over the past 60 years in applications involving target detection, characterization, imaging, and tracking. It also discusses the novel sig-nal processing concepts that take advantage of recent developments in hardware realiza-tion and implementation.

Chapter 10, "Prototype UWB Radar Object Scanner and Holographic Signal Processing," has been written by Lorenzo Capineri, Timothy Bechtel, Pierluigi Falorni, Masaharu Inagaki, Sergey Ivashov, and Colin Windsor. They developed a different approach to short-range material-penetrating radar using holographic radar, which measures the phase of the reflected signal. Their demonstration holographic radars operating at 2 GHz and 4 GHz can produce useful images with a high spatial resolution over a variety of subsurface impedance contrasts. They show how using higher frequencies (>4 GHz) in subsurface imaging is sustained by the availability of integrated electronic devices operating in this frequency range. These new technologies provide improved power consumption and signal-to-noise ratios, and high resolution analog to digital converters.

Chapter 11, "Ultrawideband Sense-through-the-Wall Radar Technology" by Fauzia Ahmad, Traian Dogaru, and Moeness Amin, presents an overview of sense through the wall (STTW) radar systems. STTW systems typically perform two main functions: obtain images of the building interior (and, related to this, reconstruct the building layout); and detect and localize targets of interest (typically humans) within buildings. Humans belong to the class of animate objects characterized by motion of the torso and limbs, breathing, and heartbeat. While both aforementioned functions may be integrated in the same sensor, the underlying operating principles are somewhat different. They examine the many problems associated with the STTW radar theory and technology.

Chapter 12, "Wideband Wide Beam Motion Sensing" by François Le Chevalier, shows how widening the bandwidth of a generic radar may improve its Doppler or angular resolution. Surveillance radars must: (1) detect, track, image, and classify all targets on the fly; (2) detect difficult targets with slow speeds (a few m/s) and low radar cross section (−20 dBm²); and (3) operate in difficult environments including urban or coastal locations and high sea states. New radar designs, involving multiple simultaneous wideband signals, must improve performance against the evolving jamming and interference threat from military and civilian radio-frequency devices. This chapter will present methods for improving radar performance and developing multiple functionalities for defense and security applications.

The contributors hope that you will find some new and useful insight into the evolving field of ultrawideband radar.

About This Book

This book presents the latest theory, developments, and applications related to high-resolution material-penetrating sensor systems. An international team of expert researchers explains the problems and solutions for developing new techniques and applications. Subject areas include ultrawideband (UWB) signal propagation and scattering, material-penetrating radar techniques for small object detection and imaging, biolocation using holographic techniques, tomography, medical applications, nondestructive testing methods, electronic warfare principles, through-the-wall radar propagation effects, and target identification through measuring the target return signal spectrum changes.

Editor

"Dictionaries are like watches: the worst is better than none at all, and the best don't run true."

Samuel Johnson (1709–1784)

James D. Taylor, Lieutenant Colonel, USAF (Retired), BSEE, MSEE, PE (Retired), SM IEEE. The siren call of ultrawideband radar has drawn James D. Taylor again into the maelstrom of writing and editing. His previous books include *Introduction to Ultra-Wideband Radar Systems* (1995), *Ultra-Wideband Radar Technology* (2000), and *Ultrawideband Radar Applications and Design* (2012) for CRC Press. He thanks the CRC Press team of Ashley Gasque, Andrea Dale, and Amber Donley for their assistance and patience in the preparation of all his books.

James D. Taylor grew up in Silver Spring, Maryland. His father, Albert L. Taylor, the director of nematology investigations for the U.S. Department of Agriculture, introduced him to science through personal example. As a result, he developed an insatiable desire to know how and why things worked. His mother, Josephine S. Taylor, assisted radar engineers as a secretary and office manager at the Johns Hopkins University Applied Physics Laboratory. She inspired his interest in radar and ensured he learned mathematics, touch typing, and clear writing at an early age.

In 1963, he earned his bachelor's degree in electrical engineering at the Virginia Military Institute and entered the U.S. Army as an air defense artillery officer. Two tours at the U.S. Army Air Defense School provided his practical and theoretical background in radar and guided missile systems. After assignments in Germany and the 101st Airborne Division in America, he transferred to the U.S. Air Force as an electronics research and development engineer in 1968.

His Air Force career started as a rocket sled test project engineer for the Central Inertial Guidance Test Facility at Holloman AFB, New Mexico. In 1977, he earned his master's degree in control theory at the Air Force Institute of Technology in Wright-Patterson AFB, Ohio. His tour as a staff engineer at the Air Force Avionics Laboratory exposed him to advanced technology and military requirements. His final assignment as director of long range technology planning in the U.S. Air Force Electronic Systems Division at Hanscom AFB, Massachusetts, led to a search for solutions to long range cruise missile detection and the new field of ultrawideband radar.

After retiring from active duty in 1991, he started consulting and writing about ultrawideband radar. To show the military possibilities, he wrote the novel *Signal Chase*. He gave short courses in Scotland, Italy, and Russia. He served as the Chairman of the IEEE Ultrawideband Radar Committee and edited the IEEE STD 1672 *Ultrawideband Radar Definitions*. In 2012, he moved to Ponte Vedra, Florida, and lives as a gentleman engineer, consultant, and editor.

"If I write a book: an illiterate condemns me; a sophomore disputes me; a parish priest accuses me of heresy; and my wine merchant cuts off my credit. Each night I pray, Dear God, deliver me from the itch to write books."

Voltaire (1694–1778)

Contributors

Fauzia Ahmad
Department of Electrical and
　　Computer Engineering
College of Engineering
Temple University
Philadelphia, Pennsylvania

Maksim Alekhin
Remote Sensing Laboratory
Bauman Moscow State Technical
　　University
Moscow, Russia

Moeness Amin
Radar Imaging Laboratory
Center for Advanced Communications
College of Engineering
Villanova University
Villanova, Pennsylvania

Lesya Anishchenko
Remote Sensing Laboratory
Bauman Moscow State Technical
　　University
Moscow, Russia

Terence W. Barrett
BSEI
Victor, New York

Angelika G. Batrakova
Kharkiv National Automobile and
　　Highway University
Department of Survey and Engineering
　　of Highways and Airfields
Kharkiv, Ukraine

Dmitriy O. Batrakov
Department of Theoretical Radiophysics
School of Radiophysics, Biomedical
　　Electronics and Computer Systems
V.N. Karazin Kharkiv National University
Kharkiv, Ukraine

Timothy Bechtel
Department of Earth and
　　Environmental Sciences
University of Pennsylvania
Philadelphia, Pennsylvania

Anatoliy Boryssenko
A&E Partnership
Belchertown, Massachusetts

Elen Boryssenko
A&E Partnership
Belchertown, Massachusetts

Lorenzo Capineri
Ultrasound and Non Destructive Testing
　　Laboratory
Department of Information Engineering
University of Florence
Firenze, Italy

Edison Cristofani
CISS Department, Royal Military Academy
Brussels, Belgium

Traian Dogaru
U.S. Army Research Lab
Adelphi, Maryland

Pierluigi Falorni
Ultrasound and Non Destructive Testing
　　Laboratory
Department of Information Engineering
University of Florence
Firenze, Italy

Fabian Friederich
Electronic Terahertz Measurement
　　Techniques
Fraunhofer Institute for Physical
　　Measurement Techniques
Kaiserslautern, Germany

Masaharu Inagaki
Walnut Ltd.
Tachikawa, Japan

Sergey Ivashov
Remote Sensing Laboratory
Bauman Moscow State Technical
    University
Moscow, Russia

Joachim Jonuscheit
Materials Characterization and Testing
Fraunhofer Institute for Physical
    Measurement Techniques
Kaiserslautern, Germany

Pavlo V. Kholod
O. Ya. Usikov Institute for Radiophysics
    and Electronics of the National
    Academy of Sciences of Ukraine
Kharkov, Ukraine

François Le Chevalier
Thales Land & Air Systems
and
Delft University of Technology
Bourg-La-Reine, France

Lanbo Liu
Department of Civil & Environmental
    Engineering
University of Connecticut
Storrs, Connecticut

Sergey A. Masalov
O. Ya. Usikov Institute for Radiophysics
    and Electronics of the National
    Academy of Sciences of Ukraine
Kharkov, Ukraine

Ram M. Narayanan
School of Electrical Engineering and
    Computer Science
The Pennsylvania State University
University Park, Pennsylvania

Gennadiy P. Pochanin
O. Ya. Usikov Institute for Radiophysics
    and Electronics of the National
    Academy of Sciences of Ukraine
Kharkov, Ukraine

Oleksandr G. Pochanin
O. Ya. Usikov Institute for Radiophysics
    and Electronics of the National
    Academy of Sciences of Ukraine
Kharkov, Ukraine

Vadym P. Ruban
O. Ya. Usikov Institute for Radiophysics
    and Electronics of the National
    Academy of Sciences of Ukraine
Kharkov, Ukraine

Oleg I. Sukharevsky
Air Force Research Center
Kharkiv Kozhedub University of Air
    Forces
Kharkiv, Ukraine

Alexander Tataraidze
Remote Sensing Laboratory
Bauman Moscow State Technical
    University
Moscow, Russia

Sergey N. Urdzik
Kharkiv National Automobile and
    Highway University
Kharkiv, Ukraine

Marijke Vandewal
CISS Department, Royal Military Academy
Brussels, Belgium

Liudmyla A. Varianytsia-Roshchupkina
O. Ya. Usikov Institute for Radiophysics
    and Electronics of the National
    Academy of Sciences of Ukraine
Kharkov, Ukraine

Vitaly A. Vasilets
Air Force Research Center
Kharkiv Kozhedub University of Air
    Forces
Kharkiv, Ukraine

Igor Vasiliev
Remote Sensing Laboratory
Bauman Moscow State Technical
    University
Moscow, Russia

Colin Windsor
Tokamak Energy, Co.
East Hagbourne, United Kingdom

Gennady S. Zalevsky
Air Force Research Center
Kharkiv Kozhedub University of Air
    Forces
Kharkiv, Ukraine

1

Introduction to Advanced Ultrawideband Radar Systems

James D. Taylor

CONTENTS

1.1 Introduction

Ultrawideband (UWB) radar systems give the user X-ray vision into the ground, solid materials, and through walls. Special purpose systems can detect small movements to remotely measure human vital signs in hospitals and hazardous environments. Large arrays of UWB radars can provide real-time imaging to search for concealed weapons on people passing through a security choke point. Special airborne sets can provide precision terrain maps and locate targets concealed by foliage. The U.S. Air Force Space Command's (AFSPC) Haystack UWB Satellite Imaging Radar (HUSIR) can make high-resolution images of satellites at synchronous orbit ranges under all weather conditions.

For practical purposes UWB radar means radiodetection and ranging systems using signal bandwidths greater than 500 MHz to measure distances with spatial resolutions $\Delta r < 30$ cm. Combining the ability to determine ranges with a fine resolution and the material's penetrating ability of electromagnetic waves can provide a wide range of remote-sensing capabilities.

Until recently, UWB radar generally implied short-range devices using *impulse* or *nonsinusoidal* wave signals in the range of 0.5–10 GHz. The GPR shown in crime dramas and archeological investigations typifies the conventional impulse signal approach.

UWB radars now work in all parts of the microwave radiofrequency (RF) spectrum using signals such as continuous wave frequency modulation (CWFM), random, and pseudorandom noise concepts based on conventional technology. Vehicle UWB radars have reserved frequency bands at 24 GHz and 77 GHz using both CWFM and impulse formats. The U.S. Air Force (USAF) HUSIR satellite imaging radar operates between 92 and 100 GHz to exploit a gap in atmospheric absorption. Chapter 6 describes nondestructive materials testing radars operating at extremely high frequencies signal frequencies from 30 to 300 GHz.

Current UWB radars measure distances by detecting reflected signals with matched filters based on the transmitted signal format. When a UWB signal encounters an object, the interaction of the surface geometry and materials changes the spectrum of the reflected waveform. Large changes of the signal spectrum will produce a weaker output from the matched filter, for example, different targets will appear more or less stronger than the others. In the past, designers selected a signal bandwidth and center frequency known to produce useful results for a given type of remote sensing. This has worked well and produced useful practical systems.

The next generation of advanced UWB radars will record the reflected waveforms from targets at different ranges. Receivers will apply time–frequency return signal analysis to identify specific target classes. An advanced UWB radar will use the target time–frequency profiles to match the radar signal to a specific target class. The matched signal will enhance the reflected energy from the selected target class and suppress other returns in the matched filtering process. The materials presented in this book can help build the next advanced UWB radar [1–5].

This chapter presents an introduction and overview to UWB radar system principles for the unfamiliar reader. Radar engineers and technical managers will find a summary of UWB radar concepts and potential future applications.

1.2 Ultrawideband Radar Applications

UWB radar systems work at ranges from under less than 1 m for material-penetrating applications out to geosynchronous orbit ranges close to 10^5 km for satellite imaging. The following examples show the scope and versatility of UWB radar applications from close ranges to space.

1.2.1 *X-Ray* Vision with Material-Penetrating Radars

Material-penetrating radars (MPRs) transmit UWB signals in the 0.5–10 GHz spectrum to penetrate dielectric materials such as soil, stone, concrete walls, structures, and composite materials. The wide signal bandwidth provides the small-range resolution measurements needed at short distances and the wavelengths allow signals to pass through solid media and reflect from objects inside. These systems can find hidden objects with a strong electrical contrast to the surrounding media such as rebars, pipes, and conduits. Special systems called ground-penetrating radar (GPR) can locate soil disturbances to find buried objects. All MPR applications depend on the electrical contrast of the target and surrounding media. Chapter 2 will discuss this in detail. [1–4,6–8] Figures 1.1 and 1.2 show examples of MPR at work.

GPR has become an important geophysical, forensic, and archeological tool for locating buried objects and measuring subsurface ground conditions. In 2012, University of Leicester archeologists and historians used GPR to locate the remains of England's King Richard III. After his death at the battle of Bosworth, Richard's followers buried his body in a Leicester, England churchyard cemetery. Five centuries later, the church had disappeared and the cemetery was put to practical use as a supermarket parking lot. In 2012, a team of archeologists and historians used a GPR survey to locate the graves under the asphalt pavement. Excavation found Richard's skeleton with his distinctive curved spine. DNA comparisons with known descendants positively verified the identity. King Richard III received proper monarch's funeral in the March of 2015 [9].

Another group of archeologists made GPR surveys around the prehistoric Stonehenge monument in England. They combined GPR, magnetometer, and 3D laser scanning to discover the remains of buildings, tombs, and other activities surrounding the megalith

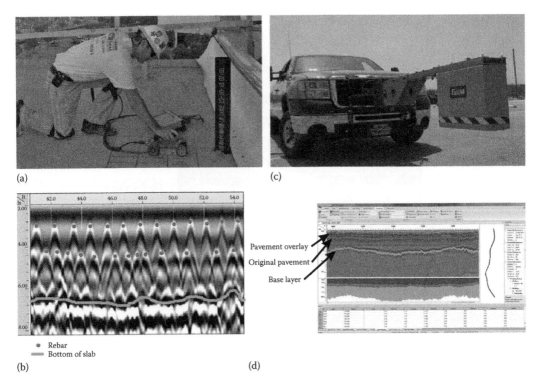

FIGURE 1.1
Examples showing material-penetrating radars can image inside dielectric materials. Performance depends on the electrical contrast between objects and the surrounding media. (a) GSSI handheld StructureScan™ standard radar, (b) image showing rebars in concrete, (c) GSSI RoadScan™ radar for road surface and bridge inspections, and (d) RoadScan™ radar image. (From GSSI, Complete GPR system for concrete inspection and analysis: StructureScan, 2015. http://www.geophysical.com/Documentation/Brochures/GSSI-StructureScanStandardBrochure.pdf [accessed September 21, 2016]; From GSSI, Complete GPR system for road inspection and analysis: RoadScan 30, 2015. http://www.geophysical.com/Documentation/Brochures/GSSI-RoadScan30Brochure.pdf [accessed September 21, 2016]. With permission.)

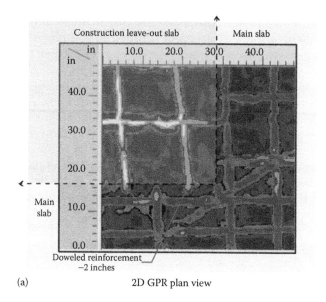

(a)   2D GPR plan view

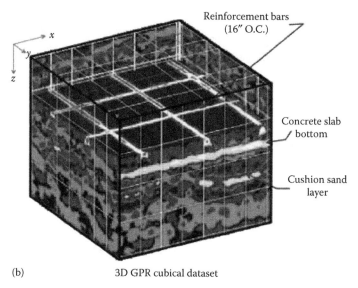

(b)   3D GPR cubical dataset

FIGURE 1.2
GPR data processing can produce (a) 2D and (b) 3D survey images showing the location of rebars, concrete slab, and the substructure of a building. (From Gehrig Inc., Gehrig state of the art noninvasive geophysical site investigations, 2014. http://gehriginc.com/ [accessed September 21, 2016]. With permission.)

stone circle. Normally this process would have required digging extensive exploratory trenches around the monument [10].

Modern highway departments use UWB radar to inspect the condition of road surfaces and bridges as shown in Figure 1.1b. Railroads use UWB radar to inspect the condition of road beds. Mining companies use GPR to inspect their mine shafts and locate mineral deposits. Military organizations use it to search for unexploded ordnance, arms caches, and mines [11]. Chapter 2 presents improvements in GPR technology for measuring pavement layer thickness.

1.2.2 Medical Remote Patient Monitoring

In 1994, Tom McEwan, the inventor of the micropower impulse radar (MIR), observed the UWB radar range display on his test bench responding to his heart rate. This stimulated engineers of the Lawrence Livermore National Laboratory to develop radar systems to remotely measure heart and respiration rates. In related applications, UWB radars can search for people trapped under avalanches, landslides, and collapsed buildings as described in Chapters 7 and 8 [12,13].

In 2008, Immoreev and Tao demonstrated how UWB patient monitoring could produce vital sign measurements comparable with conventional contact methods [14]. In 2011, Sensiotec Inc., Atlanta, GA developed the Virtual Medical Assistant® (VMA) UWB radar system for monitoring patient's vital signs. Figure 1.3a shows how the VMA fits under the patient's mattress and sends information to central stations and medical caregivers. The U.S. Food and Drug Administration (FDA) has approved the VMA for hospital use [15].

Other potential UWB radar medical applications include internal imaging and tumor detection without the use of ionizing radiation as shown in Figure 1.3b. Malignant growths and breast cancer tumors have a strong electrical contrast with the surrounding tissue, which suggests the possibility of using radar scanning to detect them. Biological tissue electrical characteristics (permittivity, permeability, and characteristic impedance) set limits on detecting tumors and radar imaging inside the body tissue. This suggests a potential

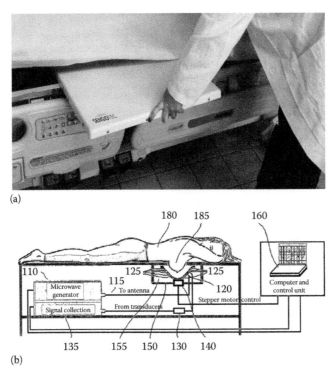

(a)

(b)

FIGURE 1.3
Medical applications of UWB radar. (a) The Sensiotec Virtual Medical Assistant® fits under a mattress and continuously measures heart rate and respiration. (Courtesy of Arkin, R., Virtual medical assistant, Sensiotec™, Atlanta, GA, 2014. http://sensiotec.com/wp/wp-content/uploads/2014/05/Brochure.pdf [accessed September 21, 2016]. With permission.) (b) UWB radar imaging of breast tissue for tumor detection and imaging. (From Li, J., Multi-frequency microwave-induced thermoacoustic imaging of biological tissue, US Patent 7,266,407 B2, September 4, 2007.)

use for reflected signal analysis and target signal matching described earlier. Figure 1.3b shows one of many concepts for breast tumor detection using UWB radar. Radar could eliminate any hazards associated with ionizing radiation [16,17].

1.2.3 Radar Safety Systems for Vehicles

Many recent automobiles and vehicles use video cameras to provide forward collision warning (FCW) and lane departure warnings (LDW). These systems process video images and alert the driver when the car starts to move out of the traffic lane or approaches another vehicle or obstacle. Although a great help and improvement, these optical systems depend on clear weather conditions to image the road ahead. The next vehicle safety evolution will use a suite of optical, acoustic sensors, and UWB radars to provide complete 360° coverage, and help the driver avoid collisions under all visibility conditions.

Research and experiments with vehicle radar started in the 1970s, but the size of components and expense limited applications. Recent advances in integrated circuits, antennas, and signal processing have provided a solution in the form of electronically scanned automotive radars. Future vehicles could use UWB frequency-modulated continuous-wave (FMCW) and impulse radar sensors to give the driver full coverage and alerts to dangerous conditions. Long- range narrow beam radars operating at 77 GHz can detect vehicles and obstacles up to 200 m away. These can provide automatic cruise control inputs to maintain a safe interval between other vehicles. Broad beam short-range radars in the 24/26 GHz and 77 GHz bands can provide 30 m range coverage for precrash warning, lane change assist, and stop-and-go assistance. Figure 1.4a shows a typical 24 GHz and 77 GHz FMCW radar systems architecture. Figure 1.4b shows how each radar can watch a different area and can warn the driver of hazards such as overtaking vehicles, obstacles to the front, and vehicles approaching from the side. Impulse UWB radars could solve many of the target detection and clutter suppression problems of broad-beam short-range systems. Kajiwara gives an excellent introduction to automotive radar technology [18]. Table 1.1 shows the characteristics of several automotive radar systems [19,20].

1.2.4 Imaging Radar for Concealed Weapons Detection

Security and police organizations must quickly and reliably search people for concealed weapons and contraband in public places such as airports and public events. Current basic search methods include visual inspections, hand pat-downs, frisking, and scanning a metal detector over each person. More sophisticated methods include thermal, X-ray, and microwave imaging systems that can only search one person at a time. This process creates major problems and delays in places such as airports and other secure facilities.

Camero-Tech of Israel developed and demonstrated a real-time radar imaging system for a concealed weapons detection (CWD) system. The CWD radar can search people as they pass through a corridor or other choke points. It uses an array of 142 UWB transmitter and receiver antennas to scan people as they walk past the hidden radar. The eight frames per second real-time imaging capability provides detailed pictures of the person from the skin outward and reveal any objects carried under their clothing or accessories. Adding special image processing software can automatically search for weapons and alert security agents. Figure 1.5a shows how two radars placed at a turn could provide complete body imaging. Figure 1.5b shows the use of reflecting plates to achieve full body imaging. Figure 1.5c shows the CWD imaging radar developed in 2010 and undergoing further refinement at the time of publication [21,22].

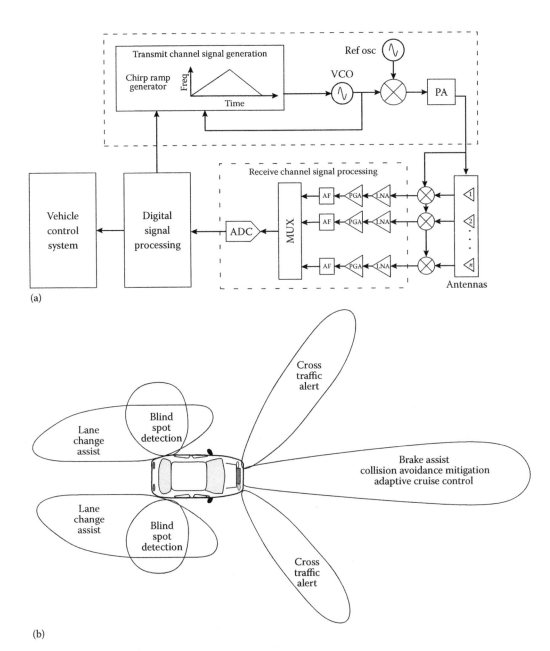

(a)

(b)

FIGURE 1.4
UWB radars can provide warning and control sensors for vehicles. (a) Typical vehicle radar architecture. (b) Radar coverage can warn the driver of dangers that take control of the vehicle to prevent collisions. (From Kajiwara, A., Ultra-Wideband automotive radar, *Advances in Vehicular Networking Technologies*, 2011. http://www. intechopen.com/books/advances-in-vehicular-networking-technologies/ultra-wideband-automotiveradar [accessed September 21, 2016]; From Autoliv Inc., Autoliv offers a wide range of industry leading radar sensors that provide a variety of active safety features. http://www.autoliv.com/ProductsAndInnovations/ ActiveSafetySystems/Pages/RadarSystems.aspx/ [accessed September 21, 2016].)

TABLE 1.1

Some Typical Automotive UWB Radar Characteristics

Radar Model	Frequency GHz	Range (m) Min/Mid/Long	Accuracy Range (m) Speed (m/s) Angle (deg)	Field of View Horizontal-Mid/Long (deg) Vertical (deg)
Delphi-ESR (Electronically Scanned Radars)	76.5	1/60/174	0.5 m 1.2 m/s 0.5°	+/−10 / +/−45 Horizontal 4.2–4.75 Vertical
Delphi SRR-2 (Side Rear Radar)	76.5	0.5 Min 80 Max	50 m closing 10 m opening	+/−75
Delphi SMS URR (Smart Micro Automotive Radar)	24	1 Min 160 Max	+/−2.5% or +/−2.5 m	+/−18 Horizontal +/−4 Vertical

Source: AutonomouStuff, Radar specification comparison chart, 2015. http://www.autonomoustuff.com/uploads/9/6/0/5/9605198/radar_comp_chart_for_web.pdf.

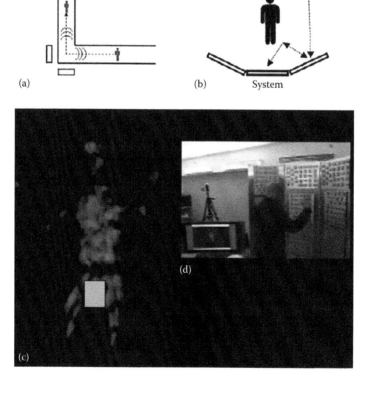

(a)   (b)   System

(c)   (d)

FIGURE 1.5

The Camero-Tech Real-time UWB radar-imaging system for concealed weapons detection. (a) Two radars can scan all people passing through a security point. (b) Adding a metal reflector can provide a 360° view of the subject for one radar. (c) The 192 array of 3–10 GHz transmitters and receivers provides the information to form images in real time. (d) Radar antenna array and subject. Automatic searches could find concealed weapons or contraband. (From Camero-Tech Ltd, Milestone in the development of Camero's ultra wide band concealed weapon detection (CWD), whole body imaging project, December. 23, 2010. http://www.camero-tech.com/news_item.php?ID=23 [accessed September 21, 2016]. With permission.)

1.2.5 Sense-Through-the-Wall Radar Systems

Sense-through-the-wall (STTW) UWB radars can show security agents what lies behind a wall and inside a room. As shown earlier in the material-penetrating discussion, a radar signal can pass through dielectric materials such as wood, drywall, brick, stone, and concrete. Taking the idea one step further, we know signals can pass through a wall, reflect from objects, and return through the wall to a receiver. With sufficient signal integration and processing, the radar can build an image showing people and inanimate objects.

Several companies have developed sense-through-the-wall radar systems to observe activities within buildings or rooms. Camero-Tech developed the Xaver™ series of through-the-wall radars that can show the location of fixed objects and track the movement of people behind concrete and stone walls. The most advanced Xaver™ 800 model shown in Figure 1.6a can look through-the-wall and give the operator a 3D image of people and objects. Figure 1.6b shows the Xaver™ 800 through-the-wall radar and typical imagery. The Xaver™ 100 one hand system can detect the presence of people behind a wall or door [23].

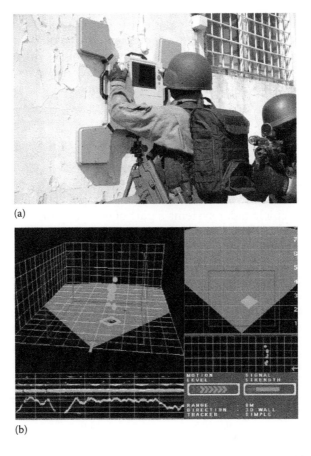

(a)

(b)

FIGURE 1.6
The Camero-Tech UWB Xaver™ 800 through-the-wall radar uses two sophisticated signal-processing technology to observe activity behind concrete walls. (a) The Xaver™ 800 in action. (b) The operators display shows the location and movement of people and location of stationary objects concealed behind walls. (From Camero-Tech Ltd, Xaver™ 800 high performance ISR portable through-wall imaging system, 2015. http://i-hls.com/wp-content/uploads/2013/11/Xavier-800-lo.jpg [accessed September 21, 2016]. With permission.)

Both the Camero-Tech CWD radar imaging and through-the-wall systems have a problem with weak reflected signals. Camero-Tech engineers solved the signal attenuation and multipath problem by collecting multiple weak radar returns in range bins for integration, signal-to-noise ratio improvement, and processing into images. Taylor et al. presented a technical explanation of the Xaver™ signal-processing algorithms in [24]. Ahmad presents the basics of sense-through-the-wall radar effects in Chapter 11, "Ultrawideband Sense-through-the-Wall Radar Technology" and discusses the many sources of signal attenuation and distortion that can occur.

1.2.6 Long-Range UWB Radar Imaging

The U.S. AFSPC must track and identify objects in space. In the 1960s, the AFSPC and the Massachusetts Institute of Technology-Lincoln Laboratory (MIT-LL) started developing UWB radars for satellite tracking and imaging under all weather conditions. Table 1.2 shows the history of AFSPC UWB radars developed for satellite imaging. Each new generation satellite imaging radar has decreased the spatial resolution and used higher operating frequencies [25].

Figure 1.7a shows the MIT-LL HUSIR radar inside the protective dome. The HUSIR radiates a 92–100 GHz signal to exploit a low attenuation portion of the atmosphere. A 35.5 m (120 ft.) diameter antenna enables the HUSIR to image satellites in geosynchronous orbits at 35,786 km (22,236 mi) above sea level. Figure 1.7b shows examples of simulated images of a satellite taken with a 30 cm (0.5 GHz) resolution and a 0.188 cm (8 GHz) resolution signals [25,26].

1.2.7 Future UWB Radar Capabilities and Applications

UWB radar system applications range from detecting buried objects to imaging satellites in space. Improvements in circuit and device technology provide systems working at frequency ranges from 0.5 to 300 GHz. Decreasing component costs will make practical applications such as vehicle radar systems a part of everyday life. The following sections will describe the basics of UWB radar technology.

Future UWB radar technology developments will include digital receivers to record the received waveform and determine the changes caused by the target. This will open new possibilities for target identification and enhancing the signals reflected by low contrast objects. Chapter 2, *Advances in Short-Range Distance and Permittivity Ground-Penetrating Radar Measurements for Road Surface Surveying* presents an approach to digital signal collection and

TABLE 1.2

History of USAF Space Command Ultrawideband Satellite Imaging Radars

Date	Radar	Frequency Range (GHz)	Bandwidth (GHz)	Spatial Resolution (cm)
1970	ALCOR C band	5.4–5.9	0.5	30
1978	LRIR X Band	9.5–10.5	1.0	15
1993	HAX Ku Band	15.7–17.7	2.0	7.5
2010	MMW Ka Band	33–37 GHz	4.0	3.75
2014	HUSIR W-band	92–100 GHz	8	1.88

Source: (From MIT LL, Haystack ultrawideband satellite imaging radar, MIT Lincoln laboratory tech notes, September, 2014; Czerwinski, M.G. et al., Development of the haystack ultrawideband satellite imaging radar, *Lincoln Laboratory Journal*, 21. With permission.)

(a)

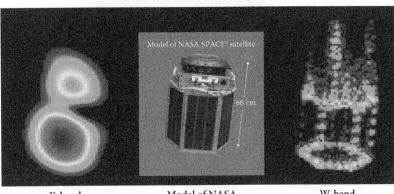

(b)

| X-band
1 GHz bandwidth
Uclassified compact
range data | Model of NASA
satellite
(Small payload access to
space experiment) | W-band
8 GHz bandwidth
Uclassified compact
range data |

FIGURE 1.7
The USAF HUSIR is part of the space surveillance network. It can image satellites at 35,786 km (22,236 mi) altitudes. (a) The 35.5 m diameter HUSIR antenna provides a very fine beam and pointing accuracy for the W-band (92–100 GHz) signal. (b) Simulated radar images of a 66 cm long satellite based on NASA data. The left and right hand images illustrate the difference in resolution between 1 and 8 GHz bandwidth signals. (From MIT LL, Haystack ultrawideband satellite imaging radar, MIT Lincoln laboratory tech notes, September, 2014; From Czerwinski, M.G. et al, Development of the haystack ultrawideband satellite imaging radar, *Lincoln Laboratory Journal*, 21. Reprinted with permission courtesy of MIT Lincoln Laboratory, Lexington, MA.)

processing which can provide new capabilities for materials penetrating, STTW, nondestructive testing and medical radar applications. Chapter 3 *Targets, Signals, and Ultrawideband Radar* presents an intuitive approach to how targets uniquely change UWB signals. Receiving and preserving the reflected signal to determine spectrum changes presents major technical challenges. Carried one step further, matching the signal to the target can substantially enhance the target returns from specific classes of objects by matching the correlation filter to the specific signal characteristics. Chapter 4 *Ultrawideband (UWB) Time–Frequency Signal Processing* will present analytical methods for identifying target signal characteristics.

1.3 UWB Radar Technology

This section presents a review of UWB radar technology including history, operating concepts, and technical definitions.

1.3.1 Ultrawideband Radar History

Starting in the 1960s many researchers developed UWB signal radar and communications systems under such names as impulse, video pulse, base-band, noise, pseudorandom noise, nonsinusoidal, wideband, and so on. Barrett shows how the development of communication systems and radars goes back at least half a century. When the terms *impulse* and *nonsinusoidal* radar created confusion, then the term UWB appeared in about 1988 to describe signals with large fractional bandwidths.

The 1990 Defense Advanced Research Projects Agency (DARPA) review board gave the term *UWB* official recognition [27,28]. The DARPA report reflected the first UWB definition, that is, a signal with a fractional bandwidth b_f greater than 25% of the center frequency [28]. This meant the signal absolute bandwidth b divided by the signal center frequency f_c, gives Equation 1.1

$$b_f = \frac{b}{f_c} = \frac{2(f_h - f_l)}{(f_h + f_l)} \tag{1.1}$$

where, f_h and f_l indicate the upper and lower frequencies of interest [29]. Since 2002, UWB has acquired several formal definitions from government regulatory organizations based on the signal absolute or fractional 10 dB bandwidth.

From 1990 to 2012, many people associated the term UWB with short-range *impulse radars* operating in the 0.5–10 GHz region. The fractional bandwidth definition became meaningless as higher frequency system use and conventional waveforms such as FMCW, linear FM modulation, random, and pseudorandom noise systems operating at higher frequency bands appeared.

In 2002, the American Federal Communications Commission (FCC) solicited comments and held public hearings regarding UWB technology. The resulting regulations for the spectra and radiated power limits on unlicensed UWB devices opened the field to commercial development by setting design goals. The FCC regulations defined UWB devices as having a bandwidth: (a) greater than 20% of center frequency at the 10 dB power point or (b) greater than 500 MHz at any center frequency. The European Union and other countries have issued similar rules regarding the spectra and power limits for unlicensed devices [30].

In 2006, the Ultrawideband Radar Committee of the IEEE AES Society published *The IEEE Std™ 1762 for Ultrawideband Radar Definitions* reflecting both the DARPA and FCC definitions [30].

Table 1.3 summarizes the different UWB definitions in terms of fractional and absolute bandwidth. (Recently many countries have started issuing regulations for UWB devices. If you plan to manufacture and sell UWB radars, check for the most recent local rules.)

In twenty-five years of writing about UWB radar, I cannot recall any case where somebody specified the fractional bandwidth of a UWB radar system. For practical purposes, we can say UWB means a radar system with spatial resolutions of the order of 30 cm (~1 ft.) or less [31,32].

TABLE 1.3

Ultrawideband Signal Definitions

Source	Signal Fractional Bandwidth	Absolute Bandwidth
DARPA 1990 [6]	>25% of center frequency	
FCC 2003	>20% of center frequency	>500 MHz
EU 2006		>50 MHz

Source: Ultrawideband Radar Committee of the IEEE AES society, *IEEE Std™ IEEE standard for ultrawideband radar definitions,* 2006; Taylor, J.D., Ch. 4, American and European regulations on ultrawideband systems, Taylor, J.D. (ed.), *Ultrawideband Radar Applications and Design,* CRC Press, Boca Raton, FL, 2012.

Figure 1.8 shows the basic concept behind short signal duration *impulse* UWB radar systems, where the signal waveform determines the system bandwidth and spatial resolution. Typical UWB waveforms have nonsinusoidal shapes such as impulses, short duration (less than 5 cycles) sinusoidal, random noise, pseudorandom noise, and frequency swept pulses.

Short-range UWB radar systems generally use single-impulse signals because of the relative simplicity of generating them. The range resolution depends on the *signal duration* as shown in Figure 1.9a for impulse systems, or the *autocorrelation time* of a particular signal waveform.

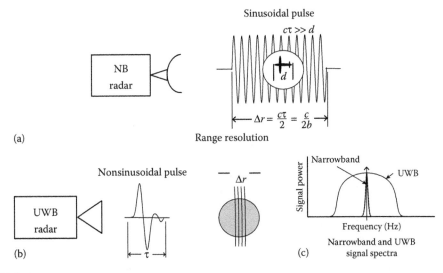

FIGURE 1.8

Comparison of narrowband (conventional) and UWB radar systems. (a) Conventional radars generally have a spatial resolution much greater than the target size. (b) UWB systems may transmit a nonsinusoidal (impulse, noise, linear FM, etc.) waveforms to achieve a short spatial resolution of the same general size, or smaller than the target. This produces multiple returns from each distance. (c) Comparison of narrowband and UWB signal spectra.

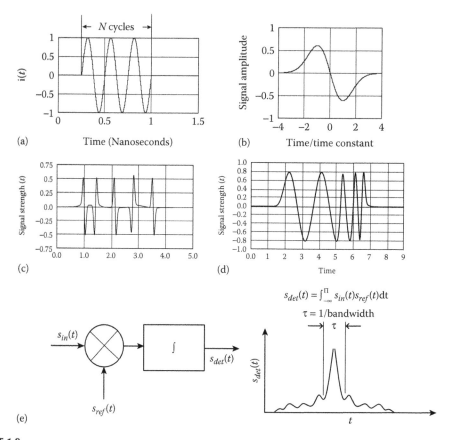

FIGURE 1.9

Wideband signals and autocorrelation detection. (a) A sinusoidal wave with less than five cycles will have an UWB width. (b) Exciting an antenna with a Gaussian-shaped DC pulse will produce a radiated Gaussian pulse. (c) A pulse train built of Gaussian pulses can provide high energy levels on target while maintaining a small-range resolution. (d) Other waveforms such as FM and random noise signals can also provide a small resolution with a high energy. (e) Autocorrelation detection produces a voltage spike of duration.

Radar developers often model impulse signals as *Gaussian* pulses generated by applying a Gaussian-shaped DC pulse to an antenna as shown in Figure 1.9b. This makes an easy analytical model for simulations and computations of waveforms and signal spectra. Impulse signal systems operate at short ranges (generally less than 50 m) because of the limited signal energy and high receiver noise levels. Legal restrictions on emitted signal power limit the signal energy of *unlicensed* devices such as GPRs, communications links, radio stethoscopes, and area surveillance systems.

Licensed and special purpose systems requiring longer ranges can use powerful pulses or long-duration nonsinusoidal signals and autocorrelation detection to detect targets as shown in Figure 1.9c–e. For example, the CARABAS airborne radar system used very high-frequency stepped frequency signals to get high quality UWB synthetic aperture radar images [3,33].

In the case of long-duration UWB signals needed for long-range applications such as the HUSIR, the receiver uses a process called correlation that compares the received signal with the transmitted signal. The UWB literature uses the terms autocorrelation, cross correlation, or matched filtering for this process [31,32].

1.3.2 Radar Bandwidth and Range Resolution

Pulse radar theory shows how the minimum spatial resolution depends on the signal pulse duration τ and the speed of light c in the particular medium. For radar analysis purposes, we can approximate the 3 dB signal bandwidth as $b = 1/\tau$, where τ indicates the pulse duration in seconds. This gives range resolution as a function of the signal bandwidth or pulse duration as

$$\Delta r = \frac{c\tau}{2} = \frac{c}{2b} \tag{1.2}$$

Note that the bandwidth b of the range resolution of Equation 1.2 applies to all radar signals. For signal formats detected by autocorrelation such as frequency swept pulse, step frequency modulation, pseudorandom and random noise, and phase shift pulse trains, the range resolution depends on the autocorrelation interval τ as shown in Figure 1.9 [32,34].

1.3.3 Unlicensed UWB Device Emission Limits

In the years before 1990, many communications and frequency allocation experts complained that UWB radar would jam every nearby electronic device. This opposition stifled commercial development of UWB radars. In 2000, the American FCC asked for formal comments from the UWB community in order to establish definitions and emission spectrum limits. The public review process resulted in FCC02-48. *Revision of part 15 of the commission's rules regarding ultra-wideband transmission systems*, which appeared in 2002. This document set emission spectrum limits for *unlicensed* UWB devices and promoted the commercial development of UWB technology [35].

The FCC regulations for *unlicensed* UWB devices set specific limits for emissions in parts of the widely used spectrum between 960 MHz and 10.6 GHz. Reference [29] provides full details regarding frequencies, emitted power levels, and measurement methods.

In 2006, the Electronic Communications Committee – European Conference of Postal and Telecommunications Administrations (ECC-CEPT) published rules for unlicensed UWB systems operating in Europe [36,37].

For *unlicensed* commercial applications, the U.S. FCC and EU ECC-CEPT have set emission power limits to avoid interference with signals in widely used spectrum regions from 960 MHz to 10,600 MHz, which includes cell phones, GPS, and Wi-Fi signals. As of 2014, the United States, United Kingdom, Canada, New Zealand, Australia, China, Japan, Singapore, Korea, and the International Telecommunication Union Radiocommunication Sector (ITU-R) have issued guidelines for unlicensed UWB devices. Each organization prescribes emission profiles similar to the one shown in Figure 1.10. As these emission profiles require careful interpretation, the UWB device developer must carefully read the descriptions of the measurement procedures for determining compliance. For example, the FCC requires measurement of the emitted radiation with a 1 MHz bandwidth filter at a specific distance. Developers need to ensure that their devices comply with the emission limits. References [38–40] summarize UWB emission limits for the United States, United Kingdom, Germany, Australia, and Canada. Other countries may follow with their own rules. The proliferation of governmental UWB device emission limits could make device compliance certification a lucrative business.

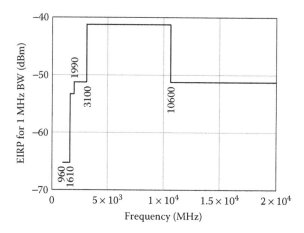

GPR and WIR radiated emissions shall not exceed the following average limits when measured with a resolution bandwidth of 1 MHZ.

Frequency (MHz)	EIRP (dBm)
960–1610	−65.3
1610–1990	−53.3
1990–3100	−51.3
3100–10600	−41.3
Above 10600	−51.3

GPR and WIR radiated emissions may not exceed these average limits when measured with a resolution bandwidth of 1 kHz.

Frequency (MHz)	EIRP (dBm)
1164–1240	−75.3
1559–1610	−75.3

FIGURE 1.10
Example U.S. FCC unlicensed UWB device effective isotropic radiated power (EIRP) radiation limits. Many countries have published legal limits for UWB devices, which follow this general format. Notice the specific instructions regarding measurement filters resolution. Developers should check the rules for the region where they intend to sell UWB devices. (From Federal communications commission [FCC] 02-48, Revision of part 15 of the commission's rules regarding ultra-wideband transmission systems, ET Docket 98–153 [released April 22, 2002].)

1.4 Conclusions

UWB radar has become an established remote-sensing technology with applications ranging from short-range X-ray vision to imaging satellites in space. The work presented in this book can help UWB radar evolve into applications for enhanced materials penetration and microwave spectroscopy.

Current radar sets treat received signals as energy bundles received with a given time delay and only indicating range to a specific point on an object. If we approach the radar as a communications system, then we can use an *impulse* signal to excite and determine the target as a transfer function, which changes the transmitted signal format.

The next step in UWB radar development will exploit the time–frequency analysis of reflected radar signals to improve the detection of specific classes of targets. The remaining chapters of this book present concepts for better performance and new applications.

Chapter 2 presents the use of return signal analysis to measure the thickness of layers in highway pavements. Chapter 3 will discuss the advanced UWB radar systems that can identify target classes and adapt the signal to enhance detection of specific types of objects. Chapter 4 will summarize the methods of time–frequency analysis. Chapter 5 will present a method for predicting how objects scatter UWB signals.

References

1. Taylor, J.D. (ed.) 1995. *Introduction of Ultra-Wideband Radar Systems*. CRC Press, Boca Raton, FL.
2. Taylor, J.D. (ed.) 2000. *Ultra-Wideband Radar Technology*. CRC Press, Boca Raton, FL.
3. Taylor, J.D. 2012. "Ch. 1 Introduction to Ultrawideband Radar Applications and Design." Taylor, J.D. (ed.) *Ultrawideband Radar Applications and Design*. CRC Press, Boca Raton, FL.
4. Sachs, J. 2012. *Handbook of Ultra-Wideband Short Range Sensing: Theory, Sensors, Applications*. Wiley-VCH Verlag & CO, Weinheim, Germany.
5. Barrett, T.W. 2012. *Resonance and Aspect Matched Adaptive Radar*. World Scientific Publishing Co. Hacksack, NJ.
6. GSSI. 2015. "Complete GPR System for Concrete Inspection and Analysis: StructureScan," http://www.geophysical.com/Documentation/Brochures/GSSI-StructureScanStandardBrochure.pdf.
7. GSSI. 2015. "Complete GPR System for Road Inspection and Analysis: RoadScan 30," http://www.geophysical.com/Documentation/Brochures/GSSI-RoadScan30Brochure.pdf.
8. Gehrig Inc. 2014. Gehrig State of the Art Noninvasive Geophysical Site Investigations. http://gehriginc.com/.
9. University of Leicester. 2015. The discovery of Richard III. http://www.le.ac.uk/richardiii/.
10. Caesar, E. "What lies beneath Stonehenge?". Smithsonian.com http://www.smithsonianmag.com/history/what-lies-beneath-Stonehenge-180952437/?no-ist.
11. Boryssenko, A. and Boryssenko, E. 2012. "Ch. 12 Principles of Materials-Penetrating UWB Radar Imagery." Taylor, J.D. (ed.) *Ultrawideband Radar Applications and Design*. CRC Press, Boca Raton, FL.
12. McEwan, T.E., Body monitoring and imaging apparatus and method, U.S. Patent No. US00557301A, November 12, 1996.
13. Taylor, J.D. and McEwan, T.E. 2000. "Ch. 6 The Micropower Impulse Radar." Taylor, J.D. (ed.) *Ultra-Wideband Radar Technology*. CRC Press, Boca Raton, FL.
14. Immoreev, I. and Tao, T-I. 2008. "UWB radar for patient monitoring." *IEEE A&E Systems Magazine*, November. pp. 11–18.
15. Arkin, R. 2014. Virtual Medical Assistant. Sensiotec™, Atlanta, GA, 2014. http://sensiotec.com/wp/wp-content/uploads/2014/05/Brochure.pdf.
16. Taylor, J.D. 2012. "Ch. 9: Medical Applications of Ultrawideband Radar." *Ultrawideband Radar Applications and Design*. CRC Press, Boca Raton, FL.
17. Li, J. and Wang, G. Multi-frequency microwave-induced thermoacoustic imaging of biological tissue, US Patent 7,266,407 B2, September 4, 2007.
18. Kajiwara, A. 2011. "Ultra-Wideband Automotive Radar." Almeida, M. (ed.) *Advances in Vehicular Networking Technologies*, InTech, http://www.intechopen.com/books/advances-in-vehicular-networking-technologies/ultra-wideband-automotiveradar.
19. Autoliv Inc., Autoliv offers a wide range of industry leading radar sensors that provide a variety of Active Safety features. http://www.autoliv.com/ProductsAndInnovations/ActiveSafetySystems/Pages/RadarSystems.aspx/.

20. AutonomouStuff. 2015. "Radar Specification Comparison Chart." http://www.autonomoustuff. com/uploads/9/6/0/5/9605198/radar_comp_chart_for_web.pdf.

21. Camero-Tech Ltd. Milestone in the Development of Camero's Ultra Wide Band Concealed Weapon Detection (CWD), Whole Body Imaging Project, December 23, 2010. http://www. camero-tech.com/news_item.php?ID=23.

22. Taylor, J.D. and Hochdorf, E. 2012. "Ch. 17 The Camero, Inc., UWB Radar for Concealed Weapons Detection." Taylor, J.D. (ed.) *Ultrawideband Radar Applications and Design*. CRC Press, Boca Raton, FL.

23. Camero-Tech. 2015. Xaver™ 800 High Performance ISR Portable Through-Wall Imaging System. http://i-hls.com/wp-content/uploads/2013/11/Xavier-800-lo.jpg.

24. Taylor, J.D., Hochdorf, E., Oaknin, J., Daisy, R. and Beeri, A. 2012. "Ch. 24 Xaver™ Through Wall UWB Radar Design Study." Taylor, J.D. (ed.) *Ultrawideband Radar Applications and Design*. CRC Press, Boca Raton, FL.

25. MIT LL. 2014. "Haystack Ultrawideband Satellite Imaging Radar." MIT Lincoln Laboratory Tech Notes, September.

26. Czerwinski, M.G. and Usoff, J.M. 2014. "Development of the Haystack Ultrawideband Satellite Imaging Radar." *Lincoln Laboratory Journal*, Vol 21 (1).

27. Barrett, T.W. 2012. "Ch. 2 Development of Ultrawideband Communications Systems and Radar Systems." Taylor, J.D. (ed.) *Ultrawideband Radar Applications and Design*. CRC Press, Boca Raton, FL.

28. OSD/DARPA Ultra-Wideband Review Panel. 1990. *Assessment of Ultra-Wideband (UWB) Technology*. DARPA, Arlington, VA.

29. Ultrawideband Radar Committee of the IEEE AES Society. 2006. *IEEE Std™ IEEE Standard for Ultrawideband Radar Definitions*.

30. Taylor, J.D. 2012. "Ch. 4 American and European Regulations on Ultrawideband Systems." Taylor, J.D. (ed.) *Ultrawideband Radar Applications and Design*. CRC Press, Boca Raton, FL.

31. Immoreev, I. 2000. "Ch. 1 Main Features of UWB Radars and Differences from Common Narrowband Radars." Taylor, J.D. (ed.) *Ultra-Wideband Radar Technology*. CRC Press, Boca Raton, FL, 2000.

32. Immoreev, I. "Chap 3: Signal Waveform Variations in Ultrawideband Wireless Systems." Taylor, J.D. (ed.) *Ultrawideband Radar Applications and Design*, CRC Press. Boca Raton, FL 2012.

33. Ulander, L., Hellsten, H. and Taylor, J.D. 2000. "Ch. 12 The CARABAS II VHF Synthetic Aperture Radar." *Ultra-Wideband Radar Technology*. Boca Raton, FL.

34. Skolnik, M.I. 1980. *Introduction to Radar Systems*. 2nd ed. McGraw-Hill, New York, NY.

35. Federal Communications Commission (FCC) 02-48. Revision of part 15 of the commission's rules regarding ultra-wideband transmission systems. ET Docket 98–153, Released April 22, 2002.

36. ECC-CEPT, Electronic Communications Committee (EDD) decision of 1 December 2006 on the conditions for use of the radio spectrum by ground-and wall-probing radar (GPR/WPR) imaging systems, ECC/DEC/(06)08 Report, December 2006.

37. ECC-CEPT, Technical Requirements for UWB LDC devices to ensure the protection of FWA (Fixed Wireless Access) systems, ECC Report 94, December 2006.

38. Rubish, Goland, "US vs Recent Canadian Rules for Ultrawideband Radio Operations." *In Compliance Magazine,* August 1, 2009. http://www.incompliancemag.com/index.php?option=com_ content&view=article&id=52:us-vs-recent-canadian-rules-for-ultra-wideband-radio-operations& catid=25:standards&Itemid=129.

39. Ministry of Economic Development. "Spectrum Allocations for Ultra Wide Band Communication Devices, Apr 2008." (Radio Spectrum Policy and Planning Group Energy and Communications Branch, Ministry of Economic Development PO Box 1473, Wellington, New Zealand. http:// www.med.govt.nz)

40. Australian Communications and Media Authority. "Planning for Ultra-Wideband (UWB) Proposals for the introduction of arrangements supporting the use of UWB devices operating in the 3.6 – 4.8 GHz and 6.0 – 8.5 GHz bands in Australia." 2010. http://www.acma.gov.au/ webwr/_assets/main/lib311844/ifc10_ultra%20wide%20band_consultation%20paper.pdf.

2

Advances in Short-Range Distance and Permittivity Ground-Penetrating Radar Measurements for Road Surface Surveying

Gennadiy P. Pochanin, Sergey A. Masalov, Vadym P. Ruban, Pavlo V. Kholod, Dmitriy O. Batrakov, Angelika G. Batrakova, Liudmyla A. Varianytsia-Roshchupkina, Sergey N. Urdzik, and Oleksandr G. Pochanin

CONTENTS

2.1 Introduction to Precise Ground-Penetrating Radar (GPR) Ranging in Multiple Transmission Media

2.1.1 The Multiple Transmission Media Problem

Expanding ultrawideband (UWB) radar system benefits requires finding new methods to increase the range measurement accuracy through multiple media. Ultimately we want to achieve small spatial resolution through innovative UWB radar systems design and signal processing [1–4]. Ground-penetrating radar (GPR) and other materials penetrating applications present special problems in range measurement because the signal must travel through several media with different electrical properties. Range measurements must account for the multiple reflections from each media layer interface and the variation of the speed of propagation in each media. The method developed for GPR has a wide range of applications in material-penetrating radar (MPR) for medical, defense, security, and nondestructive testing [5–7].

Taken from a signal analysis perspective, the UWB radar reflected impulses $e^{(s)}(t,\theta,\varphi)$ can provide significant information such as the shape and materials about the reflecting object. Properly analyzed and applied, this information can help to identify objects and improve the radar spatial resolution. From a mathematical point of view, the reflected signal waveform is the result of the following convolution operation: [2]

$$e^{(s)}(t,\theta,\varphi) = \int_{-\infty}^{+\infty} h(t-\tau,\theta,\varphi) \cdot e^{(i)}(\tau)\,d\tau \qquad (2.1)$$

which represents the impulse response of the reflecting (scattering) object $h(t-\tau,\theta,\varphi)$. The reflected signal waveform will depend on the amplitude and waveform of the incident field impulse $e^{(i)}(\tau)$, where t is time, θ, φ are directions in a spherical coordinate system from which the scattered signal arrives, and i and s denote the incident and scattered fields correspondingly. From a physical point of view the Equation 2.1 is the causality principle (Figure 2.1).

For example, Figure 2.1.b. shows the case of an δ-impulse signal reflected from a dielectric object of Figure 2.1.a. The object has a homogeneous layer of thickness d with relative permittivity ε, that means part of the normal incident wave will penetrate the object and travel through the bottom layer. The surrounding media have permittivity $\varepsilon_0 = 1$. When the impulse wave reaches the bottom layer, then part will pass through and part will reflect to the top interface. Again part of the impulse will penetrate and return to the receiver as the return signal $e^{(s)}(t,\theta,\varphi)$. The signal will continue to reflect back and

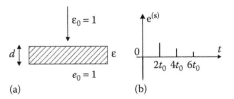

FIGURE 2.1

Reflected impulses from a dielectric layer. (a) The geometry of the dielectric reflector and (b) the signal reflected from the dielectric layer when probing, using a δ-impulse.

forth between the media interfaces and the delayed return signals will have the form shown in Figure 2.1.b [8]. The first negative impulse in Figure 2.1.b. at $t = 0$ corresponds to reflection from the surface. Next, a sequence of impulses with period $2t_0 = 2d \cdot \sqrt{\varepsilon}/c$ follows, where c is the velocity of propagation of the electromagnetic wave in free space. Partial reflection at each media interface means each succeeding return impulse will have smaller amplitude. The reflection polarity depends on ratio between permittivity of adjacent layers. It is positive if wave goes from layer with bigger ε to layer with smaller ε and it is negative in reverse case.

Glebovich described this target return sequence in Equation 2.2 as

$$e^{(s)}(t) = \Gamma_1 e^{(i)}(t) + \Gamma_2 K_1 K_2 e^{(i)}(t - 2t_0) + \Gamma_2^3 K_1 K_2 e^{(i)}(t - 4t_0) + \cdots, \tag{2.2}$$

where:

$\Gamma_2 = (\sqrt{\varepsilon} - 1)/(\sqrt{\varepsilon} + 1)$ describes the reflection coefficient from the surface between the layers $K_1 = 1 - \Gamma_2$ and $K_2 = 1 + \Gamma_2$ are transmission coefficients through the first and second boundaries respectively such that $\Gamma_1 = -\Gamma_2$; $K_1 K_2 = 1 - \Gamma_2^2$ [8].

2.1.2 Radar Measurements of Media Impulse Response

By definition, the reaction of a medium to the δ-impulse signal is the impulse response of the medium, then Figure 2.1.b shows the impulse response for the case in the Figure 2.1.a. Note how the impulse response of such a structure is a sequence of δ-impulses of infinitesimal duration. It means that by finding the impulse response, we can precisely determine the time corresponding to the reflected wave interaction with each media interface.

As shown in Figure 2.1.a, one way to find the impulse response of the structure could use the cepstrum data processing algorithm [9]. The cepstrum results from taking the inverse Fourier transform (IFT) of the logarithm of the estimated spectrum of a signal. The cepstrum process involves finding the complex spectrum of the probing and reflected signals respectively $\dot{e}^{(i)}(\omega)$ and $\dot{e}^{(s)}\omega$, and then calculating the complex reflection coefficient from

$$\dot{\Gamma}(\omega) = \dot{e}^{(s)}(\omega)/\dot{e}^{(i)}(\omega). \tag{2.3}$$

To calculate the cepstrum of power $C(e(t))$ of Equation (2.2), use the following formula (Equation 2.3):

$$C(e(t)) = \int_0^\infty \ln[\dot{\Gamma}^2(\omega)] \cdot \exp(j\omega t) d\omega, \tag{2.4}$$

and the data processing result will look like plot that is shown in (Figure 2.2).

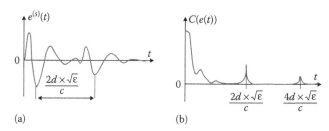

FIGURE 2.2
Analysis of the signal shown in Figure 2.1(b): (a) The received signal and (b) the cepstrum of power.

Finkelstein et al noted that such a data processing method requires a high signal-to-noise ratio (SNR) and shapes of reflections which differ only by constant coefficient for amplitude [9]. In addition to the cepstral data processing, using the well-known methods of inverse filtering and deconvolution could solve a similar problem. However, in all cases the high requirements of noise and measurement accuracy of waveforms remain unchanged for the achievement of high precision measurements of the boundaries coordinates.

Using the reflected signal waveform for probing and analyzing the return for the frequency content could significantly improve the accuracy of radar measurements. However this means the UWB radar must have very high power signal, a wide dynamic range and a receiver capable of accurately recording the signal waveform for further analysis. Meeting these conditions can provide greater range measurement accuracy in a signal passing through layers of dielectric media.

UWB electromagnetic field pulses can penetrate into a variety of dielectric media and permits the design of GPR for measuring the subsurface structure of materials such as road pavement [10–12]. Road network management quality depends on the availability and widespread use of operational methods and tools for monitoring the pavement conditions. Currently many European institutions have research programs on pavement measuring systems [13,14]. The authors have presented their research results in the cited works [15,16].

2.1.3 GPR Pavement Measurement Objectives

GPR probing can solve two major problems in road maintenance: (1) estimate the thickness of the pavement structural layers and detect subsurface defects and (2) determine any inhomogeneities in the subgrade which affect the stability and road pavement lifetime. These problems will arise during the new road construction and quality control. During the operational phase GPR measurements can evaluate the state of the road pavement. The thickness and structural defects of pavement layers and deformation characteristics of pavement material layers are the key parameters that determine its load-bearing capacity (strength). The highway construction industry standards demand precision instruments for measuring those parameters.

GPR remains rarely used for continuous nondestructive testing of extended structures in spite of the obvious benefits for two reasons. First, the GPR needs a highly skilled operator to interpret the complex raw radar data. The reflected GPR signal depends on the structural thickness and the electrophysical characteristics that affect the signal velocity in each structural layer. This phenomenon leads to significant errors in determining the thickness of pavement structural layers. To guarantee the accuracy claimed by the GPR equipment, manufacturer requires ways to reliably interpret the return signal results.

The second reason discouraging GPR use comes from comparable size of the pavement thickness of about 5 cm and for the pavement upper layers is about the same size as GPR probe pulse spatial resolution. Solving this problem requires improved GPR equipment and the development of specialized algorithms to accurately measure each pavement layer thickness.

Designing GPR instruments to accurately measure the thickness of a media layer is directly connected with the sophisticated problem of detecting and localizing the signals reflected by inhomogeneities in low-contrast conditions.

This problem becomes more complicated when the distance between the inhomogeneities in the medium is less than the probing signal spatial length. Under these conditions the pulse electromagnetic field reflected from the previous boundary is overlapped in time with the pulse reflected by the next boundary. This makes it practically impossible to separate echo signals one from another by means of only visually examining the received signal. Therefore it becomes difficult to determine the exact arrival times of the reflected signals. Determining the layer thickness requires accurately measuring the material permittivity at a known point. These problems of spatial resolution and unknown pavement characteristics turn the accurate measurement of the distances between the layer boundaries into a complex task.

Section 2.2 will present a signal return processing algorithm that can simultaneously evaluate the permittivity of pavement layers by analyzing the signals reflected from the road surface and layers beneath. The mathematical basis makes use of the Hilbert transform to determine the exact time of the return signal arrival [17].

The proposed solution uses a special algorithm to determine the permittivity and thickness of the layers. The algorithm uses the solution of the auxiliary direct problem of scattering of a plane monochromatic wave by the layered medium. The direct problem consists of finding ways to calculate the coefficients of reflection and transmission waves at the layer boundaries.

Using the GPR signal return data and Hilbert transform can determine the delay times and amplitudes of the signals reflected by the subsurface boundaries between the layers. The proposed method determines the time delay by calculating the Hilbert transform using a discrete Fourier transform. It then constructs an analytical signal and then uses it to find the delay time as the distance on the time axis between the positive peaks of the auxiliary function. It then must determine the permittivity of the top layer. Thereafter, the GPR signal processor can calculate the transmission and reflection coefficients at the first border, and it can then calculate the permittivity of the second layer. After that the signal processor can determine the layer thicknesses based on the calculated results of the permittivity and the time of passage through the layer of the probing signal. This measurement procedure can estimate the pavement thickness layer-by-layer through all of the layers of interest from top to bottom.

2.1.4 GPR Development Objectives

This general measurement scheme seems simple and physically transparent. However, the practical implementation presents serious difficulties due to the problem specifics and GPR hardware requirements. Signal transmission physics will determine the radar design requirements as described below:

1. Signal transmission physics: Measuring multiple layers with different, but close permittivities means a low reflection coefficient and small reflected signal amplitudes at the junction of each layer. Reception of very weak reflected signals will

require developing a GPR, which has a very large power budget and dynamic range. If we start with a strong probing signal, then the return signal spectrum will contain information about the electrical properties of each layer. To exploit this information, the receiver must accurately record the reflected waveforms. It needs a strong signal as a precondition for the sophisticated digital signal-processing algorithms that can note and consider every detail of the return signal. Therefore a large power output and a wide receiver dynamic range will give the capability for deeper sounding, higher resolution, higher detection probability, and lower false alarm probability for detecting certain objects.

2. GPR radar design requirements: Astanin and other researchers showed how the shape of reflected UWB signals carries information about the reflector's electrical and geometric properties [2]. In this regard, the GPR designer must ensure the collection of undistorted radar data and proper processing of multiple signal returns. Obviously accurate data collection will give better accuracy of measurement.

Received signal distortions usually occur because of following main reasons:

- Bad matching of characteristics of the transmitting and receiving antennas with parameters of probing signal
- Unsuitable antenna system configuration
- Clutter from nearby reflecting objects
- Noise
- Inaccurate collection of reflected signal amplitude data during the analog-to-digital conversion (ADC) process
- Variations in the sampling time interval, or jitter, that means an inaccurate sampling time interval setting when collecting signals with a stroboscopic receiver.

The radar data processing software can include operations such as filtering or background removal to partly correct some of the signal distortions. However attempts to correct the influence of jitter can lead to additional distortions.

These problems mean the designer must ensure very accurate GPR reflected signal data collection and suppression of signal interference by the following methods.

3. Overcoming short-range signal interference: Because the GPR works at short-ranges, this requires locating the receiving antenna close to the transmitting antenna. As a result the antenna coupling sends high power pulses from the transmitting to the receiver antenna with each pulse. In some cases the reflected signal may arrive at the receiving antenna and overlap the trailing edge of the transmitted pulse. Under these conditions the measured weak reflected signal appears on a background of intense direct coupling signal. This phenomenon limits the dynamic range of the UWB short pulse GPR and makes measurement and subsequent radar data processing more difficult. On one hand, the transmitted pulse degrades the detail and accuracy of received signal waveform and measurement. On the other hand, the strong background signal makes it difficult to record and measure the weak reflection from low-contrast boundaries [18,19]. The presence of a powerful directly coupled signal between the transmitting and receiving antennas complicates received data processing and risks damage to the sensitive receiver input circuitry.

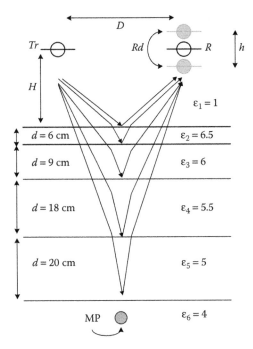

FIGURE 2.3
Model of a plane surface with five media layers with different dielectric constants and a metal cylinder (MP) beneath the lower layer. (From Varyanitza-Roshchupkina, L. A. et al., Comparison of different antenna configurations for probing of layered media, *8th International Work-shop on Advanced Ground Penetrating Radar (IWAGPR 15)*, 7–10 July 2015, Florence, Italy. © 2015 IEEE.)

As an example of interference effects, consider the probing of the multilayer medium comprising with a metal cylinder in the lower layer as shown in Figure 2.3 [19]. The radiator *Tr* forms the GPR probing signal. The reflected scattered field is recorded at the reception point *R* at distance *D* from the radiator. In this case both *Tr* and *R* locations have a height *H* above the surface. The GPR receiver collects signals for further analysis as the radiator-observation point moves along the surface.

The results of a subsurface probing simulation using specially developed software gives the result shown in (Figure 2.4) [20].

Mathematical background removal can mitigate the influence of the powerful direct coupling signal. This operation can remove the background as an average signal, remove the signal trend [21], or remove one of the signals recorded in the radar data set. Each of these cases forms the data set for subtracting, by averaging over either the set or a selected signal. The averaging process assumes the absence of reflections from the scattering objects, but includes the directly coupled signal. The processing subtracts the averaged signal from each of the analyzed signals of the data set. This helps to form a clear radar image because the processing largely eliminates the directly coupled signal, and increases the probability of detecting the reflecting boundaries.

The average signal calculation is based on the all signals composing the range profile including the reflections from local objects. As a result the averaged signal has information about these reflections in the form of pulses. Subtracting this signal from the signals of that part of the profile where no reflectors exist gives the difference signal, which has weak pulses corresponding to nonexistent objects. Thus by using the background removal we can insert the additional signals (artifacts) in the radar image.

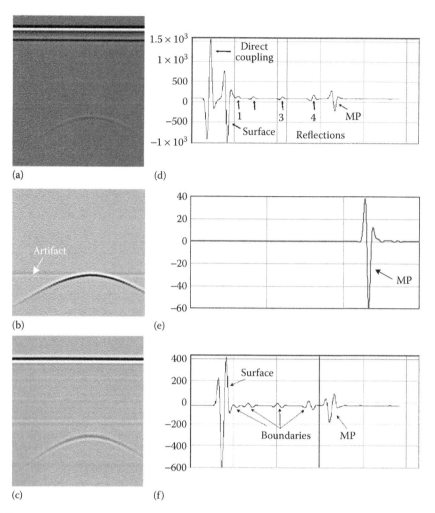

FIGURE 2.4

Subsurface B-scan simulations for the buried metal cylinder in Figure 2.3: (a) The raw data, (b) data after background removal, (c) the raw data for the differential antenna, (d) the raw data taken from the B-scan shows reflections from the interface of the different media, (e) removing the background produces the clean reflection of the buried metal cylinder, and (f) the raw data for the differential antenna (From Varyanitza-Roshchupkina, L. A. et al., Comparison of different antenna configurations for probing of layered media, *8th International Work-shop on Advanced Ground Penetrating Radar (IWAGPR 15)*, 7–10 July 2015, Florence, Italy. © 2015 IEEE.)

The same artifact problem holds true for subtracting the selected signal. Indeed, it is impossible to guarantee in advance that at some point of the probing path below the surface the reflecting object or scattering boundaries are completely absent.

As a further example, consider the case of a signal corresponding to the horizontal layer of the boundary between layers of asphalt pavement. During averaging process, the media boundary reflected signals remain in the subtrahend signal, so further subtraction of the averaged signal from the signals of the entire profile leads to the situation when reflections are removed and information about these layers will be lost. Thus the background removal process outwardly improves the radar image, but at the same time it introduces distortions in the information contained in the profile by removing the useful information about layers, and can add to the signals corresponding to nonexistent objects.

Zhuravlev et al found a way to remove an interfering signal from direct coupling between the transmitting and receiving antennas [22]. Their approach took a portion of the driving signal and sent it forward to the receiver to compensate for the interference. This principle can be used in narrowband systems, where the operating signal is a sine wave. Technically speaking, it is difficult to achieve full compensation because it requires additional adjustment of the amplitude and phase for each frequency of operating range. The presence of reflecting objects around the radar antenna system, the instability of the signal generator and the receiver can greatly complicate the adjustment procedure.

The authors invented new ways to achieve full frequency independent electromagnetic decoupling, as described in the patent [23]. In this case, the receiving antenna consists of two dipoles Rd shown in Figure 2.3 placed at different heights over the surface symmetrically to the plane of the transmitting dipole. The distance between the receiving dipoles is h. A useful signal is a difference between signals received by each of receiving dipoles. Advantage of this antenna configuration consists in the possibility of complete mitigation of powerful direct coupling signal at the input of receiver and at the same time in saving all data about all existing scattering boundaries and objects of Figure 2.4.c and f. Section 2.3 presents the details of the transmitting-receiving (TR) antenna system design and operational principles. Section 2.4 contains notes relating to ADC technology and GPR receivers for precision range measurements.

Another way to increase the power budget and GPR dynamic range uses the accumulation (summation) of the energy from multiple signals reflected from the object. Furthermore using signal accumulation can increase the SNR proportionally to square root of the number of accumulated signals. This method is particularly convenient when the UWB radar uses a stroboscopic converter to record the received signal waveforms. However the stroboscopic system must provide a very low jitter for accurate signal waveform collection. A large instability in the accumulation process will lead to distortion of the received signal shape and create difficulties for radar data processing, that degrades the resolution. Section 2.5 will describe ways for overcoming the synchronization problem.

Usually high-resolution sampling receivers require the widest possible working frequency band. However a wider working frequency band means less SNR and leads to losses of useful information. The designer needs to find an optimal working frequency band that can provide a better SNR and minimal signal shape distortion. Section 2.5 presents our analysis of the influence of the sampling gate duration on the distortion of the transformed waveform, and a special data acquisition method with an adjustable working frequency [24].

Sections 2.6 and 2.7 gives short descriptions of software for data processing and prototype of the radar.

Section 2.8 contains the results of testing the UWB GPR developed by the authors.

Section 2.9 presents results of the radar application for road inspection followed by general conclusions.

2.2 GPR Data Processing and Interpretation Using Layer-by-Layer Decomposition

Performing roadbed surveys presents several major problems that affect GPR capabilities and limit performances.

The first limitation comes from the dielectric constant and the thickness of the roadbed structural layers that will affect the performance. As a rule, the analysis of subsurface sounding results will depend on the GPR operator's experience and requires direct involvement in GPR data processing and evaluation. This approach to data results in poor GPR performance and cannot obtain quantitative characteristics describing the structure and electrical parameters of the medium under investigation.

The second complication comes from the large GPR spatial pulse duration in comparison to the thickness of the structural layers of road pavements. On the other hand using very short probing signals presents other difficulties due to the significant signal attenuation and scattering of electromagnetic pulses propagating in the road construction materials.

The third limitation comes from the low contrast (differences) between the electrophysical characteristics of each pavement layers. For example, the relative permittivity of gravel (dry) is in the range from 3.5 to 4.4, that depends on the fractional composition and degree of compaction. In a GPR probe, the relative permittivity of gravel may only be slightly different, or in some cases identical to the dielectric constant of dry sand which may vary from 3.0 to 3.7. Developing better GPR technology for road inspection requires ways to accurately measure pavement layers with a low contrast between the two paving materials. Improving road management quality means finding new measurement methods and GPR tools to distinguish between low contrast media. Ultimately we want to build a data display that tells the operator the thickness of each of several pavement layers. The mathematical approaches presented below will help solve these problems for GPR and other MPR applications in medicine, nondestructive testing, and security.

2.2.1 Methods for Solving Inverse Problems

GPR methods for measuring the dielectric constant and thickness of structural pavement layers will require ways to solve the inverse problem of determining the medium parameters. If we know the media parameters, then analytical methods can reconstruct the scattered radar electromagnetic fields in the roadbed.

We can divide these inverse problems into the following two classes according to the structure of the desired solution:

1. Continuous class: Assumes that the unknown is a continuous distribution function of the desired parameters. This works best for remote sensing to determine the distribution of soil moisture and its dependence on depth.

2. Piecewise class: Assumes that the distribution function of the unknown parameters is piecewise continuous. This works best for nondestructive quality control and the condition of layered products such as road pavement, laminated structures, and so on.

The continuous case requires finding a continuous distribution that is an infinitely dense set of numbers. For the piecewise case, the desired result is a finite set of numbers. For the continuous distribution case the task is ill-posed because (1) the solution is not unique due to the incompleteness of the original data set and (2) unsustainable because small perturbations in the input data (measurement error) may lead to significant deviations in the solution. To partially overcome these difficulties, Batrakov and Zhuck proposed various methods including the Newton–Kantorovich scheme [25–27].

To explain these methods, consider the problem of UWB GPR subsurface sounding of the pavement made of a set of dielectric layers, where the road pavement model takes the form of a layered medium.

The initial data for the inverse problem solution comes from the GPR data obtained in the first step in a pavement survey done in the steps shown in Figure 2.5 [28].

Step 1
Sense and obtain the space-time samples of fields reflected by the subsurface objects and inhomogeneities that are boundaries of layers. (In addition, the GPR data files store information about the size of the study area, a sampling step and, if recorded reflection of mutually orthogonal component of the scattered field, the polarization of the received signal.)

Step 2
Image the probed subsurface (echo profile) and make an approximate estimation of pavement layer boundaries and the local inhomogeneities in the observation area. These provide a preliminary analysis and define the geometrical and electrical parameters of the object.

Step 3
Set the initial approximate values of the geometrical and electrical parameters of the subsurface region which is approximated by a plane-layered structure.

Step 4
Find the inverse problem solution to get a quantitative estimate of the structure parameters (thickness and permittivity of layers, depth and nature of detected defects [29,30,31,8])

FIGURE 2.5
GPR pavement examination procedure.

We can use a method of the minimization of a smoothing functional to precisely measure thickness and permittivity of each layer $F[\eta]$ [25,32,33] (Equation 2.5):

$$F[\eta] \equiv \sum_{j=1}^{N} w_j \left| \mathbf{E}^{(j)}(z) - \mathbf{E}_{aux}^{(j)}(z) - \int_{V_P} L^{(j)}(z,z')\, \eta(\vec{R})\, dz' \right|^2 + \alpha \int_{V_P} w(z') \left| \eta(z') \right|^2 dz'. \quad (2.5)$$

where:

$w_j, w(z')$ are nonnegative weight coefficients and weight functions respectively

$L^{(j)}(z,z')$ is a kernel of the integral operator that defines the relationship between the measured and simulated values of experimentally measured quantities (electric field $\mathbf{E}^{(j)}(z), \mathbf{E}_{aux}^{(j)}(z)$ respectively)

j is the index of the informative parameter (frequency, polarization state, etc.), $\eta(z') \equiv \varepsilon_{aux}(z) - \varepsilon(z)$ is a correction to the value of distribution function of the unknown permittivity

z' is the coordinate of integration that must be done over whole region where the permittivity is unknown.

The smoothing functional consists of the residual between the measured samples of the pavement scattered electromagnetic field and the results of the direct problem solution of the model, as well as a stabilizing functional that actually *selects* from several solutions only one that is the least *deviated* from zero. Due to such specific problem statement this procedure becomes very effective.

Ultimately the problem is reduced to finding the solution of a system of linear algebraic equations and defining $\eta(z)$ using the direct formulas (Equation 2.6):

$$\eta(z) = \frac{1}{aw(z)}\left[q(z) + \sum_{j=1}^{N} w_j L^{(j)^*}(z,z')x_j\right],\qquad(2.6)$$

where:

$$x_j = \text{const} = \int_{V_z} dz' L^{(j)}(z,z')\eta(z'), j = 1,2,\ldots,N$$

$$q(z) = \sum_{j=1}^{N} w_j L^{(j)^*}(z,z')\left[U_{\text{aux}}^{(j)}(z) - U^{(j)}(z)\right]$$

In several studies, Goncharsky et al have attempted to calculate the thickness of the pavement layers automatically [18, 34]. To take into account possible measurement errors, the authors introduced two nonlinear parameters (p_1, p_2) and one linear parameter K. The result expressed the problem by the following equation (Equation 2.7):

$$u(x,t) = K \cdot f(t,p_1) + \int_{0}^{\infty} A(t-z,p_2) \cdot k(x,z)dz,\qquad(2.7)$$

where:
$A(t-z,p_2)$ is a function of reflection from the boundary
$k(x,z)$ is a reflection coefficient as a function of depth z

As a result, to solve the problem it is necessary to use the known radar image $u(x,t)$, the kernel $A(t-z,p_2)$, and the probing wave $f(t,p_1)$. First a set of unknown parameters (p_1,p_2) and K was defined by minimizing the residuals θ by searching the corresponding values from a certain range (Equation 2.8):

$$\theta = \left\|u(x,t) - Kf(t,p_1)\right\|^2,\qquad(2.8)$$

where, $\|\cdot\|^2$ is the square of the norm in L^2. Then it is necessary to solve the integral Equation (2.6).

Finally, in the last stage, radar image processing must be done to determine the change in thickness of layers. To solve this problem Goncharsky et al suggested the use of mathematical methods of image processing and pattern recognition [18]. To do this, we need to examine each track and choose the brightest points (their coordinates are denoted as (x,z)) in an amount determined by the parameters of the desired layers. In this case, x is the number of tracks and z is a reference number. Next the weighting function is calculated for these points (Equation 2.9):

$$W(x,z) = \sum_{i=-n}^{n}\sum_{j=-m}^{m} k(x+i, z(i)+j),\qquad(2.9)$$

where:

$z = z(0)$

$k(x+i, z(i)+j)$ is the intensity value at the coordinates $(x+i, z(i)+j)$

Next we use the obtained weighting functions to test the hypothesis that a point (x, z) belongs to boundary of one of the layers. However, the authors point out some of the problems encountered in the practical implementation of this approach that are associated with both the complexity and imperfection of the model. They also mentioned the significantly required computing resources.

To solve the problem of estimating the thickness of the coating layers, Cao and Karim applied other approaches based on other initial integral equations (for the electric and magnetic fields) and other optimization techniques (e.g., neural networks). The interested reader will find the results in the cited works [35–37].

The development of UWB impulse GPR gave new possibilities for solution of the inverse scattering problems. Such GPRs using the principle of radiation and reception of UWB signals have significant advantages in comparison with narrowband probing. Unlike earlier methods for solving inverse problems that assume probing, using narrowband monochromatic signals, the GPR radars opened up the possibility of using UWB pulses. After passing through the investigated structure these short duration pulse signals and their reflections from internal imperfections contain information about the spatial and electrical properties of the probed medium. Therefore in addition to electrodynamic models of media and objects, it is necessary to develop methods and algorithms for the solution of direct problems for simulation of the propagation of electromagnetic waves in a layered medium to enable the interpretation and use of this information. For the construction of efficient computational algorithms, these methods should be based on a universal technique that will be briefly considered.

2.2.2 UWB Impulse Signal-Processing Methods

One of the fundamental aspects of time domain signal processing and analysis is to determine the time corresponding to the pulse position on a time axis. Even with its apparent simplicity, Glebovich and Yelf note that, finding this correspondence is not a trivial task [8,38]. In our case, an obstacle to solving this problem becomes more complicated because of the previously mentioned relatively large spatial duration of the probe pulse and low contrast boundaries between layers. Among the known options to determine the position of the pulse on the time axis Krylov et al and Batrakov et al decided to apply the Hilbert transform for GPR signal processing [39,40]. This transformation is defined by the following relationships (Equations 2.10 and 2.11):

The direct Hilbert transform

$$\tilde{x}(t) = \frac{1}{\pi} \int_{-\infty}^{\infty} \frac{x(\tau)}{t-\tau} d\tau, \tag{2.10}$$

where, the function $1/(t-\tau)$ is called the kernel of the Hilbert transform and the inverse Hilbert transform:

$$x(t) = \frac{1}{\pi} \int_{-\infty}^{\infty} \frac{\tilde{x}(\tau)}{t-\tau} d\tau. \tag{2.11}$$

The integrals of the transformations have a singular point $a = (t - \tau) \to 0$, in which the principal value of the Cauchy is used for the calculation: $\lim\limits_{a \to 0} \left[\int_{-\infty}^{t-a} + \int_{t+a}^{\infty} \right]$. The reason for applying this transformation to pulses lies in its property of an *ideal phase shifter*, that is, implementing a phase rotation of the signal. In this case, the problem reduces to the evaluation of the model (modulus) of a complex analytic signal constructed from the original signal by using the Hilbert transform. Such a function is smooth and unipolar. It has only one peak corresponding to one boundary. Therefore the analysis of a signal is reduced to finding the maximum of the modulus of this function.

Then, to determine the signal delay we propose to use a discrete Fourier transform to calculate the Hilbert transform. We can then construct an analytical signal and to find the delay, as the distance on the time axis between the peaks of the positive function $S(t)$ (Equation 2.12):

$$S(t) = |h(t)| = \sqrt{x^2(t) + \tilde{x}^2(t)}. \tag{2.12}$$

We can determine the polarity of the reflected pulses from one of the properties of the transformation: the maximum modulus $|h(t)|$ that corresponds to the maximum modulus of the real alternating function [40]. To find the maximum of the conversion module, it is necessary to analyze the sign of the function $x(t)$ at the appropriate time.

2.2.3 How to Find the Thickness of Structural Pavement Layers Using Transfer Functions

In most modern GPR sets, the spatial durations of the probing impulses and dimensions of antennas are small compared with the radii of the road surface curvatures. Usually the size of the pavement surface irregularities (roughness) is substantially less than the wavefront curvature. Therefore the most effective pavement model assumes the plane-layered medium as shown in Figure 2.6. This model generally characterizes each nth layer by three parameters: permittivity, conductivity of the material, and the layer

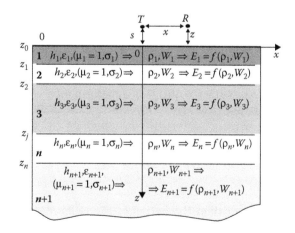

FIGURE 2.6

The pavement model has multiple layers with different electrical characteristics. The bottom $n + 1$ layer extends to infinite depth. (From Batrakov, D. O. et al., Determination of thicknesses of the pavement layers with GPR probing, *Physical Bases of Instrumentation*, 3, 46–56, 2014. With permission.)

thickness. In this case, we can assume the thickness for the bottom of the underlying layer is infinite [17].

To determine the values of permittivity and thicknesses of the layers, we can use the solution of the auxiliary problem of reflection of a plane monochromatic wave by the layered medium. This method computes the reflection and transmission coefficients at the layer boundaries.

Provided the parameters of the medium are known, there are fundamental formulas given in references [7,29,41,42,5] to calculate the reflection (R) and transmission (T) coefficients through the interface between dielectrics as shown in Figure 2.7 (Equations 2.13 and 2.14):

$$R_{n-1,n} = \frac{A_{n,n-1}}{A_{n-1,n}} = \frac{\sqrt{\varepsilon_{n-1}} - \sqrt{\varepsilon_n}}{\sqrt{\varepsilon_{n-1}} + \sqrt{\varepsilon_n}}, \tag{2.13}$$

$$T_{n\mp1,n} = \begin{cases} \dfrac{2\sqrt{\varepsilon_{n-1}}}{\sqrt{\varepsilon_{n-1}} + \sqrt{\varepsilon_n}} \\[3mm] \dfrac{2\sqrt{\varepsilon_{n+1}}}{\sqrt{\varepsilon_n} + \sqrt{\varepsilon_{n+1}}} \end{cases}, \tag{2.14}$$

where:

$n-1$, n, and $n+1$ are numbers of layers

$A_{n-1,n}$, $A_{n,n-1}$ are the amplitudes of the waves that incident on the boundary between the medium with ε_{n-1} and medium with ε_n and reflected into $n-1$th medium

$A_{n-1,n} = A_0$, $T_{n-1,n}$ is a transmission coefficient from $n-1$st layer into nth (upper line)

$T_{n+1,n}$ is a transmission coefficient from $n+1$th layer into nth

$R_{n,n-1}$ is a reflection coefficient from the boundary of the $n-1$th and nth layers back into the $n-1$th layer.

We can consider the problem in the single-scattering approximation, that is, without taking into account multiple reflections of the signal within the layers of the structure. For

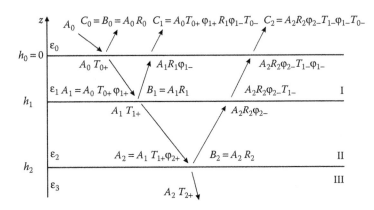

FIGURE 2.7

Diagram of the pavement layer structure and notation. I - layer #1, II - layer #2, III - layer #3. (From Batrakov, D. O. et al., Determination of thicknesses of the pavement layers with GPR probing, *Physical Bases of Instrumentation*, 3, 46–56, 2014. With permission.)

clarity and brevity, we denote the reflection coefficients from the bottom of the nth layer as $R_n = R_{n,n+1}$, and transmission coefficients through the lower boundary nth layer downward as $T_{n,+} = T_{n,n+1}$, and upward as $T_{n,-} = T_{n+1,n}$ as shown in Figure 2.7. The additional factors φ_{n+} and φ_{n-}, thus are responsible for the attenuation of the signal amplitudes during propagation downward and upward, back through the layer (in the direction of the axis OZ) in the absorption conditions. For brevity, we introduce the intermediate parameters $B_{n,n+1}$ for the reflection from the media shift from layer n to $n+1$.

Despite the fact these coefficients are equal, in general, they are reserved for the symmetry of equations.

Since at this stage, the main objective is to define the reflection coefficient from the bottom boundary of, for example, the 1st layer by the known (measured experimentally) amplitude of the signal at the output of the structure (in the upper semispace, i.e., in the air). We can introduce the concept of a layer's transfer function P_n, as the factor for the reflection coefficient R_n in the equation relating the amplitude of the wave that reflected from the lower boundary of the layer, with the amplitude of the wave passing back to the top layer (for the first layer the top layer is free space). For example, if we evaluate the passage of the wave through the border between the layers with indices 0 and 1 in the forward and reverse directions, then the signal transfer function P_1 is determined by the product of the corresponding transmission coefficients, which according to Equation 2.14, equals:

$$P_1 = T_{0,1} \cdot T_{1,0} = \frac{2\sqrt{\varepsilon_1}}{\sqrt{\varepsilon_0} + \sqrt{\varepsilon_1}} \quad \frac{2\sqrt{\varepsilon_0}}{\sqrt{\varepsilon_0} + \sqrt{\varepsilon_1}} = 4 \frac{\sqrt{\varepsilon_1} \cdot \sqrt{\varepsilon_0}}{\left(\sqrt{\varepsilon_0} + \sqrt{\varepsilon_1}\right)^2}. \tag{2.15}$$

We now designate the partial signal amplitudes reflected from the lower bounds and registered by the receiving antenna as C_n. Then for the lower boundary of the first layer there is an obvious relation as shown in Figure 2.7 (Equation 2.16):

$$C_1 = P_1 \cdot B_1 = \left(T_{0,1} \cdot T_{1,0}\right) \cdot A_0 \cdot R_{1,2}. \tag{2.16}$$

This approach has the advantage of obtaining the transfer function of the system of n layers by multiplying their own transfer functions. For example, if the layer with the number n, then we have a simple equation (Equation 2.17):

$$R_{n,n+1} = A_0^{-1} \cdot C_n / \left(P_1 \cdot P_2 \cdots P_n\right) = A_0^{-1} \cdot C_n / \prod_{n=1}^{N} P_n, \tag{2.17}$$

where: $P_n = T_{n-1,n} \cdot T_{n,n-1}$.

Next, using the definition of the reflection coefficients $R_{n,n+1} = A_{n+1,n}/A_{n,n+1}$, and using obtained from the Equation (2.13), values $R_{n,n+1}$ one can calculate the permittivity of the lower $(n+1)$th layer shown in Figure 2.6 by the equation (Equation 2.18):

$$\sqrt{\varepsilon_{n+1}} = \sqrt{\varepsilon_n} \frac{A_{n,n+1} - A_{n+1,n}}{A_{n,n+1} + A_{n+1,n}}. \tag{2.18}$$

With the Equation 2.18, one can then calculate the transmittance through the lower boundary of the nth layer. Following that, Equations 2.15 through 2.18 should be repeated.

Finally, by knowing the permittivity of the $(n+1)$th layer, and the delay Δt_n, then we can calculate thickness of the layer h_n by formulas (Equation 2.18):

$$h_n = \frac{v_n \cdot \Delta t_n}{2},$$
(2.19)

where:
$v_n = c/\sqrt{\varepsilon_n}$ is the velocity of the signal propagation in the nth layer
ε_n is the permittivity of the nth layer

Thus, with the procedure considered above, one can determine the permittivity and the thickness of the road structural layers simultaneously and without additional laboratory measurements. We can get the one-sided data by radar probing from only the space above the road surface. This opens many possibilities for measurements of layered materials of all kinds.

2.3 Measurement Accuracy of the Thin Layer Thickness

The last section described a method for distance measurements in layered media based on measurement of permittivity and thickness of layers. In order to achieve the accurate radar measurements the following preconditions should be provided:

- Simultaneous arrival of the signals reflected by all points of the boundary between layers with different electric properties, surface to the receiving antenna and
- Absence of distortion of signal shape at reflection by other object.

With these conditions, it is easy to recognize signals reflected by boundaries and to calculate electrical parameters of the layers and time intervals that probing impulses spend between propagation and reflection. As shown in Section 2.2, these preconditions correspond to the task of probing a plane-layered medium using a plane electromagnetic wave which is normally incident to the surface of the medium. To meet these conditions it is necessary to consider how the antenna configuration affects the measurement performance.

2.3.1 Comparison of Bistatic and Monostatic Antenna System Configurations

At the short ranges used by GPR systems, antennas radiate near field nonplanar waves. Moving the antenna away from the probed surface will make the wavefront closer to a plane as required for accurate measurements. The greater distance and radius of curvature of the wavefront, the better it approximates a plane wave. However, moving the antenna away from the structure requires a larger radiated power because the amplitude of signals rapidly decreases with distance.

If we compensate for these decreased signal amplitudes by increasing the GPR transmitted signal power, then because of the coupling between the bistatic transmitting and receiving antennas, we increase the amplitude of signal coming directly into the receiver.

We can decrease the direct antenna coupling by increasing the distance between the transmitting and receiving antennas. Unfortunately increasing the antenna separation will result in a propagation path, where the wavefronts are no longer normal to the pavement surface and the boundaries between layers. This will lead to less accuracy in calculating the layer thickness. We have an additional problem from the frequency dependence of reflection coefficients on the angle of incidence that will distort the sounding signal wave shape.

In the bistatic antenna system shown in Figure 2.3, the probing signal-propagation path from the transmitting antenna Tr to the boundary layer and then to the receiving antenna R, depends on the distance between the antennas D and the antenna height above the road H.

Attempts to find direct analytical formulas for calculating the path length of the probing signal in the layer, directly dependent on the known distance D and H failed. Therefore to estimate the achievable measurement accuracy we used the results of computer simulations of layered medium probing using the bistatic antenna system shown in Figure 2.3. Our simulation used the software *SEMP* [20] based on the method of finite-difference time-domain (FDTD) [43]. To simplify the analysis, we made a probing signal source of a current filament directed perpendicular to the plane of the drawing. We made the layers of this structure homogeneous with zero conductivity. The probing signal time dependence is the first-time derivative of the Gaussian pulse of 0.4 ns duration at the 0.5 level of its amplitude. This approach to the analysis of the achievable accuracy allows us to consider only the influence of the GPR geometry problem on the measurements accuracy. Our approach avoided the need to consider any other reason, for example noise, antenna characteristics, and waveform distortion during the propagation loss due to dispersion and transmission coefficient.

Let us consider the bistatic antenna system of Figure 2.3 with the antennas located at $H = 30$ cm, $D = 10$ cm, 20 cm, 30 cm, 40 cm, 50 cm, and 60 cm (6 values). Comparing the bistatic antenna configuration with quasi-monostatic configuration, where the recorded signal is the difference between signals in the receiving points R_d spaced at distance $h = 4$ cm in Figure 2.3. This quasi-monostatic differential configuration has additional very useful property. It helps to mitigate a direct coupling signal [19]. The next paragraph presents the differential antenna properties in detail.

The analyzed parameter Δt is the deviation of signal-propagation time in the layer from the propagation time for the reference case that is normally incident plane wave, for which $t_n = (2/c) d_n \sqrt{\varepsilon_n}$. The deviations of signal-propagation time Δt in the layers in the distance between the transmitter and the receiving antennas for the case of (a) bistatic and for (b) quasi-monostatic configuration are shown in Figure 2.8. From the figures, it is evident that with increasing separation D the deviation Δt increases and the accuracy decreases. Accuracy can also degrade with greater distance from the surface, but for bistatic case this degradation is much more significant as shown in Figure 2.8.a.

Note how for the analysis of the bistatic sounding we had to specifically calculate the field of direct coupling from radiator to the observation point with the absence of a layered medium for subsequent subtraction of the received signal from the simulated data set. Otherwise, a strong background of a direct coupling signal makes it practically impossible to accurately recognize the position on the time axis of the signal, reflected by the surface of the medium being probed.

Some types of antennas (e.g., TEM-horn antennas or various types of shielded antennas) radiate and/or receive signals whose waveform depends on the direction. Therefore, when using such antennas in the GPR, the designer must take into account this dependence. Usually the minimal distortions correspond to the direction along the axis of symmetry

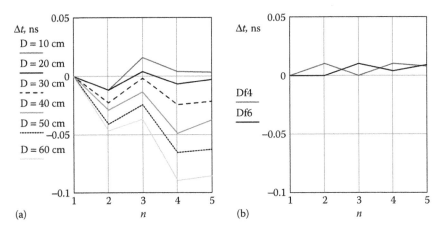

FIGURE 2.8
Deviation of the signal propagation time in the layer resulting from bistatic antenna separation. (a) Bistatic and (b) quasi-monostatic (differential) configuration ($D = 10$ cm). "n" is the number of the interface between layers. Df is distance h between receiving dipoles 4 and 6 cm. (© 2015 IEEE. Reprinted, with permission, from Varyanitza-Roshchupkina, L.A. et al., Comparison of different antenna configurations for probing of layered media, *8th International Work-shop on Advanced Ground Penetrating Radar (IWAGPR 15)*, 7–10 July 2015, Florence, Italy.)

of the antenna. The above factors lead to the conclusion that greater distortion results from increasing the distance D between the Tr and R antennas. Therefore, we need more sophisticated algorithms of radar data processing.

Considering these antenna factors, we concluded that the most accurate measurements antenna should have a monostatic configuration. In this case *monostatic* means the transmitting and receiving antennas are the same, or are aligned along the perpendicular to the surface of the probed region.

There are antenna switches for protecting of sensitive receiver input circuit from the powerful transmitting impulse during the radiation time interval (another words for duration of whole probing signal). But these antenna switches usually have switching times of tens of nanoseconds and more. In this case, a monostatic configuration would only work if the antenna system was located far enough from the surface of road to give the switch time to react and defend the receiver. But moving the antenna system away from the road surface produces a large electromagnetic field illuminating spot. This will prevent recognizing the local changing of layer thickness because of the measurement averaging of results over the spot area. It also means losing the measurement accuracy in the direction along the road surface.

Thus, we have to use a bistatic antenna configuration and locate the transmitting and receiving antennas close to each other and ensure a high electromagnetic decoupling between antennas. And bearing in mind that we use the UWB probing signal, we have to ensure this decoupling in whole working frequency band. Section 2.3.3 will explain the antenna systems operation.

To detect low contrast boundaries we need a higher GPR power output, that will increase the reflected signal and probability of detecting low contrast objects and boundaries. Thus, uncoupling the antennas allows adjusting the GPR power output depending on the object or boundary radar contrast.

Most significantly, the full frequency independent electromagnetic decoupling allows putting the transmitting and receiving antennas closer to one another to get the desirable monostatic configuration providing more accurate measuring.

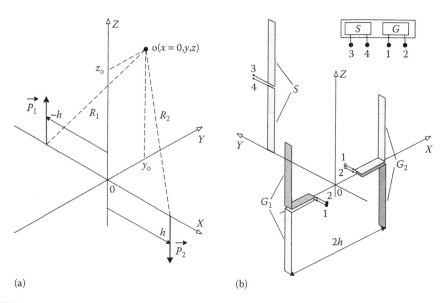

(a)   (b)

FIGURE 2.9
The differential antenna system using electric fields compensation to decouple the GPR antennas. (a) The principle of achieving decoupling. (b) The dipole antenna system. (From Kopylov, Yu. A. et al., Method for decoupling between transmitting and receiving modules of antenna system, Patent UA 81652, January 25, 2008.)

To meet these requirements, the authors suggest ways to achieve full frequency independent electromagnetic decoupling in the Ukrainian patent *Method for decoupling between transmitting and receiving modules of antenna system* [23]. The following section presents details of the transmitting-receiving (TR) antenna design and operational principles.

2.3.2 The Differential Antenna System Operating Principle

The decoupled antenna configuration has two dipoles placed symmetrically with respect to the YZ plane and excited by sources P_1 and P_2 with mutually opposite polarizations as shown in Figure 2.9.a. This antenna placement generates an electromagnetic field with only E_x and H_y components in the YZ plane. This means that if we place a plane conductor in the YZ plane, the pair of radiating dipoles does not induce any current in this conductor. Thus the receiving antenna located in the YZ plane does not receive any electromagnetic field generated by the two dipoles of the transmitting antenna. There is an absolute mutual compensation of components of the electromagnetic fields E_y, E_z, H_x, and H_z generated by the pair of transmitting dipoles. Moreover this holds independently of the exciting signal waveform and independently on the frequency band that is equal to the infinite frequency band.

Very high and frequency independent electromagnetic decoupling between the transmitting and receiving modules of the antenna system is possible if the radiating antenna is a single dipole which is located in lane YOZ and the receiving antenna is a pair of dipoles placed symmetrically with respect to the YOZ plane. In this case, it is only necessary to connect the outputs of the receiving dipoles in an appropriate way as shown in Figure 2.10.

The differential antenna decoupling works because the radiated electromagnetic field induces electromotive forces with the same waveform and amplitude in the receiving dipoles. Subtracting the signals coming from the outputs of the receiving dipoles in the summing unit produces a minimal signal at the antenna output U_{out}. Because the signal at

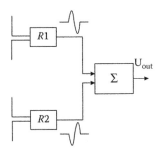

FIGURE 2.10
The summation of signals from the outputs of the differential receiving antenna module. (With kind permission from Springer Science+Business Media: *Unexploded Ordnance Detection and Mitigation,* Some advances in UWB GPR, NATO Science for Peace and Security Series –B: Physics and Biophysics, Jim Byrnes (ed.), 2009, pp. 223–233, G.P. Pochanin.)

the receiving antenna output is the difference of signals received by its two elements then it is called a *differential antenna system.*

Next we show how the differential antenna system can receive radar signals reflected by the object. Let the antenna system shown in Figure 2.11 have one transmitter dipole Tr and two receiving dipoles $R1$ and $R2$ located above the object at a distance H. The dipole axis and the direction of the currents $\vec{I}$ in them are perpendicular to the plane of the figure. Since the distances a and b are equal and the direct coupling signals passed paths a and b and induced in the receiving dipoles simultaneously and their amplitudes are equal, then by subtracting one signal from another they mutually cancel each other. This provides isolation between the transmitting and receiving modules of the antenna system. The signals reflected by an object and received by the receiving dipoles by ways c and d are different in amplitude and arrival time. The difference between these signals is not equal to zero. This ensures the reception of signals reflected by the object.

The receiving module in the differential antenna system works as a low-pass filter. In this case, the lowest working frequency depends on the relative delay between the signals received by the elements of the receiving module.

The antenna decoupling is equivalent to the background removal. In contrast to common digital procedure, the differential antenna works in an analog way and leads only to the subtraction directly coupled signals.

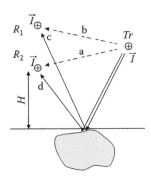

FIGURE 2.11
Signal paths for the differential antenna system located above the object. Signals through path "a" and "b" will cancel each other. Signals following paths "c" and "d" will not cancel each other and appear in the receiver.

2.3.3 The High Electromagnetic Decoupling Antenna System for UWB Impulse GPR

Let us theoretically consider how to measure the thickness of the top layer of an asphalt pavement with permittivities $\varepsilon_2 = 6$, $\varepsilon_3 = 5$ and thickness 5 cm as shown in Figure 2.12. This will help to estimate the effects of using antenna systems with and without direct electromagnetic coupling. The case without coupling is estimated by subtracting the coupling signal obtained as a result of a simulation of the task when dielectric layers are absent. In this case we put the probing impulse source and observation point at an height of 30 cm over the ground surface and spaced $D = 16$ cm as shown.

Probing the multilayer pavement by a first derivative Gaussian pulse of 0.4 ns duration at the −3 dB amplitude level will produce the signals shown in Figure 2.13 at the observation point. In the figure signal $E1$ corresponds to the case with coupling between the transmitting and receiving antennas and signal $E2$ without coupling.

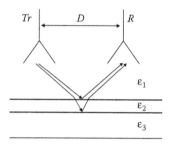

FIGURE 2.12
GPR bistatic antenna geometry of upper pavement layer thickness measurements.

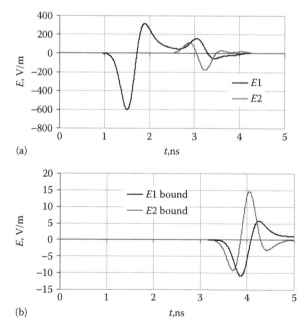

FIGURE 2.13
Synthetic GPR data from the antenna system of Figure 2.12 with above mentioned parameters. $E1$ shows the signal from bistatic antenna. $E2$ shows the signal from the differential antenna. (a) The whole reflected signal and (b) the signals reflected only by boundary between layers with $\varepsilon_2 = 6$ and $\varepsilon_3 = 5$.

Figure 2.13 shows the maximum signal amplitude at the observation point and receiver input in the configuration with coupling corresponding to $E1 \approx 600$ V/m. In the configuration without coupling, the maximum amplitude of signal at the input of receiver depends on the electric properties of the road surface and in our case is $E2 \approx 150$ V/m. It is clear that at the same radar receiver dynamic range the configuration without coupling can enlarge the GPR power budget by $4^2 = 16$ times in our case by permitting a strong probing signal, approximately 4 times larger without damaging or overloading the receiver input.

According to Figure 2.13 (b) the maximum amplitude of signal reflected by the boundary between layers with $\varepsilon_2 = 6$ and $\varepsilon_3 = 5$ in the configuration without coupling is a third more than in the configuration with coupling. This creates conditions for more accurate boundary location detection.

Another advantage of the configuration without coupling is that the allowable GPR transmitted power depends on the radar contrast of the boundaries between road structure layers. It follows from the fact that the maximum signal at the receiver input is determined only by the above contrast.

2.3.4 The Arrangement of the Transmitting-Receiving Antenna System

The differential transmitting-receiving antenna implements a method of complete frequency-independent isolation between the transmitting and receiving parts of the antenna system as described by Kopylov et al [23]. This method provides isolation by the use of a single wideband dipole as a radiating antenna and a pair of dipole receiving antennas. Pochanin et al developed several prototypes of the differential antenna systems and reported the experimental results based on this method [44–46].

The antenna system of Figure 2.14 consists of a transmitting antenna in the form of a pair of ellipses on the middle plate. The receiving module is the pair of metallic elliptical shaped dipoles one above and one below the plane of symmetry above and below the middle plate. Using dipoles as the antenna system, elements give us additional advantage because the shape of the transmitted impulse signal does not depend on the direction of the plane that is orthogonal to the electric dipole. The same reasoning applies to signal reception.

The receiving dipole outputs connect to the receiver so that when the signal received by one dipole is added to the signal received by another dipole, the signals induced by the transmitting antenna mutually cancel each other. In this case, the signals coming from

FIGURE 2.14
The differential antenna system arrangement showing the elliptical dipole transmitter (middle plate) and receiving antennas (top and lower plates).

the external objects are not compensated because of the time delay between them. Their sum at the receiver input gives information about the existence of radar objects.

The distance between the antenna elements of the receiving module is 80 mm in Figure 2.14. Owing to this separation, the antenna effectively receives the electromagnetic impulses coming from the vertical direction along the line that goes across both receiving antennas, and the typical rise time is less than 0.5 ns. A high voltage short pulse (SP) generator with an avalanche transistor switch drives the radiating antenna [4,47,48]. Transmitting and receiving dipoles are connected to the feeder lines by matching baluns.

Theoretically the transmitter and receiver antenna isolation should be absolute and frequency independent. Very accurate fabrication of the antennas and supporting structure assures an isolation better than −64.8 dB. Using a 75.7 V driving signal amplitude resulted in a direct coupling amplitude of 0.038 V and a voltage standingwave ratio (VSWR) of the transmitting and receiving antennas less than 1.6 in the working frequency band from 800 MHz to 1.6 GHz [15].

An antenna system built using this principle can also provide directional radiation and reception of UWB SP radar signals [44,45]. Its whole radar antenna radiation pattern is the product of those of the transmitting and receiving modules shown in Figure 2.15. Thus it has only two peaks along the perpendicular to the main plate in Figure 2.14. It also has a property unusual for UWB antennas that is a zero radiation pattern shown in Figure 2.15. The pattern has nulls in the transmitting dipole (symmetry) plane in any direction. It is unresponsive to clutter coming from objects and other sources of electromagnetic (EM) radiation situated in the antenna symmetry plane [45].

When using the differential antenna system, the largest signal coming from the receiving antenna to the receiver is the amplitude of signal induced in antenna by the electromagnetic wave reflected from the nearest reflecting object. If it is located quite far from the antenna system, the reflected wave has a low amplitude, therefore this permits using a higher driving voltage for the transmitting antenna excitation. This allows a substantially higher transmitted power from the UWB radar.

2.3.5 Differential Antenna Modes of Operation

A GPR with the differential antenna system can operate in the following two modes:

1. Horizontal scanning mode: The antenna system moves above the ground in this commonly used mode, as shown in Figure 2.11. The antenna pattern has two peaks in the nadir and zenith directions, and a null in the horizontal plane. Using the GPR system in the horizontal scanning mode is similar to the usual GPR technique but using a pair of differential dipoles for the receiving antenna. The higher transmit-receive antenna decoupling lets the GPR operate at higher radiated power levels than with conventional antennas.

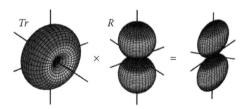

FIGURE 2.15
The differential antenna system pattern (schematically) is a product of the radiating and receiving modules patterns.

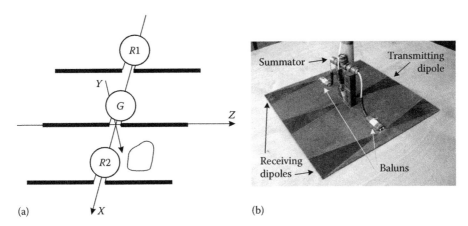

FIGURE 2.16
The GPR differential antenna scanning for small objects. (a) Accurate definition of small object location and (b) antenna system. (With kind permission from Springer Science+Business Media: *Unexploded Ordnance Detection and Mitigation,* Some advances in UWB GPR, NATO Science for Peace and Security Series –B: Physics and Biophysics, Jim Byrnes (ed.), 2009, pp. 223–233, G.P. Pochanin.)

2. Small object precision location mode: In order to provide the accurate definition that is needed for small object location, the antenna system must be rotated around the Z-axis by 90 degrees as shown in Figure 2.9 while moving the antenna along the X-axis [45,44]. The resulting configuration is shown in Figure 2.16.

If the antenna system in Figure 2.16.b moves along the axis OX over a small object as in Figure 2.16.a, then the received signal changes its waveform as shown in Figure 2.17 during the ground probing process.

Objects located far from the antenna system produce a small amplitude output signal. Figure 2.17 shows how the receiver antenna output increases as the distance between the antenna and object decreases. The output reaches a maximum, when one of the elements of the receiving antenna is over the object as shown in the left trace of Figure 2.17.a. The amplitude goes to its minimum when the object is in the antenna symmetry of the YZ plane as shown in the middle trace in of Figure 2.17.b. At further antenna displacements the signal amplitude increases again. The received signal changes polarity and reaches a maximum when the other receiving antenna element goes directly over the object as shown in the right side trace of Figure 2.17.c. Thus, when the radar moves over a small object, the signal amplitude at the output of the antenna system will go through zero and change polarity. The location of the antenna symmetry plane at the minimal output signal corresponds to the object location.

Figure 2.18 shows a GPR receiver output taken during a test to accurately locate small objects. The left side shows the initial differential antenna GPR profile without applying any data processing procedures taken over a 1 m path. The right side shows the results of GPR measurements of the same search using an ordinary bistatic antenna system.

The GPR measured the object's horizontal location with about a 2 cm accuracy. Note that in contrast to the bistatic antenna horizontal scanning mode, the differential antenna provides more accurate definition in the small object location mode and a better horizontal resolution at shallow depths.

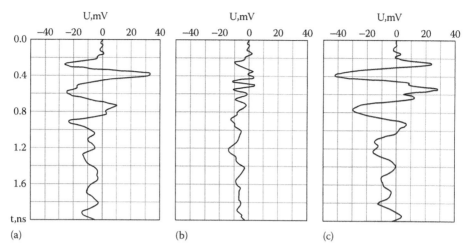

FIGURE 2.17
Differential antenna output signals for a GPR passing over a distant target. (a) Maximum signal output occurs when one antenna passes directly over an object. (b) Minimum signal output occurs when the object is in YZ plane antenna symmetry. (c) Maximum signal output with a polarity change occurs when the other antenna passes over the object. (With kind permission from Springer Science+Business Media: *Unexploded Ordnance Detection and Mitigation*, Some advances in UWB GPR, NATO Science for Peace and Security Series –B: Physics and Biophysics, Jim Byrnes (ed.), 2009, pp. 223–233, G.P. Pochanin.)

2.3.6 Microwave Tomographic Imaging with a Differential Antenna System

In addition to GPR, the differential antenna can also collect data for post processing with the microwave tomographic analysis methods developed by Soldovieri et al and Persico et al [49–51].

The first paper compares the imaging capabilities of a differential antenna systems composed of a receiving antenna-the transmitting antenna-receiving antenna. This configuration symmetrically offsets the receiving antenna relative to the transmitting antenna along three orthogonal directions [49]. Figure 2.19 shows the results of a FDTD simulation of radarograms for each measurement configuration. Further, the method of microwave tomography processing applied to the differential scattering data fields, show the capability to get a realistic picture of scattering objects. The scattering object simulations included a metal and a dielectric bar, a metal and dielectric sphere pair separated by a small distance by metallic and dielectric bars.

2.4 GPR Data Acquisition

2.4.1 GPR Analog-to-Digital Signal Data Conversion

Modern GPRs collect radar return data in digital form for computer signal processing and analysis. This requires adequate ADC characteristics in the receiver in terms of resolution and speed in ADC operations per second.

ADC resolution means the minimum change in the value of the analog signal needed to change the output digital code by one bit. The ADC resolution expressed in bits indicates

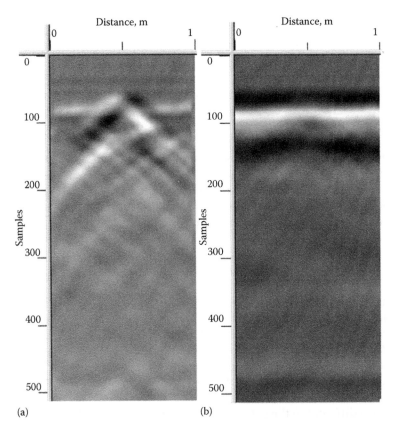

FIGURE 2.18
Experimental GPR received signal profiles showing: (a) the differential antenna system and (b) a conventional bistatic antenna system. (With kind permission from Springer Science+Business Media: *Unexploded Ordnance Detection and Mitigation*, Some advances in UWB GPR, NATO Science for Peace and Security Series –B: Physics and Biophysics, Jim Byrnes (ed.), 2009, pp. 223–233, G.P. Pochanin.)

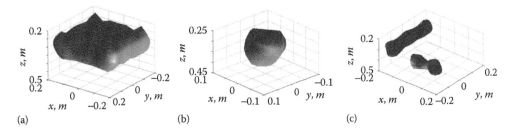

FIGURE 2.19
Radar images of scattering metallic objects taken with the differential antenna and microwave tomographic processing. (a) Bar, (b) sphere, and (c) pair of bars. (From Varianytsia-Roshchupkina, L.A. et al., Analysis of three RTR-differential GPR systems for subsurface object imaging, *Radiophysics and Electronics* 19, 48–55, 2014. With permission.)

whole number of discrete values that the converter can provide at the output. For example, a 12-bit ADC can give 4096 discrete output values (0 … 4095). Voltage resolution is equal to the voltage difference corresponding to the maximum and minimum output code, divided by the number of output discrete values. If the input range is from 0 to 1 Volt, the each 12-bit ADC increment represents an input voltage step of $1/4096 = 0.000244$ V $= 0.244$ mV.

2.4.2 Identification of GPR-Reflected Signals

If the amplitude of the recorded signal reflected by the object differs only by 1–2 discrete steps, the GPR can only determine that something reflected a signal from a given range. Based on this information the operator could not identify the reflecting object. Providing a signal identification capability requires that the ADC must cover many discrete values over the signal duration to recover the waveform. Chapter 3 will discuss receiver signal registration methods and Chapter 4 will present methods of signal time–frequency analysis.

To illustrate the possibilities for target identification, we can examine the cases of reflected signals and compare the results for cases of strong and weak electrodynamic coupling between the transmitting and receiving antennas. As initial signals, we use solutions of model problem of probing asphalt pavement with cracks inside the layer and on the surface, a practical inspection problem for inspecting a heavily traveled highway. In this case, the maximum signal will determine the ADC input range. We can select the largest of the signals resulting from either (a) a direct coupling signal from the radiator to the receiver, or (b) a reflection from the surface of asphalt. The noise level will be taken equal to U_N ~1 mV.

Table 2.1 shows the reflected signal amplitudes and analytical results.

Table 2.1 shows how with strong antenna coupling conditions the interior layer crack has an amplitude of only 4 discrete ADC units. The small reflection does not provide enough information to restore the shape of the signal and make a decision about the presence of a reflection. Considering the presence of noise, clutter, and instability of the equipment, it becomes clear that under strong coupling, the reflection from the crack inside the layer is invisible.

The situation is different in the case of weak or no antenna coupling. Reflection from the crack inside the asphalt amounts to 39-bits of ADC that is quite sufficient to distinguish the reflected signal even in the presence of noise.

This shows how we can provide the capability to detect low contrast objects by reducing the amplitude of the direct coupling signal from the transmitting antenna to the receiving antenna to the level lower than the amplitude of the signal reflected from the surface of the road. Certainly a decrease in the coupling between the transmitting and receiving antennas is possible not only by using a differential configuration but also by applying shielding.

TABLE 2.1

Pavement Probing Signal Conditions and Results

	Weak Coupling	Strong Coupling
Amplitude of the maximum signal U_{max}	0.022 V	0.95 V
Amplitude of the reflected signal		
• Surface crack	0.001 V	0.02 V
• Interior layer crack	0.002 V	0.001 V
Sampling ADC in amplitude $(U_{max} - U_N)/4096$	0.0000051 V	0.00023 V
Number of discrete ADC bits in the amplitude of the reflected signal from:		
• surface crack	196	86
• interior layer crack	39	4

2.5 Data Acquisition Methods with a Low Jitter Receiver

Unfortunately there are no multi-bit ADCs that can directly convert nano-or subnanosecond duration signals into digital form. Existing ADCs work too slowly to provide this capability. Therefore, we must use techniques developed for stroboscopic receivers used in sampling oscilloscopes. These methods are usually applied for converting the time scale from nano-or subnanosecond durations to microseconds where ADCs with many bits are available [52–55].

Because we cannot digitize an individual reflected signal, we must find a way to record it over a number of repetitions. This assumes each reflected signal, s, will closely resemble the preceding and following signals, that is, $s((n-1)t) \approx s(nt) \approx s((n+1)t)$ and each repetition will occur within an exact interval T_s.

Figure 2.20 shows the essence of stroboscopic digital conversion that consists of a sequential, step by step sampling with a small shift in time in respect to the beginning of the period T_s and sampling the amplitude of a repetitive signal in very small time gates $\Delta\tau$. The process then holds a selected measure of the amplitude during the repetition period and digitizes the measured portion using an ADC. After collecting enough partial samples of s it is possible to reconstruct the shape of the observed repetitive signal. The stroboscopic process assumes a signal with enough repetitions and time stability to ensure the collection of samples. The next sections will discuss ways to meet these assumptions in short-pulse radar applications.

2.5.1 Noise and Signal Shape Distortions versus the Sampling Gate Duration

The need for greater receiver sensitivity requires new ways to decrease the receiver noise level and increase the SNR in the ADC sampling process. Ruban and Pochanin examined this in "sampling duration for noisy signal conversion" [56].

We know that the noise level of the sampler depends on the working bandwidth of the receiver signal sampler, that depends on the sampling gate duration. The sampling gate duration of the sampler must take into account the cutoff frequency of the received signal spectrum and the measurement error requirements [57]. Precise measurements for waveform reconstruction require smaller sampling gate durations, that means a wider working

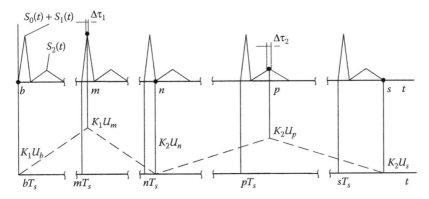

FIGURE 2.20

Stroboscopic signal conversion provides a way to accurately digitize repeating signal waveforms. b, m, n, p, s are numbers of periods; K are amplitude conversion coefficients. Input signals (above). Output signal (below). (From Zhuravlev, A. et al., Shallow depth subsurface imaging with microwave holography, *Proc. of SPIE Symposium on Defense and Security*, 9072, 90720X-1...9, 2014; G. P. Pochanin, Patent UA 81652, Jan 25, 2008.)

sampler bandwidth. At the same time, the noise level at the receiver input increases with the widening of the working bandwidth, that leads to a less sensitive sampler. This means at low SNRs, the measurement accuracy becomes significantly worse than theoretically possible, despite good matching of the signal spectrum and the working bandwidth. Thus, the extension of the working bandwidth during weak signal reception leads to the degradation of the measurement accuracy instead of the desired increase.

Our ultimate goal is to accurately determine the waveform of weak received signals. Therefore, we need to define a selection criterion for the sampler working bandwidth (sampling gate duration).

2.5.1.1 Signal Sampler Modeling and Analysis

The practical approach uses an approximate (simplified) model of the sampler, described by formula for periodic signal conversion $u(t)$ [52] (Equation 2.20):

$$U(\theta) = \int_{-\tau/2}^{\tau/2} u(t + nT_R + \theta + T_0)\,dt. \tag{2.20}$$

where:
$U(\theta)$ is the average signal amplitude at the moment θ
τ is the sampling gate duration
T_R is the repetition period
T_0 is the time delay
n is the sampling number

To show effectiveness, we shall consider the recording of impulses of different shapes including the Gaussian impulse and its first and second derivatives. In the model, the detected signal is a sum of the clear signal and the noise voltage – $u(t) = u_i(t) + u_n(t)$.

2.5.1.2 ADC Sampling and Conversion Errors

The ADC process gives a set of approximations of the received waveform. We need to ask how good is the approximation for a given sampling rate and interval? Correlation of the original and sampled signal can provide a measure of ADC effectiveness. This section will examine the effects of ADC conversion errors by comparing shapes of the converted signal $U(\theta)$ and the true detected signal $u_i(t)$. This process assumes that $\theta = t$ and uses the correlation coefficient $R_{U,ui} = \text{cov}(U, u_i)/\sigma_U\sigma_{ui}$, where $\text{cov}(U, u_i)$ is the covariance of signals $U(t)$ and $u_i(t)$, σ_U and σ_{ui} are the standard deviation of $U(t)$ and $u_i(t)$, respectively.

Judging by the maximal value of the function $R_{U,ui}$, this indicates the degree of resemblance between the ideal noise-free signal and a signal converted by the sampler with a sampling duration τ. Therefore, the correlation coefficient can estimate the degree of converted signal distortion.

The graphs of Figure 2.21 show the dependencies of signal correlation on the sampling duration for chosen noise levels at the sampler input. Each plot shows the calculated correlation coefficient for the Gaussian pulse and two derivative signals. In these calculations the noise level σ was normalized to the peak signal amplitude u_{max}. The horizontal axis shows the sampling duration τ normalized to the pulse duration δ at the level $0.5u_{max}$.

The graphs in Figure 2.22 show an obvious maximal correlation coefficient for each noise level. This is due to the fact that in a wide working frequency band of the sampler,

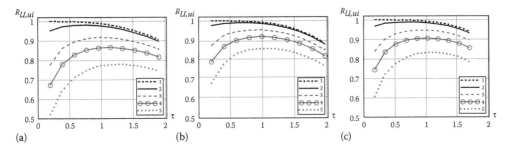

FIGURE 2.21

The correlation coefficient $R_{U,ui}$ of the converted signal and the source signals depends on the sampling interval and the noise level in these plots for (a) Gaussian pulse signal, (b) first derivative of the Gaussian pulse, and (c) 2nd derivative of the Gaussian signal. The plots show how the correlation depends on the sampling duration τ normalized to the pulse duration δ and for noise levels normalized to the peak signal level u_{max}. In each case, trace 1 = 0 u_{max}, trace 2 = 0.2 u_{max}, trace 3 = 0.5 u_{max}, trace 4 = 0.7 u_{max}, and trace 5 = 1 u_{max}. (From Ruban, V.P. and Pochanin, G.P., Sampling duration for noisy signal conversion, *Proc. of 5th Int. Conf. on Ultra Wideband and Ultra Short Impulse Signals*, September 6–10, Sevastopol, Ukraine, pp. 275–277. © 2010 IEEE.)

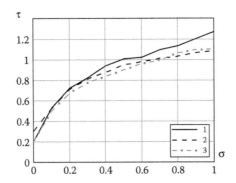

FIGURE 2.22

Dependence of optimal normalized sampling duration τ on the noise level σ: Trace 1-the Gaussian signal, Trace 2- for the 1st derivative of Gaussian signal, Trace 3- for the 2nd derivative of Gaussian signal. (From Ruban, V.P. and Pochanin, G.P., Sampling duration for noisy signal conversion, *Proc. of 5th Int. Conf. on Ultra Wideband and Ultra Short Impulse Signals*, September 6–10, Sevastopol, Ukraine, pp. 275–277. © 2010 IEEE.)

where $\tau \ll \delta$, the noise essentially distorts a received waveform. In a more narrow band, where $\tau > \delta$, the signals are essentially distorted because of the converting process. The amplitude of the converted signal is averaged over the sample duration. Also fast amplitude changes in the time interval of the sample are smoothed and converted to a single value of the amplitude. The graphs also show how increasing the noise level decreases the signal correlation, and the maximum of correlation curve moves towards a long sampling duration.

Obviously, at a certain level of noise the maximum of the correlation curve will correspond to the certain length of the sample.

Based on this criterion, we plotted graphs of the dependence of optimal sampling duration on the noise level for three signal types shown in Figure 2.23.

The graphs show how, when the noise level drops to zero, the optimal sampling duration tends to values that correspond to classical criteria of one $2B$ samples per second, where, B indicates the highest frequency of the wave.

Maximal deviation. Another way to estimate the signal conversion error is to calculate the relation between the amplitudes of the original signal U_m and a converted signal U_c at the areas of maximal distinction between them. As it turns out, these areas of maximal

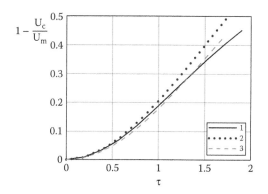

FIGURE 2.23
Signal conversion errors versus the normalized sampling interval τ of: 1-the Gaussian signal, 2- the Gaussian 1st derivative, and 3-the second derivative. (From Ruban, V.P. and Pochanin, G.P., Sampling duration for noisy signal conversion, *Proc. of 5th Int. Conf. on Ultra Wideband and Ultra Short Impulse Signals,* September 6–10, Sevastopol, Ukraine, pp. 275–277. © 2010 IEEE.)

distinction coincide with extreme values of signals. Therefore, we propose to estimate the measurement error of the signal amplitude using relative units (relative to the peak-maximum) at time points, that are coincident with signal extremes values.

Dependencies of the conversion error in the area of peaks of signals on sampling duration for three signal types are shown in Figure 2.23. These graphs can help to determine the maximum discrepancy of the converted signal when selecting an optimal sampling duration. For example, to detect a Gaussian signal derivative waveform with a sampler input noise level of $\sigma = 0.7$, then we determine an optimal sampling duration of $\tau = 1$ from the plot in Figure 2.22. At this sampling duration, the maximum converted signal amplitude will differ by 20% from a signal at sampler input. A signal converted by the sampler, where sampling duration is determined by the traditional condition (sampling gate duration equals 0.1), is shown in Figure 2.24.a [52]. A signal converted by the sampler where sampling duration equals to 1, is shown in Figure 2.24.b.

The diagrams clearly have shown how the increase of sampling duration leads to a sufficient increase of SNR. Under those conditions, the converted waveform has changed very slightly.

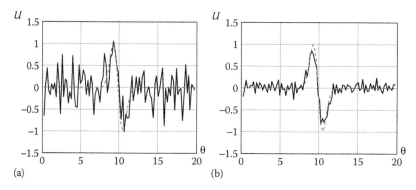

(a)   (b)

FIGURE 2.24
Comparison of input and output converted signals (a) At a normalized sampling duration 0.1; and (b) at a normalized sampling duration 1. In each case the input signal is marked by dashed line. Increasing the sample duration increases the SNR. (From Ruban, V.P. and Pochanin, G.P., Sampling duration for noisy signal conversion, *Proc. of 5th Int. Conf. on Ultra Wideband and Ultra Short Impulse Signals,* September 6–10, Sevastopol, Ukraine, pp. 275–277. © 2010 IEEE.)

The examined approach permits

- Evaluating the degree of distortion of the noisy signal being detected with a given noise level at the optimal sampling duration.
- A way to determine a sampling duration of the sampler, judging by a permissible error of signal conversion.
- A way to evaluate the noise level, at which the cross-correlation of original and converted signals is maximal.

2.5.1.3 Data Acquisition by a Stroboscopic Receiver with a Variable Sampling Gate Duration

The previous section showed there is an optimal sampling gate duration that can provide a better SNR with minimal signal shape distortion. It depends on the radar transmitted pulse duration, but more importantly on the duration of the received pulses. Choosing an optimal sampling gate width can give the required resolution and to raise the sensitivity as high as possible [24].

Reflected objects buried deeply in the ground have smaller reflected signal amplitudes than other closer objects. At the same time, long duration pulse signals have longer rise and fall times. Therefore at the beginning of the scan we can choose a small *sampling gate* duration, where the reflected signals correspond to boundaries close to the ground surface. In this case, the probing signal-propagation path is small, therefore the attenuation is insignificant and the SNR is large. This condition means that the small sampling gate duration works better because it ensures better waveform reconstruction, a higher radar measurements accuracy, and better GPR spatial resolution.

If receiving reflected signals from deep layers or buried objects, then the probing wave path of propagation is long and has a high attenuation due to soil properties, and the probing wave reflection by upper layers. This means the GPR has a small SNR at longer ranges that makes it reasonable to extend the sampling gate duration. As a result, a longer sampling gate duration ensures an increased SNR and improves the received waveform reconstruction. These conditions give better radar range measurement accuracy and GPR spatial resolution.

Therefore, we can change the sampling gate duration to make it optimal for a certain depth. This will improve the receiver sensitivity that ensures better waveform reconstruction and achieves the highest of possible resolution along the whole A-scan.

Because the optimal sample duration is a critical characteristic for UWB SP radar, our prototype GPR included a computer software function to adjust the sample duration during radar measurements [58,59]. In addition, the radar had special modes that shortened the sample duration to minimize waveform distortion in high SNR cases for close and high contrast objects. For deeper or low contrast targets, the computer could increase the sample duration to optimally record the signal waveform.

2.5.2 Received Signal Accumulation Approaches to Decrease Jitter

For signal measurement without noise, the ADC amplitude resolution is directly determined by the number of bits. In practice, the input signal SNR limits the ADC resolution. When high-intensity noise is applied to the ADC input, the discrimination of adjacent levels of input signal becomes impossible and degrades the ADC resolution.

A possible way to decrease the noise level and to increase power budget of UWB radars, is accumulating the energy of a number of signals reflected by an object [46]. Accumulating or summing a certain number of signals, increases the SNR proportionally as the square root of number of accumulated signals.

Note that conventional SP UWB systems such as the Novelda nanoscale impulse radar and Camero, through the wall, and imaging radars use a system of accumulating and integrating received energy in range bins to improve performance for range measuring systems [4]. In these cases, narrow band systems perform signal accumulation in analog form in selective oscillating circuits that permit selective reception of the chosen signal and suppression of other signals including noise and clutter.

Using stroboscopic transformation as a method of receiving reflected pulse signals promotes accumulation. The result of such an accumulation process is the same as filtering with a digital comb filter [60].

Our GPR uses both digital and analog accumulators. The term digital accumulation means (1) the recording of signals (which includes reception, stroboscopic conversion, analog-digital conversion, the transmission of numerical data in the computer) for a certain number of signals (2) then summing of the numbers corresponding to the signals amplitudes at the same points in time and (3) then dividing the amplitude by the number of averaged signals. After signal accumulation, the processor records the results to a file and displays them. Many GPR systems use digital accumulation.

Analog accumulation means the summation of the received signal energy during the stroboscopic conversion in the sample and hold. This analog accumulation process performs a sequential selection of amplitude samples of the received signal corresponding to the same time point for a predetermined number of times. After the analog-to-digital conversion, the signal processor transfers the signal file to a PC to record and to process the radar probing results. In addition to increasing the SNR, the analog accumulation allows a significant expansion of the operating GPR receiver bandwidth owing to the additional charging of the sample and hold capacity [61].

The effective use of accumulation in UWB radar systems requires applying highly stable measuring equipment with very low jitter. If the temporal instability is large, the accumulation can distort the received signal shape, and increase its duration. As a result, the UWB radar range resolution becomes worse. Signal shape distortions create additional difficulties for radar data processing. The investigation by Kholod and Orlenko has shown the dependence of an amount of signals that could be accumulated on jitter [62]. According to their results, it is possible to determine the optimum quantity of the pulses that may be accumulated as (Equation 2.21)

$$N = \frac{0.2\sqrt{3}}{n \cdot \zeta} \cdot T,$$

(2.21)

where:

    n is the ratio of maximum frequency in the spectrum of the signal to the minimal frequency

    T is pulse repetition period

    ζ is the root mean square (rms) deviation of the pulse repetition period.

    For example, for an impulse radar signal with the upper frequency 1 GHz, with the pulse repetition frequency 1 MHz, and the quantity of the stored pulses $N = 10$, then the pulse repetition period instability should not exceed the value $\zeta = 35$ ps. For an accumulation of 35 pulses, the radar requires providing instability of $\zeta \leq 10$ ps.

    Thus, higher stability measuring equipment means the signal process can accumulate a larger number of signals without damaging the reflected signal information and could achieve a larger power budget.

At least two reasons cause the temporal instability in UWB SP radars with stroboscopic converters. These are (1) synchronizing jitter of the power probing pulse generator and (2) sampling pulse generator jitter in the synchronizing pulses going to the stroboscopic converter. This especially concerns stroboscopic converters using the analog method of forming of time shift for the sampler.

We can improve the equipment stability by using the following approaches:

- Replacement of the analog circuit providing time shift between samples in the stroboscopic converter by a digital delay line (DDL) with a fixed time offset between samples
- Using a small part of the energy of a powerful probe pulse to synchronize the DDL as shown in Figure 2.25.

Using this method of forming of time delay in the microchip, makes it possible to provide extremely low jitter. Currently chip digital delay lines in increments of 10 ps or even 5 ps are available. Accordingly, the instability cannot exceed a few picoseconds. Synchronization of the receiver by means of the emitted signals solves the problem of choosing a temporary moment corresponding to the beginning of the recording of the received signal.

Using fast comparators, we can get a total jitter that is approximately 12 ps as shown by the plot in Figure 2.26, where 500 signals are placed in the same plot and the quickest changing part of the signals are magnified to get the ability to see the time interval, where all 500 signals are fitted in the time interval equal to only 12 ps. Calculation of the root mean square of time-base-errors of synchronization, $\delta\tau$ has shown the value approximately equal to 2.4 ps for the distribution as shown in Figure 2.27.

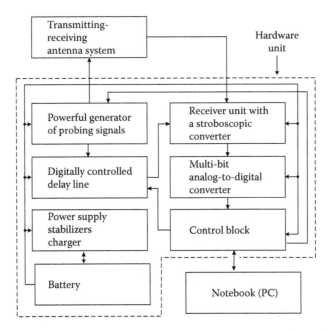

FIGURE 2.25
Block diagram of the UWB radar for recording complete received waveforms.

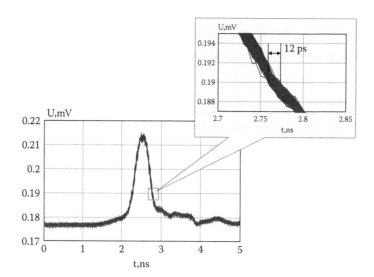

FIGURE 2.26
Falls of 500 signals in the same plot showing the effects of comparator jitter. The *x*-axis shows the time scale and the *y*-axis shows the voltage.

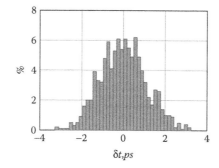

FIGURE 2.27
Distribution diagram for the time-interval-errors of synchronization.

2.6 Software for Radar Data Processing

We created three different software modules to use all the possible advantages of the radar. (1) The *SignalProcessorEx* module controls radar data acquisition. (2) The *GPR ProView* module performs initial GPR data processing and (3) The *GeoVizy* module can accurately determine the depth of the pavement layers boundaries.

2.6.1 The *SignalProcessorEx* Data Acquisition Module

The *SignalProcessorEx* module controls all of the GPR functions. These include the commonly used functions such as time scale selection, determining the number of samples in A-scans, averaging received signals, background removal, and others. The module provides the following possibilities to select characteristics for optimum radar system performance:

- The stroboscopic receiver sample gate duration.
- The number of accumulated analog signals.
- The transmitted signal repetition rate.

These new functions allow us to optimize the receiver parameters such as the SNR and thereby extend the GPR dynamic range.

The radar module and PC data exchange protocol has great flexibility. The operator can add new radar functions by adding blocks of new computer programs.

2.6.2 The *GPR ProView* Module

The *GPR ProView* module has a number of signal-processing operations intended to improve quality of the radar data. The software module includes: (1) different filters such as frequency, spatial, mask and others and (2) transforms including the Hilbert, Hough, [4,44,63] and Fourier.

2.6.3 Computational Algorithm of the Software *GeoVizy*

The *GeoVizy* algorithm computation uses the reasoning of the Section 2.2.3 to determine the thickness of pavement layers by using transfer functions. It consists of the following 7 steps:

Step 1: Based on primary radar data and using the Hilbert transform, the algorithm determines the time delay Δt_i of signals coming from the lower boundaries and signals amplitudes [39,40]. It also uses this information as an algorithm input.

Step 2: It applies Equation 2.18 to determine the upper layer permittivity $\sqrt{\varepsilon_1}$ and then ε_1. It determines the incident signal amplitude from the registered reflection from a metal sheet lying on the surface of the test coverage area immediately prior to the measurements.

Step 3: It uses Equations 2.14 through 2.17 to calculate the times $T_{0,1}$, $T_{1,0}$, and amplitudes $A_{0,1}$, $A_{1,0}$. It does this to use those quantities again in Equation 2.18 to calculate the second layer permittivity ε_2.

Step 4. It repeats the calculations until it determines the dielectric constant of the lowest layer – substrate (soil).

Step 5: It then calculates the layer boundary coordinates according to the formula (Equation 2.22):

$$Z_I = \sum_{n=1}^{I} \frac{(t_n - t_{n-1}) \cdot c}{2 \cdot \sqrt{\varepsilon_n}} = \sum_{n=1}^{I} \frac{\Delta t_n \cdot c}{2 \cdot \sqrt{\varepsilon_n}}. \tag{2.22}$$

where:

Z_I is the coordinate of the *I*th boundary (the highest boundary has index 0)

t_n, t_{n-1} are the time of the signal passing the boundaries nth and $(n-1)$th determined in the stage 1

ε_n is the permittivity of the nth layer that has been calculated at the stages 2 or 3

Step 6: It calculates the thicknesses of the layers h_n using the formula (Equation 2.23):

$$h_n = Z_n - Z_{n-1}; \quad Z_0 = 0, n = 1, 2, \dots, I. \tag{2.23}$$

Step 7: It records the data for later use.

Thus, the *GeoVizy* procedure can determine the thickness and the permittivity of pavement structural layers by direct surface measurements and without additional laboratory tests.

2.7 The Ground-Penetrating Radar Prototype

2.7.1 Radar Equipment Components

The pavement monitoring radar *ODYAG* shown in Figure 2.28, consists of the transmitting-receiving antenna, the block of electronics including the sounding pulse generator, control unit, sampling receiver, DDL, ADC, power supply with a battery (6V, 7A·h), and charger. It also has a power supply cord adapted for the vehicle electrical system, a feeder line, and a computer control cable. The *ODYAG* system includes computer software for GPR control and data acquisition, the software, the radar data processing, and interprets the results [15].

2.7.2 *ODYAG* Specifications

The sounding pulse generator:

- Amplitude of the sounding pulse at the load 50 Ohms is more than 75.7 V
- Repetition rate is less than 500 kHz
- Pulse rise time is less than 0.4 ns

The transmitting – receiving antenna:

- Operating frequency band of the transmitting and receiving parts of the antenna has a range from 0.8 GHz to 1.6 GHz
- Attenuation of the signal passing directly from the transmitting antenna to the receiver input is more than 64.8 dB

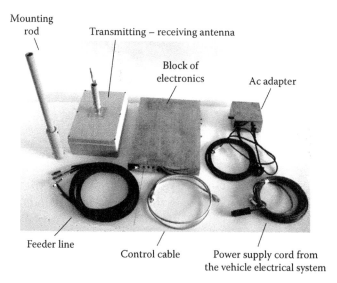

FIGURE 2.28
The GPR *ODYAG* components. (From Pochanin, G.P. et al., GPR for pavement monitoring, *Journal of Radio Electronics*, 2013. http://jre.cplire.ru/alt/jan13/8/text.pdf [accessed from September 21, 2016]. With permission.)

The strobe receiver unit:

- The level of noise at the input of the stroboscopic converter is less than 200 μV
- The gating step is 10 ps
- Rise time characteristics of the stroboscopic converter is less than 0.2 ns
- The jitter is less than 3 ps

The ADC has 16-bit resolution.
    The control unit:

- Communicates with the computer by the Ethernet port via a standard cable connection (100 Mb/s).

The scan ranges: 5 ns, 10 ns, 20 ns, and 40 ns.
    The software provides GPR control and data transfer. It displays the sounding results on the computer monitor in the form of the received waveforms or profile and calculates the pavement structural layers thickness.

2.8 *ODYAG* GPR Test Results

Figure 2.29 shows *ODYAG* GPR tests performed at two locations on recently repaired roads, drilling to provide a comparison of core sample measured by radar results.

FIGURE 2.29
GPR *ODYAG* tests performed on newly repaired roads. After the measurements, a core sample was taken to evaluate the radar accuracy. (From Pochanin, G.P. et al., GPR for pavement monitoring, *Journal of Radio Electronics*, 2013. http://jre.cplire.ru/alt/jan13/8/text.pdf [accessed from September 21, 2016]. With permission.)

(a)

(b)

FIGURE 2.30
GPR investigation of highway M18: (a) scanning in a longitudinal direction and (b) cores of GPR data for verification of the obtained results.

First road in Figure 2.29 was M03 Kyiv–Kharkov–Dovzhansky, the other in Figure 2.30, M18 Kharkov–Simpheropol–Alushta–Yalta. The goal of investigations was to assess the thickness of the road pavement structural layers, evaluating subsurface inhomogeneities in the pavement structure, and evaluation of subgrade soil moisture.

Figure 2.31 shows the measurement results processed with the *GeoVizy* software. Table 2.2 shows the results of GPR measurement of the pavement layer thickness.

2.9 Example GPR Applications

The highway pavement measurements in the previous section demonstrated how the GPR can accurately determine layer thickness on a stretch of newly paved road. This section will show several more results of GPR road inspection applications. We will show further examples of the practical application of the proposed algorithm for measuring the thickness and electric parameters of the upper pavement layers on the

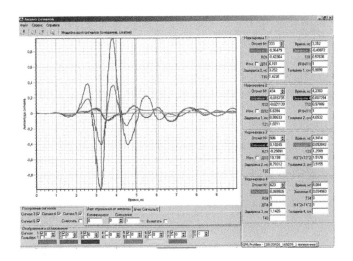

FIGURE 2.31
Computer display operating window for the GeoVizy software module (Russian version). The measurement determined the pavement layer thickness. (From Pochanin, G.P. et al., GPR for pavement monitoring, *Journal of Radio Electronics*, 2013. http://jre.cplire.ru/alt/jan13/8/text.pdf [accessed from September 21, 2016]. With permission.)

TABLE 2.2

GPR Pavement Layer Thickness Measurement Results

	Depth	
GPR Measurement	**The Actual Value from the Core**	**Estimated Value Using the Results of GPR Surveys**
Location: M03 Kyiv–Kharkov-Dovzhansky at 519 km		
Package thickness of asphalt layers	10.5 cm	10.124 cm
The thicknesses of the first and second layers	The upper layer: 5.5 cm The lower layer: 5.0 cm	The upper layer: 5.174 cm The lower layer: 4.9496 cm
Location: M03 Kyiv–Kharkov-Dovzhansky at 528 km		
Thicknesses of the first, second and third layers	The upper layer: 6.0 cm The second layer: 4.0 cm The third layer: 4.0 cm	The upper layer: 5.989 cm The second layer: 4.09 cm The third layer: 3.915 cm

section of the repaired road. The measurement accuracy met the requirements of the corresponding standard.

As it was mentioned above, we tested our newly developed GPR to inspect 5 km of recently repaired road section between the cities of Simferopol and Alushta on the Crimean peninsula. Figures 2.32 [16], 2.33 and 2.34 show examples of raw radar probing, radarograms of the results after preprocessing, and radar post processing results. As it follows from the Figure 2.34, the thickness of upper asphalt layer, that is stone mastic asphalt concrete, equals 5 cm ± 0.5 cm. The total coating thickness varies from 10 to 37 cm. Figure 2.35 shows the mobile laboratory used to evaluate the highway.

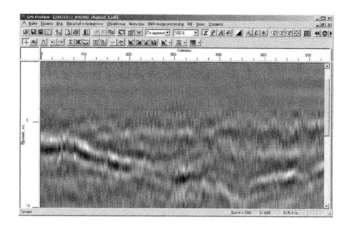

FIGURE 2.32
GPR pavement inspection data collected over a 5 km stretch of highway and displayed by the *GPR ProView* software. (From Pochanin, G.P. et al., Measuring of thickness of asphalt pavement with use of GPR, *Proc. of the 15th International Radar Symposium*, University of Technology, Warsaw, Poland, pp. 452–455, 2014). With permission.)

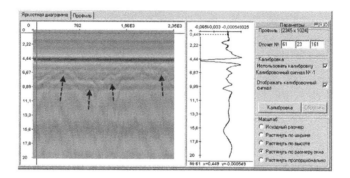

FIGURE 2.33
Radarogram of the results of the scan in the longitudinal direction of the section 680 + 800 − 681 + 000. The arrows indicate the cross sections of the subsurface irregularities (cracks in the old surface, filled with bitumen) at a depth of 15–30 cm.

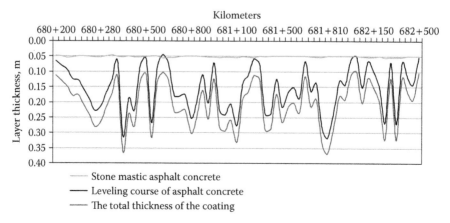

FIGURE 2.34
Post processing results of 5 km road survey in Figure 2.32.

FIGURE 2.35
The mobile laboratory for pavement monitoring.

2.10 Conclusions

This chapter presented possible ways to increase the power budget of a UWB SP GPR and to make more precise measurement of thickness layers of the road pavement.

The GPR shown above used an antenna system with complete frequency independent decoupling between the transmitting and receiving modules, which makes:

- It possible to significantly gain probing signal power,
- The radar less sensitive to obstructive objects including the radar hardware located in the antenna system symmetry plane.

Producing a high stability of the transmitter synchronization, enables the use of coherent analog energy accumulation of a large number of reflected signals that increased receiver SNR.

The accompanying software provided an adjustable sampling gate duration that allowed optimizing the duration during the probing process. This adjustable gate duration minimizes distortion of the received signal and a sufficient SNR for accurate waveform determination.

These GPR and software innovations now help the users to quickly determine thickness of the pavement structural layers with the high precision needed for road monitoring. The radar measurements are compared favorably with measurements taken from pavement samples (cores).

When used for road monitoring, the GPR showed good accuracy and high productivity. Users could install the GPR on a mobile laboratory and monitor pavement conditions at a speed of 30 to 40 km/h.

Editors Epilogue

The methods described here take the GPR from a simple range measuring device to a valuable instrument for determining the properties of media and objects. The example of pavement layer measurement represents just one potential application for the SP GPR and signal-processing methods. The methods have potential applications in other MPR problems such as nondestructive testing, security, medical imaging work, and others to be determined.

References

1. Harmuth, H.F. 1981. *Nonsinusoidal Waves for Radar and Radio Communication.* Academic Press, New York.
2. Astanin, L.Yu. and Kostylev, A.A. 1997. *Ultrawideband Radar Measurements: Analysis and Processing.* The Institute of Electrical Engineers, London.
3. Taylor, J.D. 1995. *Introduction to Ultra-Wideband Radar Systems.* CRC Press, Boca Raton, FL.
4. Taylor, J.D. 2012. *Ultrawideband Radar Applications and Design.* CRC Press, Boca Raton, FL, 536 p.
5. Finkelstain, M.I. 1994. *Subsurface Radar.* Radio i Svjaz, Moscow. (in Russian)
6. Daniels, D.J. 2004. *Ground Penetrating Radar.* The Institution of Electrical Engineers, London.
7. Grinev, A.Yu. [ed] 2005. *Problems of Subsurface Radiolocation.* Radiotekhnika, Moscow. (in Russian)
8. Glebovich, G.V. (ed). 1984. *Investigation of Objects Using Picosecond Pulses.* Radio and Communication, Moscow. (in Russian)
9. Finkelstein, M.I., Kutev, V.A. and Zolotarev, V.P. 1986. *The Use of Radar Subsurface Probing in Engineering Geology.* p. 128, Finkelstein, M.I. (ed), NEDRA, Moscow. (in Russian)
10. Ground penetrating radar for road structure evaluation and analysis. http://www.geophysical.com/roadinspection.htm.
11. Concrete & Pavement. //http://www.sensoft.ca/Applications/Concrete-and-Pavement.aspx
12. Engineering geophysical surveys for highways. URL: http://www.geotechru.com/en/file-manager/download/446/road_inspection.pdf.
13. Pajewski, L. and Benedetto, A. (eds). 2013. *Civil Engineering Applications of Ground Penetrating Radar, Booklet of Participants and Institutions.* Aracne, Rome, July 194 pp.; ISBN 978-88-548-6191-6.
14. Cost action TU1208. Civil engineering applications of ground penetrating radar http://gpradar.eu/onewebmedia/BOOKLET%20TU1208%20WEB.pdf.
15. Pochanin, G.P., Ruban, V.P., Kholod, P.V., Shuba, A.A., Pochanin, A.G., Orlenko, A.A., Batrakov, D.O. and Batrakova, A.G. 2013. "GPR for pavement monitoring," *Journal of Radio Electronics.* No.1. Moscow, http://jre.cplire.ru/alt/jan13/8/text.pdf.
16. Pochanin, G.P., Ruban, V.P., Batrakova, A.G., Urdzik S.N. and Batrakov, D.O. 2014. "Measuring of thickness of asphalt pavement with use of GPR," *Proc. of the 15th International Radar Symposium.* (Gdansk, Poland, June 16–18, 2014), University of Technology, Warsaw, Poland, pp. 452–455.
17. Batrakov, D.O., Batrakova, A.G., Golovin, D.V., Kravchenko, O.V. and Pochanin, G.P. 2014. "Determination of thicknesses of the pavement layers with GPR probing," *Physical bases of instrumentation,* Vol. 3, No. 2, p. 46–56. (in Russian)
18. Goncharsky, A.V., Ovchinnikov, S.L. and Romanov, S.Yu. 2009. "The inverse problem of wave diagnostics of roadway," *Computational Methods and Programming.* Vol. 10, pp. 275–280. (in Russian)
19. Varyanitza-Roshchupkina, L.A., Pochanin G.P., Pochanina I.Ye. and Masalov S.A. 2015. "Comparison of different antenna configurations for probing of layered media," *8th International Work-shop on Advanced Ground Penetrating Radar (IWAGPR 15), Conference Proceedings,* Florence, Italy, July.

20. Varyanitza-Roshchupkina, L.A. 2006. "Software for image simulation in ground penetrating radar problems," *III International Workshop "Ultra Wideband and Ultra Short Impulse Signals" (UBUSIS 2006): proc.* Sevastopol, pp. 150–155.

21. Golovko, M.M., Sytnic, O.V. and Pochanin, G.P. 2006. "Removing of the trend of GPR data," *Electromagnetic Waves and Electronic Systems.* Vol. 11, No. 2–3, pp. 99–105. (in Russian)

22. Zhuravlev, A., Ivashov, S., Razevig, V., Vasiliev, I. and Bechtel, T. 2014. "Shallow depth sub-surface imaging with microwave holography," *Proc. of SPIE Symposium on Defense and Security. Radar Sensor Technology XVIII Conference.* Baltimore, MD, May 5–7, Vol. 9072. pp. 90720X-1...9.

23. Kopylov, Yu.A., Masalov, S.A. and Pochanin, G.P. Method for decoupling between transmitting and receiving modules of antenna system. Patent UA 81652. Jan 25, 2008. (in Ukraine)

24. Pochanin, G.P. and Ruban, V.P. Stroboscopic method of recording signals. Patent UA 96241, Dec. 07, 2010. (in Ukraine)

25. Batrakov, D.O. and Zhuck, N.P. 1994. "Solution of a general inverse scattering problem using the distorted Born approximation and iterative technique," *Inverse Problems.* Vol. 10, No. 1. pp. 39–54. Feb.

26. Batrakov, D.O. and Zhuck, N.P. 1994. "Inverse scattering problem in the polarization parameters domain for isotropic layered media: solution via Newton-Kantorovich iterative technique," *Journal of Electromagnetic Waves and Applications.* June Vol. 8, No. 6. pp. 759–779.

27. Zhuck, N.P. and Batrakov, D.O. 1995. "Determination of electrophysical properties of a layered structure with a statistically rough surface via an inversion method," *Physical Review* B-51, No. 23, June 15. pp. 17073–17080.

28. Batrakova, A.G. 2011. *Recommendations for the determination of the thickness of the structural layers of the existing pavement,* Ukraine State Road Agency (Ukravtodor) normative document, R V.2.3-218-02071168–781. (in Ukraine).

29. Brekhovskikh, L.M. 1973. *Waves in Layered Media.* Nauka, Moscow, p. 343. (in Russian)

30. Masalov, S.A. and Puzanov, A.O. 1997. "Diffraction videopulses on layered dielectric structures," *Radio Physics and Radio Astronomy,* Vol. 2, No. 1. pp. 85–94. (in Russian)

31. Masalov, S.A and Puzanov, A.O. 1998. "Scattering videopulses on layered soil structures," *Radio Physics and Radio Astronomy.* Vol. 3, No. 3/4. pp. 393–404. (in Russian)

32. Batrakov, D.O. 1995. Development of radio physical models applied to the problem of inhomogeneous media sensing. Dissertation of. Dokt. of Sciences, Kharkov. 297 p. (in Russian)

33. Batrakova, A.G. and Batrakov, D.O. 2002. "Application of electromagnetic waves for the analysis of hydrogeological conditions and diagnostic properties of road pavement," *Bulletin of Kharkov National Automobile and Highway University.* No. 17. pp. 87–91. (in Russian)

34. Goncharsky, A.V., Ovchinnikov, S.L. and Romanov, S.Yu. 2009. "Solution of the problem of pavement wave diagnostics on supercomputer," *Computational Methods and Programming* Vol. 10. pp. 28–30. (in Russian)

35. Cao, Y., Guzina, B.B. and Labuz, J.F. 2008. *Pavement evaluation using ground penetrating radar, Final Report,* University of Minnesota. 102 p.

36. Cao, Y., Guzina, B.B. and Labuz, J.F. 2011. "Evaluating a pavement system based on GPR full-waveform simulation," *Transportation Research Board (TRB) 2011 Annual Meeting,* pp. 1–14.

37. Karim, H.H. and Al-Qaissi, A.M.M. 2014. "Assessment of the accuracy of road flexible and rigid pavement layers using GPR," *Eng. & Tech. Journal,* Vol. 32, Part (A), No. 3, pp. 788–799.

38. Yelf, R. and Yelf, D. 2006. "Where is the true time zero?," *Electromagnetic phenomena.* Vol. 7, No. 1 (18), pp. 158–163.

39. Krylov, V.V. and Ponomarev, D.M. 1980. "Definition signal delay by Hilbert and methods of its measurement," *Radiotechnika i elektronika.* Vol. 25, No. 1. pp. 204–206. (in Russian)

40. Batrakov, D.O., Batrakova, A.G., Golovin, D.V. and Simachev, A.A. 2010. "Hilbert transform application to the impulse signal processing," *Proceedings of UWBUSIS'2010,* Sevastopol, Ukraine, September 6–10. pp. 113–115.

41. Born, M. and Wolf, E. 1973. *Principles of Optics,* Nauka, 1973. 720 p. (in Russian)

42. Finkelstein, M.I., Mendelssohn, V.L. and Kutev, V.A. 1977. *Radiolocation of layered Earth's surface.* Sov. Radio, Moscow. 176 p. (in Russian)

43. Taflove, A. 1995. *Computational Electrodynamics: The Finite-Difference Time-Domain Method*. Artech House, Boston-London.
44. Pochanin, G.P. 2009. "Some Advances in UWB GPR," *Unexploded Ordnance Detection and Mitigation*, NATO Science for Peace and Security Series –B: Physics and Biophysics, Jim Byrnes (ed), Springer, Dordrecht, (The Nederland). pp. 223–233.
45. Pochanin, G.P. and Orlenko, A.A. 2008. "High decoupled antenna for UWB pulse GPR "ODYAG"," *4ᵗʰ Int. Conf. on "Ultra Wideband and Ultra Short Impulse Signals"* September 15–19, Sevastopol, Ukraine, pp. 163–165.
46. Pochanin, G.P., Ruban, V.P., Kholod, P.V., Shuba, A.A., Pochanin, A.G. and Orlenko, A.A. 2013. "Enlarging of power budget of ultrawideband radar," *6th International Conference on "Recent Advances in Space Technologies-RAST2013"* June 12–14, 2013. Istanbul (Turkey). pp. 213–216.
47. Lukin, K.A., Masalov, S.A. and Pochanin, G.P. 1997. "Large current radiator with avalanche transistor switch," *IEEE Trans. Electromagnetic Compatibility*. Vol. 39, May (2), pp. 156–160.
48. Keskin, A.K., Senturk, M.D., Orlenko, A.A., Pochanin, G.P. and Turk, A.S. 2014. "Low cost high voltage impulse generator for GPR," *30th International Review of Progress in Applied Computational Electromagnetics (ACES 2014)*, Jacksonville, FL, March 23–27.
49. Varianytsia-Roshchupkina, L.A., Gennarelli, G., Soldovieri, F. and Pochanin, G.P. 2014. "Analysis of three RTR-differential GPR systems for subsurface object imaging." *Radiophysics and Electronics*, Vol. 19, No. 4. pp. 48–55.
50. Varianytsia-Roshchupkina, L.A., Soldovieri, F., Pochanin, G.P. and Gennarelli, G. 2014. "Full 3D Imaging by Differential GPR Systems," *7th International Conference on "Ultra Wideband and Ultra Short Impulse Signals"* September 15–19, Kharkiv, Ukraine, pp. 120–123.
51. Persico, R., Soldovieri, F., Catapano, I., Pochanin, G., Ruban, V. and Orlenko, O. 2013. "Experimental results of a microwave tomography approach applied to a differential measurement configuration," *7th International Work-shop on Advanced Ground Penetrating Radar (IWAGPR 13), Conference Proceedings*, Nantes, France. pp. 65–69.
52. Ryabinin, Yu.A, 1972. *Sampling Oscillography*. Sov. Radio, Moscow, 1972. (in Russian).
53. TEK Sampling Oscilloscopes, Technique Primer 47W-7209, http://www.cbtricks.com/miscellaneous/tech_publications/scope/sampling.pdf.
54. TEK Sampling Oscilloscopes, Technique Primer 47W-7209, http://stilzchen.kfunigraz.ac.at/skripten/comput04/47w_7209_0.pdf.
55. Jol, Harry M., (ed), 2009. *Ground Penetrating Radar Theory and Applications*. Elsevier.
56. Ruban, V.P. and Pochanin, G.P. 2010. "Sampling duration for noisy signal conversion," *Proc. of 5th Int. Conf. on "Ultra Wideband and Ultra Short Impulse Signals"* September 6–10, Sevastopol, Ukraine, pp. 275–277.
57. Astanin, L.Yu. and Kostylev, A.A. 1989. *Basic Principles of Ultrawideband Radiolocation Measurements*. Radio i Svyaz', Moscow. (in Russian)
58. Ruban, V.P. and Shuba, O.O. 2012. "Sampling pulse width versus forward current in the step recovery diode," *Proc. of 6th Int. Conf. on "Ultra Wideband and Ultra Short Impulse Signals"* September 17–21, Sevastopol, Ukraine, pp. 72–74.
59. Ruban, V.P., Shuba, O.O., Pochanin, O.G., Pochanin, G.P., Turk, A.S., Keskin, A.K., Dagcan, S.M. and Caliskan, A.T. 2014. "Analog signal processing for UWB sounding," *7th International Conference on "Ultra Wideband and Ultra Short Impulse Signals"* September 15–19, Kharkiv, Ukraine, pp. 55–58.
60. R.E. Bogner and A.G. Constantinides (eds). 1975. *Introduction to Digital Filtering*. Wiley, London.
61. Ruban, V.P., Shuba, O.O., Pochanin, O.G. and Pochanin, G.P. 2014. "Signal sampling with analog accumulation," *Radiophysics and Electronics*. Vol. 19, No. 4, pp. 83–89. (in Russian)
62. Kholod, P.V. and Orlenko, A.A. 1998. "Optimum synthesis of transmitting-receiving sections of subsurface radar," *Proc. of the Third Int. Kharkov Symp. "Physics and Engineering of Millimeter and Submillimeter Waves", MSMW-98*, Vol. 2, Kharkov, Ukraine, September 15–17, pp. 546–548. (in Russian)
63. Golovko, M.M. and Pochanin G.P. 2004. "Application of Hough transform for automatic detection of objects in the GPR profile," *Electromagnetic Waves and Electronic Systems*, Vol. 9, No. 9–10, pp. 22–30. (in Russian).

3

Signals, Targets, and Advanced Ultrawideband Radar Systems

James D. Taylor, Anatoliy Boryssenko, and Elen Boryssenko

CONTENTS

3.1 Introduction to the Advanced Ultrawideband Radar Concept

Imagine a radar that can identify targets by their reflected signal spectra and adjusting its transmitted signal to match a specific class of targets. The advanced ultrawideband (UWB) radar evolves the technology from a passive detector of reflected electromagnetic signals to an active sensor for finding special targets. This chapter introduces the concepts and technology needed to build this system.

The advanced UWB radar will transmit a *virtual-impulse* signal. The receiver will first detect the return and then it will convert the received signal to a digital format. The signal processor will analyze the return to determine the time–frequency characteristics of each target. It can then use the time–frequency profile to identify the target. The operator can then adjust the transmitted signal spectrum to match the target and enhance detection of a particular target class.

Classical communications system theory says we can determine the transfer function of a system (object/target) by exciting it with a Dirac-delta impulse of an infinite bandwidth and for an infinitesimally short duration. A radar target will respond with signals determined by its resonating elements that depend on the geometry and material properties. Thus we can say targets have *resonance* characteristics that produce a unique time-frequency signature when excited.

Such formal studies have been done for narrowband and wideband time-domain radar systems to exploit target-specific unique signal signatures produced by natural resonances and use them for target identification. The singularity expansion method (SEM) was formally developed by C.E. Baum in the 1970s for transient electromagnetic scattering from targets illuminated by pulsed electromagnetic (EM) radiation (Baum, C.E., 1976). SEM is based on some observations that the transient responses of complex electromagnetic scatterers are dominated by a small number of damped sinusoids. From a formal mathematical perspective, it is backed by the analytic properties of the electromagnetic responses that are expressed as a function of the two-sided Laplace transform complex frequency, variable s. Singularities of the Laplace transform are used to characterize the electromagnetic response of a structure to incident radiation or a driving source, in both the time and complex frequency domains (Baum, C.E et al. 1991).

A wave of new interest in the SEM emerged a decade ago and was correlated with interests in UWB applications including subsurface characterization of targets like landmines and similar obscured objects (Baum, C.E., 1997). Such SEM representations have been developed for both induced currents and scattered fields including their aspect-independent forms. Processing signal SEM signatures would give some indications on the general shape and the constitution of the illuminated targets. It was successfully demonstrated in simulations and with less success for measured data because of sensitivity of SEM techniques to noise level. For example, in the case of landmine detection, the strong soil attenuation lowers the Q factor of the observable resonances and, thus, reduces the detectability and signal-to-noise ratio (SNR). Other close approaches are associated with E-pulse and K-pulse (excitation pulse) techniques (Kennaugh, E., 1981; Martin, S. 1989; Rothwell, E.J et al. 1987). For example, the K-pulse technique exploits a different approach. Instead of illuminating a target with a pulse that excites all inherent target resonances, the target is illuminated by a *matched* pulse that produces a short possible detected response and so on.

The advanced UWB radar system will transmit a wide spectrum *virtual-impulse* signal that simulates a Dirac-delta *impulse* signal. The virtual-impulse spectrum bandwidth will cover the probable resonances in a class of targets. Distinct resonant elements of the target will appear in the return signal. Demonstrations have shown how the different objects uniquely modify the waveform (and spectrum) of UWB signals. Note that the advanced UWB radar can operate at any place in the microwave spectrum. The designer must assure that the UWB modulation bandwidth covers the likely resonances of a target class. (Barrett, T.W. 1996; Barrett, T.W. 2012; VanBlairicum, M.L. 1995)

The advanced UWB radar can carry the process one step further by changing the transmitted waveform to match a specific target class. When using correlation detection, this waveform tailoring will enhance the response of specific classes of targets. Targets outside of the selected class will have weaker returns because their signature does not match the correlation reference signal.

The advanced UWB radar will require a receiver capable of converting short duration impulse signal returns to a digital form for storage and time–frequency analysis. For special applications, it may require a transmitter, that can change signal waveforms to match the target. Knowledge of the target characteristics means the radar can modify its signal waveform to enhance the returns from specific classes of objects through correlation detection. This target probing and signal matching concept has broad applications for medical, security, nondestructive testing, and other radar applications.

The advanced UWB radar concept resulted from my discussions with Terrence W. Barrett about his book *Resonance and Aspect Matched Adaptive Radar (RAMAR)*. Barrett patented the basic concepts in 1996 and 2006. He demonstrated the basic technology in 2011. (Barrett, T.W. 1996; Barrett, T.W. 2012)

This chapter will present a conceptual guide to the UWB impulse signal and target interaction phenomenon. It will show how to use the signal spectrum change to identify and enhance detection and tracking of certain target classes. After developing the performance objectives, it will present an advanced UWB radar architecture and concept of operation. The final sections present UWB radar digital receivers and transmitters implementations.

3.2 Advanced UWB Radar Applications

3.2.1 Benefits of UWB Radar Return Signal Analysis

Capturing UWB radar return signals can provide a way to examine the spectrum change between the transmitted and the returned signal. The reflected return signal waveform (spectrum) depends on the band of the transmitted signal frequencies and physical characteristics of the target geometry and materials. If the radar target contains resonances in the spectrum of the UWB impulse signal, then analysis of the return signals into time and frequency components opens possibilities for the following:

- Target identification: Comparing the return signal spectrum against a database of known returns could provide a way to identify the target. The identification process could determine if the target belongs to a specific class of objects.

- Target return enhancement: The time–frequency signal-processing techniques described in Chapter 4 can determine the target resonance characteristics. This can let the radar modify the transmitted signal to match the target and enhance the detector correlation output from the specific target class return signal. Providing the target class signal spectrum as the reference signal for a correlation filter will enhance tracking of that target class and reduce the correlated detection signal from other targets. Barrett developed and demonstrated this technique in *Resonance and Aspect Matched Adaptive Radar* (Barrett, T.W. 2012).

- Intelligent remote sensing: Suppose the advanced radar operator could select a signal spectrum based on either a priori knowledge, or from immediate measurements and signal analysis. This can help to continually modify the transmitted spectrum to search for specific classes of objects. For example, a GPR could search for specific classes of buried objects, for example, land mines, low contrast objects (plastic mines), and structures. A medical radar could search for low contrast tumors or tissue pathology with harmless, low power EM radiation. Chapter 2 described detecting pavement layers using return signal analysis methods.

3.2.2 Advanced UWB Radar Technology Requirements

Building an advanced UWB radar system will require several major components including:

- Time-domain receiver: This means an antenna and a wideband receiver that can record the received signal waveform in a digital format for time–frequency analysis. Barrett described such a receiver used in the Resonance and Aspect Matched Adaptive Radar (RAMAR) demonstrations (Barrett, T.W. 2012). Pochanin et al. described a time-domain receiver based on a stationary signal and stroboscopic digitization methods in Chapter 2. This requires capturing the received waveform with enough resolution for signal processing to accurately analyze the time frequency components using the techniques of Chapter 4.

- Signal sampling and digitization: The sampling methods will depend on the capabilities of analog-to-digital converters (ADC) and the bandwidth of the needed signal spectrum as determined by the target resonance characteristics. The short received signal duration and low power levels may require techniques for increasing the amplitude for accurate digitization and time–frequency analysis.

- Time–frequency signal processing: The signal-processing system can apply one of the several methods described in Chapter 4 to determine the time–frequency profile of the signal. The resulting profile can provide the basis for target identification and transmitted signal waveform synthesis.

- Variable waveform transmitter: The system must transmit a wide variety of waveforms for target probing and/or return signal enhancement. This will require a digitally controlled transmitter that can synthesize a signal from the target time-frequency profile.

- Specialized system architecture: All advanced systems will share a common architecture, but operate in different frequency ranges depending on the type of targets of interest. Artificial intelligence signal processing will play important roles in locating and tracking specific type of targets if the time–frequency characteristics change due to movement.

3.3 UWB Radar Signals and Targets

Past UWB radar designs have found better ways to find objects by transmitting a fixed waveform signal and detecting the reflected energy from a given class of objects. Generally these approaches used receivers with matched filter detection based on the transmitted signal and assuming no return signal spectrum changes. Signal processing stored the received energy from multiple returns for integration to higher SNR levels. The target RCS for that particular signal spectrum set the system performance limits. These simple and effective designs work well in many applications.

Researchers have demonstrated the distinct UWB radar signatures of targets and suggested their application for passive identification. Astanin, Barrett and others have proposed matching signals to target characteristics. Immoreev et al. have shown how the waveform of a reflected UWB signal varies depending on the bistatic angle from the target. This section will give an intuitive explanation of how to exploit target resonance effects with a *virtual impulse* signal (Astanin, L.Y. and Kostylev, A.A. 1997; Astanin, L.Y. et al. 1994; Immoreev, I. 2000; Immoreev, I. 2012).

3.3.1 UWB Signal and Target Interactions

Most radar books include an illustration similar to Figure 3.1(a) that shows the normalized reflected energy from a perfectly conducting metallic sphere of radius a plotted against the wavelength λ divided by the sphere circumference $2\pi a/\lambda$. The reflected energy falls into three regions as follows:

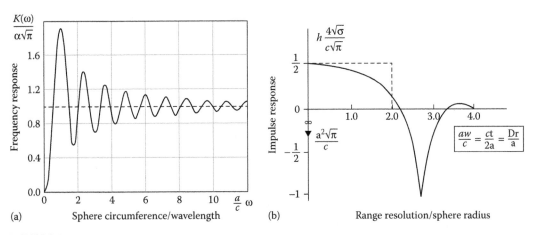

(a) Sphere circumference/wavelength

(b) Range resolution/sphere radius

FIGURE 3.1

Signal reflection from a perfectly conducting sphere. (a) The classic Mie resonance response results when the incident wavefront diffracts around the sphere and falls in phase with the directly reflected signal to produces an enhanced RCS and (b) the reflected energy response for a UWB impulse signal of duration τ produces this response. (From Astanin, L.Y. et al., *Radar Target Characteristics: Measurements and Applications*, CRC Press, Boca Raton, FL, 1994).

1. The Rayleigh region with long wavelengths so that $((2\pi a/\lambda) < 1.0)$. The wavefront passes around the sphere with no significant bending or diffraction. The sphere has a low effective reflecting area, or radar cross section (RCS) compared to its physical size.

2. The Mie resonance region reflection occurs when the wavelength approaches the same size as the sphere circumference, that is, $(1 < (2\pi a/\lambda) < 10)$. The resonance model assumes a single continuous frequency signal acting according to wave mechanics. This assumes that the wave front travels around the sphere and directly back to the source. This delayed diffracted wave front produces reinforcements and cancellations when it combines with the directly reflected wave. The largest reflection enhancement (+3 dB) occurs when $(2\pi a/\lambda) = 1$. The maximum RCS enhancement happens when the wavelength matches the path length around the sphere. In this condition the diffracted wavefront reinforces a directly reflected wavefront reflected by the front half of the sphere. Any deviations from this condition produce less RCS enhancement depending on the ratio of $2\pi a/\lambda$. (Knott, E.F. 2008)

3. The optical region occurs when the wavelength is much smaller than the sphere circumference so $((2\pi a/\lambda) > 10)$. This provides a convenient narrowband RCS model based on the target size (cross section area) and shape. Note that most radar systems operate in the optical region for practical considerations of antenna size or frequency allocations.

Astanin et al. examined the impulse signal response spherical target. His analysis used an UWB impulse signal of duration τ and plotted the response against the range resolution $\Delta r = c\tau/2$ divided by the sphere of radius a or $c\tau/2a$. The impulse response plot shown in Figure 3.1(b) implies an optimal *physical size* of an impulse signal for detecting targets with some characteristic dimension close to signal spatial resolution (Astanin, L.Y. et al. 1994).

Note that the model does not cover the case where the impulse signal bandwidth covers natural target resonances.

Immoreev developed the idea that objects have unique responses depending on the physical length of the UWB signal $c\tau_a$, where τ_a indicates the signal autocorrelation time. For signal lengths about the same size as some target dimension, that is, $d \approx c\tau_a$, the resulting target return signal will change waveform. His analysis assumed the transmitted signal induced currents in the target, which then radiated a new signal in all directions. He further showed how a reflected signal waveform will change depending on the bistatic angle from the target, or with the aspect angle of the target with respect to the arriving wavefront (Immoreev, I. 2000; Immoreev, I. 2012).

For the opposite case of $d < c\tau_a$, the overresolved target case, the target becomes a series of returns from each reflecting part with different, but close time delays between them (Immoreev, I. 2000; Immoreev, I. 2012). Figure 3.2 shows the practical effects of under-and overresolved targets in UWB radar design. For applications such as radar imaging, the system needs the overresolved target return of Figure 3.2(b) for a fine grain target picture. For other applications the system can work with the under-resolved signal of Figure 3.2(c) Table 3.1 summarizes the continuous wave and impulse conditions.

Sachs et al. pointed out how UWB radar detection resembles determining the *impulse response* of an electrical system (Sachs, J. 2012). Although the analytical Dirac-delta impulse has a theoretically infinite spectrum, a signal with a spectrum covering the resonances of the object could achieve the same effect. Examining the radar return signal permits

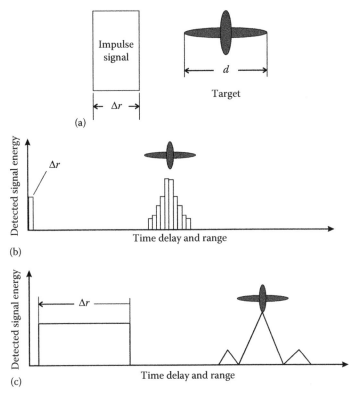

FIGURE 3.2
UWB radar reflections depend on the spatial resolution for impulse waveform matched filter detection. (a) The ratio of UWB signal-range resolution to the target size will determine the type of target response, (b) a spatial resolution much smaller than the target will result in returns from the target at each down-range increment. The overresolved target case works well for UWB imaging, and (c) the underresolved target cases detects large object, the same general size as the target, and produces single response.

TABLE 3.1

Spherical Radar Target Resonance and Spatial Resolution Conditions

Continuous Wave		Impulse Signal	
Sphere circumference/wavelength		Range resolution/sphere radius	
$2\pi a/\lambda$		$\Delta r/a$	
$2\pi a/\lambda < 1$	Rayleigh scattering	$\Delta r/a > 1$	Over-resolved
$1 < 2\pi a/\lambda < 10$	Mie resonance	$\Delta r/a \sim 2.7$	Optimal impulse response for a sphere
$2\pi a/\lambda > 10$	Optical scattering	$\Delta r/a < 1$	Under-resolved

modeling the target as a set of multiple resonating points. Figure 3.3(a) shows the concept *impulse response* measurement using a signal spectrum covering the major resonances of the object. With appropriate signal processing, a UWB radar could determine the presence and electrical characteristics of an object as shown in Figure 3.3(b). In this case, the return nonsinusoidal signal is made up of sum of waveforms from the characteristic object resonances as shown in Figure 3.3(c). Time frequency analysis could find a distinct target time frequency profile. Astanin and Immoreev have established the basic theory for understanding UWB signals waveform shifting. Relating the waveform shift to the target

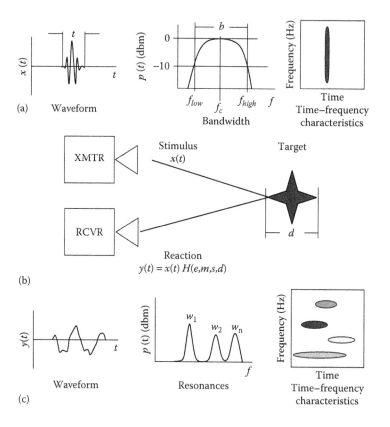

FIGURE 3.3
The target electrical and geometrical characteristics will modify the UWB radar impulse return signal. If the transmitted impulse signal bandwidth covers the target resonances, then return signal analysis can determine the target radar characteristics or transfer function. (a) The UWB signal approximates an *impulse function* covering the target frequencies of interest, (b) the target material and geometry determine the response to the *impulse function* signal, and (c) the return signal will have a different waveform. Time–frequency analysis will produce a unique target signature for different objects.

characteristics will take the first step to a new field of radar spectroscopy and tomography. Barrett demonstrated this in Chapter 4 (Barrett, T.W. 2008; Barrett, T.W. 2012).

Each target will have multiple resonating components that will remain the same. Barrett observed that changing the target aspect will change the return signal waveform by shifting the time sequence of each arriving signal component. Depending on the aspect angle with respect to the virtual-impulse wavefront, some resonating elements will have a stronger response while others diminish. To understand the concept, visualize the radiation pattern of a half-wave dipole. Some resonant responses will arrive at different times with respect to others. Although these aspect angle related delays will change the composite waveform, they will not change the overall frequency content in the reflected signal (Barrett, T.W. 2012).

3.3.2 Virtual-Impulse Signal

Suppose we want to build an advanced UWB radar for the purpose of detecting a specific class of aircraft targets such as modern fighters. Assume we know the maximum target size d as shown in Figure 3.4(a). This means the target will have several resonating

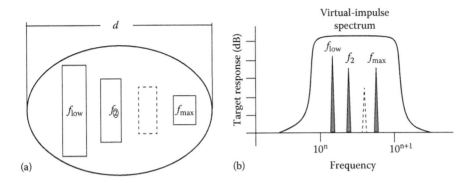

FIGURE 3.4
The virtual-impulse signal has a spectrum covering the range of resonators in a given class of targets.
(a) A simplified target model with multiple resonators. The overall physical size of the object will determine
the lowest frequency of interest. The maximum frequency of interest will depend on the target structure.
(b) The virtual-impulse signal spectra covering the range of resonances associated with a particular class of
objects. The reflected signal will consist of strong returns at each resonator frequency.

parts with frequencies $f_1, f_2, \ldots, f_n$. The lowest resonance frequency of interest will have a
frequency $f_{low} < c/2d$. The advanced UWB radar will use a virtual-impulse signal as shown
in Figure 3.4(b) that will excite the resonating elements of the target, for example, wings,
tail, and fuselage.

For practical purposes radar targets such as mines, trucks, aircraft, artillery shells, and
missiles have distinct resonance signatures depending on the materials, physical size,
and aspect angle with respect to the radar line of light. The work of Baum, VanBlairicum,
Astanin, and Immoreev predicted this interaction of UWB signals and targets, and suggested the potential applications to radar target identification.

Barrett's RAMAR experiments demonstrated the target resonance concept on real
world objects. The RAMAR used a broad spectrum signal modulation on a carrier signal
to determine target resonances. After return signal time–frequency analysis, the demonstration RAMAR synthesized a signal modulation tailored to the target. In Chapter 4,
Barrett presents methods for the time–frequency analysis of UWB signals. The resulting time–frequency profiles can identify targets and provide information for signal
resonance modeling. He demonstrated methods for analyzing the returns from a UWB
signal to provide an optimal signal *tuned* to a particular target class. As expected from the
theoretical studies, field test results showed how the target resonances depended on the
wavefront aspect angle with the object. RAMAR demonstrations showed how matching
signals to the target resonances can substantially increase the return signal amplitude
(Barrett. T.W. 2012).

The advanced UWB radar will operate as shown in Figure 3.5. The radar transmits
an *impulse* signal covering the time–frequency characteristics of multiple target configurations. For example, this impulse signal could use broadband modulation of a carrier signal, such as linear frequency, step frequency, random noise, and pseudorandom
noise modulations with a 3 GHz bandwidth on a 30 GHz radar signal. In this figure, the
radar receives and digitizes returns from two different target geometries. The signal-
processing system performs a time–frequency analysis that breaks the return into several distinct frequencies of a specific time duration. Each target class has distinct target
characteristics shown in Figure 3.5(b). Comparing target signatures with a database of
radar signatures from previous observations, or predicted responses, could provide the

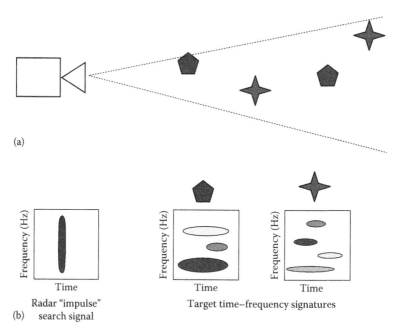

FIGURE 3.5
UWB radar target identification process. (a) The radar searches the field view in an *impulse* search mode to locate targets falling under the search signal frequency characteristics and (b) signal processing determines the time-frequency characteristics of each return and sorts them into categories. A priori data may indicate the class of target to the operator.

operator with a target identification. For example, the star reflectors could indicate a specific class of mines, aircraft, rocket, vehicles, and material defect. The pentagonal symbols represent a benign object.

To illustrate the advanced UWB radar possibilities, imagine a security surveillance radar network for detecting and locating pickup trucks with machine guns. The operator could first search the field of view and determine the classes of targets visible. Signal processing will produce a time frequency signature of each return as shown in Figure 3.5(b). This would produce a display of each return signature *file* organized by range and direction as shown in Figure 3.6(a). Either the radar operator, or preprogrammed signal controls could then synthesize a signal matched to armed pickup trucks. The new display of Figure 3.6(b) will then enhance the returns from trucks and suppress other returns because they do not match the expected target return waveform set in the correlation detectors.

3.3.3 Target and Signal Matching Possibilities

The advanced UWB radar can use a virtual-impulse signal to determine the time–frequency profile and characteristic resonances of specific target classes with resonances included in the impulse signal bandwidth. This target-signal matching feature can expand the radar capabilities for searching for specific target classes. The concept has potential applications across a wide range of radar functions and opens new possibilities for practical applications.

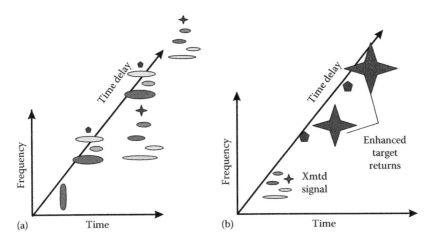

FIGURE 3.6
UWB radar modes and signal returns. (a) The initial search determines the classes of objects in the radar field of view and (b) changing the transmitted signal to match the target class time–frequency analysis will produce enhanced returns from the selected target class and suppress unwanted returns.

Barrett's Chapter 4 discusses the various time–frequency methods available. Sukarevsky's Chapter 5 presents a method for predicting how an object will scatter UWB impulse signals. Chapter 6 describes nondestructive testing that could benefit from signal analysis. Chapters 7 and 8 present biolocation applications that could benefit from these target return analysis techniques. In Chapter 12, Francois LeChevalier presents another approach to tailoring signals to targets to enhance specific performance objectives.

3.4 The Advanced UWB Radar

Many researchers have advanced the idea of *intelligent* or *active* radio signal processing to enhance target detection or communications. I as well as other researchers have suggested the basics of this concept in earlier books and papers (Taylor, J.D. 2012; Taylor, J.D. 2013). This section includes ideas similar to those of Joseph Guerci's *Cognitive Radar* (Guerci, J.R. 2010) and Simon Haykin's *Cognitive Dynamic Systems: Perception-Action Cycle, Radar, and Radio* (Haykin, S. 2011). The advanced UWB radar concept draws heavily on Barrett's RAMAR for the methods and benefits of matching signal waveforms to specific targets (Barrett, T.W. 2012).

3.4.1 Performance Objectives

This section presents an intuitive overview of the performance objectives, technical issues, and an advanced UWB radar system that can modify signals to match specific target classes.

3.4.2 Fixed Signal UWB Radar Limitations

Why build an advanced UWB radar for target and signal matching? Currently UWB radar systems operate in a passive fixed signal mode with respect to their targets. These systems transmit a single UWB waveform signal appropriate to the particular radar design objective and probable target characteristics. Conventional UWB receivers and signal processors detect reflected energy and the signal arrival time to determine the range for each reflector. Range resolution depends on the signal duration or autocorrelation time of the particular signal. The correlation (matched filter) detection process assumes a strong resemblance between the transmitted and reflected signals for a maximum detector output. Collected received energy returns recorded against the arrival time produces a matrix of high-resolution range information about targets in the field of view. The material-penetrating radar (MPR) shown earlier in Figures 1.1 through 1.3 illustrate this type of signal processing.

For large targets, the target data matrix shows the returns from the multiple parts of a large object, as shown in Figures 1.4 through 1.7. Signal-processing methods can produce an output for a specific purpose. In each case, the fixed signal waveform and target characteristics set inherent limits on system performance. The fixed signal approach provides practical benefits and economical solutions to specialized requirements. The advanced UWB radar with target return time–frequency analysis could provide a remote sensor for special applications.

3.4.3 The Advanced UWB Radar Architecture

The signal target to matching concept described earlier could improve radar performance for a wide range of special applications. A radar with return signal analysis could adjust the transmitted signal format to achieve maximum returns for detecting specific target classes. For example, the radar could match the signal to targets such as classes of mines, underground objects, diseased biological tissues, imperfections in dielectric composite materials, and flying, or ground based objects.

An advanced UWB radar architecture would have the same functional elements for all applications, but different configurations depending on the particular operational purpose (Taylor, J.D. 2013). For example:

- Ground-penetrating radar (GPR) signal processing could search for specific objects, soil differences, and other useful conditions. In Chapter 2, Pochanin described a GPR that analyzes the return signal waveform to achieve precise ranging and to identify layers of asphalt pavement in highways. In Chapter 7 Liu describes a radar for searching victims buried under collapsed buildings. Both these applications could potentially benefit from an advance UWB radar approach with target signal matching to increase detection performance.

- Nondestructive testing radars could search for changes in material layers indicating flawed or damaged dielectric parts or structures. In Chapter 6 Cristofani describes applications of radar to inspecting composite aircraft parts that could benefit from signal to target matching.

- Medical radars as shown in Figure 1.3 could search a patient for changes in tissue characteristics and reflectivity. Doctors operating in austere conditions, such as a field hospital or remote clinic, could use the advanced UWB radar to quickly image internal tissues and evaluate injuries without X-rays.

- Through-the-wall and security imaging radars as shown in Figures 1.5 and 1.6 could use optimal signals for contraband and weapons detection. Chapter 11 presents the basics of through-the-wall radar systems.
- A long-range surveillance and tracking radar could modify its signal to enhance the detection of specific target classes. Barrett's RAMAR could evolve to this way.

An advanced UWB radar architecture would look like Figure 3.7 and the signal optimization process like Figure 3.8. The architecture assumes overall control using artificial intelligence with active user inputs to determine the types of targets emphasized. For example, the user could direct the radar to switch back and forth between target types.

The target identification signal process would follow the general process shown in Figure 3.8. In this case the term *impulse signal* refers to any signal format that can provide a range of frequencies covering the targets of interest. As Barrett points out, the *impulse* could include the modulation of some powerful carrier signal. For high power systems, the carrier signal could exist in some designated frequency band with a UWB modulation appropriate to the target (Barrett, T.W. 2012).

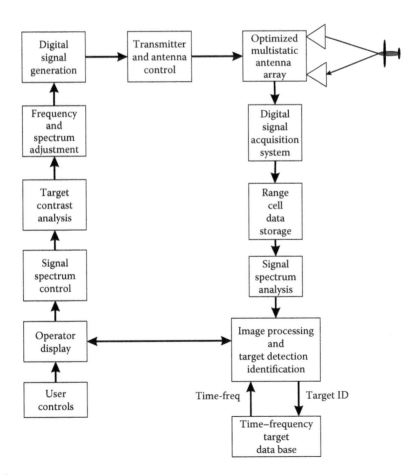

FIGURE 3.7
Notional advanced UWB radar architecture for determining target time–frequency characteristics, identifying targets, and adapting the signal for a maximum target return.

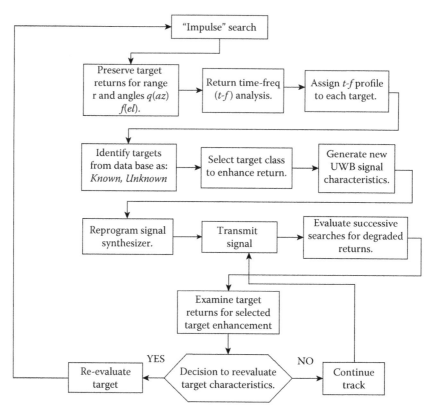

FIGURE 3.8
The advanced UWB radar could use an active probing process that begins with an impulse signal to determine the characteristics of targets in the field of view. After analyzing all target returns, it will select a class of targets for enhancement. It then adjusts the signal spectrum to the selected target class. This will give an optimum correlation detection output for those targets and diminished outputs for other targets. It could periodically evaluate target return strength/quality. If the target return quality deteriorates, then it will reevaluate the target and adjust the signal format.

3.5 Advanced UWB Radar Technology Requirements

The advanced UWB radar shown in Figures 3.7 and 3.8 for locating and measuring the target characteristics will require the following components:

- A receiver with both frequency and time-domain signal-processing capabilities.
- A digital receiver which can preserve the signal waveform. The nonsinusoidal UWB waveform will present special problems for accurate analog to digital conversion. Section 3.6 UWB signal registration sections will discuss potential analog-to-digital converter (ADC) solutions. Chapter 2 discussed the received signal digitization problem and presented one solution.
- A range bin memory to store and integrate multiple weak signal returns to accurately reconstruct the nonsinusoidal (multiple frequency) waveform for time–frequency analysis.

- Return signal time–frequency analysis processing to determine characteristics of targets in the field of view. Chapter 4 presents a summary of signal analysis techniques for UWB signals.
- A library of known or predicted target time–frequency signatures for target class identification.
- A search mode using different signal waveforms to search for previously unknown target signatures.
- Capability for storing unknown signatures for further processing and identification.
- Signal modification software to optimize returns from a class of targets.
- A digital waveform synthesizer to translate the target time–frequency profile into a transmitter input.
- A signal generator that converts the selected target class time–frequency characteristics and produces a signal to match the target and control the transmitter signal. This will involve special digital-to-analog converter (DAC) technology and a special transmitter power amplifier.
- Digitally controlled transmitter and an antenna.
- Artificial intelligence system controller to adapt to changing target conditions.

In conventional radars, frequency domain processing locates targets based on the reflected energy time delay. Practical range sensors use many techniques based on matched filtering and return signal integration in range bins for detection.

Time-domain processing requires recording the reflected signal waveform for further analysis. This could mean some single pass digitization, or the collection and integration of multiple signal digital conversions to get the waveform details needed for successful operation. The notional drawings of Figures 3.1 through 3.3 showed how target material and physical characteristics can modify a UWB radar return waveform. Analysis of the time-domain response can identify the electrical and geometrical characteristics of radar targets for identification.

The following sections will discuss signal ADC methods and UWB transmitter types.

3.6 UWB Signal Registration

Current UWB radar receivers treat target reflections as energy bundles received with a given time delay indicating the range to a specific point on a target. The advanced UWB radar can overcome the limits of systems using one signal format for all targets. If we approach the radar design using communications system theory, then we can use an *impulse* signal to excite and determine the target transfer function which shapes the return signal spectrum. Although the ideal Dirac-delta pulse has an infinite spectrum extent, a *practical* Dirac-delta pulse will have a limited spectrum because only such signals can be generated and radiated. At the same time, target-specific probing pulses need to cover the spectrum of natural resonances and not waste energy beyond that spectrum.

The receiver capabilities will limit the radar performance by its ability to capture the target return from the impulse probing signal for time–frequency analysis. Adequate ADC of the complicated target waveform can use many approaches including special methods and approaches based on compressive sensing and other mixed analog-to-digital methods such as time lenses. We can make a reasonable stationarity assumption that each received return from a given target will repeat periodically, which means we can integrate successive returns or use lower ADC sampling applied to different parts of successive signals. Some systems will need real-time or *near real-time* operation to acquire the waveform within a few successive signal repetitions.

3.6.1 Technical Considerations in Digitizing Received UWB Signals

The advanced UWB radar receiver performance will depend on an ADC for signal data acquisition, measurements, and automation. All ADCs process continuous real-world signals into discrete digital formats for efficient storage. The ADC output will go to central processing units (CPUs), digital signal processors (DSP), field-programmable gate arrays (FPGAs), or graphical processor unit (GPU) to support major signal-processing routines required to implement advanced radar concept. Now we need to examine how well an ADC can convert a nonsinusoidal short period signal.

Modern ADCs have many architectures including the sigma-delta, successive approximation register (SAR), high-resolution, and high-speed ADCs (Kester, W. 2004; Pelgrom, M.J.M., 2010). Commercial electronics in the audio and video bands for communication applications such as 3G, software-defined radio (SDR) and other systems now use ADCs. The radio frequency (RF) and microwave bands for UWB radar applications have much higher data throughput and dynamic range requirements which limit ADC applications for real-time operations.

This section focuses on real-time operations with high-speed and high-dynamic range ADCs. Applying high-speed ADCs front-ends in broadband and UWB EM systems has many technical issues dominated by the following:

1. Performance trade-offs between conversion speed and conversion accuracy (resolution) while mitigating associated excessive power consumption. For example, some ADCs have high sampling rates but not the best signal-to-noise ratio (SNR). ADCs with a good SNR typically have lower sampling rates. The capability to simultaneously achieve good SNR and high speeds is mutually exclusive for *single-core* ADCs. This condition forces advances in multi-core ADCs or an ADC made from cascaded and properly clocked low-rate ADC units (Rolland, N. et al. 2005).

2. Implementation complexity and integration with other subsystems. The engineer must consider the radar performance objectives and architecture as a whole, when designing the digital time-domain receiver. The overall systems architecture must integrate both the hardware and software parts of the entire system into a functioning unit.

3. Economic factors could make the cost of high-speed ADCs prohibitive for many applications. The systems engineer must clearly understand and consider all the above factors separately and in their mutually conflicting relations when planning the radar architecture. The lack of systematically published references means the designer must consult many datasheets, application notes, webcasts, and other sources. The final selection of a particular ADC architecture and components will come to a question of systems cost versus user benefits.

This section will present the technical considerations of ADC devices in a compact and comprehensive form.

3.6.2 ADC Operating Principles and Performance Metrics

3.6.2.1 ADC Theory and Limitations

To understand the problem of UWB signal digitizing we need to understand how ADC converters work and the design assumptions behind them. We must also understand the nature of reflected UWB impulse signals modified by the target.

Our discussion of the advanced UWB radar receiver will treat the ADC as a *black box*. Functionally the ADC takes in a *continuous* analog input signal (voltage) and emits a stream of digital words indicating the measured input value at given discrete time intervals (Baker B., 2011; Pearson, C. 2011). Signal digital conversion requires two consecutive processes: (1) signal sampling and (2) digitization (quantization) as shown in a high-level time domain functional scheme in Figure 3.9.

First, the ADC samples the input analog signal at the sampling frequency F_s which is at least twice the known signal frequency, or maximum expected signal frequency of the target response, F_{max} according to the Nyquist theorem. Then the sampled signals are transformed into a digital format as illustrated in Figure 3.9 for the 3-bit case when a quantized number of a closed digital bit is assigned at the output. The nomenclature of the major parameters in this presentation level includes the following:

1. n is the number of output bits (resolution)
2. A_{IN} is the analog input voltage that must not exceed V_{max}
3. V_{REF} is the reference voltage (or current) used to compare the input signal

More ADC operational features can be assessed in the frequency domain. This type of information is provided in datasheets for a set of single harmonic input signals as sketched in Figure 3.10 for frequency F_1. Table 3.2 shows the key ADC parameters.

The ADC SNR is the first important characteristic. There are two SNR interpretations, namely the ideal (or theoretical) and the practical (or real). The ideal SNR for an n-bit ADC is the ratio of the root-mean square (RMS) full-scale, digitally reconstructed, analog input V, that is $0.5V/\sqrt{2}$, to its RMS quantization error (i.e., V LSB$/\sqrt{12}$, where LSB is the least significant bit Equation 3.1):

$$SNR = 2^n \sqrt{3}/\sqrt{2} = 1.225 \cdot 2^n \qquad (3.1)$$

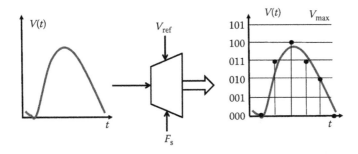

FIGURE 3.9
The ADC works in the time-domain to sample the broad spectrum input voltage at the sampling frequency F_s. This 3-bit ADC provides an output of the input signal. Accurate signal digitizing requires prior knowledge of the signal amplitude and frequency.

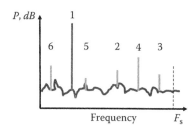

FIGURE 3.10
ADC frequency domain operation with the input signal tone numbered by 1 (fundamental) and a set of harmonic products numbered from 2 to 6, among which the harmonic 4, in this particular example, shows highest power and, thus, would be used in SFDR (3.6) calculations.

TABLE 3.2

Key ADC Operational Features

Feature	Meaning
Signal power	Power P_s associated with a single tone signal F_1
Noise floor power	The inherent noise P_n within the ADC input channel caused by thermal noise, interference, and so on. Limits the smallest resolvable increment
Power of harmonics	Signal power $P_i (i = 1, 2, 3 \dots 6)$ of corresponding harmonic frequencies $F_i = F_1$
Power of highest spur	Signal power P_H associated with the highest power harmonic. (4th harmonic in Figure 3.10)

or on the logarithmic decibel (dB) scale (Equation 3.2)

$$\text{SNR}_{dB} = 6.021 \cdot n^n + 1.763 \tag{3.2}$$

This quantity can be considered also as the theoretical dynamic range (DR) limited only by quantization noise. Other noise contributions involved degrade the DR as further shown. For the given signal spectral presentation in Figure 3.10(2), the SNR is defined as (Equation 3.3)

$$\text{SNR}_{dB} = 10 \log 10 (P_S / P_N) \tag{3.3}$$

SNR and DR are both functions of input signal frequency F_1, because the ADC operates across certain bands. This single tone signal is often called the *carrier* and SNR is then characterized in dBc units which means *dB to carrier*.

Most ADC datasheets provide the frequency-dependent SNR(F) and DR(F) for the several operational conditions defined as follows:

1. For a number of characteristic frequency points below the Nyquist limits.
2. At several input signal magnitudes below its maximum permitted extent, V_{max} in Figure 3.9.
3. For one or a few conversion rates.

The ADC is by its nature a nonlinear device that outputs a number of harmonics which are unwanted byproducts that contribute to the overall output noise as shown in Figure 3.10.

Several such harmonics, labeled from 2 to 6 are typically specified in datasheets and contribute the total distortion power (Equation 3.4)

$$P_D = P_2 + P_3 + P_4 + P_5 + P_6 \tag{3.4}$$

Then the total harmonic distortion (THD) is defined as (Equation 3.5)

$$\text{THD} = 10 \log 10 (P_S / P_D) \tag{3.5}$$

The spurious-free dynamic range (SFDR$_{dB}$) is defined as (Equation 3.6)

$$\text{SFDR}_{dB} = 10 \log 10 (P_S / P_H) \tag{3.6}$$

where, P_H is next highest spur, for example $P_H = P_4$ in Figure 3.10. The ratio of the signal power to the overall power of noise and harmonic distortions defines the signal-to-noise and distortion (SINAD) (Equation 3.7):

$$\text{SINAD} = 10 \log 10 \left(\frac{P_S}{P_N + P_D} \right) \tag{3.7}$$

An important ADC performance measure is the effective number of bits (ENOB), which is derived by combining Equations 3.2, 3.3, and 3.6 as (Equation 3.8):

$$\text{ENOB}_{\text{bits}} = \frac{(\text{SINAD} - 1.763)}{6.021} \tag{3.8}$$

An ADC can have several composite figures of merits (FOM). The one shown below gets the most consideration:

$$\text{FOM} = F_s \cdot 2^{\text{ENOB}} \tag{3.8a}$$

This FOM suggests that adding an extra bit to an ADC is just as hard as doubling its bandwidth (Walden, R.H. 1999).

The sampling frequency stability is important for the ADC performance. Any instability, often called collectively jitter, contributes to degradation of the key ADC parameters, for example, SNR, SINAD, and ENOB.

3.6.3 UWB Signal ADC Time Interleaved Digitizing Strategies

In many cases, ADC technology may not support single pass UWB signal digitization. In this case we can apply digitizing techniques based on the known or assumed signal repeatability. Pochanin described a stroboscopic approach in Chapter 2. Time interleaved digitizing provides a fast hardware approach to solving the problem (Pelgrom, M.J.M., 2010).

Using several ADCs running in an interleaved mode might resolve the tradeoff between achieving a good SNR and high sampling rates. As shown in Figure 3.11, a modern multicore time interleaving ADC combines several ADCs, say N units, of lower sampling F_s rate within the same package. All ADC cores operate in parallel being clocked at the same rate F_s with mutually adjusted time shifts to enable resulting higher sampling rate NF_s. The samples produced by each core are then combined into one data stream at the output. This increases the power consumption for N converters by the factor N. A block-diagram for $N = 4$ is illustrated in Figure 3.11(a) and timing diagram is shown in Figure 3.11(b),

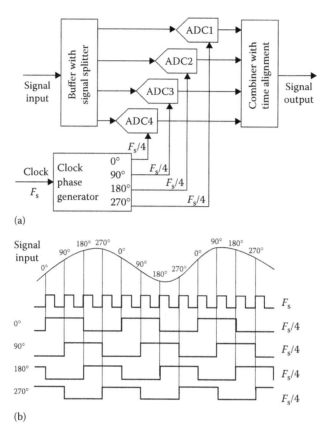

(a)

(b)

FIGURE 3.11
A four-core interleaved ADC converter can achieve a good SNR and high sampling rates. (a) The four-core
time-interleaving ADC block-diagram and (b) the four-core interleaving ADC clock timing diagram.

shows the clock phase shift of 90 degrees with respect to each consecutive ADC unit
(Hopper, R.J. 2015).

In an ideal case, the composite SNR has a performance roughly equivalent to an individual ADC core. However, the real hardware of multicore ADCs introduces errors that
degrade the overall spurious-free dynamic range (SFDR). Figure 3.12 shows three such
potential analog errors including misalignment in channel gain, DC offset, and timing
shift for a two-core ADC used for a simple example. Their combined effect translates to
spurious products in the captured signal spectrum as shown in Figure 3.13. In particular,
the offset error introduces a discrete spurious tone and their quantity depends on the
number of interleaved cores. For a four-core interleaved ADC, the interleaving spurs are
located at $F_s/4$ and $F_s/2$ in Figure 3.13. The signal dependent errors of gain and clock phase
yield images that are centered on the discrete frequencies $F_s/4$ and $F_s/2$ also shown in
Figure 3.13. (Hopper, R.J. 2015)

It is potentially possible to construct an interleaved ADC from several off-the-shelf
chips that will require quite advanced treatment of all involved signal integrity and board
design issues (Rolland, N. et al. 2005).

However, there are some commercial ADC chips that are already based on the
time-interleaved principle discussed above. The dual ADC ADS54J60 chip from Texas
Instrument uses four interleaved cores per channel to achieve a 1 Gigasample per second

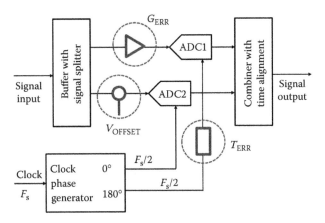

FIGURE 3.12
Three major error sources in a two-core ADC caused by gain error G_{ERR}, DC offset, V_{OFFSET}, and clock imbalances T_{ERR}.

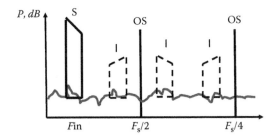

FIGURE 3.13
The channel errors in a four-core ADC from misalignment in channel gain, DC offset, and timing produces a typical performance spectrum with spurious outputs where S – input signal, I – input images, and OS – offset spurs.

(GSPS) output sampling rate. This converter employs a proprietary digital interleaving correction block to adjust for the core imbalances. This correction scheme always works in the background so there is never an interruption to the output data stream and achieves better than 80 dBc correction. The Texas Instrument ADC12J4000 chip uses four interleaved cores to achieve a 4 GSPS output sampling rate. This device operates with several options embedded for interleaving correction that can generally keep interleaving spurs better than 70 dBc at room temperature. (Hopper, R.J. 2015; TI ADC083000. 2015; TD ADC 12J000. 2015).

3.6.4 Nonconventional ADC Front-Ends

There are some cases that apply ADCs for RF, although benefiting from running them in *anomalous* modes that seem to violate the general basis of ADC operation dictated by proper selection of the Nyquist sampling frequency. The sampling theorem states that the sampling rate must be at least twice the largest bandwidth of the signal. However this can be changed to have it lower (undersampling) and higher (oversampling) with interesting practical advantages as discussed in the following subsections (Hopper, R.J. 2015).

3.6.4.1 Aliasing an ADC Mixer with Undersampling

A common way to perform ADC sampling at the Nyquist frequency to avoid aliasing as illustrated for a sinusoidal signal in Figure 3.14(a). In Figure 3.14(b), the signal frequency is increased six times, although the same sampling frequency is preserved and the original signal still yet to be sampled at the same temporal sample points as in Figure 3.14(a). Normally this situation is treated as aliasing, when the higher frequency will alias down to the ADC's capture bandwidth. From another perspective, the ADC in the case of Figure 3.14(b) acts like a conventional RF mixer implemented with such a properly set under-sampling ADC technique. This can substantially simplify the receiver architecture by providing both a down-conversion mixing and digitization functions performed in a single down-conversion digital mixer built on a single ADC (Hopper, R.J. 2015).

To better understand this approach, that exploits benefits from aliasing, the overall signal spectrum is split into separate Nyquist zones in Figure 3.15. The first Nyquist zone represents the maximum sample bandwidth, that is equal to the sampling rate, F_s, divided by two. The higher Nyquist zones represent the adjacent spectrum bands with equivalent

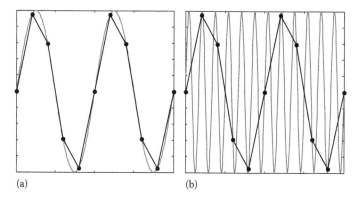

(a)   (b)

FIGURE 3.14

Higher-frequency aliasing used to operate as a digital down-conversion mixer. (a) Signal sampled at the Nyquist frequency and (b) signal of (a) at six times the frequency, but still sampled at the same time intervals as the lower frequency.

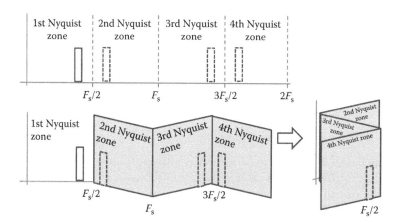

FIGURE 3.15

Nyquist spectrum zones with the first Nyquist zone for the maximum sample bandwidth and a sequence of higher Nyquist zones folded back onto the first Nyquist zone due to aliasing.

bandwidth. The down-conversion that happened in the aliasing mode can be interpreted through the spectrum folding back onto the first Nyquist zone like an accordion. Each signal that ultimately resides in the first Nyquist zone has a counterpart located in a higher Nyquist zone. With proper analog filtering, the ADC can capture a desired signal in one of the higher Nyquist zones, equivalent to a higher RF frequency (Hopper, R.J. 2015).

3.6.4.2 ADC Oversampling

In some practical cases such as a GPR with sequential sampling, the bandwidth or the output frequency is not excessive and could be in the audio band. For normal practice, a low-speed ADC device with matched Nyquist sampling needs to be employed. However, there is still an advantage to utilizing the higher sampling rate capabilities of RF sampling converters. This case is called oversampling (also called averaged sampling) aimed to improve the SNR. This method measures the relative level between the desired signal power and the entirety of the noise power within the first Nyquist zone that represents the entire bandwidth of the device. The Nyquist zone bandwidth is the sampling rate divided by two, namely $F_s/2$ in Figure 3.16. As discussed above, all signals and noise from higher Nyquist zones will fold back into the first Nyquist zone of Figure 3.16 (Hopper, R.J. 2015).

There are several practical benefits to oversampling in RF systems. Most evident is that the image signal components in the higher Nyquist zones are separated farther in frequency space for higher sampling frequency as in Figure 3.16. If so, then we can implement antialiasing analog filtration with a simpler and substantially smaller roll-off to eliminate interfering signals that can alias down into the captured bandwidth. Another benefit of oversampling is improving the SNR performance beyond the theoretical quantization noise limitations. In general, the quantization noise is equally distributed across the Nyquist bandwidth. By increasing the sampling rate, the same quantization noise is spread over a larger Nyquist bandwidth while the desired signal remains fixed. To the end, downsampling (decimation) is typically performed in a digital filter to return to the

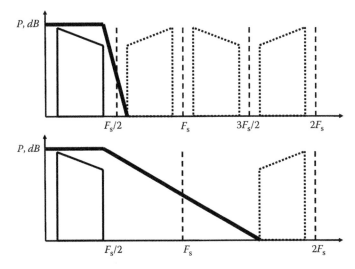

FIGURE 3.16
Two-sampling cases: one signal at the top, sampled near the Nyquist rate and one, at the bottom, that is oversampled with wider separation between Nyquist zones.

original bandlimited spectrum of the signal of interest and decrease the noise bandwidth (Hopper, R.J. 2015).

3.6.4.3 Signal Voltage Dithering Improves ADC Performance

Dithering means the addition of some white noise to analog signals before their digitization. This method helps resolve the trade-off between the bandwidth and dynamic range in ADCs. Figure 3.17(a) shows the usual method of adding external noise at the ADC input. This might look like a counterintuitive and wrong approach since it might degrade the SNR. However it does not degrade the SNR. Furthermore, the dither can be interpreted as a method of statistical linearization, namely smoothing the staircase-like ADC transfer functions. This technique originated in the audio field, where it improved quality of the sound of analog audio signal by extending the dynamic range in magnetic tape recorders back in the 1930s. Similarly this approach works for ADCs to enhance dynamic range by spreading the spurious spectral contents of the signal over its spectrum. Dithering gives this positive effect because of decorrelation noise in time and decorrelation between noise and signal. Applications where the spectral distortion caused by the quantization of low level signals is particularly undesirable will especially benefit from applying dither. Apparently a level of white random noise needs proper setting to maximize the SNR while avoiding rising of the spurious- free dynamical ranges (SFDR) as shown in Figure 3.16(b). For ideal converters, the optimum dither is white noise at a voltage level of about 1/3 LSB rms (Analog Devices, 2004; Baker B. 2011; Hopper, R.J. 2015; Leon Melkonian, L. 1992; Pearson, C. 2011; Pelgrom, M.J.M. 2010).

3.6.5 Compressive Sensing and Sampling

Moving more wireless systems operational, signal processing into the digital domain has produced promising advances. Both software radio (software defined radio) and cognitive radio rely on a number of features observed in signal channels. Exploiting the sparse nature of real signals in the time or frequency domain, or both, can work greatly to improve the major functionality of wireless and sensing systems.

3.6.5.1 Compressive Sensing Signal Acquisition and Analog to Information Converters

Nyquist sampling for modern broadband systems requires the expensive, fast, and power-hungry ADCs discussed earlier in this chapter. In addition this will produce a tremendous flood of data forwarded to the digital processing elements. This situation results

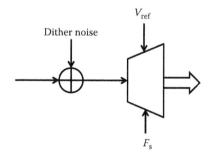

FIGURE 3.17
Dithering, the addition of white noise to the ADC input can improve the ADC SNR.

from the traditional *brute force* approach to capture and digitize radar return signals of UWB radar systems.

To understand the significance, we can consider signals in a broader sense using 2D signal spectrograms in the time–frequency coordinates shown in Figure 3.18. In this diagram, signals appear only in one or a few localized areas while most of the diagram is virtually blank. In the traditional *brute force* approach, the whole area with dominant *blank* spaces must be acquired, digitized, and processed.

Digitizing blank spaces produces many redundant computations with no significance to the system performance. *Brute force* ADC also produces apparent penalties of increased cost, size, weight, and power consumption. These hardware oriented approaches can produce a flood of data that will overwhelm downstream processing and information extraction algorithms. Resolving the signal-processing bottleneck with more computer power, these hardware based approaches will fail to produce an optimal system with regard to the bandwidth-bits product figure of merit.

For many, if not most, of the radar and sensor operational scenarios, the full signal spectrograms are similar to those in Figure 3.18 with signals that are inherently sparse in the sense that only a few spots are occupied across limited bands during limited periods of time. This means, for example, from a mathematical perspective, that the total time–bandwidth product (TBWP) occupancy might be much smaller than the total collection TBWP area $\mathrm{TBWP_{MAX}} = (T_{max} - T_{min})(F_{max} - F_{min})$.

For example, if $F_{max} - F_{min} = 6$ GHz, this instantaneous bandwidth will traditionally define the Nyquist sampling rate that must be as high as $F_s \geq 12$ GHz and the ADC needs at least a 12 GSPS capability for this case. If the receiver works at the 12 GHz Nyquist sampling frequency, then only a fraction of the recorded data $O(\mathrm{TBWP}/\mathrm{TBWP_{MAX}})$ will have any meaning and empty digital recordings will take most of the data storage volume because of the sparse signal nature. Hence, we need a smart approach to resolve the issue by exploiting new hardware architecture designed to avoid the redundancy indicated above, that will much improve the size, weight, and power (SWAP), functionality, and cost (Davenport, M.A. et al. 2010; Donoho, D.L. 2006; Ender, J. 2013).

Many attempts have been made to properly record the real-world sparse radar signals pictured in Figure 3.18. One such successful attempt employs compressive sensing (CS), that this study considers only partially from a relevant perspective of subNyquist sampling. In other words, CS provides simple and efficient signal acquisition method at a low rate below the Nyquist limit followed by computational reconstruction. Using CS signal-processing techniques enables signal sampling with fewer randomized samples than Nyquist sampling (Laska, J.N. et al. 2006; Liu, Y. and Wan, Q. 2011).

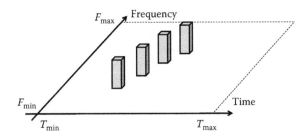

FIGURE 3.18
Signal occupancy in the time–frequency plane shows a sparse structure. Compressive sensing eliminates digitizing and processing the areas with no information.

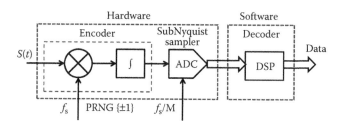

FIGURE 3.19
Analog to Information converter (AIC) with low rate subNyquist sampling and full signal recovery in digital domain.

CS assumes that the signals have a sparse representation. This also assumes that the signals are compressible in some mathematical sense, for example, by the $O(\text{TBWP}/\text{TBWP}_{\text{MAX}})$ sparsity measure simultaneously in the frequency and time domains. Thus the CS signal sampling framework enables sampling at subNyquist rates as small as F_s/M where $M \approx \text{TBWP}/\text{TBWP}_{\text{MAX}}$. Practically speaking, the number M can be in the range of tens and hundreds depending on the particular operational specifics. A number of theoretical developments made in the CS field can leverage the structure of sparse signals into reduced sampling rates to implement so-called compressive sampling for direct analog to information conversion (AIC). A typical AIC combines the hardware-software architecture and consists of a front-end hardware encoder and a back-end software decoder with a low-rate subNyquist shown in Figure 3.19 with low-rate ADC sampling in between (Liu, Y. and Wan, Q. 2011).

The hardware portion of the AIC in Figure 3.19 includes an analog mixer (modulator) operating on a pseudorandom number generator (PRNG) ±1 sequence clocked at the Nyquist sampling rate. The sampled signals are fully recoverable in the CS framework if certain conditions are met and the recovery performed in the digital domain using a number of reported algorithms. Figure 3.20 shows some results for AIC performance prediction simulated in MATLAB®. In this model, the 12-bit ADC used in the AIC converter, samples a test signal composed of two continuous wave (CW) signals of 1.85 GHz and 6 GHz. This represents two signals that are sparse in the frequency domain and sampled in time-domain, making them treatable in the CS framework. The Nyquist sampling requires 12 gigabits per second (GBPS), but the subNyquist sampling is performed at two frequencies: (i) four times lower, that is, 3 GSPS, shown at the left in Figure 3.20 and 10 times lower, that is, 1.2 GSPS, shown at the right in Figure 3.20. The original and recovered signals are depicted in Figure 3.20 by their power spectral densities (PSD) that allow for estimation of the resultant SFDR, defined similarly to an ADC. In this simulation, signal recovery was performed via convex programming supported by the l1-magic MATLAB® package (l-1 Magic, 2006). The SFDR range is defined earlier in Figure 3.14(b) as the difference between the original signal amplitude and the highest spur. It turns out that sampling at 3 GHz (four times lower the Nyquist frequency) preserves nearly the full dynamical range of the 12-bit ADC used in the model Figure 3.20 at the left, while the lower rate subNyquist sampling degrades the SFDR in Figure 3.20 at the right (Baraniuk, R. and Steeghs, P. 2007).

3.6.6 UWB Signal Real-Time *Time Stretching, Time Lens,* or *Time Imaging*

Beyond the standard techniques for digital real-time UWB signals registration exists an alternative approach based on *time stretching*, time lens, or *time imaging* (Kolner, B. 1994; Kolner, B.H. and Nazarathy, M. 1989). This approach originated from the optical domain

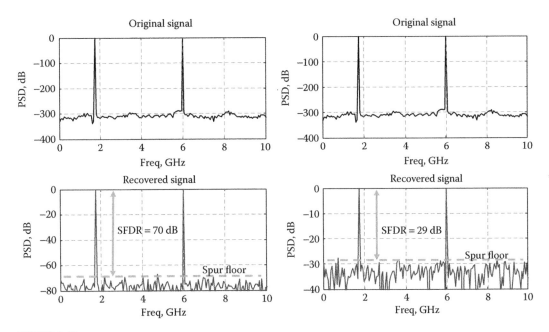

FIGURE 3.20
Simulated AIC sampling of two CW signals at 1.85 GHz and 6 GHz with their original PSD, at the top, and recovered, at the bottom, from subNyquist samples: (left) sampling is performed at 3 GSPS that is four times lower the Nyquist 12 GSPS with 70 dB SFDR signal recovering; (left) sampling is performed at 1.2 GSPS, that is, ten times lower the Nyquist 12 GSPS with 29 dB SFDR signal recovering.

for measurement of ultrafast optical waveforms. Such measurements made by using high speed photodiodes and sampling oscilloscopes are typically limited to a resolution of several picoseconds. To achieve femtosecond resolutions, the ultrafast optical waveforms can be stretched to timescales compatible with high speed electrical instruments (Coppinger, F., et al. 1999; Jalali, B. and Han, Y. 2013). This explains in part the meaning of *temporal imaging* in Figure 3.21. Additional meaning comes from the analogy between the spatial problem of diffraction and the temporal problem of dispersion and general *time-space* duality in Figure 3.22.

An interpretation of the *time-space* duality uses simplified solutions to the general wave propagation problem. For such solutions, the slowly varying envelope equations corresponding to modulated plane waves in dispersive media have the same form as the paraxial equations describing the propagation of monochromatic waves of finite spatial extent (diffraction). There is a correspondence between the time variable in the dispersion problem and the transverse space variable in the diffraction problem.

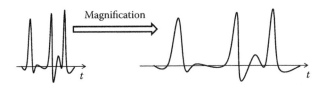

FIGURE 3.21
Signal temporal magnification by stretching (zooming) in time.

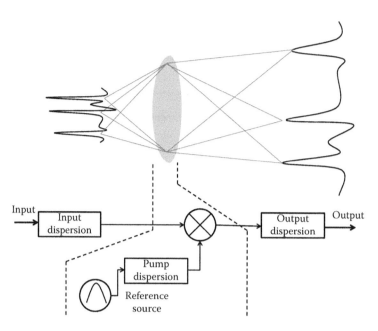

FIGURE 3.22
Optical and time microscopes use similar basic operational principles. The optical lens magnifies (spreads) the image in space for accurate measurements. The pump and reference source turn a short signal into a long one for sampling and digitizing at lower sampling frequencies.

Figure 3.22 shows a simplified overview of the optical analogy. In this case, an optical microscope uses a convex lens to magnify the image of the original object. A signal pump, for example, a chirped pulse and a nonlinear multiplier (mixer) operates similarly to the optical lens. Before mixing, the input signal goes through the *input dispersion* block and the pump signal passes through the *pump dispersion* block. The mixer output then passes through the *output dispersion* block to get a magnified version of the input signal. A magnification factor as for the lens, is mainly defined by the dispersion factor of the three dispersion blocks, although the pump source spectrum needs to agree with the initial signal spectrum. Simple math shows how this signal conversion works. Its practical implementation in real-world hardware introduces some restrictions and certain imperfections compared to just ideal mathematical case.

Optical band dispersion is made using dispersive fibers (Coppinger, F. et al. 1999). In the RF and microwave bands stretching applies two typical approaches: (1) transition to the optical band, perform stretching in optical band, and convert the optical signals back to their electrical counterpart; (2) perform all functions in the electrical domain. Both approaches have been demonstrated so far for RF and microwave bands with some degree of success (Schwartz, J.D. et al. 2007). Stretching such signals in time, enables sampling with slower rate ADCs below their initial Nyquist limits without loss of information (Kolner, B. 1994; Kolner, B.H. and Nazarathy, M. 1989).

Note how time compression of pulses is also possible by changing the order of the pump dispersion that applies dispersive lines. This can work for signal shape/spectrum control in addition to the methods considered in this chapter (van Howe, J. et al. 2004). For example, a waveform can be generated using a slow rate direct digital synthesis (DDS) as described further and then transformed to higher band by its compression in time.

3.7 Target Adaptive UWB Radar Transmitter Requirements

3.7.1 Targets and Signal Modulation Objectives

Target-adaptive UWB radars must generate signals with wavelength twice the size of the target of interest, as discussed in Section 3.6.2.2. The target matched signal will cover the range of resonances expected for a particular class of targets, for example, mines, vehicles, and tumors. The transmitted target matched signal may have two formats:

1. A short duration UWB impulse with components between f_{low} and f_{high} determined by the generating pulse width and shape, for example, Gaussian and square wave. This will be a typical case for short-range systems.

2. Modulation of a carrier about a center frequency as found in frequency-modulated continuous-wave and similar radars. Long-range systems will probably use this approach, as exemplified by the Haystack UWB Satellite Imaging Radar (HUSIR) described in Chapter 1. In Chapter 9 Ram Narayanan describes progress made in using noise signal radars.

In either case, the signal bandwidth must cover the resonances of the target class (Barrett, T.W. 1996; Barrett, T.W. 2012). The target matched signal will have components with wavelengths twice the electrical size of the target d. The target size will determine f_{low} and either the spatial resolution or the highest target resonance of interest will determine f_{high}. Figure 3.23 shows the frequency as function of target size based on a half wave dipole resonance frequency $f_{res} = c/2d$. Actual objects may have more complicated frequency relations with the physical size.

The advanced UWB radar systems transmitter will need special hardware implementations to provide reconfigurable diverse pulse shapes and frequency spectra. Each radar application will have special requirements to achieve more functional capabilities for lower power, lower complexity, and lower cost. Reviewing the modern literature and

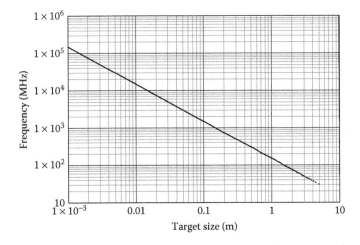

FIGURE 3.23
Estimated target resonant frequency versus size based on $f_{res} = c/2d$ where, d indicates some characteristic target component size, for example, diameter, length, wing, and span. A real object can respond with multiple frequencies higher than the lowest shown here for the maximum dimension.

Internet sources leads to the following classifications of available methods to generate such tunable UWB signals:

1. Traditional Gaussian pulse generation
2. Frequency up-conversion
3. Analog filtration
4. Digital logic
5. Analog–digital synthesis

3.7.2 Direct Digital Synthesis

Some relevant technical developments have been demonstrated for UWB communication systems, where tunability means *modulation* (Win, M.Z, and Scholtz, R.A. 2000). Note that such generated signals need the necessary power for successful transmission. This requires high power amplifiers that will not distort the UWB pulses, or operate with controllable and acceptable distortions.

The transmitter must transmit two types of signals: the *virtual-impulse* covering the range of potential target resonances and the *target matched* signal consisting of the target resonance frequencies. The main technical issues will center on the specialized signal generation and power amplification needed for adequate radar function levels. Such devices will have the general form shown in the block diagram of Figure 3.24.

The following sections describe methods for generating UWB signals with specific power spectral characteristics.

3.7.2.1 Traditional Signal-Generation Methods

Early techniques for generating Gaussian-like impulse signals cannot meet the operational requirements of advanced UWB radar systems. These methods based on semiconductor active elements such as step recovery diode, drift step recovery diode, avalanche transistors and so on have demonstrated limited flexibility for generating tailored signals. (Protiva, P. 2007)

3.7.2.2 Frequency Up-Conversion

Analog carrier-based methods for creating UWB pulses exploit two major generation techniques (1) heterodyning as shown in Figure 3.25 and (2) time-gated oscillators as shown in Figure 3.26.

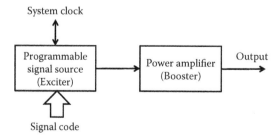

FIGURE 3.24
A general block diagram of a target matched signal transmitter for *virtual-impulse* and *target matched* signals. The *virtual-impulse* signal code will depend on the expected electrical size of the target. Time frequency analysis for *virtual-impulse* reflected signals will determine the spectrum of the target matched signal.

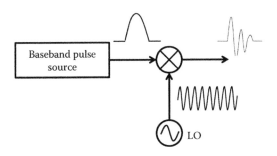

FIGURE 3.25
The heterodyning transmitter generates the baseband pulse at a lower frequency for ease of implementation. Mixing the pulse generator output with a higher frequency local oscillator (LO) produces at output pulse at the sum of the baseband and LO frequencies.

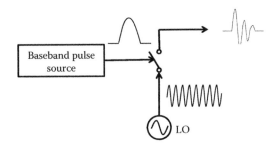

FIGURE 3.26
Alternative time gating for pulse generation. This method uses a short duration sine wave to generate a UWB pulse signal. The resulting output has a sinc like spectrum.

In heterodyning as shown in Figure 3.25, a pulse is first generated at baseband with a low-pass spectrum, that is typically easier to do. In turn, a baseband signal source can be made using one of the options above from 4 to 6 as described later. Secondly, the baseband signal is up-converted to a higher target frequency band using a local oscillator (LO) and a mixer (multiplier). The LO can use either fixed or variable CW sources. Obviously, frequency up-conversion relaxes to some extent requirements for the baseband signal source. However, conversion efficiency might be low and require power boosting as shown in Figure 3.28. This approach can use several available tuning options including: (1) initial baseband signal adjustments and (2) center frequency adjustments controlled by LO (voltage controlled oscillator, etc.).

Short duration sinewaves (less than 5 cycles) have UWB characteristics. Figure 3.26 shows how to generate a short duration sinewave with the time gated oscillator. In this case, a continuously running LO provides the basic sinewave. A time gated switch controlled by a broadband pulse turns the continuous wave signal on and off to form short duration sinewave pulses. Another approach switches the LO itself on and off to achieve the same output (Song, F. et al. 2008). A typical output in this case is made of a few cycles of the LO having a rectangular amplitude envelope in time domain and sinc-like frequency spectrum. The resulting output signal will have the power spectral density with the bandwidth shown. This approach has narrow tuning capabilities changed by the LO frequency and the number of cycles permitted by the on/off control (Song, F. et al. 2008).

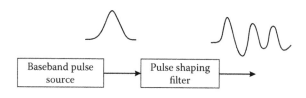

FIGURE 3.27
Generation of baseband pulses followed by filtration. This can simplify the design by controlling the pulse shaping filter to achieve a desired signal spectrum. The pulse shaping filter design must compensate for the antenna characteristics.

3.7.3 Analog Filtering

Another established method generates baseband pulses and then analog filters them to shape the desired pulse as shown in Figure 3.27. This approach has a very limited tuning capability defined by the baseband source made, using one of the options above from 4 to 6. The baseband source mostly defines the center frequency and bandwidth. Note that such a filter transformation often happens unavoidably in antennas when baseband signals need to radiate (Rajesh, N. and Pavan, S. 2015).

The antenna frequency response characteristics will also shape the pulse. The engineer must account for pulse distortion by the antenna in the exciting pulse shaping process. This requires considering the antenna and transmitter antenna characteristics together (Boryssenko, A. and Schaubert, D.H. 2006). Optical means can provide another approach to pulse shaping with low energy efficiency (Hedayati, H. et al. 2011).

3.7.4 Digital Logic Signal Generation and Transmitters

Tunable UWB pulse generation using digital logic shows many possibilities. Pulse generators have many specific features, but all employ several delays, inverters, and NOR/NAND gates in their signal generating circuits (Bourdel, S. et al. 2010; Chang, K.C. and Mias, C. 2007). Figure 3.28 shows a simplified typical scheme made from a single inverter delay and a single NOR gate. The example pulse generator trigger splits a pulse into two signal paths with a delay in one of them. The signals from both paths with a proper mutual delay enter the NOR gate that produces output impulses on every falling edge of the trigger pulse. NAND gates can be employed instead of NOR gates. This enables impulses with the opposite polarity on every rising edge of the trigger. The circuit of Figure 3.28 shows only two signal phases for simplicity. Practical circuits of this type will have many signal phases and many logic functional elements. Such pulse generators can be built using discrete digital logic (Schwoerer, J. et al. 2005). However their functionality would be limited because of many technical challenges. Those challenges could be collectively related to signal integrity issues due to tolerances in the printed circuit board, time mismatching, and other effects associated with discrete-built electronics compared to integrated circuit (IC)-based electronics. IC-based digital logic elements can provide more accuracy, functionality, bandwidth, and performance. Two approaches exist here with custom IC logic using complementary metal-oxide semiconductor process and reprogrammable off-the-shelf FPGA logic. Their advantages and disadvantages present the designer several pros and cons. The custom IC logic approach has a major drawback in higher development costs. A lack of bandwidth and special features could become a major bottleneck of reprogrammable off-the-shelf FPGA logic (Strackx, M. et al. 2013).

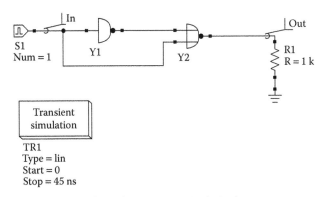

Digital impulse generator using logic elements.

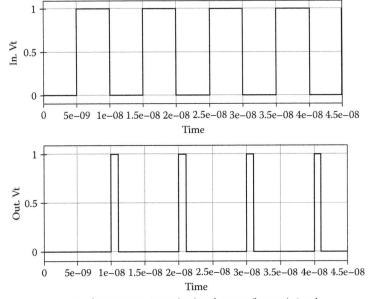

Impulse generator input (top) and output (bottom) signals.

FIGURE 3.28
The digital logic impulse generator uses an inverter as a time delay to determine the output pulse time duration. The output pulse width and bandwidth will depend on the inverter delay.

3.7.5 Analog–Digital Signal Synthesis

Analog–digital synthesis generates signals by combining multiple copies of baseband pulsed signals, that are individually delayed and weighted to get a given magnitude and polarity. Figure 3.29 shows how an array of N baseband pulse generators (BPG) generates the signal copies in a *distributed waveform generator*. The process starts by sending a trigger to a digital delay line made of cascaded T_D delays with a sampling frequency of $f_s = 1/T_D$. This approach provides flexibility for pulse shaping and spectral tuning if the necessary BPG are available or can be built. BPG arrays seems the most realizable way to use custom mixed analog–digital ICs. Figure 3.30 shows another version of this analog–digital signal synthesizer that avoids BPGs by directly weighting the delayed edges of the trigger signal and combining them for a proper pulse shaping (Cutrupi, M. et al. 2010; Zhu Y. 2007; Zhu Y. 2009; Zhu, Y. et al. 2009).

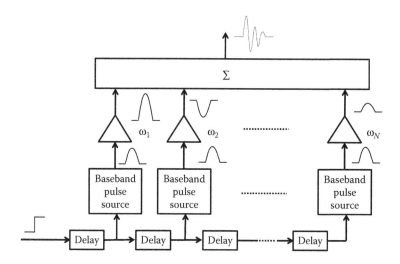

FIGURE 3.29
The distributed waveform generator uses baseband pulse generator (BPG) arrays to synthesize special UWB waveforms by combining multiple broadband impulse combinations.

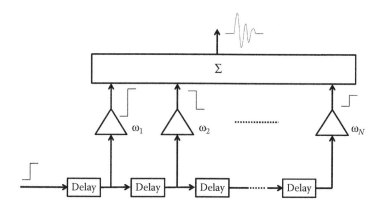

FIGURE 3.30
The combined trigger edge UWB pulse generator eliminates the BPGs of Figure 3.29. It uses multiple time delays to generate a series of fast rise time trigger pulses. It achieves pulse shaping by directly weighting the delayed edges of the trigger signal and combining them into a single output pulse.

3.7.6 Direct Digital Synthesis (DDS)

A more promising and universal approach for UWB pulse generation uses high speed DAC waveform synthesis. Figure 3.31 shows a generic DAC waveform synthesizer. In theory, a high speed DAC with good resolution could directly generate signals with necessary time-spectral features, although being fully reconfigurable (reprogrammable). However, like the ADCs considered earlier, such DACs need to operate with high sampling rates. For example, the DAC will must have a sampling frequency of at least 20 GHz for a 3–10 GHz UWB pulse and with 6-bit resolution or better. The high sampling rate poses a performance challenge for the DAC and generating the input digital data stream from a digital memory as shown in Figure 3.31. This approach increases the design complexity and power consumption. If the UWB pulse usually lasts a few nanoseconds, then

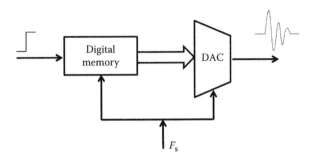

FIGURE 3.31
UWB pulse generation by direct digital waveform synthesis.

there are only tens of samples in a single pulse. Since the pulse shape does not need to change at high speed, a time interleaved DAC seems a promising path similar to time-interleaved ADC to obtain high sampling rates. In general, the overall technical challenges to get the necessary DAC capabilities is comparable to those of advanced UWB radar systems ADC.

There are off-the-shelf DAC chips that operate in a few GHz bands. For example, DAC3482 from Texas Instruments, Dallas, Texas is a very low power, high dynamic dual-channel, 16-bit DAC with a sample rate as high as 1.25 GSPS. The AD9912 from Analog Devices, Norwood, Massachusetts is a 1 GSPS DDS chip with an integrated 14-bit DAC. The AD9119/AD9129 from Analog Devices are high performance, 11-/14-bit RF DAC that can support data rates up to 2.85 GSPS (Baranauskas, D. and Zelenin, D. 2006).

A variant of the DDS approach associated with software-defined radio (SDR) can operate up to a few GHz but with a signal bandwidth not exceeding 100 MHz (e.g., the Analog Devices AD936x SDR on chip) (Pu. D, et al. 2015).

3.8 Advanced UWB Radar Conclusions

The advanced UWB radar concept offers many possibilities for expanded remote-sensing capabilities. The full development of the advanced UWB radar including signal and target interactions phenomenon, digital receiver design, signal-processing methods, variable waveform transmitter design, and the control architecture would require a dedicated book. This summary provides an overview of the concept, the potential technical issues, and suggestions for further innovations.

Special UWB radar systems will evolve to fully exploit the possibilities of return signal analysis and signal matching. The remaining chapters of this book present concepts for better performance and new applications.

This chapter presented an introduction to the major technical considerations required for building an advanced UWB radar that can

1. Identify target classes through time–frequency analysis of signals reflected by targets.
2. Change the transmitted waveform to match the target characteristics and enhance the returns from a specific target class.

Building an advanced UWB radar for a specific function will require the following:

1. Understanding the target class and virtual-impulse signal interaction.
2. Tailoring the signal bandwidth and spatial resolution to the target class.
3. Building a time-domain receiver that can digitize each reflected signal for time-frequency analysis. This will require matching of ADC capabilities with techniques such as selective sampling, compressive sensing, and other techniques.
4. Building a signal synthesizer that can build a target matched signal from the time-frequency analysis information.
5. Building a transmitter and antenna that can synthesize and transmit the target matched signal.

Advanced UWB radar performance will depend on the available ADC technology capabilities. The ability to synthesize target matched UWB signals will restrict applications to specific signal modulation bandwidths and classes of targets. Further ADC technology evolution will increase the range of applications. Although the general principles remain the same, each new realization will depend on the functional objective. For example, an advanced GPR for plastic mine location will have a much different design than a radar for area surveillance against specific classes of aircraft, vehicles, or watercraft.

We have presented the major performance objectives and technical considerations for building an advanced UWB radar. Depending on the target class and functional performance objectives, each case will require understanding the interaction of the signal and target at a new level. Chapter 12, *Wideband Wide Beam Motion Sensing* will discuss targets and signal tailoring from a different perspective.

References

Analog Devices. 2004. *Data Conversion Handbook*. Newnes, Oxford, UK.

Astanin, L.Y. and Kostylev, A.A. 1997. *Ultrawideband Radar Measurements Analysis and Processing*, The Institution of Electrical Engineers. London, UK.

Astanin, L.Y., Kostylev, A.A., Zinoviev, Yu.S. and Pasmurov, A.Ya. 1994. *Radar Target Characteristics: Measurements and Applications*. CRC Press, Boca Raton, FL.

Baker B. 2011. A Glossary of Analog-to-Digital Specifications and Performance Characteristics, *Texas Instrument Application Report Texas Instrument Application Report SLAA510*, January.

Baranauskas, D. and Zelenin, D. 2006. A 0.36W 6b up to 20GS/s DAC for UWB wave formation. *IEEE International Solid-State Circuits Conference*, February pp. 2380–2389.

Baraniuk, R. and Steeghs, P. 2007. Compressive Radar Imaging. *Radar Conference, 2007 IEEE*. pp. 128–133.

Barrett, T.W. 1996. *Active Signaling Systems*. US Patent No. 5,486,833 dated January 23.

Barrett, T.W. 2008. *Method and application of applying filters to n-dimensional signal and image in signal projection space*. US Patent No. 7,496,310, September 16, 2008.

Barrett, T.W. 2012. *Resonance and Aspect Matched Adaptive Radar*. World Scientific Publishing Co. Hacksack, NJ.

Baum, C.E. 1976. *Transient Electromagnetic Fields,* Ch. The Singularity Expansion Method, pp. 128–177, Springer-Verlag, New York.

Baum, C.E. 1997. Discrimination of buried targets via the singularity expansion, *Inverse Problems*, vol. 13, no. 3, pp. 557–570.

Baum, C.E., Rothwell, E.J. and Chen, K.M. 1991. The singularity expansion method and it application to target identification. *Proc. IEEE, 79* (10), pp. 1481–1482.

Boryssenko, A. and Schaubert, D.H. 2006. Electromagnetics-Related Aspects of Signaling and Signal Processing for UWB Short Range Radios. *J. VLSI Signal Process. Syst.* Vol. 43(1), pp. 89–104.

Bourdel, S., Bachelet, Y., Gaubert, J., Vauche, R., Fourquin, O., Dehaese, N. and Barthelemy, H. 2010. A 9-pJ/Pulse 1.42-Vpp OOK CMOS UWB pulse generator for the 3.1–10.6-GHz FCC band. *IEEE Transactions on Microwave Theory and Techniques,* Vol. 58(1), pp. 65–73.

Chang, K. C. and Mias, C. 2007. *Pulse generator based on commercially available digital circuitry. Microwave and Optical Technology Letters.* Vol. 49(6). pp. 1422–1427.

Coppinger, F., Bhushan, A.S. and Jalali, B. 1999. Photonic time stretch and its application to analog-to-digital conversion. *Microwave Theory and Techniques, IEEE Transactions on.* Vol. 47(7), pp. 1309–1314.

Cutrupi, M., Crepaldi, M., Casu, M.R. and Graziano, M. 2010. A flexible UWB Transmitter for breast cancer detection imaging systems. *Design, Automation & Test in Europe Conference &* Exhibition 8–12 March 2010, pp. 1076–1081.

Davenport, M.A., Schnelle, S.R., Slavinsky, J.P., Baraniuk, R.G. and Wakin M.B. 2010. A Wideband Compressive Radio Receiver'. *2010 Military Communications Conference,* pp. 1193–1198.

Donoho, D.L. 2006. Compressed Sensing, *IEEE Trans. on Information Theory,* Vol. 52(4), pp. 1289–1306, April.

e2V Semiconductor, 2001, Dithering in Analog to Digital Conversion, SAR 20017.

http://www.e2v.com/shared/content/resources/File/documents/broadband-data-converters/doc0869B.pdf

e2V Technologies, 2016, Broadband data converters, http://www.e2v.com/products/semiconductors/broadband-data-converters/

Ender, J. 2013. A Brief Review of Compressive Sensing Applied to Radar, *14th International Radar Symposium,* pp. 3–22.

Guerci, J.R. 2010. *Cognitive Radar: The Knowledge-Aided Fully Adaptive Approach.* Artech House, Boston, MA.

Haykin, S. 2011. *Cognitive Dynamic Systems: Perception-Action Cycle, Radar, And Radio.* Cambridge University Press, Cambridge, UK.

Hedayati, H., Arvani, F., Noshad, M., Mir-Moghtadaei, V. and Fotowat-Ahmady, A. 2011. Design of an optical UWB pulse leading to an in-band interference tolerant impulse radio UWB transceiver, *Ultra-Wideband (ICUWB), 2011 IEEE International Conference on,* pp. 25–28.

Hopper, R.J. 2015. TI E2E Community, RF sampling series blogs. http://e2e.ti.com/tags/RF%2bsampling%2bseries

Immoreev, I. 2000. "Ch. 1: Main Features of UWB Radars and Differences from Common Narrowband Radars." Taylor, J.D. (ed.) *Ultra-Wideband Radar Technology.* CRC Press, Boca Raton, FL.

Immoreev, I. 2012. "Ch. 3: Signal Waveform Variations in Ultrawideband Wireless Systems." Taylor, J.D. (ed.) *Ultrawideband Radar Applications and Design.* CRC Press, Boca Raton, FL.

Jalali, B. and Han, Y. 2013. Tera Sample-per-Second. Ch 9. *Microwave Photonics, Second Edition,* Lee, C.H. (ed). Taylor & Francis. Boca Raton, FL, pp. 307–361.

Kennaugh, E. 1981. "The K-pulse concept." *IEEE Transactions on Antennas and Propagation.* Vol. 29(2) pp. 327–331, March.

Kester, W. 2004. *Data Conversion Handbook.* Analog Devices Inc.

Kim, Tae Wook. 2015. *Ultrawideband radar,* US Patent No. 9,121,925, September 1, 2015.

Knott, E.F. 2008. "Ch. 14 Radar Cross Section." Skolnik, M.I. *The Radar Handbook. 3rd Edition,* McGraw-Hill, New York, NY.

Kolner, B., 1994. Space-time duality and the theory of temporal imaging. *Quantum Electronics, IEEE Journal of,* Vol. 30(8). pp. 1951–1963.

Kolner, B. H. and Nazarathy, M. 1989. Temporal imaging with a time lens. *Optics Lett.,* 14:630–632.

l-1 Magic 2006. http://users.ece.gatech.edu/justin/l1magic/

Laska, J.N., Kirolos, S., Massoud, Y., Baraniuk, R., Gilbert, A., Iwen, M. and Strauss, M. 2006. Random Sampling for Analog-To-Information Conversion of Wideband Signals. *IEEE Dallas Circuits and Systems Workshop (DCAS) 2006.* pp. 119–122.

Li, Yong, and Tian, Y. Gui. 2008. "A Radio-frequency Measurement System for Metallic Object Detection Using Pulse Modulation Excitation." *17th World Conference on Nondestructive Testing.* 25–28 October 2008, Shanghai, China.

Liu, Y. and Wan, Q. 2011. Anti-Sampling-Distortion Compressive Wideband Spectrum Sensing for Cognitive Radio, *Int. Journal of Mobile Communications.*

Martin, S. 1989, A Comparison of the K-Pulse and E-Pulse Techniques for Aspect Independent Radar Target Identification, Naval Postgraduate School Monterey Ca ADA220172.

Melkonian, L. 1992. Improving A/D Converter Performance Using Dither, AN-804, National Semiconductor Application Note 804, February.

Pearson, C. 2011. High-Speed, Analog-to-Digital Converter Basics, *Texas Instrument Application Report SLAA510*, January.

Pelgrom, M.J.M. 2010. *Analog-to-Digital Conversion*, Springer Science & Business Media.

Protiva, P., Mrkvica, J. and Machacm, J. 2007. Universal Generator of Ultra-Wideband Pulses. *Radioengineering* Vol.17(4), pp. 74–78, December.

Pu. D., Cozma, A. and Hill T. 2015. Four Quick Steps to Production: Using Model-Based Design for Software-Defined Radio, Analog Dialogue, Analog Devices. http://www.analog.com/library/analogdialogue.

Rajesh, N. and Pavan, S. 2015. Programmable analog pulse shaping for ultra-wideband applications, *Circuits and Systems (ISCAS), 2015 IEEE International Symposium on*, pp. 461–464.

Rolland, N., Benlarbi-Delai A. Ghis, A. and Rolland, P.A. 2005. 8-GHz bandwidth spatial sampling modules for ultrafast random-signal analysis, *Microwave and Optical Technology Letters,* Wiley, 44, pp. 292–295.

Rothwell, E.J., Chen, K. M. Nyquist, P. and Sun, W. 1987. "Frequency domain E-Pulse synthesis and target discrimination." *IEEE Trans. Antennas Propagation,* vol.AP-35, No. 4, pp. 426–434, April.

Sachs, J. 2012. *Handbook of Ultra-Wideband Short Range Sensing: Theory, Sensors, Applications.* Wiley-VCH Verlag & CO, Weinheim, Germany.

Schwartz, J.D., Azaa, J. and Plant, D.V. 2007. A Fully Electronic System for the Time Magnification of Ultra-Wideband Signals. *Microwave Theory and Techniques, IEEE Transactions.* Vol. 55(2), pp. 327–334.

Schwoerer, J., Miscopein, B., Uguen, B. and El-Zein, G. 2005. A discrete fully logical and low-cost sub-nanosecond UWB pulse generator. *Wireless and Microwave Technology, 2005. WAMICON 2005. The 2005 IEEE Annual Conference*, pp. 43–46.

Song, F., Wu, Y., Liao, H. and Huang, R. 2008. Design of a novel pulse generator for UWB applications. *Microw. Opt. Technol. Lett.* Vol. 50(7), pp. 1857–1861.

Strackx, M., Faes, B., D'Agostino, E., Leroux, P. and Reynaert, P. 2013. FPGA based flexible UWB pulse transmitter using EM subtraction, *Electronics Letters*, Vol. 49(19) pp. 1243–1244, September 12.

Taylor, J.D. 2012."Ch. 1 Introduction to Ultrawideband Radar Applications and Design." Taylor, JD., *Ultrawideband Radar Applications and Design.* CRC Press, Boca Raton, FL.

Taylor, J.D. 2013. "Ultrawideband Radar Future Directions and Benefits." *Progress In Electromagnetics Research Symposium Proceedings,* Stockholm, Sweden, August 12–15, pp. 1575–1578.

Texas Instruments, ADC12J4000, 2015. 12-Bit, 4.0 GSPS RF sampling ADC with JESD204B interface. http://www.ti.com/product/ADC12J4000.

Texas Instruments, ADC083000, 2015, 8-Bit, 3 GSPS, High Performance, Low Power A/D Converter. http://www.ti.com/product/adc083000.

VanBlairicum, M.L. 1995. "Ch. 9 Radar Cross Section and Target Scattering." Taylor, James D. (ed.) *Introduction to Ultra-Wideband Radar Systems*, CRC Press, Boca, Raton, FL.

van Howe, J., Hansryd, J. and Xu, C. 2004. Multiwavelength pulse generator using time-lens compression. *Opt. Lett.* No. 29, pp. 1470–1472.

Walden, R.H. 1999. Analog-to-digital converter survey and analysis, *IEEE Journal on Selected Areas in Communications*, Vol. 7(4), April. pp. 539–549.

Win, M.Z and Scholtz, R.A. 2000. Ultra-wide bandwidth time-hopping spread spectrum impulse radio for wireless multiple-access communications. *IEEE Transactions on Communications,* No. 48:679–689, April.

Zhu Y. 2007. A 10 Gs/s distributed waveform generator for subnanosecond pulse generation and modulation in 0.18um standard digital CMOS, *Radio Frequency Integrated Circuits (RFIC) Symposium, 2007 IEEE*, pp. 35–38.

Zhu, Y. 2009. Distributed waveform generator: A new circuit technique for ultra-wideband pulse generation, shaping and modulation. *IEEE Journal of Solid-State Circuits*, Vol. 44(3), pp. 808–823.

Zhu, Y., Zuegel, J.D., Marciante, J.R. and Wu, H. 2009. Distributed Waveform Generator: A New Circuit Technique for Ultra- Wideband Pulse Generation Shaping and Modulation. *IEEE Journal of Solid State Circuits*, Vol. 44(3), pp. 808–823, March.

4

Ultrawideband (UWB) Time–Frequency Signal Processing

Terence W. Barrett

CONTENTS

4.1 Introduction and Objectives

4.1.1 Overview

Shannon's theorem tells us that reliable communication in a communication channel depends on the bandwidth (BW) and the signal-to-noise ratio (SNR). Applying the same concept to radar implies that a reflected signal with a large signal BW and a high SNR radar return will contain information about the target. The problem lies in finding ways to analyze the return signal to access relevant information about a target in whatever form it might be and determine its characteristics in a usable format so that a description of the target is obtained upon which relevant decisions can be made. As explained in Chapter 1, the significance of these new signal analysis methods requires understanding the difference between ultrawideband (UWB) and conventional *narrowband* radar systems. Briefly stated, conventional narrowband radar uses *frequency domain* (energy) collection methods, while advanced UWB radar for target identification uses *time-and-frequency* (time–frequency) collection and analysis approaches.

We know that a UWB radar return will contain features unique to a particular target's size and geometry. Target identification through return signal analysis depends on finding ways to mathematically describe the target return signal and bring out the time–frequency characteristics. UWB signals' analysis requires applying techniques suited to the short time duration and wide range of frequencies contained in a reflected transient waveform. In many instances, if the radar signal processor can compare the target return time–frequency characteristics with known returns, then it can identify the target type. A smart-adaptive radar can then change signal waveform changes to improve target detection as described in Barrett (1996, 2012). This opens multiple possibilities for UWB radar applications in medical, nondestructive testing, security, and other applications.

This chapter shows analytical techniques for determining the target return time–frequency characteristics of UWB signals.

4.1.2 Modern UWB Radar Time–Frequency Signal-Processing Requirements

This chapter will describe some new ways to analyze a UWB target signal to determine a distinct time–frequency signature which the radar image processor can match with known signatures to *identify* the target and its aspect. The following sections will demonstrate ways to analyze short-duration impulse returns from a military Humvee truck and a Dodge Ram pickup truck with similar overall dimensions, as well as other targets. The signal analysis results will show how the different geometries produce different UWB radar returns.

Time–frequency target detection and identification require a radar receiver with a high analog-to-digital conversion (ADC) speed to accurately record the return signal waveform to retain both the timing and the frequency content. To retain the frequency content, the receiver must sample the return signal and record the amplitudes at time intervals much smaller than the signal duration. The full exploitation of time-domain methods assumes a good SNR and analog-to-digital (ADC) capabilities to reconstruct the received signal waveform. ADC-digitizing rates have limited time-domain receiver development to experimental laboratory radars. Barrett's 2012 – RAMAR book describes two time–frequency receivers used in these tests.

Note: *For this chapter, assume we have a digitized UWB signal from sample targets.* Now we need to find ways to process those received signals and get useful information.

4.1.3 Time and Frequency Domain Signal-Processing Methods and Objectives

Target identification means the chosen signal-processing system must detect a significant feature or features in a target's return signal waveform. This applies to a single return or a synthetic aperture radar (SAR) map, in order to make a decision concerning a target's detection, identification, and classification. (The same procedures apply to multiple targets.) Each signal-processing method has presuppositions about the received signal, and these will influence the approach to the unprocessed signal. These unknowns complicate selection of the best methods to bring some desired features in the signal to prominence. Selecting a method requires a trade between enhancing the desired signal features and obscuring less important ones. Therefore the chosen signal processing method, together with its presuppositions of valid application, will depend on the sometimes unknown features of the signal that is most useful for the chosen decision process. The signal-processing problem is also complicated by whether the signal is a long duration or quasi-steady state signal, or of short duration and a transient signal. These observations reduce to the single conclusion that you will require a repertoire of signal-processing methods to explore and, bring to prominence and consideration, useful signal features for decisions.

All these considerations apply particularly to *time–frequency* UWB radars, as well as RAMARs (resonance and aspect matched radars) – Barrett (2012), which by definition address transient, not quasi-steady state, signals. Appropriately addressing this class of signals requires time–frequency methods, rather than merely time or frequency domain methods. Therefore in the following sections, we shall examine a variety of signal-processing methods. You will learn about the signal presuppositions and requirements concerning their valid application. These methods have a particular relevance to transient signals such as UWB signals and time–frequency analysis methods.

We can group signal-processing methods into the following general classes:

1. Nonlocal or global signal-processing methods which examine the whole signal. For example, the well-known Fourier transform (FT) can provide precision in frequency, but not in time.

2. Local methods which process parts of the signal separately, as in wavelet methods (Meyer, 1993; Mallat, 1999). These methods provide precision in time but imprecision in frequency.

3. Hybrid time–frequency bilinear methods such as the Wigner–Ville Distribution (WVD) and the ambiguity function (AF).

All other competing methods and approaches to signal analysis and decomposition start with different assumptions and drawbacks. Some examples include the following:

1. The singular value decomposition (SVD) eigensignal/eigenimage approach uses basis signals/images defined by properties of the signal/image itself. However, this is not an efficient way to treat signals/images because those basis signals/images change from one signal/image to the next, hence the extracted properties change. Therefore, the eigenvalues so obtained cannot be used for compression and transmission.

2. The Karhunen–Loève (or Hotelling) transform assumes the signal/image to be processed is ergodic, that is, the spatial or temporal statistics of a single signal or image are the same over an ensemble of signals or images. But this is not always the case – signals and images are not the result of simple outcomes of a random process, there being always a deterministic underlying component.

3. Basis signal/images are constructed from the conventional FT. These assume the signal or image is repeated periodically in all directions, or that the signal or image is bandwidth, but not space or time limited.

4. The Weber–Hermite transform (WHT) discussed here does not require these assumptions. Furthermore, while the conventional FT provides no local information, that is information about a signal at defined times, the WHT is designed to address such information and the signal frequencies at such times.

5. Wavelet transforms and the WHT trade minimum imprecision in frequency detection for local information about the signal at a certain time.

In the following sections, we will examine these leading signal analysis methods, indicate their advantages/disadvantages, and introduce new techniques such as the WHT and carrier frequency-envelope frequency (CFEF) spectral methods, which are suitable for UWB return signal analysis.

We will commence with an examination of bilinear methods which include the WVD and AF.

Having established our objectives, we can examine each method, show examples of the results, and compare the relative merits of each.

4.2 The Wigner–Ville Distribution

Time–frequency methods, including the well-known spectrogram, are designed for linear, but nonstationary, data. Time–frequency distributions are obtained from the product of a signal at a *past time* and the signal at a *future time*, or, equivalently, from the product of a signal's higher frequency components and a signal's lower frequency components. Therefore, time–frequency analyses are *nonlocal or global transformations*. Because the analyzed signal is used twice in these definitions, the time–frequency distribution is said to be *bilinear* or *quadratic* (Hlawatsch & Boudreaux-Bartels, 1992). Time–frequency analysis is effective in analyzing nonstationary signals whose frequency distribution and magnitude vary with time and is phase-shift invariant. But these methods have a drawback through the potential cross-term "contamination," which can occur when analyzing multi-component signals. However, using window functions, the cross-terms can be mitigated, or reduced, with the penalty of some loss of resolution. Time–frequency methods are suitable for signals with narrow instantaneous BW and specifically, signals from linear time varying (LTV) systems.

The *general class, or Cohen's class, C,* for all time–frequency representations is (Cohen, 1989, 1995) shown as follows:

$$C(t,\omega) = \frac{1}{4\pi^2} \iiint_{-\infty}^{+\infty} s^*(u - \tau/2)s(u + \tau/2)\varphi(\theta,\tau)\exp\left[-i\theta t - i\tau\omega + i\theta u\right] du\, d\tau\, d\theta \qquad (4.1)$$

or

$$C(t,\omega) = \frac{1}{4\pi^2} \iiint_{-\infty}^{+\infty} S^*(u + \theta/2)S(u - \theta/2)\varphi(\theta,\tau)\exp[-i\theta t - i\tau\omega + i\tau u] d\theta\, d\tau\, du \qquad (4.2)$$

where:

$\varphi(\theta, \tau)$ is a function called the kernel* (Claasen & Mecklenbräuker, 1980a-c; Janssen, 1981, 1982, 1984)

s is a time-domain signal

S is a frequency domain signal

---

* The term "kernel", as used here, refers to the part of an integral transform, or an operator, that maps between two spaces–in the present instance, the original space is the returned UWB transient signal defined in the time domain with A-to-D conversion of sufficient speed to retain frequency information, and the target space is a time–frequency space. In the target space, the returned UWB transient signal is defined with respect to vectorial components known as a "basis". Thus, if the kernel in the transform/operator is changed, the basis in the target space is changed, and the depiction of the UWB transient signal is changed.

t is time

ω is radial frequency

u is a general variable addressing time or frequency that positions the kernel along the signal length

τ is the time lag variable used in autocorrelation analysis, that is, the kernel maps signals by τ

θ is the Doppler frequency, that is, the kernel maps signals by θ and τ

We can define the general signal class in terms of the *characteristic function, M as follows*:

$$C(t,\omega) = \frac{1}{4\pi^2} \iint\limits_{-\infty}^{+\infty} M(\theta,\tau)\exp[-i\theta t - i\tau\omega]d\theta d\tau, \tag{4.3}$$

where:

$$M(\theta,\tau) = \varphi(\theta,\tau)\int\limits_{-\infty}^{+\infty} s^*(u-\tau/2)s(u+t/2)\exp[i\theta u]du \tag{4.4}$$

$$= \varphi(\theta,\tau)A(\theta,\tau)$$

and $A(\theta,\tau)$, the *symmetrical* AF. This allows classifying time–frequency distributions with respect to the kernel function $\varphi(\theta,\tau)$ (Cohen, 1989, 1995).

The deterministic autocorrelation function is defined as follows:

$$R(\tau) = \int\limits_{-\infty}^{+\infty} s^*(u)s(u+\tau)du \tag{4.5}$$

and a *deterministic generalized local autocorrelation function* is defined as follows (Choi & Williams, 1989; Cohen, 1989):

$$R_\tau(\tau) = \iint\limits_{-\infty}^{+\infty} s^*(u-\tau/2)s(u+\tau/2)\varphi(\theta,\tau)\exp\left[i\theta(u-t)\right]d\theta du. \tag{4.6}$$

This autocorrelation is dependent on time.

Turning now from the general class, we will specifically examine the WVD (Wigner, 1932; Ville, 1948) which is defined as

$$WVD(t,\omega) = \frac{1}{2\pi}\int\limits_{-\infty}^{+\infty} s^*(t-\tau/2)s(t+\tau/2)\exp[-i\tau\omega]d\tau$$

$$= \frac{1}{2\pi}\int\limits_{-\infty}^{+\infty} S^*(\omega-\theta/2)S(\omega+\theta/2)\exp[-it\theta]d\theta, \tag{4.7}$$

where s is a time-domain signal and real. The WVD is the FT with respect to τ of an auto-correlation specifically defined as

$$R(t,\tau) = s^*(t-\tau/2)s(t+\tau/2) \tag{4.8}$$

or

$$R(t,\tau) = \int_{-\infty}^{+\infty} WVD(t,\omega)\exp[i\tau\omega]d\omega = s^*(t-\tau/2)s(t+\tau/2) \tag{4.9}$$

As $\varphi(\theta,\tau) = 1$ in the case of the WVD, the WVD characteristic function is as follows:

$$M(\theta, \tau) = A(\theta, \tau) \tag{4.10}$$

Figure 4.1 shows the results of analyzing sinusoid and pulse test signals of Figure 4.1a and b with the WVD to the appearance of the cross-terms. This figure shows that the WVD can highlight sinusoids and test pulses, that is, simple signals without multiple components and cross-interference, but in the case of a multiple component signal shown in Figure 4.1c cross terms appear as light areas on the three traces.

To demonstrate a practical use of the WVD transformation, we can analyze the UWB radar return from a Humvee with approximate dimensions of 4.7 m length, 2.16 m width, and 1.83 m height. The transmitted pulse was a 2 nsec envelope-modulated pulse with a 35.3 GHz (Ka-band) carrier of BW 1 GHz, and of power < 0.5 W. The targets for these series of tests were at a range of 150 m.

The one-sided UWB WVD time–frequency spectra for a Humvee target at aspect angles such as 0, 45, 90, and 180° are shown in Figure 4.2. The Choi–Williams (1989) modification of the WVD was used. Nevertheless, it is difficult to determine which components of the WVD obtained are signal components, and which are residual cross-terms.

4.3 The Ambiguity Function

The AF is a time–frequency transform related to the WVD transform. The symmetric AF is defined as follows:

$$AF(\theta, \tau) = \int s^*(t-\tau/2)s(t+\tau/2)\exp[i\theta t]di \tag{4.11}$$

and is generally complex. The AF is thus the FT with respect to t of an autocorrelation also introduced and used above in defining the WVD, and defined as follows:

$$R(t,\tau) = s^*(t-\tau/2)s(t+\tau/2). \tag{4.12}$$

or

$$R(t,\tau) = \int_{-\infty}^{+\infty} AF(\theta, \tau)\exp[-i\theta t]d\theta = s^*(t-\tau/2)s(t+\tau/2). \tag{4.13}$$

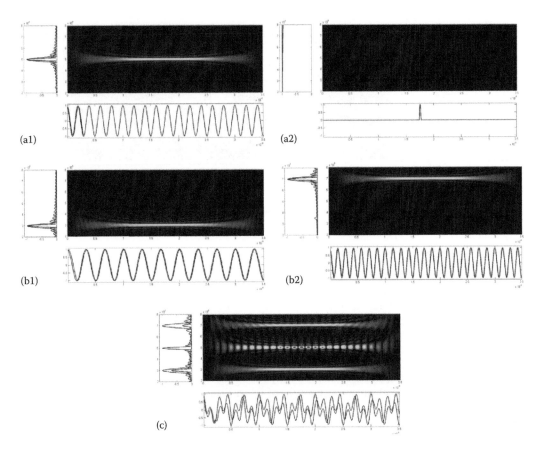

FIGURE 4.1
The one-sided Wigner-Ville Distribution (WVD) time–frequency spectra for a sinusoid (a1), a pulse (a2); a low-frequency sinusoid (b1); a high-frequency sinusoid (b2); the low- and high-frequency sinusoids combined (c). In the marginal in all cases, the blue lines indicate the WVD marginals, and the red lines the Fourier transform marginals. It is apparent that, whereas there are no cross-terms in (b1) and (b2), they are a "contaminant" in (c). The horizontal axis indicates time and the vertical axis frequency spectrum.

The AF was first introduced by Ville (1948) and Moyal (1949), and extensively applied to conventional radar signals by Rihaczek (1969, Rihaczek & Hershkowitz, 1996; 2000). The relation of the AF to matched filters was described by Woodward and Davies (1950) and Woodward (1953), and the AF is the characteristic function of the WVD. Specifically, the WVD is the double FT of the symmetric AF, which is shown in the following:

$$WVD(t,\omega) = \int\int_{-\infty}^{+\infty} AF(\theta, \tau)\exp[-i(\omega\tau + \theta t)]d\theta d\tau. \tag{4.14}$$

However, the AF differs from the WVD in that, while the autoterms are concentrated *around* the origin and the cross-terms are concentrated *away from* the origin. Therefore, it is possible to suppress the cross-terms by applying a low-pass 2D filter in the ambiguity domain. The AF is considered to extend the notion of autocorrelation to nonstationary signals (Flandrin, 1998).

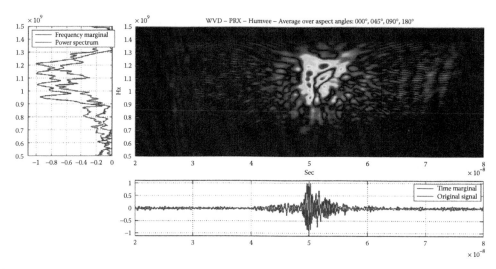

FIGURE 4.2
The Wigner-Ville Distribution (WVD) time–frequency spectra of the UWB return signal from a Humvee truck averaged over the return signals for the target at aspect angles, 0, 45, 90 and 180°. In the marginals, the blue lines indicate the WVD marginals, and the red lines the Fourier transform marginals.

One-sided AF lag time (τ) time–frequency spectra are shown for test signals in Figure 4.3 and for a Humvee truck at an aspect angle of 0° in Figure 4.4. The cross-terms are away from the origin and the time symmetry of the AF is due to the chosen symmetry of the autocorrelation time lag range.

4.4 The Weber–Hermite Transform and Weber–Hermite Wave Functions

Weber–Hermite wave functions have both a global and local application and provide a solution to the well-known "energy confinement problem," or time–bandwidth product (TBP) limiting problem, addressed by Slepian and collaborators (Landau & Pollak, 1961, 1962; Slepian, 1964, 1978; Slepian & Pollak, 1961).

Slepian et al's solution to the energy confinement problem was the prolate spheroidal wave function (PSWF) series, which can be recursively, but not analytically, generated. An alternative solution is the parabolic cylinder or Weber–Hermite wave function (WHWF) series (Barrett, 1972, 1973a,b), which provide analytic functions. The PSWFs provide a slightly better (i.e., smaller) BW occupancy, but with a slightly inferior (i.e., larger) time occupancy. As the WHWFs are analytic and can be analytically generated to any level, they are preferred here.

The WHWFs are related to *Weber's equation* (Weber, 1869) as follows:

$$\frac{d^2\psi_n(x)}{dx^2} + \left(n + \frac{1}{2} - \frac{1}{4}x^2\right)\psi_n(x) = 0 \tag{4.15}$$

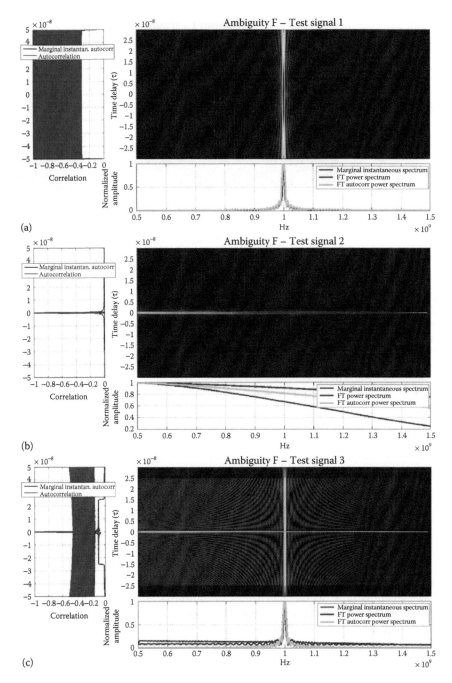

FIGURE 4.3
The one-sided ambiguity function (AF) time (lag)-frequency spectra for (a) a single frequency sinusoid (b) a pulse, and (c) the sinusoid and pulse. In the left marginal are the AF marginal and also the autocorrelation. In the bottom marginal are the Fourier transform derived spectrum and also the Fourier transform/autocorrelation derived spectrum. The time (lag) symmetry of the AF is due to the chosen symmetry of the autocorrelation time lag range.

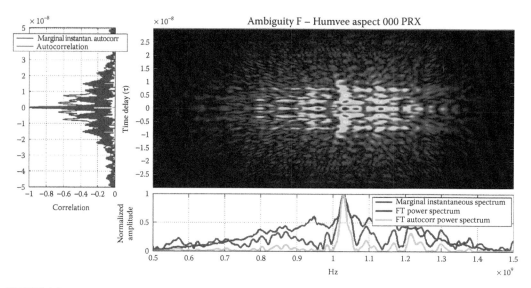

FIGURE 4.4
One-sided UWB ambiguity function (AF) time (lag)-frequency spectra for a Humvee truck at an aspect angle of 0°. In the left marginal are the AF marginal and also the autocorrelation. In the bottom marginal are the Fourier transform derived spectrum and also the Fourier transform/autocorrelation derived spectrum. The time (lag) symmetry of the AF is due to the chosen symmetry of the autocorrelation time lag range. The transmitted pulse was a 2 nsec envelope-modulated pulse with a 35.3 GHz (ka-band) carrier of bandwidth 1 GHz, and of power <0.5 W. The targets for these series of tests were at a range of 150 m.

for which the *general Weber equation*, or *parabolic cylinder differential equation*, is (Abramowitz & Stegun, 1972) as follows:

$$\frac{d^2\psi_n(x)}{dx^2} + (ax^2 + bx + c)\psi_n(x) = 0 \tag{4.16}$$

The solutions of this equation are the parabolic cylinder or WHWFs:

$$\psi_n(x) = 2^{-n/2}\exp[-x^2/4]H_n\left(x/\sqrt{2}\right) \quad n = 0, 1, 2, \ldots \tag{4.17}$$

where the H_n are Hermite polynomials. When n is an integer, the Weber–Hermite functions become proportional to the Hermite polynomials. (Caution: Other terminology is used for the WHWFs, for example, Hermite–Gaussian functions.) The present author prefers the designation WHWFs because (a) Hermite–Gaussian implicates Gaussian in all polynomials, $n = 0, 1, 2, \ldots$ but the Gaussian is only just the first, for the case $n = 0$; (b) Weber's equation is more general than Hermite's equation; (c) the name "Weber–Hermite" follows the *Mathematical Encyclopedia* (Hazewinkel, 2002) usage; and (d) other texts, for example, Morse & Feshbach, 1953, vol. 2, p. 1642; Jones, 1964, p. 86 also use the terminology "Weber–Hermite".

The WHWFs are given a physical representation as follows. The one-dimensional wave equation is written:

$$-\frac{1}{2m}\frac{\partial^2\psi}{\partial x^2} + V(x)\psi = E\psi, \tag{4.18}$$

with spring potential:

$$V(x) = \frac{1}{2}kx^2 = \frac{1}{2}m\omega^2 x^2,$$

(4.19)

where:

$\omega = \sqrt{k/m}$ is the angular frequency
k is the stiffness constant
m is the mass
x is the field deflection of the oscillator

The wave equation can be written in dimensionless form by defining the independent variables $\xi = \alpha x$ and an eigenvalue, λ and requiring:

$$\alpha^4 = mk, \quad \lambda = 2E\left(\frac{m}{k}\right)^{1/2} = \frac{2E}{\omega}.$$

(4.20)

The dimensionless form is then written as

$$\frac{\partial^2 \psi}{\partial \xi^2} + \left(\lambda - \xi^2\right)\psi = 0$$

(4.21)

which is also a form of Weber's equation and permits solutions as a function of $n = E/2\beta - 1/2$. In order for the solutions to be quadratically integrable, it is necessary that n takes on integer values $n = 0,1,2,\dots$ (Morse & Feshbach, 1953, p. 1641). With normalization factors, the solutions are the Weber–Hermite or parabolic cylinder functions as given below:

$$\psi_n(t) = \frac{1}{\sqrt{2^n n!}}\left(\frac{\alpha}{\pi}\right)^{1/4} \exp\left(-\frac{\alpha t^2}{2}\right) H_n\left(t\sqrt{\alpha}\right)$$

(4.22)

where:
$\alpha = m\omega$
H_n are Hermite polynomials
α is a time–frequency trade parameter/variable
If in the case of a function $f(x)$, an expansion of the following form:

$$f(x) = a_0\psi_0(x) + a_1\psi_1(x) + \cdots + a_n\psi_n(x) + \cdots$$

(4.23)

exists, and if it is legitimate to integrate term-by-term between the limits $-\infty$ and $+\infty$, then:

$$a_n = \frac{1}{\sqrt{(2\pi)^{1/2} n!}} \int_{-\infty}^{+\infty} \psi_n(t)f(t)dt.$$

(4.24)

The first six WHWFs (i.e., WHWF 0–5) are shown in the time and frequency domains in Figure 4.5. It is apparent that the WHWFs increase in temporal length × BW size, that is, in time-bandwidth product (TBP), as n increases. Just as a FT decomposes a signal into an amplitude sum of sinusoidal functions, so a Weber–Hermite (WH) transform decomposes a signal into an amplitude sum of WHWFs shown here. A major difference between

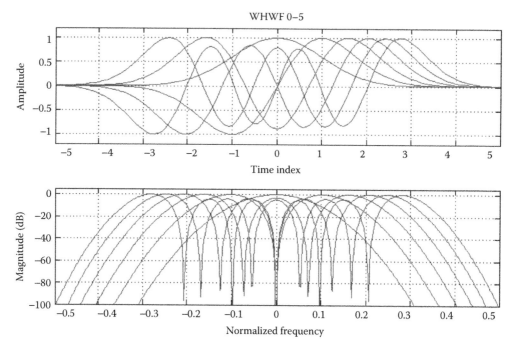

FIGURE 4.5
The first six Weber–Hermite wave functions (WHWF) (upper) and their log magnitude spectra (lower) are labeled according to an increasing n (n = 0, 1, 2, 3, 4, 5). As n increases, the temporal length and bandwidth increase, that is, the time–bandwidth product increases.

the two transforms is that, whereas the sinusoidal functions of the FT technically have no beginning and no end, the WHWFs (a) have a beginning and end – they are finite or "compactly supported," and (b) each WHWF has a finite and precisely defined TBP. The WHWF transform is thus more suitable for decomposing transient signals, which also have a beginning and end.

In a manner similar to that of the FT, a WHT can be constructed by matrix methods. Figure 4.6a shows a 128 × 256 WH (magnitude) matrix. If the complex WH matrix is designated, W, then, as W is a unitary matrix, $WW\dagger = I$, where $W\dagger$ is the conjugate transpose (Hermitian adjoint) of W, and I is the identity matrix shown in Figure 4.6b. The inverse of W, or W^{-1}, is equal to the conjugate transpose: $W^{-1} = W\dagger$.

We can examine the effects of the WHWF waveform on truck and Humvee targets. Figure 4.5 shows the waveform generated using the second WHWF in the series WHWFn, n = 0, 1, 2, …., that is, WHWF1. This provides a differentiating filter with the filter Q set at 4. Figure 4.7 shows the WHT time–frequency spectra for the return signals from a 2 ns UWB pulse (RX) from a Humvee and Dodge Ram pickup truck targets at various aspect angles. Appendix A develops the relationship of WHWFs to the fractional Fourier transform (FRFT) and additional derivations of WHWFs.

The WHWFs – of increasing TBP – can be given both a *globally distributed* or a *locally distributed* matrix form. As shown in Figures 4.8 and 4.9, both matrices are unitary and forward, and inverse global and local WH transforms of 1D and 2D signals are possible as shown in Figure 4.10. Figures 4.11 and 4.12 indicate the complementarity of Fourier and WH transforms. As continuous wave (CW) signals are the basis functions of the FT, the Fourier transform-based power spectrum well discriminates the CW signals and the

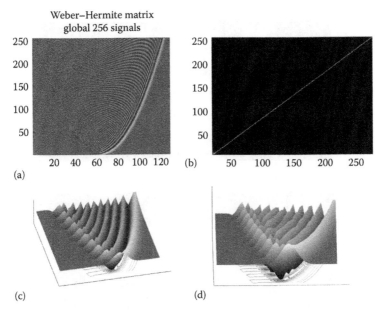

FIGURE 4.6
Example Weber–Hermite matrices: (a) 128 × 256 Weber–Hermite (WH) (global) matrix (magnitude). If the complex WH matrix is designated, W, then, as W is a unitary matrix, $WW^\dagger = I$, where $W^\dagger$ is the conjugate transpose (Hermitian adjoint) of W, and I is the identity matrix shown in (b). The inverse of W, or W^{-1}, is equal to the conjugate transpose: $W^{-1} = W^\dagger$. (c) and (d) show exploded views of the first 20 signals in (a). The WHWF matrix when multiplied with its conjugate transpose results in an identity matrix and indicates that a signal can be transformed into a WH domain, and an inverse WH transform will reconstitute the signal, that is, forward and inverse transform exists and no information is lost.

WH transform does so poorly. In contrast, as WH signals are the basis functions of the WH transform, the WH transform discriminates well the WH packet signals and the Fourier transform-based power spectrum does so poorly.

The ability to reconstruct corrupted signals by separation of the corruption and the signal and the removal of the corruption in the transform domain is an indication of the appropriateness of a transform for analysis of specific families of signals. Figures 4.13–4.15 indicate that the WH transform has this capability in the case of transient wave packet signals, but the FT does not. Therefore, the WH transform has a more important role to play in the characterization of UWB return signals.

Applying the WH transform to process the UWB returns form an airborne UWB radar from two similar targets, the Humvee and Dodge truck shown in Figure 4.16 demonstrates how each target produces a unique signature. In all of the following analyses of return signals, the transmitted pulse was a 2 nsec envelope-modulated pulse with a 35.3 GHz (ka-band) carrier, of BW 1 GHz, and of power < 0.5 W. The targets for these series of tests were at a range of 150 m. Figure 4.17 shows a clear difference in the processed UWB returns for the two targets at ~600 MHz in the Fourier spectra. Significantly, a WH transform of the UWB returns amplified the difference to 31.2 dB as shown in Figure 4.18.

Another illustration of WH transform used the artillery shell shown in Figure 4.19 as a target. A time–frequency analysis of the UWB return – but, significantly, not a Fourier spectral analysis – revealed a resonance ~600 MHz as shown in Figure 4.20.

There is a clear indication that the WH transform provides time–frequency UWB return signal target signatures that facilitate decisions concerning the identification of the target.

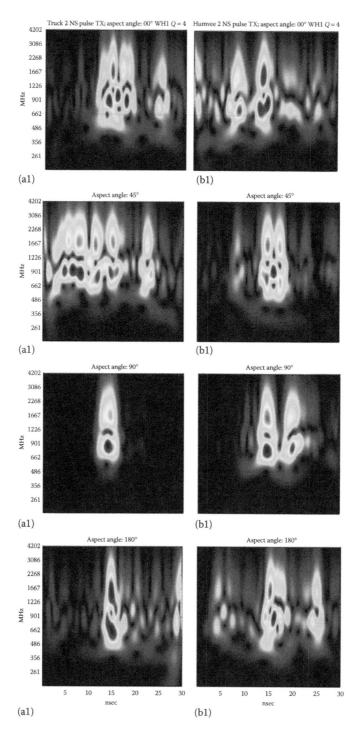

FIGURE 4.7
Time–frequency spectra using Weber–Hermite wave function 1 (WHWF 1) – a differentiating wavelet filter – with filter $Q = 4$. Left column (a1) shows the truck target. Right column (b1) shows the Humvee target. UWB TX is 2 nsec pulse. Target aspect angle, top to bottom, 0, 45, 90, and 180°.

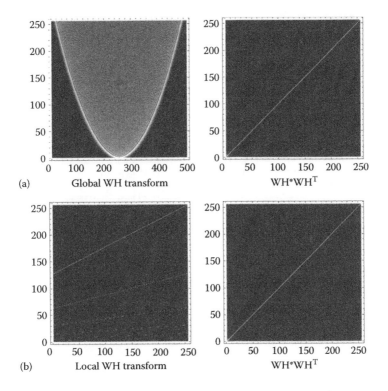

(a) Global WH transform   WH*WH^T

(b) Local WH transform   WH*WH^T

FIGURE 4.8
Examples of Weber–Hermite (WH) transforms. (a) Left: *Global* WH transform matrix. (b) Left: *Local* WH transform matrix, in both of which the time–bandwidth product (TBP) increases from bottom to top according to $\Delta f \Delta t = 1/2(2n+1)$, $n = 0,1,2,\ldots$ In the case of both left matrices, which are unitary, the product of each matrix with its conjugate transpose results in an identity matrix – both right matrices. Therefore, both global and local WH transforms, forward and inverse, of 1D and 2D signals are indicated. These are general matrices and the axes are general, but can, of course, be time and frequency, as used here. The major point is that the global or local WH matrices when multiplied with their appropriate conjugate transpose result in an identity matrix, indicate that a signal can be transformed into a WH global or local domain, and the appropriate inverse WH transform will reconstitute the signal, that is, that forward and inverse transforms exist and no information is lost.

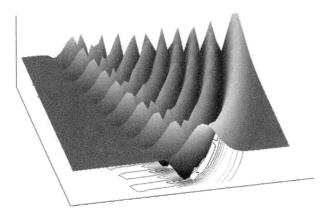

FIGURE 4.9
A perspective of the general global Weber–Hermite (WH) matrix for 20 WHWF signals (basis elements) which are seen in the above Figure 4.8 – left.

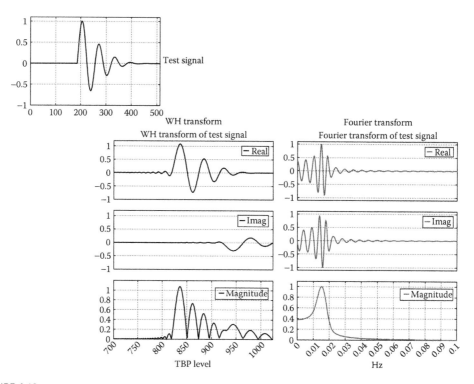

FIGURE 4.10
A transient test signal (upper), and the real, imaginary, and magnitude components of the Weber–Hermite (WH) global transform and the Fourier transform.

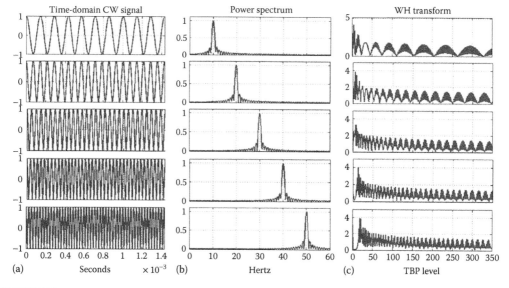

FIGURE 4.11
Comparison of signal power spectra and Weber–Hermite (WH) transform. (a) Time-domain constant wavelength (CW) signals, (b) signal power spectra, and (c) global WH transform. As CW signals are the basis functions of the Fourier transform, the Fourier transform-based power spectra discriminate well the CW signals and the Weber–Hermite (WH) transform does poorly.

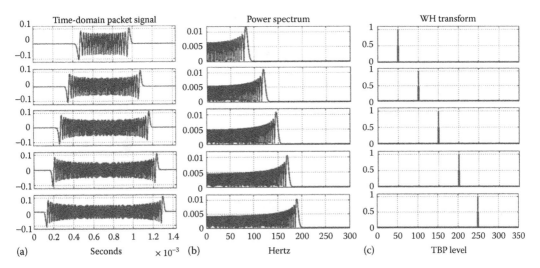

FIGURE 4.12
Comparison of (a) time domain Weber–Hermite (WH) packet signals, (b) WH signal power spectra, and (c) global WH transform. As WH signals are the basis functions of the WH transform, the WH transform discriminates well the WH packet signals and the Fourier transform-based power spectrum does poorly.

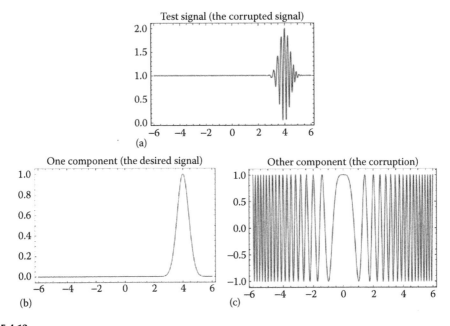

FIGURE 4.13
The WH transform can aid in recovering signals by identifying the signal and corrupting components.
(a) A corrupted signal is as follows:

$$s(t) = \exp\left[-\pi(x-4)^2\right]\left(\exp[-i\pi t^2] rect[t]\right),$$

$$rect[t] = 1 \quad (-8 \le t \le +8)$$

$$rect[t] = 0 \quad (x < -8; x > +8).$$

(b) The desired or designated signal is $\left[-\pi(x-4)^2\right]$
(c) The corruption is $\exp[-i\pi t^2] rect[t]$

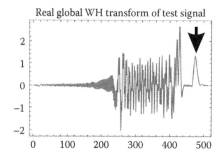

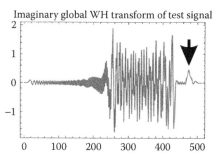

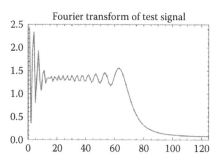

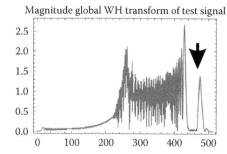

FIGURE 4.14
The real, imaginary, and magnitude of the global Weber–Hermite (WH) transform of the corrupted signal are shown in Figure 4.13. The signal corruption component and the desired signal component of the transform (indicated by arrow) are separated. However, the Fourier transform of the corrupted signal – lower left – does not permit separation of the two components.

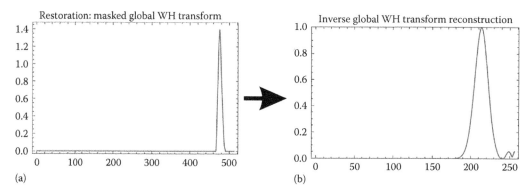

FIGURE 4.15
Signal capture without corruption using the Weber–Hermite "transform." (a) After masking the corrupted signal in the WH domain, the transform of the desired signal component remains. (b) An inverse WH transform captures the desired signal component in the original domain without the corruption.

0° aspect   180° aspect   0° aspect   180° aspect

90° aspect Humvee   90° aspect Dodge Ram

FIGURE 4.16
Truck UWB radar targets Humvee (4.7 m length, 2.16 m width, and 1.83 m height) and Dodge Ram pickup (5.5 m length, 1.9 m width, and 1.7 m height).

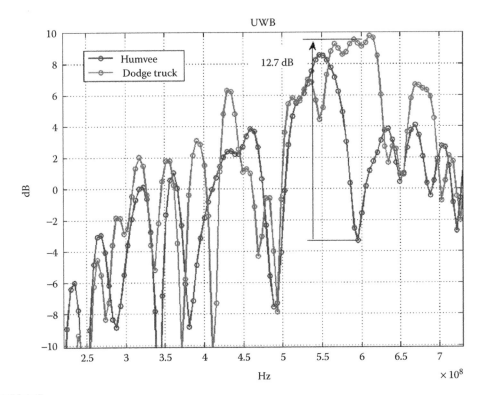

FIGURE 4.17
The Fourier spectra of UWB returns from Humvee and Dodge trucks. The Fourier transform clearly shows a spectral difference around 600 MHz for the Dodge truck.

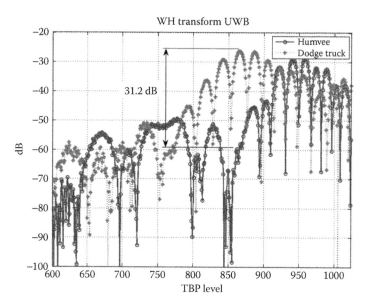

FIGURE 4.18
Weber–Hermite (WH) transform spectra of the same UWB returns from Humvee and Dodge trucks as shown in Figure 4.17. There is a difference of 31.2 dB at a Weber–Hermite wave functions (WHWFs' filter time–bandwidth product (TBP) of around 850. These return signals were obtained from UWB transmission from a light aircraft flying above the targets.

FIGURE 4.19
The artillery shell (155 × 667 mm) radar test target. Expected resonance is 3e8/(667e-3) = 4.4978e8 or approximately 450 MHz – neglecting the effect of shell's taper. Empirical testing showed the main resonance at around 600 MHz.

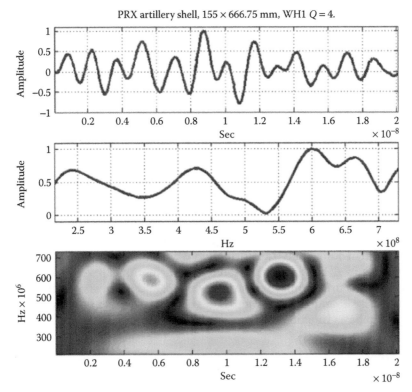

FIGURE 4.20

The UWB signal return (RX) for a 155 mm artillery shell. Top: time-domain RX signal. Middle: Fourier spectrum. Bottom: WH time–frequency spectrum. Whereas, the Fourier spectrum (sufficient for a linear time-invariant (LTI) system response description) indicates multiple resonant peaks, the WH time–frequency spectrum (appropriate for linear time varying (LTV) system response description) clearly reveals the major resonance at 600 MHz.

4.5 The Fractional Fourier Transform

The quadratic FRFT is a generalization of the conventional FT (Namias, 1980; Dickinson & Steiglitz, 1982; McBride & Kerr, 1987; Ozaktas & Mendlovic, 1993; Lohmann, 1993; Ozaktas et al., 1994, 2001). Conversely, the linear canonical transforms (LCT), or special affine Fourier transform (SAFT), or ABCD transform (Bernardo, 1996) generalize the FRFT (Papoulis, 1977; Pei & Ding, 2001; Saxena & Singh, 2005). The FRFT leads to a generalization of the time (or space) and frequency domains. Here, we address the FRFT, but other well-known transforms can be fractionalized (Lohman et al., 1996; Herrmann, 2011). A fundamental property of the FRFT is that performing the ath FRFT operation followed by a WVD corresponds to rotating the WVD of the original signal by an angle parameter of α ($\alpha = a\pi/2$; $a = \alpha2/\pi$ both defined below) in the clockwise direction (Lohmann, 1993; Pei & Ding, 2001), or a decomposition of the signal into chirps. Whereas the conventional FT acting on a function can be interpreted as a linear differential operator acting on that function; the FRFT generalizes this definition by having the differential operator depend on the continuous order parameter, a. Therefore, the ath order FRFT is the ath power of the FT operator.

The conventional FT is a special case of the FRFT, the FRFT adding an additional degree of freedom to signal analysis with introduction of the fraction or order parameter, a. The solution to a signal analysis problem can then be optimized depending on the objective at a specific a and the chosen optimization criterion.

A physical picture is conveniently based on diffraction optics. It is well-known that the far-field diffraction pattern is the FT of the diffracting object, thus the generalization is that FRFTs are the field patterns at closer distances. An alternative picture is an optical path involving many lenses separated by a variety of distances, with the FRFTs being the amplitude distributions of a wave propagating through the total system of lenses. As a wave propagates through the system, the amplitude distribution evolves through FRFTs of increasing order.

Whereas the conventional FT utilizes cisoidal (complex exponential) and steady state, basis functions, the FRFT (except at $a = 1.0, 2.0, ...$) utilizes frequency modulated or chirp basis functions that convolve maximally with signals changing in frequency – as in wave propagation through a dispersive medium – and such wave dispersion can occur on the surface of targets. In the following set of equations, the order parameter variable, a ranges from -2 to $+2$. When $a = +1$, the transformation is the conventional forward Fourier transformation; when $a = -1$, the transformation is the conventional inverse Fourier transformation. When $a = 0$, the time-domain signal is regained; and when $a = +2$, the time-domain is regained, but parity is changed. At other values of a, the ath order FRFT results with the corresponding chirp basis functions. The relations are described in the following equations:

$$FRFT_a = \int_{-\infty}^{+\infty} K_a(u,t) f(t) dt \tag{4.25}$$

where:

$$K_a(u,t) = A_a \exp\left[i\pi\left(u^2 cot(\alpha) - 2ut\, csc(\alpha) + t^2 cot(\alpha)\right)\right], \tag{4.26}$$

$$A_a = \sqrt{1 - i cot(\alpha)} \tag{4.27}$$

$$\alpha = a\pi/2 \tag{4.28}$$

Note how the kernel, $K_a(u,t)$, functions in the manner of a Green's function as in the well-known Schrödinger's equation.

In the case $a = 1$, then $\alpha = \pi/2$, $u = f$, and

$$FRFT_1(f) = FT(f) = \int_{-\infty}^{+\infty} \exp[-i2\pi ft] f(t) dt, \tag{4.29}$$

that is, the conventional *forward* FT.

In the case $a = -1$, then $\alpha = -\pi/2$, $u = f$ and

$$FRFT_{-1}(f) = FT(f) = \int_{-\infty}^{+\infty} \exp[+i2\pi ft] f(t) dt \tag{4.30}$$

that is, the conventional *inverse* FT.

Using operator formalism, the FRFT is generalized to the special affine Fourier transform (SAFT) (Pei & Ding, 2000). For example, let

$$O_F^\alpha(f(t)) = FRFT_\alpha \qquad (4.31)$$

then the SAFT, or canonical transform (Moshinsky & Quesne, 1971; Abe & Sheridan, 1994) is as follows:

$$O_F^{(a,b,c,d)}(f(t)) = \sqrt{\frac{1}{|b|}} \exp\left[i\pi\left(u^2(d/b) - ut(1/b)\right) + t^2(a/b)\right]f(t)dt \text{ when } b \neq 0, \qquad (4.32)$$

$$O_F^{(a,b,c,d)}(f(t)) = \sqrt{d}\exp\left[i\pi(u^2(cd))\right]f(du)dt \text{ when } b = 0, \qquad (4.33)$$

and where $ad - bc = 1$ must be satisfied.

The FT is then described as a rotation by 90° and represented as follows:

$$\begin{bmatrix} a & b \\ c & d \end{bmatrix} = \begin{bmatrix} 0 & 1 \\ -1 & 0 \end{bmatrix}, \qquad (4.34)$$

and the FRFT is described as a rotation by an arbitrary angle and presented as follows:

$$\begin{bmatrix} a & b \\ c & d \end{bmatrix} = \begin{bmatrix} \cos\theta & \sin\theta \\ -\sin\theta & \cos\theta \end{bmatrix}. \qquad (4.35)$$

The SAFT has the additive and reversible properties which are as follows:

$$O_F^{(d,-b,-c,a)}\left(O_F^{(a,b,c,d)}(f(t))\right) = f(t). \qquad (4.36)$$

The FRFT is periodic in a (or α) with period 4 (or 2π). Therefore, the transform is fully defined for $a \in (-2, 2]$ or $\alpha \in (-\pi, \pi]$. The following relations result where $\mathcal{F}^i$ is the FRFT for $a = i$, $\mathcal{J}$ is the identity matrix, and $\mathcal{P}$ indicates a parity change (Candan et al., 2000; Ozakta et al., 2001):

$$\mathcal{F}^0 = \mathcal{J}, \qquad (4.37)$$

$$\mathcal{F}^1 = \mathcal{F}, \qquad (4.38)$$

$$\mathcal{F}^2 = \mathcal{P}, \qquad (4.39)$$

$$\mathcal{F}^3 = \mathcal{F}\mathcal{P} = \mathcal{P}\mathcal{F}, \qquad (4.40)$$

$$\mathcal{F}^4 = \mathcal{F}^0 = \mathcal{J}, \qquad (4.41)$$

$$\mathcal{F}^{4j+a} = \mathcal{F}^{4k+a}, \qquad (4.42)$$

where j and k are arbitrary integers.

The FRFT operations are additive. For example, the 0.3th FRFT of the 0.6th FRFT is the 0.9th FRFT. The inverse, $(\mathcal{F}^a)^{-1}$ of the ath order FRFT operator, $\mathcal{F}^a$ is equal to the operator $\mathcal{F}^{-a}$ because $\mathcal{F}^{-a}\mathcal{F}^a = \mathcal{J}$

A coordinate multiplication operator, $\mathcal{U}$, can be defined for which impulse signals, δ, are the eigensignals; and also a differentiation operator, $\mathcal{D}$, for which harmonic signals are the eigensignals (Candan et al., 2000). Then $\mathcal{U}$ and $\mathcal{D}$ are Hermitian which are shown as follows:

$$[\mathcal{U},\mathcal{D}]=\frac{i}{2\pi}\mathcal{J} \tag{4.43}$$

with the following properties:

$$\mathcal{U}f(u) = u(f(u)), \tag{4.44}$$

$$\mathcal{D}f(u) = \frac{1}{2\pi i}\frac{d}{du}f(u). \tag{4.45}$$

Under an FRFT, the Hamiltonian Hermitian operator, $\mathcal{H}$, is invariant;

$$\mathcal{H}_F = 2\pi\frac{1}{2}(\mathcal{D}^2 + \mathcal{U}^2) \equiv 2\mathcal{J}, \tag{4.46}$$

and a chirp multiplication becomes:

$$\mathcal{H}_m \equiv 2\pi\frac{1}{2}\mathcal{U}^2, \tag{4.47}$$

a chirp convolution is:

$$\mathcal{H}_c \equiv 2\pi\frac{1}{2}\mathcal{D}^2, \tag{4.48}$$

and the FRFT defined in these terms is:

$$\begin{aligned}\mathcal{F}^a &= \exp[-\alpha\mathcal{H}] = \exp\left[-i\alpha\left(\pi(\mathcal{D}^2 + \mathcal{U}^2) - 1/2\right)\right]\\ &= \exp\left[-i\pi(\csc\alpha - \cot\alpha)\mathcal{U}^2\exp[-\pi(\sin\alpha)]\right]\\ &\quad \times \mathcal{D}^2\exp\left[-i\pi(\csc\alpha - \cot\alpha)\mathcal{U}^3\exp[i\alpha/2]\right].\end{aligned} \tag{4.49}$$

therefore, the FRFT can be described as follows:

1. a chirp multiplication, followed by
2. a chirp convolution, followed by
3. another chirp multiplication
4. a product by a complex amplitude factor

Alternatively, as the FRFT is a rotation of the WVD (Lohmann, 1993), the FRFT can be described as follows:

1. a conversion of a, for example, 1D signal to a 2D WVD, followed by
2. a rotation of the WVD, followed by
3. a 1D FT

With this description, the rotation (2) of the WVD is composed of three shearing processes, such as left, down, and right.

The FRFT can be described in terms of FT eigenfunctions (Pei & Ding, 2002). Let ψ_n, $n = 0,1,2,3\ldots$ denote eigenfunctions of the ordinary FT operation, then depending on the order parameter, a, the FRFT will either provide a predominately chirp multiplication (averaging) ($a = 0, 2, 4, 6, 8$, and so on), or predominately chirp convolution (differentiation) ($a = 1, 3, 5, 7, 9$, and so on), and the relation of the averaging, differentiation, and Hamiltonian operators is as follows:

$$\pi(\mathcal{D}^2 + \mathcal{U}^2) = \psi_n(u) = (n + 1/2)\psi_n \tag{4.50}$$

$$\mathcal{H}\psi_n(u) = \lambda_n\psi_n(u) \tag{4.51}$$

Computing the FRFT of a signal corresponds to expressing it in terms of an orthonormal basis formed by chirps – complex exponentials with linearly varying instantaneous frequencies (Almeida, 1994). While there is an FRFT definition for all classes of signals (Cariolario et al., 1998), there are many definitions of the discrete FRFT (DFRFT), and none of these definitions presently satisfy all the properties of the continuous FRFT (Ozaktas et al., 1994; Pei & Ding, 2000). However, Pei & Ding (2000) proposed a closed form under appropriate sampling constraints that provide all properties of the continuous FRFT, including reversibility, except the additive property, but a conversion operation was proposed as a substitute. The sampling operations are as follows:

$$O_{DFRFT}^{-\alpha,\Delta u,\Delta t} = \left(O_{DFRFT}^{\alpha,\Delta t,\Delta u}\left(f(t)\right)\right) = f(t). \tag{4.52}$$

That is, the DFRFT of order $-\alpha$ with the sampling interval Δu in the input and Δt at the output is the inverse of the DFRFT of order α with the sampling interval Δt in the input and Δu at the output (Erseghe et al., 1999; Pei & Ding, 2000; Li et al., 2007).

The FRFT spectra are a order-parameter versus frequency spectra. As the order-parameter, a, changes, the orthogonal basis functions change. Therefore, the location of the maximum amplitude in the FRFT spectra will depend on *whether the signal analyzed has frequency-modulated components, that is, a changing instantaneous frequency*. Figure 4.21 demonstrates this with three test signals as follows: (i) a 500 MHz, CW; (ii) a 0–1 GHz LFM; and (iii) a square wave. In the case of the first signal, which is CW and has no frequency modulation, the FRFT shows a maximum amplitude at $a = 1.0$, that is, the conventional FT. In the case of the second signal, which is LFM, the maximum amplitude is at $a = 1.4386$, a value of a which does *not* provide the conventional FT. In the case of the square wave, which, again, has no frequency-modulated components, the maximum amplitude is at $a = 1.0$, that is, again the conventional FT.

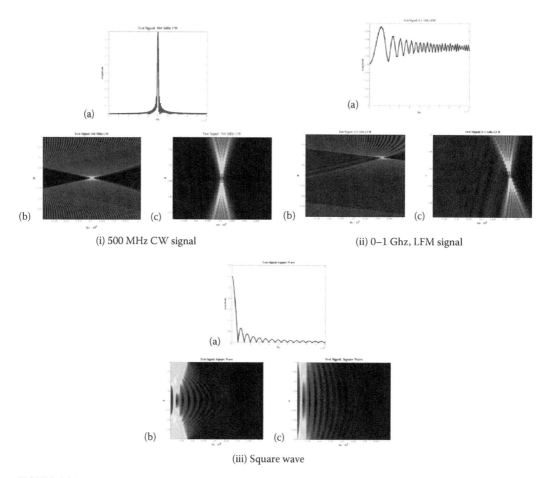

(i) 500 MHz CW signal (ii) 0–1 Ghz, LFM signal

(iii) Square wave

FIGURE 4.21
Three test signals are as follows: (i) 500 MHz, CW; (ii) 0–1 GHz, LFM; and (iii) Square Wave. (a) is the FT of test signal; (b) is the FRFT a order-parameter frequency spectrum of test signal; and (c) is the amplification of the (b) spectrum. The maximum amplitude in (i) is at $a = 1.0$, that is, the conventional FT; in (ii), the maximum amplitude is at $a = 1.4386$, which does *not* give the conventional FT; in (iii), the maximum amplitude is at $a = 1.0$, that is, again the conventional FT.

Figure 4.22 (i)–(iii) shows cuts through the FRFT a-frequency spectra of three test signals shown in Figure 4.21 at the position of the maximum amplitude. Here, in Figure 4.22 (i)–(iii), the (a) cut is a cut at $a = 1.0$ providing the conventional FT. The (b) cut is a cut at $a = 1.4386$ providing the maximum amplitude but *not* the conventional FT. The (c) cut is a cut at $a = 1.0$ again providing the conventional FT, indicating that the FRFT provides optimum characterization of FM signals.

The question of whether real targets act in a manner similar to frequency dispersive media is addressed in Figures 4.23 and 4.24. Figure 4.23 shows the FRFT order-parameter frequency spectra for the UWB return signals from a truck and Humvee target. It is noteworthy that there is asymmetry in these spectra for $0 < a < 0.995$ versus $1.005 < a < 2$. In the

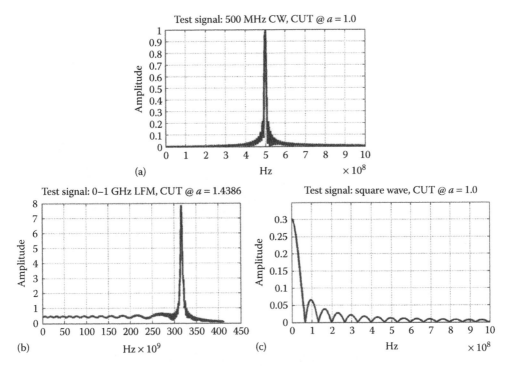

FIGURE 4.22
Examples of cuts through the fractional Fourier transform (FRFT) *a*-frequency spectra of test signals (i)–(iii) of Figure 4.21 (4.1) at the position of the maximum amplitude. (a) A cut at $a = 1.0$ providing the conventional FT, (b) a cut at $a = 1.4386$ providing the maximum amplitude but *not* at the conventional FT, and (c) a cut at $a = 1.0$, again providing the conventional FT.

FRFT for both targets, there is asymmetry around $a = 1$, – the conventional FT – indicating that the targets act as if differential dispersive media. In the marginal below, each figure is plotted as the conventional FT (red); and the differential marginal spectrum, which is the average of the spectra for $1.005 < a < 2$ minus the average of the spectra for $0 < a < 0.995$ (blue). The differential marginal spectrum provides a more complex and detailed signature of the target based on chirp basis functions rather than cisoidal basis functions as in the case of the conventional FT. Figure 4.24 shows that the differential marginal spectra of Figure 4.23 plotted in overlap distinguish the two targets.

FRFT kernels corresponding to different values of the order parameter, *a*, are related to wavelets (Onural, 1993; Ozaktas et al., 1994). Thus, filtering operations using the FRFT *a* domain can also be interpreted as filtering in the corresponding wavelet domain. The orthogonal bases of the FRFT are WHWFs[*] (Namias, 1980; Ozaktas et al., 1995; Candan et al., 2000; Cariolaro et al., 1998). As a class, however, FRFTs need not be based on WHWFs,

---

[*] As mentioned above, there are other appellations for the WHWFs, for example, Hermite–Gaussian functions. The present author prefers the designation WHWF because (a) the name "Hermite–Gaussian" implicates Gaussian in all polynomials, $n = 0, 1, 2, \ldots$ instead of just the first for $n = 0$; (b) Weber's equation is more general than Hermite's equation; (c) the name "Weber–Hermite" follows the *Mathematical Encyclopedia* usage (Hazewinkel, 2002); and (d) other texts, (e.g., Morse & Feshback, 1953, vol. 2, p. 1642; Jones, 1964, p. 86), have used the terminology "Weber–Hermite".

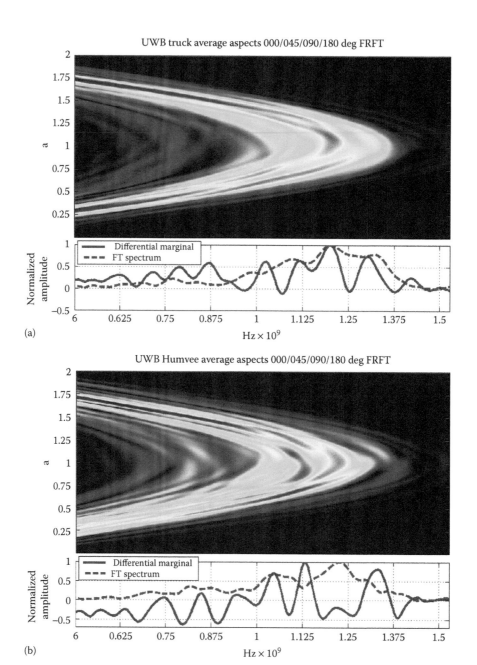

FIGURE 4.23
Fractional Fourier transforms of UWB signal returns averaged over target aspects 000, 045, 090, and 180° from (a) Truck target; and (b) Humvee target; over the order parameter $0 < a < 2$. In the FRFT for both targets, there is asymmetry around $a = 1$ – the conventional Fourier transform – indicating that the targets act as if differential dispersive media. In the marginal below, each figure is plotted as the conventional Fourier transform (red); and the differential marginal spectrum, which is the average of the spectra for $1.005 < \alpha < 2$ minus the average of the spectra for $0 < a < 0.995$ (blue). The differential marginal spectrum provides a more complex and detailed signature of the target based on chirp basis functions rather than cisoidal basis functions as in the case of the conventional Fourier transform.

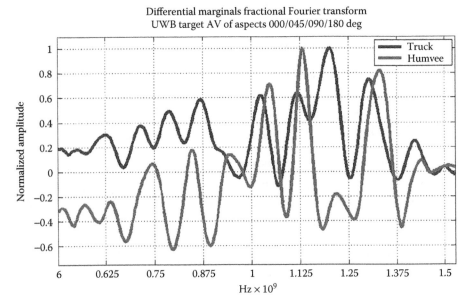

FIGURE 4.24
The differential marginal spectra of Figure 4.23 are plotted in overlap. The two targets are clearly distinguishable.

as Li (2008) has recently shown how different varieties of FRFTs can result from using orthogonal bases based on various different wavelets. The relationship of WHWFs and the FRFT is described in more detail in Appendix B.

4.6 Multiple Window Spectral Analysis

The averaging of LTV radar signals is not straightforward due the problem of constant change or nonstationarity. The multiple window spectral analysis method provides a way to address random, nonstationary signals and to provide a low bias and variance spectral estimator (Thomson, 1982; Martin & Flandrin, 1985; Frazer & Boashash, 1994; Xu et al., 1999). The following section shows how WHWFs were used to obtain the multiple-window spectra. The procedure has two steps as mentioned in the following:

1. Process the UWB return signals using, for example, the WHWF series, the series being pairs of scaling functions (averaging filters) and wavelets (differentiating filters). For example, all even numbered filters are scaling functions, or averagers, for example, WHWF0, WHWF2, WHWF4, WHWF6, and so on; and all odd numbered filters are wavelets or differentiators, for example, WHWF1, WHWF3, WHWF5, WHWF7, and so on. Arbitrarily, only the eight filters, WHWF0–7, were used here. Scaling these filters provided eight localized time–frequency spectra as shown in Figure 4.25.

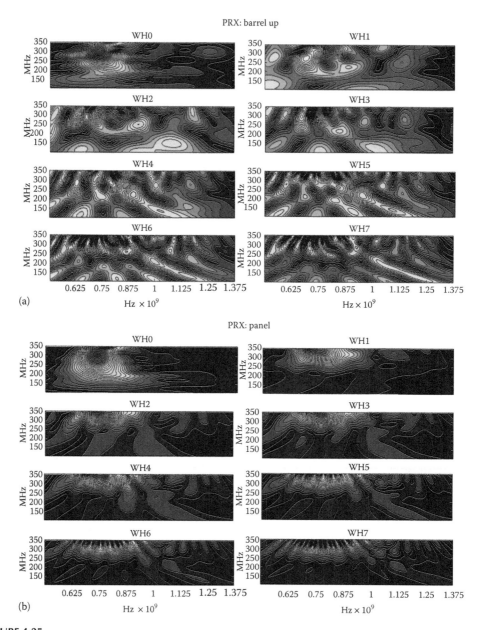

FIGURE 4.25
The transmitted signal was an amplitude-modulated transient signal of center frequency 472.5 MHz and a bandwidth of 515 MHz (215 – 730 MHz). On the receive side, the sampling rate was 10 GSPS. WH transforms of the UWB return signals calculated using the WHWF series which is composed of pairs of scaling functions (averaging filters) and wavelets (differentiating filters) are shown here. All even numbered filters are scaling functions, or averagers, for example, WHWF0, WHWF2, WHWF4, WHWF6, and so on, and all odd numbered filters are wavelets, or differentiators, for example, WHWF1, WHWF3, WHWF5, WHWF7, and so on. Arbitrarily, only the eight filters, WHWF0-7, were used here. Scaling these filters provided eight localized time–frequency spectra. (a) UWB signal return for a barrel (22.5 in (57.15 cm) diameter, and 33.5 in (85 cm) height) target showing eight spectra (WHWF0-7) and (b) UWB signal return for a roof panel (5 m × 0.91 m) target showing eight spectra (WHWF0-7).

2. Average these time–frequency spectra to provide the aggregate-localized MWTFAs such as those that are shown in Figure 4.25. Another example is shown in Figure 4.26 which shows the time-domain UWB return signal from a barrel target in Figure 4.26a. and its MTWFA spectrum in Figure 4.26b.

The MWTFA technique removes a deficiency in wavelet transforms. This deficiency is that while wavelet analysis provides excellent time location of high frequencies, there is poor time location at low frequencies. This is known as the bias/variance dilemma. To remove this dilemma, Thomson suggested using a different set of windows to compute several spectra of the entire signal and then averaging the resulting spectra to construct a spectral estimate (Thompson, 1982). To obtain an estimate at low bias and low variance,

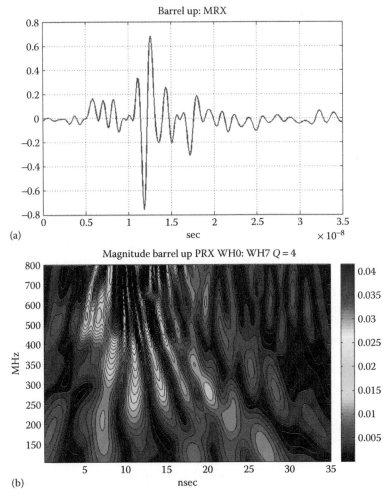

(a)

(b)

FIGURE 4.26
Example UWB signal returns from a barrel target. (a) A time domain averaged UWB return from a target barrel (22.5 in (57.15 cm) diameter, and 33.5 in (85 cm) height) in upright position and (b) the multiple window time-frequency analysis (MWTFA) spectrum of the signal return based on eight WHWFs is WHWF0:WHWF7, that is, eight time–frequency spectra were obtained from the UWB return and then averaged.

the windows must be orthogonal (to minimize variance) and optimally concentrated in frequency (to reduce bias).

The MWTFA shown in these examples is a WHWF expression of Thomson's multiple window method (Thomson, 1982), which is a time–frequency distribution estimator for a random process. It can substitute for a FT periodogram, which makes an inappropriate estimator for short, time-limited signals. Thomson's original approach to spectral estimation of short, time-limited signals computed several periodograms using a set of orthogonal windows that are locally concentrated in frequency and then averaged (Xu et al., 1999).

4.7 The Hilbert–Huang Transform

The Hilbert transform (HT) provides a signal's instantaneous spectral profile and a time versus instantaneous frequency representation of a signal as shown in Figure 4.27. The Hilbert–Huang transform (HHT) gives an *energy-time–frequency* method of signal analysis for nonstationary and nonlinear signals (Huang et al., 1998, 1999, 2003; Huang & Shen, 2005; Huang & Attoh-Okine, 2005). The HHT uses *empirical mode decomposition* (EMD) together with HTs of those EMDs, and is designed for the analysis of *signals that are neither linear, nor stationary*. The motivation for the method is that analyses in terms of any *a priori* chosen basis cannot fit the signals arising from all systems (Huang & Shen, 2005). In other words, one basis set does not fit all (mathematically or physically). Therefore, the HHT takes an adaptive approach to the empirically derived breakdown of a nonlinear and nonstationary signal into its multiple components. This adaptive approach is called a *sifting process* and yields an empirical basis set of intrinsic mode function (IMF) components. The basis set is derived from the data itself. In some instances, the HHT can provide more accurate identifications of nonstationary systems than wavelet methods (Shan & Li, 2010).

As the HHT is relatively new, there are still problems with the transform that continue to be addressed, such as the choice of signal envelope generation method (Chen et al., 2006), the stopping criteria for the sifting process, and the production of mode mixing for some signals (Datig & Schlurmann, 2004; Huang & Wu, 2008). The HHT method offers a meaningful insight into physically defined (as opposed to mathematically defined) instantaneous frequencies.

The HT, which accesses a signal's instantaneous frequencies is as follows:

$$y(t) = \frac{1}{\pi} P \int\limits_{-\infty}^{+\infty} \frac{x(\tau)}{t-\tau} d\tau, \qquad (4.53)$$

where P is the Cauchy principle value of the singular integral. The analytic signal is defined as follows:

$$z(t) = x(t) + iy(t) = a(t)\exp[i\varphi t], \qquad (4.54)$$

where:
$a(t) = \sqrt{x^2 + y^2}$ is the instantaneous amplitude
$\varphi(t) = \arctan(y/x)$ is the instantaneous phase function

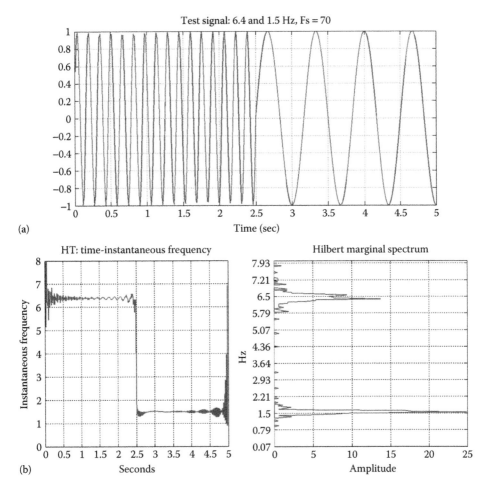

FIGURE 4.27
Hilbert transform of a test signal with two frequency components, 6.4 and 1.5 Hz, (Fs = 70 SPS). (a) A time domain representation of test signal. (b) Left – Hilbert transform of test signal. Right – Hilbert marginal spectrum.

The instantaneous frequency is then defined as follows:

$$\omega = \frac{d\varphi}{dt},\tag{4.55}$$

which provides only physically meaningful results for mono-component signals (Huang et al., 1998), Figure 4.27.

Alternatively, the HHT method involves Hilbert transforming a signal's IMF components, which are extracted by a process called EMD, and seeks the results for multicomponent signals. The decomposition method is as follows:

The necessary condition for IMF extraction are as follows: (i) the signal must be symmetric with respect to the local zero mean and (ii) have the same number of zero crossings or extrema. The decomposition used the following steps:

1. Identify the local maxima of $x(t)$ and connect them with a (for example, cubic) spline, to provide the upper envelope, $e_1(t)$.
2. Identify the local minima of $x(t)$ and connect them with a (for example, cubic) spline, to provide the lower envelope, $e_1(t)$.
3. Calculate the local mean, $m_1(t) = (e_1(t) - e_1(t))/2$.
4. Calculate the difference, $c_1(t) = x(t) - m_1(t)$.
5. Repeat steps (1)–(4) with $c_1(t)$ replacing $x(t)$ to obtain $c_{11}(t) = x(t) - m_{11}(t)$.
6. Repeat this *sifting process* k times:

$$c_{1k}(t) = c_{1(k-1)}(t) - m_{1(k-1)}(t),$$
(4.56)

resulting in the first IMF:

$$C1(t) = c_{1k}(t).$$
(4.57)

An IMF is subject to the following constraints:

a. The number of extrema and the number of zero-crossings must either be equal or differ by utmost one.
b. At any point, the mean value of the envelope defined by the local maxima and the envelope defined by the local minima must be zero.

7. k is defined by a stopping criterion. One criterion for stopping the sifting process is when $SD = 0.2$–0.3, where SD is defined:

$$SD = \sum_{t=0}^{N} \left[\frac{\left| c_{1(k-1)}(t) - c_{1k}(t) \right|^2}{c_{1(k-1)}^2(t)} \right].$$
(4.58)

Another criterion used is that the number of zero-crossings and the number of extrema must be equal.

8. Subtract $C1(t)$ from the signal according to the following:

$$r_1(t) = x(t) - C1(t),$$
(4.59)

replace $x(t)$ with $r_1(t)$ and repeat steps (1)–(6).

9. The result is a series of IMFs, $Ci(t)$, $i = 1, 2, \ldots, n$ with a final residue $r_n(t)$ becoming a monotonic function such that (Huang et al., 1998):

$$x(t) = \sum_{i=1}^{n} Ci(t) + r_n(t)$$
(4.60)

10. As the $r_n(t)$ are monotonic and can be neglected, the HT of $Ci(t)$ results in the *Hilbert spectrum*:

$$X(t) = Re\left(\sum_{i=1}^{n} a_i(t)\exp\left(i\int \omega_i(t)dt\right)\right). \tag{4.61}$$

In comparison, the FT of the original data is as follows:

$$F(\omega, t) = Re\left(\sum_{i=1}^{\infty} a_i \exp(-i\omega_i t)\right), \tag{4.62}$$

with a_i and ω_i as constants. Therefore, the IMF is a generalized Fourier expansion.

We first demonstrate the method using a test signal as an example (Flandrin & Gonçalvès, 2004). This test signal is a more difficult signal to deconstruct than will be commonly encountered empirically. We construct the test signal as follows (Figure 4.28). The first two component signals are oscillatory, frequency-modulated signals. It is these types of signals that one might want to process separately further using time–frequency methods.

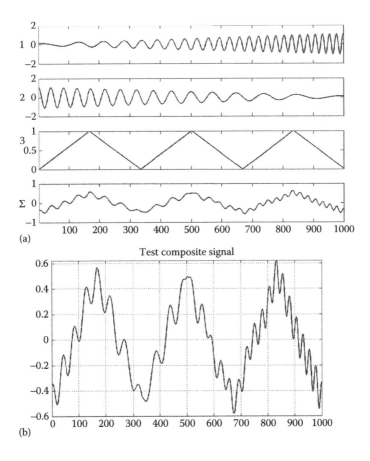

FIGURE 4.28
Test signal components are described as follows: (a) The first two signals are oscillatory, frequency-modulated signals. The third signal is a slow DC component, and the fourth signal is the combination or composite of the three preceding and represents an RX signal, before processing and (b) the fourth composite signal is amplified. Model signal due to Flandrin & Gonçalvès (2004).

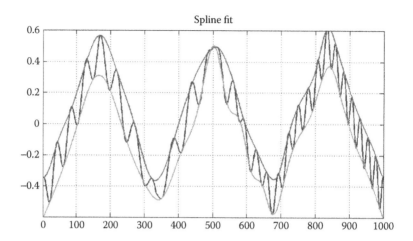

FIGURE 4.29
The composite signal together with the envelopes calculated by a spline fit. The instantaneous mean is calculated from the difference of the envelopes and subtracted from the original composite signal, which is designated as C0. The resultant from the subtraction is the first intrinsic mode function and is designated C1 (or the high-frequency component of C0), and the remainder is designated as R1 (or the low-frequency component of C0).

The third component signal is a slow DC component, and the fourth signal is the combination of the three preceding components and is considered to represent an example received signal (RX).

Calling this test RX signal, C0, the signal envelopes (Figure 4.29) are calculated and the sifting process begun. After 50 sifting iterations – this choice of 50 is arbitrary and signal-dependent – the first IMF, C1 is obtained, and from the relation, C0 = C1 + R1, it should be noticed that R1 captures almost the entire slow DC component of the composite signal (Figure 4.30). C1 contains almost all of the high-frequency components of the original signal (Figures 4.31 and 4.32), and both C1 and R1 can be separately processed using, for example, time–frequency methods. This example recovered the signal components after one stage of sifting. In most cases, more stages are required.

This example indicates that the HHT, used as a pre-processing method, decomposes the RX signal for separate spectral analysis in a manner similar to a filter bank (Flandrin et al., 2003, 2004).

The same procedure can be applied to empirical (MAP) RX data. Figure 4.33 shows the calculated envelopes (upper and lower) for the PRX signal from a barrel target positioned up. The PRX is designated as C0. The instantaneous average is calculated from these envelopes, which is then subtracted from the PRX, or C0, giving C1 (the high-frequency IMF). The remainder is R1 (the lower frequency remainder IMF). The process is repeated using C1 to obtain C2 and R2, and so on.

Figure 4.34 shows the time domain C0 – C8 IMFs for the target barrel up and barrel side PRXs. The spectra for the C0 – C4 and the R1 – R4 IMFs are shown in Figure 4.35. Separation of the high-frequency C IMFs and the lower frequency R IMFs progressively shifts downwards – by subtraction of the $C_{(i+1)}$ from the R_i to obtain the $R_{(i+1)}$ IMF – the remaining signal $R_{(i+1)}$ toward the low frequencies. In fact, Figure 4.35 shows that the R's

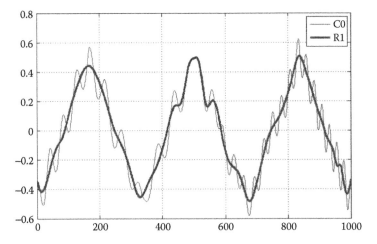

FIGURE 4.30
The original composite signal, C0, and the first low-frequency intrinsic mode functions (IMFs), R1. C1 is obtained by the recursive subtraction of the instantaneous mean, permitting the extraction of the high-frequency components in the signal. This process is known as "sifting". The remainder is R1 = R0 – C1. Here, R1 captures almost the entire DC component in the C0 signal.

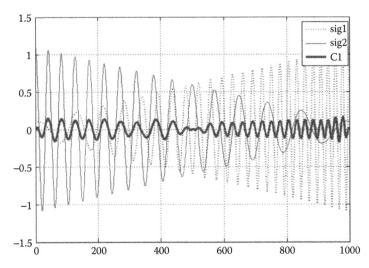

FIGURE 4.31
The two separate oscillatory components of the original composite signal, signals 1 and 2 in Figure 4.28 and the high-frequency components, C1, of the original signal, C0, obtained by the sifting process. C0 = C1 + R1.

and the C's at each level, *i*, form the output of low–high pass frequency filter pairs. This is reminiscent of wavelet multiresolution analysis (Flandrin et al., 2003).

The correlations of the high-pass C1–4 and the low-pass R1–4 IMFs of Figure 4.35 from two targets, Barrel positioned side up and Barrel position sideways, are shown in Table 4.1. The original PRXs are from the same barrel target, but in two different aspects, such as up and side.

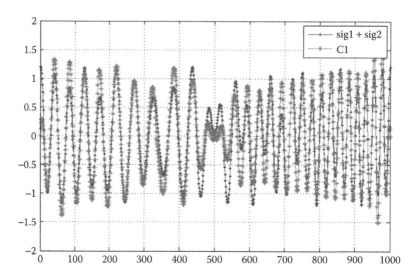

FIGURE 4.32
The two added oscillatory components of the composite C0 signal: signal 1 + signal 2, and the high-frequency IMF signal, C1, obtained by the sifting process. C0 = C1 + R1. It is clear that the sifting process has extracted almost all of the oscillatory, high-frequency components, C1, from the composite signal, C0.

4.8 Carrier Frequency-Envelope Frequency Spectra

After reception and processing into the frequency domain, if the UWB return signal information in the time domain is of no importance, that is, information about the orientation of the target is not required, then the following frequency–frequency or CFEF spectrum provides significant information (Figure 4.36). The CFEF of Figure 4.36 is obtained by the Fourier transformation of each line in a multiple window time–frequency analysis (MWTFA) (Figure 4.36a). The CFEF y-axis is identical to the multiple window time–frequency spectrum analysis (MWFTA) y-axis, but the CFEF x-axis is also a frequency axis and the influence of temporal jitter has been removed. To obtain an averaged signal, the multiple CFEF is averaged and the jitter is not an influence on the average. The average is obtained for a penalty of loss time of RX arrival information. The CFEF method provides detailed spectral information concerning a target return. As shown in Figure 4.36b, information concerning a target's return signal's "envelope" frequencies as well as "carrier" frequencies is revealed.

Pre-processing of real signals into their WH transforms provides a signal decomposition in complex number form. The return signal is decomposed into a hierarchical TBP basis more suited to LTV signal analysis (as opposed to LTI), than is decomposition into constant frequency sinusoidal basis functions as in the case of conventional Fourier decomposition.

As described above, the WHWFs series are a hierarchy of pairs of scaling functions and wavelets. All even numbered filters are scaling functions, or averagers, for example, WHWF0, WHWF2, WHWF4, WHWF6, and so on, and all numbered filters are wavelets, or differentiators, for example, WHWF1, WHWF3, WHWF5, WHWF7, and so on. Arbitrarily, only the eight filters, WH0–7, were used in this analysis. Scaling these filters provided eight localized time–frequency spectra. Summing these spectra

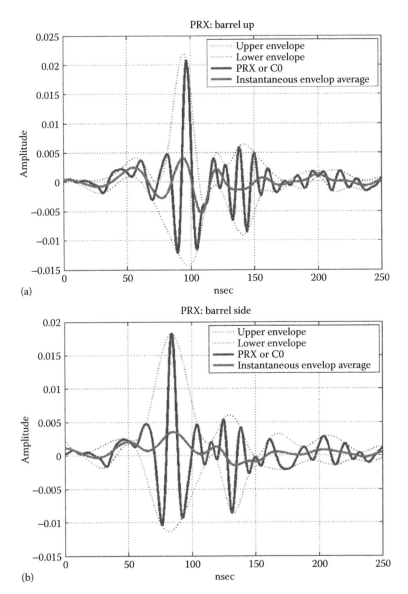

FIGURE 4.33
UWB signal returns from a barrel (22.5 in (57.15 cm) diameter, and 33.5 in (85 cm) height). (a) C0 = UWB return signal barrel target positioned up and (b) C0 = UWB return signal barrel target positioned on its side. A snapshot of the first instantaneous average which is calculated from the difference of the envelopes and updated recursively in the sifting process. The final instantaneous average, not shown here, is then subtracted from the UWB return signal (PRX), or C0, giving C1. The remainder is R1. The sifting process is repeated using C1, to obtain C2 and R2, and so on.

provided the multi-window aggregate-localized magnitude time–frequency spectra as shown in Figure 4.36a. The real part of these spectra was used in the following additional analysis.

The FT was taken of each frequency line of a multi-window spectrum, resulting in a frequency × frequency spectrum, or a CFEF spectrum (e.g., Figure 4.36b) identifying both the frequencies of the modulating envelopes of the carrier signal frequencies. There are

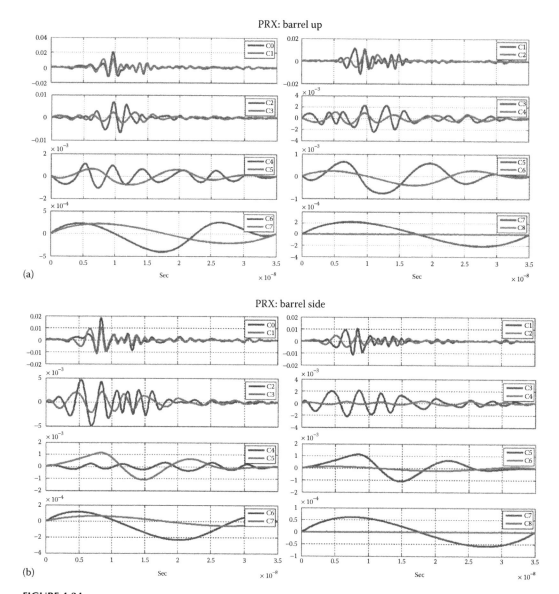

FIGURE 4.34
Time-domain signal returns from the same barrel. The calculated (sifted) UWB C1, C2, C3, C4, C5, C6, C7, and C8 IMFs for targets: (a) Barrel positioned up and (b) barrel positioned sideways.

peaks in Figure 4.36b at the "carrier frequencies" indicating amplitude envelope modulations of the carrier frequency components of the return signal (RX) at the same frequency as the RX carrier frequency components – *a double modulation of both carrier and envelope at the same frequency.* There are also lower envelope frequency peaks indicating steady bursts cross all carrier frequencies, that is, a spike, that are indicated on the left as vertical bands.

The rationale for these plots is to remedy the fact that a time comparison (or a time average) of time–frequency plots requires that the compared plots represent signal arrival at the receiver at precisely the same time, that is, the same position in all time–frequency spectra. By Fourier transforming at each frequency line of a MWFTA spectrum, the impediment of

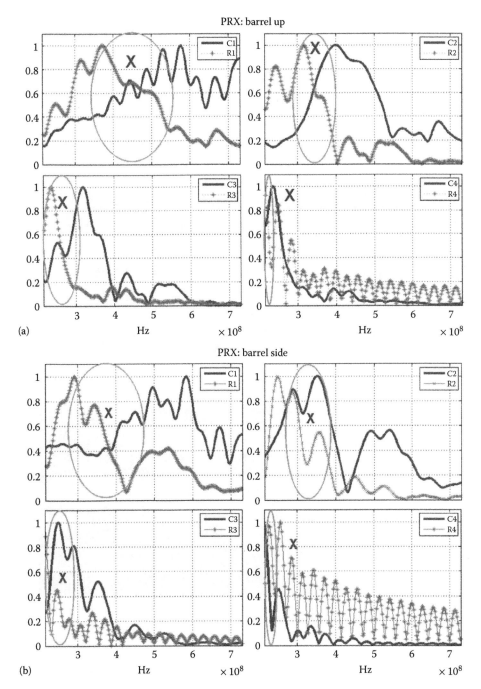

FIGURE 4.35
Barrel reflected signals in the frequency domain. The calculated (sifted) UWB C1, R1, C2, R2, C3, R3, and C4, R4 IMFs for (a) Barrel positioned up and (b) barrel (22.5 in (57.15 cm) diameter, and 33.5 in (85 cm) height) positioned sideways. Notice that the C(i) and R(i) form high-pass, low-pass pairs, analyzing the previous R(i-1).

TABLE 4.1

Barrel Up & Barrel Side Cross-Correlations
High-pass & Low-pass Intrinsic Modes

H-P	Correlation	L-P	Correlation
C1s	0.8627	D1s	0.9683
C2s	0.9122	D2s	0.7735
C3s	0.7894	D3s	0.8443
C4s	0.9030	D4s	0.8290

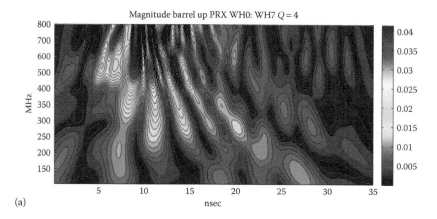

(a)

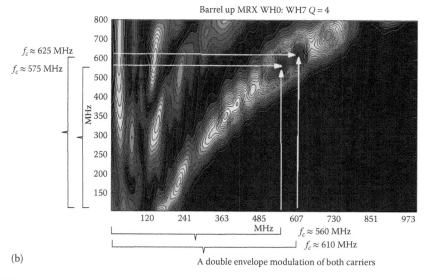

(b)

FIGURE 4.36

Examples of the return signals processed with multiple window time–frequency spectrum analysis (MWFTA). (a) A multiple window time–frequency spectrum for a return signal, target Barrel (22.5 in (57.15 cm) diameter, and 33.5 in (85 cm) height) positioned up, calculated by the average of eight time–frequency spectra using filters – WHWF0-7, $Q = 4$ and (b) a *Carrier* Frequency-*Envelope* Frequency (CFEF) spectrum. These plots are the Fourier transforms of each frequency line of the multiple window time–frequency spectrum, providing the spectra, 100 MHz – 1 GHz, of the modulating envelopes of the carrier signal frequencies of the multiple-window spectrum. There are peaks in these spectra at "carrier frequencies" and "envelope" modulations at specific frequencies. There are also lower envelope frequency peaks indicating steady bursts cross all carrier frequencies, that is, a spike indicated on the left as a vertical band.

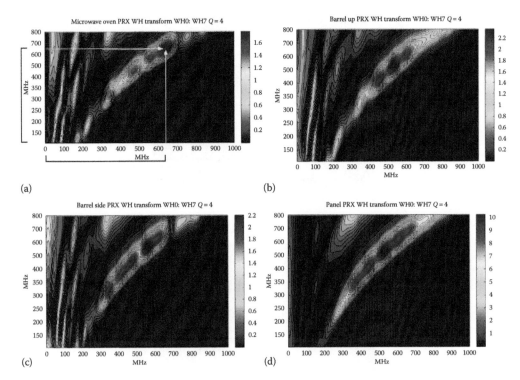

FIGURE 4.37
Carrier frequency-envelope frequency (CFEF) received UWB signal spectra for targets: (a) Microwave oven (top 0.54 m², side 0.047 m²), (b) upright barrel (22.5 in (57.15 cm) diameter, and 33.5 in (85 cm) height), (c) barrel positioned sideways, and (d) roof panel (5 m × 0.91 m).

different signal arrival times is remedied, and provides additional information concerning the envelope modulation. There is, of course, the cost that all time-of-RX-signal arrival information is lost.

As illustration, Figure 4.37 shows UWB CFEF spectra for the targets, such as microwave oven, barrel up, barrel sideways, and a roof panel; and Figure 4.38 shows the three truck UWB signal returns.

The magnitudes of the FTs of the cuts at each frequency through the time–frequency MWTFA spectrum as shown in Figure 4.36a. The spectra shown indicate both the spectra of the return signal carrier frequencies as well as of the modulating envelopes of those carriers. It can be seen that, although there are peaks in these spectra at approximately the carrier frequencies – by reading across the y-axis – that indicate short, almost monocycle, bursts at the appropriate frequency; by reading down to the x-axis, it can be seen that the frequencies of the envelopes modulating the carriers are also indicated. On the far left, a broad band carrier signal is indicated that is modulated by a low-frequency envelope. Any long vertical band indicates a broad band spike burst. In the middle, there are "islands" indicating short-duration envelopes modulating "carrier frequencies", that is, wave packets. There are also wave packets located at approximately the same frequency on the y- and x-axes indicating wave packet envelope modulations of the "carrier frequency" components of the return signal at the same frequency as those "carrier frequency" components – *a double modulation involving both a "carrier" and envelope at the same frequency.*

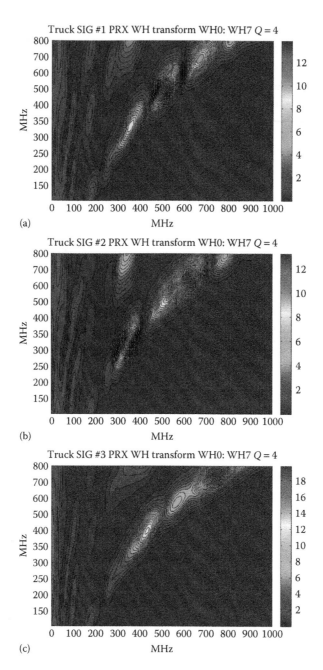

FIGURE 4.38
Carrier frequency-envelope frequency (CFEF) spectra for a truck target at three aspect angles: (a) 0°, (b) +6°, and (c) –6°.

The motivation for such CFEF spectra is to circumvent the difficulty that a comparison of different time–frequency spectra requires that the compared spectra represent signals with receiver arrivals at precisely the same time, that is, time frequency spectra are only comparable if the time axis is aligned, which requires that the different targets must be at precisely the same distance from the transmitter. By Fourier transforming at each frequency line of the time–frequency spectrum shown in A, the link to a specific RX receiver time of signal arrival is broken. There is, of course, the cost that all time-of-RX-signal arrival information is lost.

4.9 The Radon Transform

Radon (1917) showed how a function can be defined in terms of its integral projections. The mapping to the projections is known as the Radon transform. Another transform, the Hough transform was designed to detect straight lines in images (Hough, 1962) employing a template. Both are mappings from image space (or source space) to parameter space (or destination space). The two approaches to the mappings can be distinguished by noting that, whereas the Radon transform considers how a data point in destination space is obtained from the data in source space, the Hough space considers how a data point in source space is obtained from the data in destination space (Ginkel et al., 2004). The original Hough transform can also be considered as a discrete version of the Radon transform. However, the mathematical formalism for both transforms is identical (Deans, 1981; Illingworth & Kittler, 1988); and Gel'fand et al., (1966) formulated the Radon transform in terms of the Dirac delta function, permitting the Radon transform to be treated as a form of template matching. Therefore, the Radon transform and the Hough transform are equivalent forms of template matching, and henceforth we shall confine ourselves to addressing the Radon transform.

Following Ginkel et al. (2004), the Radon transform can be described in terms of generalized functions as follows:

$$(\mathcal{L}_C \mathcal{J})(p) = \int_R C(p, x) \mathcal{J}(x) dx, \tag{4.63}$$

where:
the integration is over the extent of the (real numbers) of the image
$\mathcal{L}_C$ a linear integral operator with kernel
C is a generalized function
$\mathcal{J}$ is a 2D image
p is a vector containing parameters
x are the spatial coordinates
the integral is a volume integral

The Radon transform of the cross WVD (or RCWVD) is related to the FRFT (Lohmann & Soffer, 1993). The RCWVD of an angle is equal to the pth-order FRFT of the first function times the pth-order conjugate FRFT of the second function (Raveh & Mendlovic, 1999).

The Radon transform and its inverse are well-known and well used in applications such as computed axial tomography (CAT scan), barcode scanners, electron microscopy of macromolecular assemblies like viruses and protein complexes, reflection seismology, and in the solution of hyperbolic partial differential equations, tomographic or image reconstruction (Deans, 1983; Barrett, 1984; Helgason, 1999; Ramm & Katsevich, 1996; Kak & Slaney, 2001). A general description is as follows: the omnidirectional (0–180 degrees) Radon

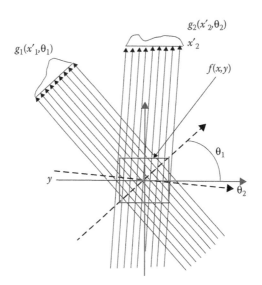

FIGURE 4.39
Radon transform: parallel projections for two angles θ_1 and θ_2. $f(x,y)$ represents an image.

transform of an (x,y) image is the collection of line integrals, $g(s,\theta)$ along lines inclined at angles θ fom the y axis and at a distance s from the origin. The line integrals are space limited in s and are periodic in θ with period 2π (Figure 4.39).

Referring to Figure 4.39, if the Radon transform at angle θ, or projection g_θ the image $f(x,y)$ is

$$g_\theta(\theta,x') = \int f\left(x'\cos\theta - y'\sin\theta, x'\sin\theta - y'\cos\theta\right)dy' \tag{4.64}$$

and if $f(x,y)$ a WVD transform of a 1D signal, then,

$$g_\theta(\theta,x')[WVD] = \left|FRFT_a\right|^2 \tag{4.65}$$

or, in other words, the 1D Radon transform or projection of the 2D WVD of a 1D signal onto an axis x_i' making an angle $\theta = a\pi/2$ (radians) with the x-axis is equal to the squared modulus of the ath order FRFT of the signal. More simply, the Radon–Wigner transform is the squared magnitude of the FRFT (Wood & Barry, 1994a,b,c).

One motivation for using Radon transforms in signal processing is to achieve omnidirectional filtering (0–180°) of 2D images. 1D signals are given a 2D image representation by the WVD and AF transforms, and the enhancement of features requires the application of filters omnidirectionally. However, conventional signal/image representation methods do not provide algorithms that filter omnidirectionally, that is, unbiased in any direction. For example, the 2D generalization of quadrature mirror (wavelet) filters, addressing the analysis of trends (averages) and fluctuations (differences), decomposes an image by computing: (1) trends along rows followed by trends along columns; (2) trends along rows followed by fluctuations along columns; (3) fluctuations along rows followed by trends along columns; and (4) fluctuations along both rows and columns. In a pyramidal filtering scheme, four arrays of coefficients (of decreasing size) are produced at each level and the filtering is performed only in the vertical and horizontal directions. This choice of procedure is biased to the right angles of rows and columns; and conventional generalization of a 1D filter to 2D form provides no detection capability for image features that lie at specific angles. However, it is possible to analyze trends and fluctuations omnidirectionally (0–180 degrees) *by applying 1D filters to the 2D projection space representation of the image*, that is, the image's Radon transform (Barrett, 2008). Thus in projection space, the image is a 2D omnidirectional representation with the image distributed along the x-axis as a function of the Cartesian angle of the image; and can be filtered through all angles (0–180 degrees), yet with 1D filters.

The same observation applies to image compression for transmission. In the case of 2D wavelet compression of 2D images by conventional methods, significant coefficients resulting from a pyramidal analysis are transmitted, together with a significance map. If an omnidirectional representation is required and using these conventional methods, the same procedure must be applied n times to compress the same 2D images at all n angles. However, using Radon transform methods, it is possible to provide methods for compressing 2D images oriented omnidirectionally (0–180 degrees) with 1D filters, for example, wavelets, but using only one procedure. The resulting coefficients are then transmitted with the significance map.

This observation also applies to image enhancement methods. In the case of 2D wavelet image enhancement of 2D images by conventional methods, a 2D, but still specifically oriented, for example, wavelet, is used to process an image. To provide omnidirectional processing of an image, the orientation of the wavelet must be progressively changed with the image being processed by the same methods at all the orientations adopted by the wavelet. Thus, processing an image at n angles requires n processing sequences. However, applying the detecting filter in projection space, it is possible to enhance 2D images in any orientation by applying 1D filtering of the rows of the Radon-transformed image and then by inverse Radon transforming the resultant of that filtering, all in one processing sequence.

Thus transforming to projection space permits the analysis of trends and fluctuations omnidirectionally by applying 1D filters to the 2D projection space representation of the image, that is, the image's Radon transform. In projection space, the image is in a linearized omnidirectional representational form and can be filtered through 0–180° in a series of but one 1D procedure. Briefly, these results are achieved by (i) a Radon transform of the image(s) or array(s); (ii) a convolution of the chosen 1D filter with, for example, a 1D Ram–Lak, or other band-limited filter; (iii) a convolution of the resultant 1D filters with each of the 1D columns of the 2D Radon transform or projection space version of the image; and (iv) an inverse Radon transform of the linearly filtered projection space image back to a Cartesian space omnidirectionally filtered form.

The inverse Radon transform is obtained by means of a back-projection operator. The back-projection operator at (r,ϕ) is the integration of the line integrals $g(x',\theta)$ along the sinusoid $x' = r\cos(\theta - \phi)$ in the (x',θ) plane. Thus the back-projection operator maps a function of (x',θ) coordinates into a function of spatial coordinates, (x,y) or (r,ϕ), and integrates into every image pixel over all angles θ. A drawback is that the resulting image is blurred by the point spread function, but the remedy lies in the fact that the back-projection is the adjoint of the inverse Radon transform, which can be (x',θ) obtained by a filtering operation that has a variety of approximations, for example, Ram–Lak, Shepp–Logan, and so on Ram–Lak being the most common (Ramm & Katsevich, 1996; Kak & Slaney, 2001).

Figure 4.40 illustrates the relation of objects in Cartesian space and Radon transform projection space using two test images in Figure 4.40a and b. Peaks in Cartesian space become lines in projection space; and lines in projection space become peaks. In Figure 4.40c, an image corrupted with Gaussian noise is restored.

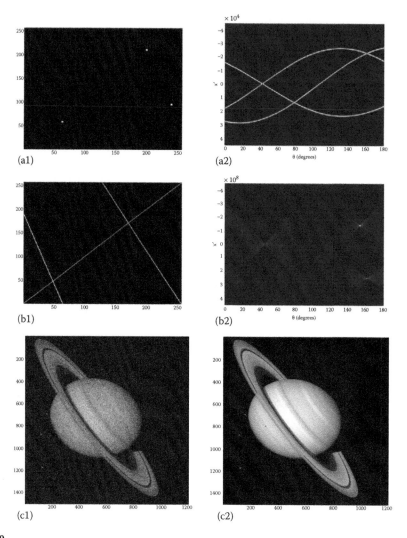

FIGURE 4.40

Examples of the relation of objects in Cartesian space and Radon transform projection space using two test images.

- (a1) Image with three points or peaks (square).
- (a2) Radon transform of image with three points or peaks. The three points (square) appear as three curved lines.
- (b1) Images with three lines (straight).
- (b2) Radon transform of image with three lines (straight). The three lines appear as three peaks in projection space. The peaks in projection space are related to the lines in Cartesian space as follows:
    - (i) A peak at angle θ on the x-axis in projection space represents a line at an inclination θ to the vertical in Cartesian space
    - (ii) A peak at x' on the y-axis in projection space represents a line located at a distance x' from the center of the Cartesian space image.
- (c1) Image of Saturn with Gaussian noise.
- (c2) Image of Saturn with noise removed by the following:
    - (i) Radon transform to projection space or domain.
    - (ii) Application of one-dimensional low-pass filter at all angles in projection space, that is, equivalent to omnidirectional filtering in the Cartesian space or domain.
    - (iii) Inverse Radon transform to Cartesian space.

(After Barrett, T.W., Method and application of applying filters to n-dimensional signals and images in signal projection space, United States Patent No 7,426,310, dated September 16, 2008.)

4.10 Target Linear Frequency Response Functions

There are three major approaches to calculating the *linear* frequency response function (FRF). The three methods are described by the following equations:

$$H(f)_1 = \frac{S(f)_{xy}}{S(f)_{xx}} \tag{4.66}$$

$$H(f)_2 = \frac{S(f)_{yy}}{S(f)_{yx}} \tag{4.67}$$

$$H(f)_3 = \frac{S(f)_{yy} - S(f)_{xx} + \sqrt{\left(S(f)_{xx} - S(f)_{yy}\right)^2 + 4\left|S(f)_{xy}\right|^2}}{2S(f)_{yx}} \tag{4.68}$$

where:
$S(f)_{xx}$ is the autospectrum or spectral density function of the transmitted signal TX
$S(f)_{yy}$ is the autospectrum or spectral density function of received signal, RX
$S(f)_{xy}$ and $S(f)_{yx}$ are the cross-spectra of TX and RX

and these FRFs, of $H(f)_1$, $H(f)_2$ and $H(f)_3$ are unbiased with respect to the output noise, to the input noise, and to both the input and output noises, respectively.

Calculation of the FRF is dependent on calculating the input to the system (i.e., the transmitted signal directly before, and exciting, the target) and the output from the system (i.e., the return signal directly after, and radiating from, the target). Therefore, the following compensations were applied to the transmitted signals at the transmitter and the received signals at the receiver as follows:

1. switching parasitics (on the TX path)
2. transmit magnitude compensation function (on the TX path)
3. inverse of receive magnitude compensation function (on the RX path)
4. inverse of cable loss (on the TX path)

The transmitted (TX) and received (RX) signals compensated in this way provide $S(f)_{xx}$, $S(f)_{xy}$, and $S(f)_{yx}$. These permitted the calculation of $H(f)_1$, $H(f)_2$, and $H(f)_3$ are shown in Figures 4.41 and 4.42.

4.11 Singular Value Decomposition and Independent Component Analysis

If can we can build an archive of target signal returns, that is, *a priori* information, then identification of a target on the basis of a specific target return can proceed in a number of ways with reference to that archive.

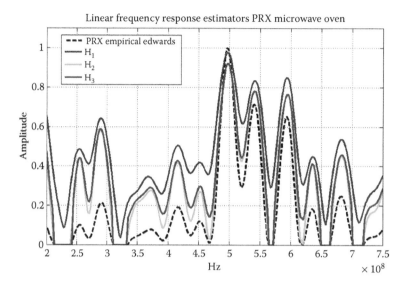

FIGURE 4.41

The frequency response functions for a microwave oven target that is estimated and empirically obtained from a returned signal. They are shown as follows:

The empirically obtained UWB return signal (PRX).
$H(f)_1$ unbiased with respect to the output noise.
$H(f)_2$ unbiased with respect to the input noise.
$H(f)_3$ unbiased with respect to both the input and output noises.

This section describes one possible method based on a model that exploits a set or matrix of unknown transmitted signals (UTX). The UTX enter a channel, and a known set of signals, (KRX) is received (Figure 4.43). The role of KRX can be played by a set of signal returns obtained from a target at a number of aspect angles. By assuming some plausible statistical properties of the channel and the UTX, the KRX can be decomposed to obtain an estimate of the common aspects of the return signals under the explicit assumptions adopted for the UTX, that is, a set of signals designated: ERX. Therefore the ERX are an estimate of the UTX, or [ETX]. Any further signal returns can then be correlated with any of the set of ERX/[UTX] signals for target identification purposes, and the correlations obtained provide a measure of identification confidence.

The role of the target in this model is as a mixing matrix. That is the basic model, but there are, however, several strategies for extracting the source signals ERX/[UTX], for example, principal component analysis (PCA), singular value decomposition (SVD), independent component analysis (ICA), Gram-Schmidt orthogonalization (GSO). But any method adopted is only as valid as the assumptions implicit in that method. We turn now to introduce some of these methods and their assumptions.

Principle component analysis (PCA) eigen-decomposition, which is related to Factor analysis[*] (FA), is a multivariate analysis or analysis of multiple variables treated as a single entity which makes manifest latent information in the original data and both PCA and FA are based on correlation techniques for source signal separation. This process can result in data reduction. Information in the data is divided into two subspaces – a signal subspace

---

[*] FA is a form of PCA with the addition of extra terms for modeling the sensor noise associated with each signal mixture.

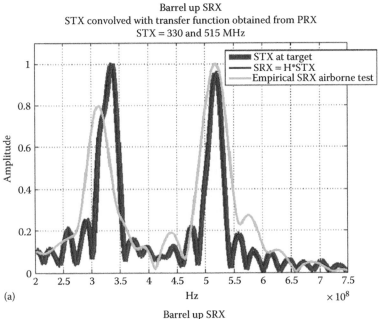

(a)

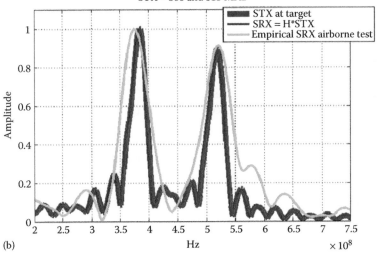

(b)

FIGURE 4.42

Transfer function calculations. Figure (a) and (b) show: (1) the calculated and compensated transmitted UWB signal (STX) *at the target*, commencing with the empirical STX at the transmitter; (2) the return signal (SRX) calculated from the target transfer function, H, that was itself calculated using the transmitted UWB signal (STX) and the signal return (SRX) *at the target*; and (3) the empirically obtained UWB SRX spectra.

and a noise subspace. The signal subspace should be twice the number of sinusoids in the data if known. PCA operates by transforming to a new set of uncorrelated variables – the principle components (PCs). The PCs are also orthogonal and ordered in terms of variability. However, only if the original variables are Gaussian are the uncorrelated PCs also independent, because PCA uses only 2nd order statistics, for example, variances.

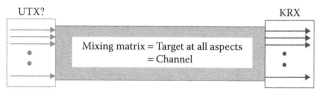

UTX = Unknown set of target aspect-independent signals.
KRX = Known set of received signals
     = Set of MRXs or PRXs at various target aspect angles.

FIGURE 4.43
In this target identification model, a set or matrix of unknown aspect independent signals, UTX, enters a channel, and a known set of signals, KRX, is received (i.e., return signals (RX) at set target aspect angles). By assuming some statistical properties of the channel, the KRX can be decomposed to obtain an estimate of the common aspects of the RX signals, that is, a set of signals designated: ERX. Therefore, the ERX are an estimate of the UTX, or [UTX]. Any later received RX signals can then be correlated with any of the ERX/[UTX] set for target identification purposes. The role of the target in this model is as a mixing matrix. There are several strategies for extracting the source signals, ERX/[UTX], for example, PCA, SVD, ICA, GSO, and so on, but any method adopted is only as valid as the assumptions implicit in that method.

(Variables with Gaussian distributions have zero statistics above 2nd order.) A generalized form of PCA is SVD.

Another description of PCA is that it is a general multivariate technique whereby signals are decorrelated and then the components extracted according to the decreasing order of their variances. PCA provides a technique for computation of the eigenvectors and eigenvalues of the, for example, return signal (RX) matrix of signals (Jolliffe, 2002). The technique involves the linear transformation of a group of correlated variables to achieve certain optimum conditions, the most important of which is that the transformed variables are uncorrelated (Jackson, 2003). PCA assumes that the UTX signals have Gaussian probability distributions.

The related SVD is a PCA matrix-based method for obtaining principal components without having to obtain the covariance matrix. SVD requires the assumption that the "important" source signals in the matrix of signals, UTX, required for target identification, are not among the smaller eigenvectors. Another method to decorrelate a set of signals is Gram-Schmidt orthogonalization (GSO). GSO depends on the initial choice among the ERX/[UTX] signals, removing that chosen signal, and projecting the remaining reduced set of signals onto a lower dimensional plane. The correct initial choice is thus critical.

ICA is a generalization of PCA, of Factor analysis (FA), and of a form of blind source separation (BSS). ICA assumes that the UTX and any noise components, are nonGaussian and statistically independent, and uses different optimization criteria. The requirement of statistical independence sets ICA apart from the other procedures. Whereas PCA and FA find a set of signals, ERX/[UTX], that are uncorrelated with each other, ICA finds a set of signals, ERX/[UTX], that are statistically independent from each other. It should be noted that a lack of correlation is a weaker property than independence: whereas independence implies a lack of correlation, a lack of correlation does not imply independence (Stone, 2004). ICA can use a PCA pre-processing step that decorrelates (i.e., whitens) the KRX. There are a number of different forms of ICA, for example, temporal, spatio-temporal, local (Hyvärinen et al., 2001; Cichocki & Amari, 2002). ICA is more noise-sensitive than PCA, and assumes that signals are the product of instantaneous linear combinations of independent sources. There are two major assumptions: the sources are independent and are nonGaussian. Unlike PCA, ICA uses higher-order statistics. Correlation is a measure of the amount of covariation between two signals, for example, x and y, and depends only on the first moment of the probability density function of x and y: pdf_{xy}. Independence – a stronger condition for signal separation – is a measure of the covariation of all the moments of the probability function pdf_{xy}.

Another method, complexity pursuit, assumes that the source signals, UTX, have informational temporal or spatial structure. Complexity pursuit uses a measure of complexity, for example, Kolmogorov complexity (Cover & Thomas, 1991), and the signal with the lowest complexity is extracted (Hyvärinen, 2001). The method can be used on mixtures of KRX that are super-Gaussian, Gaussian, and sub-Gaussian.

These methods are also related to blind equalization, as system identification is similar to the problem of equalizing a linear channel (Ding & Li, 2001). The objective of blind equalization in this case is to recover the underlying target response characteristics based solely on the probabilistic and statistical properties of the *a priori* data.

4.11.1 Singular Value Decomposition

The following Figure 4.44 indicates that under the assumption required by the SVD method – that the UTX signals are Gaussian - the SVD method can identify models targets of P-51 and C-160 aircraft on the basis of their returns signal (RXs) and SVD-originated ERX/[UTX]s.

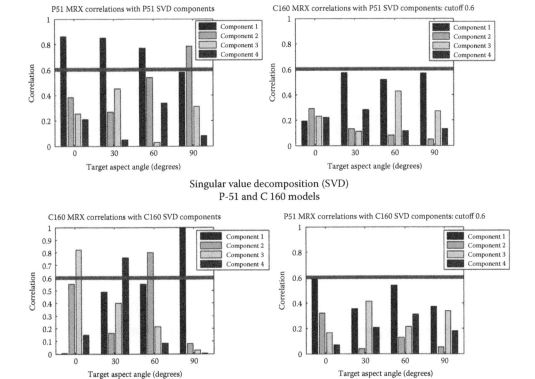

FIGURE 4.44
Examples of singular value decomposition (SVD) target identification through correlations of target RXs for 0, 30, 60 & 90° aspect angles, with Target [UTX]s = SVD components. Upper left: correlations of P-51 model RXs with P-51 model [UTX]s. Upper right: correlations of C-160 model RXs with P-51 model [UTX]s. Lower left: correlations of C-160 model RXs with C-160 model [UTX]s. Lower right: correlations of P-51 model RXs with C-160 model [UTX]s.

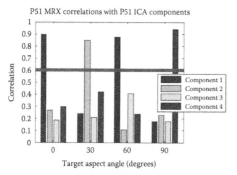
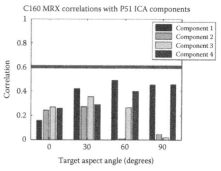

Independent component analysis (ICA)
P-51 and C-160 models

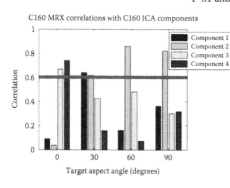
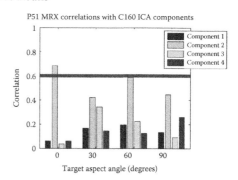

FIGURE 4.45
Examples of ICA target aircraft model identification: correlations of target MRXs for 0, 30, 60, and 90° aspect angles, with target [UTX]s = ICA components. Upper left: correlations of P-51 model RXs with P-51 model [UTX]s. Upper right: correlations of C-160 model RXs with P-51 model [UTX]s. Lower left: correlations of C-160 model RXs with C-160 model [UTX]s. Lower right: correlations of P-51 model RXs with C-160 model [UTX]s.

4.11.2 Independent Component Analysis

The following Figure 4.45 shows that under the assumption required by ICA – that the UTX signals are nonGaussian and statistically independent – the same P-51 and C-160 aircraft model targets can also be identified on the basis of their RXs and ICA-originated [UTX]s. Figure 4.46 demonstrates similarly that Truck and Humvee targets can be identified.

4.12 Blind Source Separation, Matching Pursuit, and Complexity Pursuit

In previous sections, the WVD transform time–frequency approach was considered as a form of analysis that decomposed a target's return signal (RX) into a time–frequency spectrum that identifies the local, or transient, features of signals, as opposed to the classical Fourier-based approaches which made to find the global or constant wavelength features.

The WVD is based on instantaneous autocorrelations and provides an energy picture of signals. For example, (a) the sum over frequency at a specific time or (b) the sum over time at a specific frequency provides the signal energy (a) at that time and (b) at that frequency,

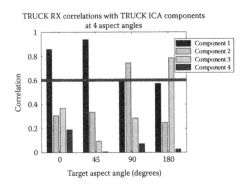

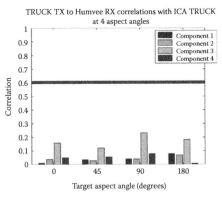

Independent component analysis (ICA)
Dodge truck and Humvee

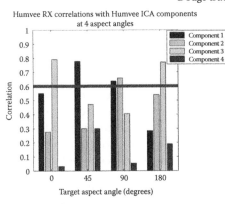

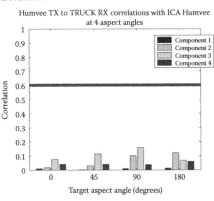

FIGURE 4.46
Examples of ICA truck target identification: correlations of target RXs for 0, 30, 60, and 90° aspect angles, with target [UTX]s = ICA components. Upper left: correlations of truck RXs with truck [UTX]s. Upper right: correlations of Humvee MRXs with truck [UTX]s. Lower left: correlations of Humvee RXs with Humvee [UTX]s. Lower right: correlations of truck MRXs with Humvee [UTX]s.

respectively. As indicated previously, WVD has major advantages, one being that it is related to the AF, the FRFT, and other measures known in optics. However, a drawback of the WVD comes from the presence of cross-terms generated in the transform. These cross terms although physically meaningful in optics are yet a hindrance when the WVD transform is used to describe and interpret nonoptical signals. The application of filters can mitigate the presence of cross-terms (Choi & Williams, 1989), but the application of *optimum* filters requires *a priori* knowledge of the frequency content of the signals being analyzed. In contrast to the WVD, all classical approaches based on the FT have deficiencies that are addressed in Marple (1987).

Other spectral estimation and analysis methods also have assumptions which, may, or may not apply in any particular instance. For example, multiple signal classification algorithm (MUSIC) is a nonparametric eigen-analysis frequency estimation procedure (Bienvenu & Kopp, 1983; Schmidt, 1986). MUSIC has a better resolution and better frequency estimation characteristic than classical methods, especially at high white noise levels. However, its performance is worse in the presence of colored noise. The eigen-decomposition produces eigenvalues of decreasing order and orthonormal eigenvectors.

The resulting spectra are not considered true power spectra estimates, as signal power is not preserved. Therefore, MUSIC spectra are considered pseudospectra and frequency estimators.

Yet another signal decomposition method, complexity pursuit, is based on the assumption that a mixture of signals is usually more complex than the simplest (least complex) of its constituent source signals (the complexity conjecture). Thus complexity pursuit is a BSS method and seeks a weight vector which provides an orthogonal projection of a set of signal mixtures such that each extracted signal is minimally complex. Complexity can be defined in different ways; but in complexity pursuit, it is defined using criteria related to Kolmogorov complexity (Cover & Thomas, 1991), and the least complex search strategy is based on gradient ascent.

There are other approaches to signal decomposition, including autoregressive (AR) parametric modeling in general, the modified covariance method – an AR method, and infinite impulse response (IIR) parametric modeling, all of which assume linearity and a correct model order.

Now, the underlying assumption of a UWB radar is that there is no multiplication of input and TX components, and RX components are statistically independent. RXs are assumed to be mixtures of independent components and the temporal complexity of a mixture is assumed to be greater than that of its simplest (least complex) source signal.

In the case of ICA, the assumptions likewise require that the sources of extracted signals are statistically independent. (It does not follow that uncorrelated sources are also statistically independent, which is a stronger requirement than correlation as noted previously.) Therefore, under this assumption, as the variance of a target's return signal is provided by changes in aspect angle, it might be supposed that sources, or ICA components, extracted from a mixture of several RXs from a responding target at various aspect angles, should roughly match. In fact, this approximate match is shown in Figure 4.47 for the case of RXs from truck and Humvee targets. The good fit of the power spectra and the spectra of ICA components supports the assumption that the RXs are: (a) the product of instantaneous linear combinations of independent sources (located on the target) and the sources are (b) independent and (c) nonGaussian – as (a), (b), and (c) are also the assumptions of ICA analysis.

Of course, extrapolation of such attributes, (a), (b), and (c) to all targets under all conditions would not be warranted, but extrapolation is warranted, to a degree, in the case of these tested targets and similar targets. One can then ask of these cases: what is the optimum basis representation of target RXs? A technique designed to answer this question is Matching Pursuit, according to which efficient decomposition can only be achieved in a dictionary containing functions reflecting the structure of the analyzed signal (Mallat & Zhang, 1993, Durka, 2007). An advantage of this approach is that both transients and constant wavelength signals can be captured in an analyzed signal, but a disadvantage is that the choice of filter dictionaries presupposes knowledge of the types of signal being processed. Furthermore, efficient decomposition can only be achieved in a dictionary containing functions reflecting the structure of the analyzed signal, as the Matching Pursuit procedure is as follows: find in the dictionary a function that best fits the signal; subtract its contribution from the signal; and then repeat on the remaining residuals.

There is much leeway in the choice of dictionary, or basis, in Matching Pursuit decomposition as shown in Figures 4.48 and 4.49. In Figure 4.48, a time-domain MRX signal is analyzed into a dictionary, or basis, of sinusoid vectors. The reconstruction from the basis or dictionary representation of sinusoid vectors is seen to be an accurate representation of

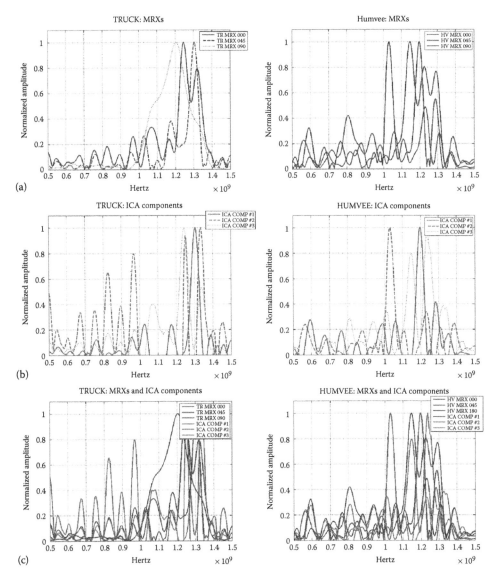

FIGURE 4.47
Received signal (RX) power spectra, ICA component spectra, and their overlap for a truck (Left column) versus Humvee (Right column). Top row: the power spectra for the RXs of the two targets at aspect angles 00°, 45°, and 90°. Middle row: ICA component spectra for a mixture of the same RXs. Bottom row: the two upper rows overlapped. The good fit of the power spectra and the spectra of the ICA components support the assumption that the RXs are (a) the product of instantaneous linear combinations of independent sources (located on the target), and the sources are (b) independent and (c) nonGaussian, as (a), (b), and (c) are the assumptions of ICA analysis. However, extrapolation of such attributes to all possible targets under all possible conditions is not justified.

the original MRX signal. However, a decomposition into WH filters/wavelets of increasing TBP provides an equally accurate reconstruction as shown in Figure 4.49; but, whereas the FT is optimum for constant wavelength signals, the WH transform is optimum for transient signals.

The criteria used in these blind source separations (BSSs) shown in Figures 4.47 through 4.49 have involved the moments of a joint probability density function (*pdf*) or

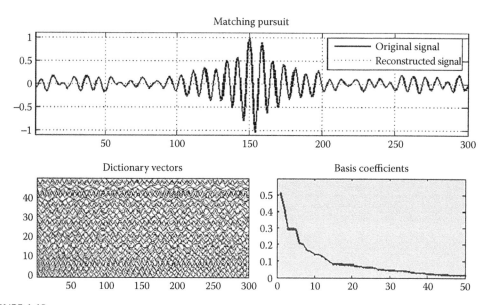

FIGURE 4.48
A time-domain RX signal, upper, is analyzed into a dictionary or basis of sinusoid vectors, lower left, with coefficients shown in lower right. The reconstruction from the basis or dictionary representation is an accurate representation of the original RX signal (upper).

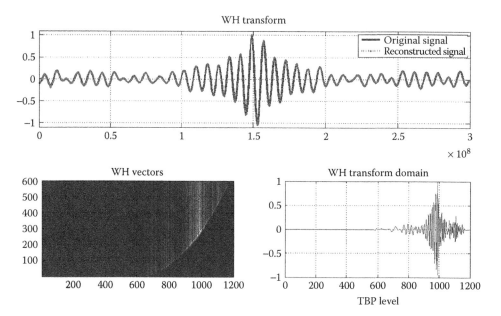

FIGURE 4.49
The same time-domain mixture of received signals (MRX) as shown in Figure 4.48, upper, is WH transformed, lower left, with a time–bandwidth product WH vector representation shown in lower right. The reconstruction from the WH transform representation is also an accurate representation of the original RX signal (upper).

the fit to an arbitrary dictionary or basis. Another criterion is that of maximizing entropy. Entropy is, in one sense, a measure of the uniformity of the distribution of a bounded set of values such that complete uniformity corresponds to maximum entropy (Cover & Thomas, 1991). The critical observation is that variables with maximum entropy distributions are statistically independent (Stone, 2004, 2005).

Finding independent signals by maximizing entropy is known as infomax (Bell & Sejnowski, 1995), which is equivalent to maximum likelihood estimation (MLE). The term complexity pursuit was coined (Hyvärinen, 2001) to describe the process of extracting from mixtures the minimally complex source signals. The supposition in this case is that maximum predictability is equal to minimal complexity (Xie et al., 2005). Thus unlike PCA and ICA which presuppose power density function models of the signal, complexity pursuit depends only on the complexity of the signal.

An example of BSS extraction of complexity pursuit source components from a mixture of return signal (RXs) is shown in Figures 4.50 and 4.51. Here four extracted source components are compared with the conventional power spectrum of the average of the mixture. It can be seen that the separated components largely agree with the average power spectrum. However, once again extrapolation of the validity of the assumptions of infomax to all targets under all conditions is not warranted.

This section reviewed the application of several approaches to BSS using a variety of assumptions. It was shown that some of these techniques and assumptions apply to vehicle targets. However, as it is (statistically) impossible to prove the null hypothesis, one cannot extrapolate the claim to all targets under all conditions. The lesson is probably that a variety of signal-processing techniques using a variety of assumptions should be applied in the BSS of mixtures of received signals from a newly encountered class of targets. Nonetheless, for the targets tested, it appears that the assumption that there is no multiplication of input and transmitted components, and that the received components are statistically independent, appears supported. The received signals can be assumed to be mixtures of independent components and the temporal complexity of a mixture is greater than that of its simplest (least complex) source signal.

4.13 Conclusions

In UWB time–frequency return signal processing, there are many choices to be made concerning what information in the signal is to be brought to center stage while relegating other information to backstage; and, at the same time, deciding whether the signal satisfies the presuppositions of a particular processing choice. Another choice is whether a unitary transform is required. A unitary transform, such as the WHWF transform, many wavelet transforms, the WVD, and the FT, all permit an inverse operation recovering the original signal. However, an inverse operation is sometimes not required, permitting reliable processing decisions to be made in a signal space for which an inverse is not possible. A further complication is that these operations assume that the transfer function of the target is linear, which, although likely the case, may not always be so. If nonlinear operations need to be addressed, then a commencement has been made in addressing nonlinear

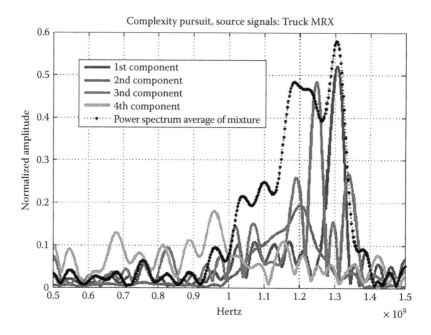

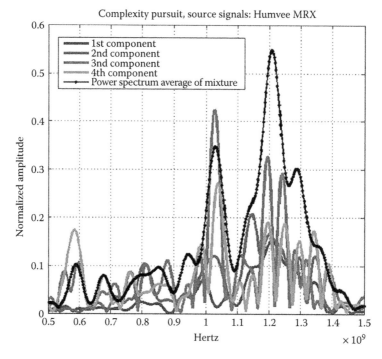

FIGURE 4.50
Source signals separated from a mixture of RX signals by a complexity pursuit algorithm. Upper: truck target, and lower: Humvee target. The four source components separated in each case are compared with the conventional power spectrum of the average of the mixture. It can be seen that the separated components largely agree with the average power spectrum.

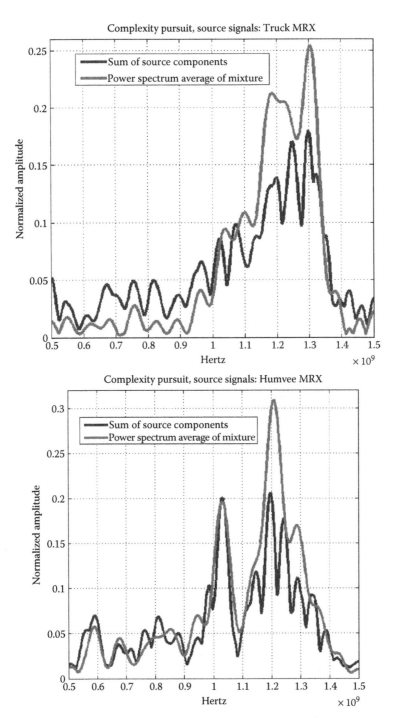

FIGURE 4.51
The same data as shown in Figure 4.50 with the four source components summed and compared with the conventional power spectrum of the average of the mixture. It can be seen that the separated components largely agree with the average power spectrum.

system analysis (Bendat, 1990; Bendat & Piersol, 1993, 2000), but there is further to go in that direction.

Given that the targets addressed are linear, another distinction arises. Conventional signal-processing treats targets as LTI systems. However in the case of UWB, the variation in time of signal arrival of returning signal components at the transceiver dictates that the targets be treated as linear time varying, or LTV, systems. The concept that targets interrogated by UWB transient pulses are LTV systems has consequences. The well-known, and ubiquitously used, FT is strictly only appropriate for LTI systems. The more appropriate techniques for LTV systems are time–frequency and wavelet techniques, and these have been addressed here.

Furthermore, UWB return signals from LTV systems differ from return signals from LTI systems. (LTI systems are addressed by Bell, 1988, 1993; Haykin, 2006; and Guerci, 2010.) However, in the case of a short-pulse transient and UWB transmit and receive pulse, the distinction between an LTV and LTI system has practical significance. For example, a variable network is defined as one in which one or more element values are dependent in a specified way upon a combination of three variables, such as time, input, and output. Whereas in LTI systems, a transfer or system function is defined as the FT of the response to a unit impulse and hence is a function of frequency and phase, but independent of time, in the case of LTV systems, a transfer or system function is defined as a function of frequency and phase with time as a parameter (Zadeh, 1950a,b). In the limiting case of time independence (which reduces to aspect angle independence), an LTV transfer function reduces to that of a LTI transfer function, but in the case of time dependence (and aspect dependence) as in UWB radars, the LTV transfer function differs from the LTI transfer function. The LTV transfer function is an instantaneous transfer function. These differences dictate different signal-processing choices in the case of UWB signal processing as compared with conventional radar signal processing.

4.14 Signal Processing and Future UWB Radar Systems

The UWB community has contended that full exploitation of UWB signals requires approaches to signal processing beyond the classical methods developed for narrowband constant wavelength signals. As demonstrated here, no single method can provide a totally satisfactory product signal-processing solution for all applications.

The signal-processing engineer must define the processing objectives and then select from the available library of time–frequency methods to find the best approach for a particular application. Some specialized cases may require new approaches derived from the available library of approaches shown in Table 4.2.

The designer must find the best method based on the transmitted signal format, target characteristics (i.e., material, geometry, size relative to the signal spatial resolution) and signal analysis objectives. Different targets and objectives will require different approaches.

Developing signal analysis methods based on the techniques presented here can greatly increase the value of UWB radar beyond simple range profiling and imaging. Practical applications of time–frequency signal analysis techniques will make UWB radars more valuable in medical, nondestructive testing, security, defense, and future unknown systems.

TABLE 4.2

Summary of Frequency and Time–Frequency Analysis Transforms

Transform	Functional Use	Comments
Fourier Transform	Determines frequency content of continuous signals and depending on constructive and destructive interference of basis elements.	• A global analysis method. • Determines global (constant) features of the waveform. • Optimum transformation for constant wavelength signals. • Trades precision in signal frequency detection for imprecision in timing of onset and offset of frequency occurrence. • Basis functions are sinusoids infinite in time. • Assumes constant wavelength signals are of infinite duration. • Misleading for analysis of transient signals–dependent on unphysical constructive and destructive interference.
Wavelet transforms	Provides the local frequency information content of a signal at defined times.	• A local analysis method. • Trades precision in timing of signal frequency onset and offset for imprecision in frequency detection. • Basis elements formed by dyadic or binary dilations and translations (in frequency and time). • Based on a constant timebandwidth product: $\Delta f \Delta t = 1/2\pi$. • Choice of a variety of possible wavelet transforms maximizing different signal elements–choice determined by the properties of the signal analyzed and the objectives of the user/operator/receiver.
Ambiguity Function	Global analysis. Two-dimensional function of time delay and Doppler frequency.	• Drawback: provides excellent time location of high frequencies but poor time location at low frequencies – the bias/variance dilemma. • A global and local analysis method. • Provides a time(lag)-frequency(Doppler) spectrum. • Related to the Wigner-Ville Distribution by a Fourier transform. • Lags and Doppler referenced to either (a) the signal itself by auto-correlation or (b) a matched filter by cross-correlation.
Wigner-Ville Distribution (WVD) transform	Form of analysis that decomposed a target's return signal into a time–frequency spectrum	• A global and local nonlinear analysis method. • The Fourier transform of the auto-correlation function expressed in terms of the averaged time and the time lag. • Makes manifest the local, or transient, features of signals. • Major drawback: appearance of cross-terms in complex signals, partially mitigated by filtering.

(Continued)

TABLE 4.2 (*Continued*)

Summary of Frequency and Time–Frequency Analysis Transforms

Transform	Functional Use	Comments
Weber–Hermite Transform based on the Weber–Hermite Wave Functions (WHWF) series	Exists in both global and local analysis forms.	Exists in both (i) Global form, and (ii) Local form. • Basis functions formed by non-dyadic or non-binary dilations and translations (in frequency and time). • Expansion based on a changing time–bandwidth product: $\Delta f \Delta t = 1/2\pi (2n+1)$, $n = 0,1,2, \ldots$. • In the Global form, signal decomposed into n WHWF basis functions. • In the Local form, choice of even n WHWFs as averaging wavelet filters, and odd n WHWFs as differentiating filters. Related to Prolate Spheroidal Waveform transform. • Provides a solution to the bias/variance dilemma
Hilbert Transform (HT)	A linear operator, or non-causal FIR filter, that maps a function, $u(t)$, to a function $H(u)(t)$ with the same domain. The original signal is phase shifted by $\pi/2$.	• The Hilbert transform is used to calculate the "analytic" signal. If the original signal is $\sin(\omega t)$, the -Hilbert transform is $-\cos(\omega t)$ and the analytic signal is $\sin(\omega t) - j\cos(\omega t)$. • The analytic signal has no negative frequency components.
Hilbert–Huang Transform (HHT)	Provides a signal's instantaneous spectral profile and a time versus instantaneous frequency representation, using an *energy–time–frequency* method of signal analysis suitable for nonstationary and nonlinear signals.	• The motivation for the method is that an analyses in terms of any *a priori* chosen basis cannot fit the signals arising from all systems, that is one basis set doesn't fit all (mathematically or physically). • An adaptive approach is used to breakdown a nonlinear and nonstationary signal into its multiple components. The adaptive approach is called a *sifting process* yielding an empirical basis set of intrinsic mode function (IMF) components. The basis set is derived from the data itself. • In some instances, the HHT can provide more accurate identifications of nonstationary systems than wavelet methods. • Problems remain regarding: choice of signal envelope generation method, the stopping criteria for the sifting process, and the production of mode mixing for some signals.
Multiple Window Time–Frequency Analysis (MWTFA)	MWTFA is an averaging of linear time-variant (LTV) radar signals addressing the problem of constant (random) change or nonstationarity providing a way to address random, nonstationary signals and to provide a low bias and variance spectral estimator.	• The MWTFA technique removes a deficiency in wavelet transforms, namely, that while wavelet analysis provides excellent time location of high frequencies, there is poor time location at low frequencies. This is known as the bias/variance dilemma.

(Continued)

TABLE 4.2 (*Continued*)

Summary of Frequency and Time–Frequency Analysis Transforms

Transform	Functional Use	Comments
		• The dilemma is addressed by using different set of windows to compute several spectra of the entire signal and then averaging the resulting spectra to construct a spectral estimate. • To obtain an estimate at low bias and low variance, the windows must be orthogonal (to minimize variance) and optimally concentrated in frequency (to reduce bias).
Carrier Frequency-Envelope Frequency (CFEF)	• Discards signal information in the time domain. • Provides the relation of signal carrier frequency to envelope frequency in a Carrier Frequency-Envelope Frequency (CFEF) double-frequency spectrum.	• A CFEF is obtained by the Fourier transformation of each line in a Multiple Window Time–Frequency Analysis (MWTFA). • The influence of temporal jitter in signal averaging is removed. • An averaged signal is obtain without the influence of temporal jitter – with the penalty of loss of return signal arrival information. • Information concerning a return signal's "envelope" frequencies, as well as "carrier" frequencies is revealed.
Radon Transform	A mapping of a signal to integral projections. A mapping from signal space (or source space) to parameter space (or destination space). • A form of template matching. • Identical in mathematical formalism to the Hough transform.	• The Radon transform makes explicit how a data point in destination space is obtained from the data in source space.
Hough	• Designed to detect straight lines in images employing a template. • A form of template matching. • Identical in mathematical formalism to the Radon transform.	• The Hough space considers how a data point in source space is obtained from the data in destination space. • A discrete version of the Radon transform.
Fractional Fourier Transform (FRFT)	The FRFT is a generalization of the conventional Fourier transform (FT).	• Provides a generalization of the time (or space) and frequency domains. • Performing the αth FRFT operation followed by a WVD corresponds to rotating the WVD of the original signal by an angle parameter α in the clockwise direction, or a decomposition of the signal into chirps. • The αth order FRFT is the αth power of the FT operator. • The conventional FT is a special case of the FRFT, the FRFT adding an additional degree of freedom to signal analysis with introduction of the fraction or order parameter α.

(Continued)

TABLE 4.2 (*Continued*)

Summary of Frequency and Time–Frequency Analysis Transforms

Transform	Functional Use	Comments
Radon–Wigner	The squared magnitude of the fractional Fourier transform	• Signal analysis can be optimized depending on the objective at a specific α and the chosen optimization criterion. • The FRFT makes explicit whether a signal's main spectral component is sinusoidal or frequency modulated (chirp).
Blind Signal Separation (BSS)	BSS seeks a weight vector which provides an orthogonal projection of a set of signal mixtures The criteria used in BSSs involve the moments of a joint probability density function (*pdf*) or the fit to an arbitrary dictionary or basis.	• BSS is carried out by a variety of techniques – for example SVD, ICA. • Given received signals and suppositions concerning the target's transfer function, BSS attempts to reconstruct the transmitted signals.
Principle Component Analysis (PCA) eigen-decomposition.	• A multivariate analysis or analysis of multiple variables treated as a single entity. • A general multivariate technique whereby signals are decorrelated and then the components extracted according to the decreasing order of their variances. • The components are orthogonal linear combinations that maximize the total variance. • In contrast, in the related Factor analysis (FA), the factors are linear combinations that maximize the shared portion of the variance, that is the underlying latent constructs. • FA uses a variety of optimization routines and the result, unlike that for PCA, depends on the optimization routine used and starting points for those routines.	• Can find latent information in the original data. • Based on correlation techniques for source separation. • Information in the signal is divided into two subspaces – a signal subspace and a noise subspace. The signal subspace should be twice the number of sinusoids in the data if known. • PCA operates by transforming to a new set of uncorrelated variables – the principle components (PCs). • The PCs are also orthogonal and ordered in terms of variability. • Drawback: only if the original variables are Gaussian are the uncorrelated PCs also independent, because PCA uses only 2nd order statistics, for example variances. (Variables with Gaussian distributions have zero statistics above 2nd order.) • Singular Value Decomposition (SVD) is a generalized form of PCA.
Factor analysis (FA)	• FA is a form of PCA with the addition of extra terms for modeling the sensor noise associated with each signal mixture.	• Based on correlation techniques for source separation. • The factors extracted are linear combinations that maximize the shared portion of the variance, that is the underlying latent constructs. In contrast, in PCA the components are orthogonal linear combinations that maximize the total variance.

(*Continued*)

TABLE 4.2 (*Continued*)

Summary of Frequency and Time–Frequency Analysis Transforms

Transform	Functional Use	Comments
		• FA uses a variety of optimization routines and the result, unlike that for PCA, depends on the optimization routine used and starting points for those routines.
Singular Value Decomposition (SVD)	A PCA matrix-based method obtains principal components without having to obtain the covariance matrix.	• SVD requires the assumption that the "important" source signals in the matrix of signals, required for target identification, are not among the smaller eigenvectors.
Independent Component Analysis (ICA).	A generalization of PCA, of Factor analysis (FA), and a form of blind source separation (BSS).	• Assumes that the transmitted signal and any noise components, are non Gaussian and statistically independent.
		• ICA is a generalization of PCA, of FA, and a form of BSS. ICA assumes that the unknown source signals and any noise components, are non Gaussian and statistically independent.
		• The requirement (assumption) of statistical independence sets ICA apart from the other procedures.
		• Whereas PCA and FA find a set of estimated source signals, that are uncorrelated with each other, ICA finds a set of estimated source signals, that are statistically independent from each other.
		• A lack of correlation is a weaker property than independence: whereas independence implies a lack of correlation, a lack of correlation does not imply independence.
		• ICA is more noise-sensitive than PCA, and assumes that signals are the product of instantaneous linear combinations of independent sources. There are two major assumptions: the sources are independent and are non Gaussian.
		• Unlike PCA, ICA uses higher-order statistics. Correlation is a measure of the amount of covariance between two signals, for example x and y, and depends only on the first moment of the probability density function of x and y.
		• Independence – a stronger condition for signal separation – is a measure of the covariation of all the moments of the probability function.

(Continued)

TABLE 4.2 (Continued)

Summary of Frequency and Time–Frequency Analysis Transforms

Transform	Functional Use	Comments
Gram-Schmidt Orthogonalization (GSO).	• Decorrelates signals.	• GSO depends on the initial choice among the signals, removing that chosen signal, and projecting the remaining reduced set of signals onto a lower dimensional plane. • The correct initial choice is thus critical.
Karhunen–Loève (or Hotelling) transform.	• Represents a stochastic process as an infinite linear combination of orthogonal functions by minimization of the total mean square error. Orthogonal basis functions are determined by a covariance function.	• Assumes the signal/image to be processed is ergodic, that is that the spatial or temporal statistics of a single signal or image are the same over an ensemble of signals or images. • But this is not always the case: signals and images are not the result of simple outcomes of a random process, there being always a deterministic underlying component.
Complexity pursuit,	• Complexity pursuit is a blind source separation (BSS) method and seeks a weight vector which provides an orthogonal projection of a set of signal mixtures such that each extracted signal is minimally complex. • Extracts from signal mixtures the minimally complex source signals.	• Provides an orthogonal projection of a set of signal mixtures such that each extracted signal is minimally complex and the signal with the lowest complexity is extracted. • Assumes that a mixture of signals is usually more complex than the simplest (least complex) of its constituent signals. • Assumes that maximum predictability is equal to minimum complexity. • Assumes that the source signals have informational temporal or spatial structure. • Uses a measure of complexity, for example Kolmogorov complexity. • Unlike PCA and ICA which presuppose power density function models of the signal, complexity pursuit depends only on the complexity of the signal.
Matching pursuit,	• Signal decomposition achieved by means of a dictionary containing functions reflecting the structure of the analyzed signal.	• Captures transient and constant characteristics. • The choice of filter dictionaries presupposes knowledge of the types of signal being processed. • Furthermore, efficient decomposition can only be achieved in a dictionary containing functions reflecting the structure of the analyzed signal. *(Continued)*

TABLE 4.2 (*Continued*)

Summary of Frequency and Time–Frequency Analysis Transforms

Transform	Functional Use	Comments
Multiple Signal Classification Algorithm (MUSIC)	• Non parametric eigen-analysis frequency estimation procedure	• Better resolution and frequency estimation characteristic than classical methods, especially at high white noise levels. • Performance worse in the presence of colored noise • The eigen-decomposition produces eigenvalues of decreasing order, and orthonormal eigenvectors. -Resulting spectra are not considered true power spectra estimates as signal power is not preserved. • Therefore MUSIC spectra are considered pseudospectra and frequency estimators.
Autoregressive (AR) parametric modeling and infinite impulse response (IIR) parametric modeling, all of which which assume linearity and a correct model order.	• Parametric modeling method.	• These methods assume linearity and a correct model order.
Infomax	• Finds independent signals by maximizing entropy.	• Equivalent to Maximum Likelihood Estimation (MLE). • Assumes that maximum predictability is equal to minimal complexity.

APPENDIX 4.A Derivations of Weber–Hermite Wave Functions (WHWFs)

The relationship of WHWFs to the FRFT is as follows. Let $\psi_i(u), i = 0,1,2,3,\ldots$ note WHWFs that are eigensignals (or eigenfunctions) of the ordinary FT operation with respect to the eigenvalues, λ_i (Wiener, 1933; Dym & McKean, 1972; McBride & Kerr, 1987), and let these functions establish an orthonormal basis for the space of well-behaved finite-energy signals (functions). The FRFT is then given by (Ozaktas et al., 2001):

$$\mathbb{F}^a \psi_i = \lambda_i^a \psi_i(u) = \exp(-ia_i)\psi_i(u) = \exp(-ia_i\pi/2)\psi_i(u), \tag{4.A.1}$$

and a given function, $f(u)$ which can be expanded as a linear superposition of WHWFs is:

$$f(u) = \sum_{i=0}^{\infty} C_i \psi_i(u), \tag{4.A.2}$$

with

$$C_i = \int \psi_i(v) f(v) dv. \tag{4.A.3}$$

applying $\mathbb{F}^a$ to both sides, we get the following:

$$\mathbb{F}^a f(u) = \sum_{i=0}^{\infty} \exp\left(-\frac{ia_i\pi}{2}\right) C_i \psi_i(u) = \int \sum_{i=0}^{\infty} \exp\left(-\frac{ia_i\pi}{2}\right) \psi_i(u)\psi_i(v) f(v) dv, \tag{4.A.4}$$

and

$$K_a(u,v) = \sum_{i=0}^{\infty} \exp\left(-\frac{ia_i\pi}{2}\right) \psi_i(u)\psi_i(v). \tag{4.A.5}$$

providing the relation of WHWFs to the FRFT.

Turning now to derivations of the WHWFs, there are a number of slightly different derivations, but each is noteworthy in presenting a different insight into the physical nature of these functions. We examine seven in the following:

1. A classical derivation is as follows. The parabolic cylinder functions, or WH functions, are solutions to Weber's equation (Weber, 1869):

$$\frac{d^2\psi_n(x)}{dx^2} + \left(n + \frac{1}{2} - \frac{1}{4}x^2\right)\psi_n(x) = 0, \tag{4.A.6}$$

for which there is a general Weber equation or parabolic cylinder differential equation (Abramowitz & Stegun, 1972, p. 686):

$$\frac{d^2\psi_n(x)}{dx^2} + \left(ax^2 + bx + c\right)\psi_n(x) = 0, \tag{4.A.7}$$

with the point $x = \infty$ strongly singular.

This equation permits two solutions derived as follows (Whittaker, 1902, 1903; Whittaker & Watson, 1927, p. 347). The substitutions,

$$\psi = x^{-1/2}W_{k,m}, \quad z = x^2/2,$$ (4.A.8)

where $W_{k,m}$ is the Whittaker function (Whittaker & Watson, 1927, p. 347; Abramowitz & Stegun, 1972, p. 505),

$$\frac{d}{zdz}\left[\frac{d\left(wz^{1/2}\right)}{zdz}\right] + \left(-\frac{1}{4} + \frac{2k}{z^2} + \frac{3}{4z^4}\right)wz^{1/2} = 0,$$ (4.A.9)

for which:

$$\frac{\partial^2 w}{\partial z^2} + \left(2k - \frac{1}{4}z^2\right)w = 0,$$ (4.A.10)

converts the Weber equation to the Whittaker equation, which is a special case of the confluent hypergeometric equation*. In particular, taking the solution for which $R(z) > 0$ the solution is as follows:

$$\psi_n = \left(2^{\frac{n}{2}+\frac{1}{4}}\right)\left(z^{-\frac{1}{2}}\right)\left(W_{\frac{n}{2}+\frac{1}{4},\frac{1}{4}}\left(1/2z^2\right)\right)$$

$$= \frac{1}{\sqrt{z}}2^{n/2}\exp\left[-z^2/4\right]\left(-iz\right)^{1/4}\left(iz\right)^{1/4} {}_1F_1\left(\frac{1}{2}n+\frac{1}{4};\frac{1}{2};\frac{1}{2}z^2\right)$$ (4.A.11)

where ${}_1F_1(a; b; z)$ a confluent hypergeometric function (Gauss, 1812; Abramowitz & Stegun, 1972, p. 503). With

$$w = z^{-1/2}W_{k,-1/4}\left(\frac{1}{2}z^2\right),$$ (4.A.12)

where W is a Whittaker function defined in the above Equation 4.A.12, Weber's equation can be separated into:

$$\frac{d^2U}{du^2} - \left(c + k^2u^2\right)U = 0,$$ (4.A.13)

---

* The hypergeometric differential equation is a second-order linear ordinary differential equation whose solutions are given by the hypergeometric series. The hypergeometric series have the following form: $\left(\sum_{n=0}^{\infty} a_n\right)\left(\sum_{n=0}^{\infty} b_n\right) = \sum_{n=0}^{\infty} c_n$, where $c_n = a_k b_{n-k}$. The confluent hypergeometric equation is a degenerate form of the hypergeometric equation.

or Weber's first derived equation,

$$\frac{d^2V}{dv^2} + \left(c - k^2 v^2\right)V = 0,$$ (4.A.14)

or Weber's second derived equation.

For nonnegative n, and after renormalization, the solution to Weber's first derived equation reduces to the following:

$$U_n(x) = 2^{-n/2} \exp\left[-x^2/4\right] H_n\left(x/\sqrt{2}\right) \quad n = 0, 1, 2, \ldots,$$ (4.A.15)

which are parabolic cylinder functions or WHWFs, and where H_n is a Hermite polynomial.

Similarly, completing the square, Weber's second derived equation can be rewritten as follows:

$$\frac{\partial^2 \psi}{\partial x^2} + \left[a\left(x + \frac{b}{2a}\right)^2 - \frac{b^2}{4a} + c\right]\psi = 0.$$ (4.A.16)

Defining:
$u = x + b/2a$; $du = dx$, and substituting, gives:

$$\frac{\partial^2 \psi}{\partial u^2} + \left[au^2 + d\right]\psi = 0,$$ (4.A.17)

where $d = -b^2/4a + c$. Again, this equation admits of two solutions, an even and an odd. Continuing with the even solution, the solution is as follows:

$$\psi(x) = \exp\left[-x^2/4\right] {}_1F_1\left(\frac{1}{2}a + \frac{1}{4}; \frac{1}{2}; \frac{1}{2}x^2\right),$$ (4.A.18)

where ${}_1F_1(a; b; z)$ is, as before, the confluent hypergeometric function, and the solutions of this equation are, again, the parabolic cylinder or WHWFs.

2. A second parallel derivation commences with the one-dimensional wave equation as follows:

$$-\frac{1}{2m}\frac{\partial^2 \psi}{\partial x^2} + V(x)\psi = E\psi,$$ (4.A.19)

with spring potential as follows:

$$V(x) = \frac{1}{2}kx^2 = \frac{1}{2}m\omega^2 x^2,$$ (4.A.20)

where:
  $\omega = \sqrt{k/m}$ the angular frequency
  k is the stiffness constant
  m is the mass
  x is the field deflection of the oscillator

or

$$-\frac{1}{2m}\frac{\partial^2 \psi}{\partial x^2} + \frac{1}{2}kx^2\psi = E\psi. \tag{4.A.21}$$

This wave equation can be written in dimensionless form by defining the independent variables $\xi = \alpha x$ and an eigenvalue, λ, and requiring:

$$\alpha^4 = mk, \quad \lambda = 2E\left(\frac{m}{k}\right)^{1/2} = \frac{2E}{\omega}. \tag{4.A.22}$$

The dimensionless form is then:

$$\frac{\partial^2 \psi}{\partial \xi^2} + (\lambda - \xi^2)\psi = 0. \tag{4.A.23}$$

which is a form of Weber's equation.

3. A derivation based on a familiar model goes as follows. The wave equation for a vibrating string is associated with the difference between the total kinetic energy of the string and its potential energy being as small as possible. If a string vibrates with simple harmonic motion, then the time dependence is expressed as follows:

$$\psi(x,t) = \psi(x)\exp[-i\varepsilon\alpha^2 t], \tag{4.A.24}$$

and the function ψ must satisfy Helmholtz's equation:

$$\frac{\partial^2 \psi}{\partial x^2} + k^2\psi = 0, \tag{4.A.25}$$

with k a real constant. When $k = 0$, he equation is a one-dimensional Laplace equation; and when k^2 is a function of the coordinates and

$$\varepsilon = (2M/(h/2\pi)^2)E. \tag{4.A.26}$$

This equation is Schrödinger's equation for a particle with constant E (Morse & Feshbach, 1953, p. 494).

When ψ is space-dependent, the wave equation is as follows:

$$\frac{\partial^2 \psi}{\partial x^2} + (\varepsilon - \alpha^2 x^2)\psi = 0, \tag{4.A.27}$$

where $\alpha = M\omega/(h/2\pi)$ and which is yet another form of Weber's equation. This equation permits solutions as a function of

$$n = \frac{\varepsilon}{2\beta} - \frac{1}{2} = \frac{E}{(h/2\pi)\omega} - \frac{1}{2} \tag{4.A.28}$$

In order for the solutions to be quadratically integrable, it is necessary that n take on integer values, such as $n = 0,1,2,...$ (Morse & Feshbach, 1953, p. 1641). With normalization factors, the solutions are the Weber–Hermite (WH) or parabolic cylinder functions (Morse & Feshbach, 1953, p. 1642):

$$\psi_n(t) = \frac{1}{\sqrt{2^n n!}} \left(\frac{\alpha}{\pi}\right)^{1/4} \exp[-\alpha t^2/2] H_n(t\sqrt{\alpha}) \tag{4.A.29}$$

where $\alpha = M\omega/(h/2\pi)$. For the classical result, we substituted $(h/2\pi) \to 1$, and α becomes a time–frequency trade parameter/variable. This is the general form of the WHWFs.

If for a function, $f(x)$ an expansion of the form:

$$f(x) = a_0\psi_0 + a_1\psi_1 + \cdots + a_n\psi_n + \cdots \tag{4.A.30}$$

exists, and if it is legitimate to integrate term-by-term between the limits $+\infty$ and $-\infty$, then:

$$a_n = \frac{1}{\sqrt{(2\pi)^{1/2} n!}} \int_{-\infty}^{+\infty} \psi_n(t) f(t) dt, \tag{4.A.31}$$

and such a function can be expanded in terms of n WHWFs with coefficients α_n.

4. There is a derivation based on the Helmholtz equation. This derivation commences with the Helmholtz equation in parabolic cylinder coordinates:

$$\frac{1}{u^2 + v^2}\left(\frac{\partial^2\psi}{\partial u^2} + \frac{\partial^2\psi}{\partial v^2}\right) + \frac{\partial^2\psi}{\partial z^2} + k^2\psi = 0. \tag{4.A.32}$$

The equation can be separated with:

$$\psi(u,v,z) = U(u)V(v)Z(z), \tag{4.A.33}$$

resulting in:

$$\frac{1}{u^2 + v^2}\left(VZ\frac{\partial^2 U}{\partial u^2} + UZ\frac{\partial^2 V}{\partial v^2}\right) + UV\frac{\partial^2 Z}{\partial z^2} + k^2 UVZ = 0. \tag{4.A.34}$$

Dividing by UVZ and separating out the Z part gives:

$$\frac{\partial^2 Z}{\partial z^2} = -\left(k^2 + m^2\right)Z, \tag{4.A.35}$$

which can be solved, permitting the derivation:

$$\left(\frac{1}{U}\frac{\partial^2 U}{\partial u^2} - k^2 u^2\right) + \left(\frac{1}{V}\frac{\partial^2 V}{\partial v^2} - k^2 v^2\right) = 0. \tag{4.A.36}$$

With

$$\left(\frac{1}{U} \frac{\partial^2 U}{\partial u^2} - k^2 u^2 \right) = c, \tag{4.A.37}$$

$$\left(\frac{1}{V} \frac{\partial^2 V}{\partial v^2} - k^2 v^2 \right) = -c, \tag{4.A.38}$$

we have again:

$$\frac{d^2 U}{du^2} - (c + k^2 u^2)U = 0, \tag{4.A.39}$$

or Weber's first derived equation,

$$\frac{d^2 V}{dv^2} + (c - k^2 v^2)V = 0, \tag{4.A.40}$$

or Weber's second derived equation
   The solutions of Weber's first derived equation reduce to:

$$U_n(x) = 2^{-n/2} \exp[-x^2/4] H_n(x/\sqrt{2}), \quad n = 0, 1, 2, \dots \tag{4.A.41}$$

or WHWFs as before.

5. A derivation can be based on the electric Hertz vector, Π_e. Hertz showed that the electromagnetic field can be expressed in terms of a single vector function (Hertz, 1889, 1893). This potential is known as *the Hertz (electric) vector, the polarization potential, or a "super-potential"*. A second Hertz vector, the *"Hertz magnetic vector potential"*, Π_m, and related to the magnetic polarization, was introduced by Righi (1901). Together, Π_e and Π_m form a six vector (i.e., an antisymmetric tensor of the second rank). Π_e and Π_m have similar transformation properties to E and B in the case of Π_e, and D and H in the case of Π_m (Born & Wolf, 1999, p. 84). Their relationship is as follows:

Assuming the Lorentz gauge, a vector, called *the polarization vector, p* is defined with respect to the actual charges and currents as (Panofsky & Phillips, 1962, p. 254):

$$\frac{\partial p}{\partial t} = J; \ \nabla p = -\rho, \tag{4.A.42}$$

where:
   J is the free current density
   ρ is the free charge density

A generalization introduces a magnetic density vector, m so that these equations become (*cf.* Chapou Fernández et al., 2009):

$$\frac{\partial p}{\partial t} = J - \frac{1}{\mu \epsilon}(\nabla \times m); \ \nabla p = -\rho. \tag{4.A.43}$$

Π_e and Π_m are then defined as two *retarded potentials* (Born & Wolf, 1999, p. 84):

$$\Pi_e = \frac{1}{\epsilon} \int_V \frac{[p]}{R} dV', \tag{4.A.44}$$

$$\Pi_m = \mu \int_V \frac{[m(r')]}{R} dV', \tag{4.A.45}$$

where:
  R is the distance from a point, r, to a volume element, dV', at a point, r'
  the brackets $[\ ]$ indicate that p is to be evaluated at the retarded time, $t - |r - r'|/v$, where
    v is the velocity of propagation.

In terms of Π_e and Π_m, the electromagnetic field can be defined by the following A vector potential field (Jones, 1964; Born & Wolf, 1999):

$$A = \mu\epsilon \frac{\partial \Pi_e}{\partial t} + \nabla \times \Pi_m, \tag{4.A.46}$$

$$\varphi = -\nabla \cdot \Pi_e, \tag{4.A.47}$$

$$\nabla^2 \Pi_e - (\mu\epsilon)^2 \frac{\partial^2 \Pi_e}{\partial t^2} = -p/\epsilon, \tag{4.A.48}$$

$$\nabla^2 \Pi_m - (\mu\epsilon)^2 \frac{\partial^2 \Pi_m}{\partial t^2} = -\mu m, \tag{4.A.49}$$

$$\nabla \cdot (\nabla^2 \Pi_e) = \nabla^2 (\nabla \cdot \Pi_e). \tag{4.A.50}$$

In a source-free region, in vacuo, an electromagnetic field can be described in terms of either the electric or the magnetic Hertzian potential.
  For a field in the absence of magnetic polarization:

$$A = \mu\epsilon \frac{\partial \Pi_e}{\partial t}, \tag{4.A.51}$$

$$\varphi = -\nabla \cdot \Pi_e, \tag{4.A.52}$$

$$E = \nabla(\nabla \cdot \Pi_e) - \mu\epsilon \frac{\partial^2 \Pi_e}{\partial t^2}, \tag{4.A.53}$$

$$B = \mu\epsilon \left(\nabla \times \frac{\partial \Pi_e}{\partial t} \right), \tag{4.A.54}$$

$$\nabla^2 \Pi_e - (\mu\epsilon)^2 \frac{\partial^2 \Pi_e}{\partial t^2} = -p/\epsilon. \tag{4.A.55}$$

For a field in the absence of electric polarization:

$$A = \nabla \times \Pi_m, \tag{4.A.56}$$

$$\varphi = 0, \tag{4.A.57}$$

$$E = -\mu\epsilon \frac{\partial(\nabla \times \Pi_m)}{\partial t}, \tag{4.A.58}$$

$$B = \nabla \times (\nabla \times \Pi_m), \tag{4.A.59}$$

$$\nabla^2 \Pi_m - (\mu\epsilon)^2 \frac{\partial^2 \Pi_m}{\partial t^2} = -\mu m. \tag{4.A.60}$$

For a field with both electric and magnetic polarization:

$$A = \mu\epsilon \frac{\partial \Pi_e}{\partial t} + \nabla \times \Pi_m, \tag{4.A.61}$$

$$\varphi = -\nabla \cdot \Pi_e, \tag{4.A.62}$$

$$E = \nabla(\nabla \cdot \Pi_e) - \mu\epsilon \frac{\partial^2 \Pi_e}{\partial t^2} - \mu\epsilon \frac{\partial(\nabla \times \Pi_m)}{\partial t}, \tag{4.A.63}$$

$$B = \mu\epsilon \left(\nabla \times \frac{\partial \Pi_e}{\partial t} \right) + \nabla \times (\nabla \times \Pi_m). \tag{4.A.64}$$

For a conductor without sources:

$$E = \nabla(\nabla \cdot \Pi_e) - \mu\epsilon \frac{\partial^2 \Pi_e}{\partial t^2} - \mu\sigma \frac{\partial \Pi_e}{\partial t} = \nabla \times (\nabla \times \Pi_e), \tag{4.A.65}$$

$$B = \mu\nabla \times \left(\epsilon \frac{\partial \Pi_e}{\partial t} + \sigma\Pi_e \right), \tag{4.A.66}$$

$$\nabla^2 \Pi_e - \mu\epsilon \frac{\partial^2 \Pi_e}{\partial t^2} - \mu\sigma \frac{\partial \Pi_e}{\partial t} = 0. \tag{4.A.67}$$

Invariance is obtained for the transformations:

$$\Pi_e' = \Pi_e + \nabla \times F - \nabla G, \tag{4.A.68}$$

$$\Pi_m' = \Pi_e - \mu\epsilon \frac{\partial F}{\partial t}, \tag{4.A.69}$$

where the vectors $\boldsymbol{F} = F_x\mathbf{i} + F_y\mathbf{j} + F_z\mathbf{k}$, and the scalar function, G, defined over three dimensions of Cartesian coordinates, and with $\mathbf{i}$, $\mathbf{j}$, and $\mathbf{k}$ the unit vectors for the x-, y- and z-axes, respectively, are solutions to the wave equations:

$$\nabla^2 \boldsymbol{F} - (\mu\epsilon)^2 \frac{\partial^2 \boldsymbol{F}}{\partial t^2} = 0, \tag{4.A.70}$$

$$\nabla^2 G - (\mu\epsilon)^2 \frac{\partial^2 G}{\partial t^2} = 0. \tag{4.A.71}$$

for which the usual definitions apply:
∇ as the gradient operator (grad):

$$\nabla F = \frac{\partial F_x}{\partial x}\mathbf{i} + \frac{\partial F_y}{\partial y}\mathbf{j} + \frac{\partial F_z}{\partial z}\mathbf{k}. \tag{4.A.72}$$

$\nabla \cdot$ as the divergence operator (div):

$$\Delta \cdot \boldsymbol{F} = \frac{\partial F_x}{\partial x} + \frac{\partial F_y}{\partial y} + \frac{\partial F_z}{\partial z}. \tag{4.A.73}$$

$\nabla \times$ as the curl operator (curl):

$$\nabla \times \boldsymbol{F} = \left(\frac{\partial F_z}{\partial y} - \frac{\partial F_y}{\partial z} \right)\mathbf{i} + \left(\frac{\partial F_x}{\partial z} - \frac{\partial F_z}{\partial x} \right)\mathbf{j} + \left(\frac{\partial F_y}{\partial x} - \frac{\partial F_x}{\partial y} \right)\mathbf{k}. \tag{4.A.74}$$

∇^2, sometimes Δ, as the Laplace operator, or the divergence of the gradient (del):

$$\Delta \boldsymbol{F} = \nabla \cdot (\nabla \boldsymbol{F}) = \nabla^2 \boldsymbol{F} = \frac{\partial F_x^2}{\partial x^2} + \frac{\partial F_y^2}{\partial y^2} + \frac{\partial F_z^2}{\partial z^2}. \tag{4.A.75}$$

and where SI units are implied:
   μ is the magnetic permeability $(\text{kg} \cdot \text{m})/(\text{s}^2 \cdot \text{A}^2)$
   ϵ is the dielectric constant $(\text{A}^2 \cdot \text{s}^4)/(\text{kg} \cdot \text{m}^3)$
   σ is the conductivity $((\text{A}^2 \cdot \text{s}^3)/(\text{kg} \cdot \text{m}^3))$
   ρ is the charge density $(\text{A} \cdot \text{s})/\text{m}^3$
   m is the magnetization or magnetic density A/m
   p is the electric polarization density $(\text{A} \cdot \text{s})/\text{m}^2$
   J is the current density A/m^2
   E is the electric intensity or electric field strength $(\text{kg} \cdot \text{m})/(\text{A} \cdot \text{s}^3)$, with A = ampere
   B is the magnetic flux density or magnetic induction, $\text{kg}/(\text{A} \cdot \text{s}^2)$
   A is the vector potential $(\text{kg} \cdot \text{m})/(\text{A} \cdot \text{s}^2)$
   φ is the scalar potential $(\text{kg} \cdot \text{m}^2)/(\text{A} \cdot \text{s}^3)$

In parabolic cylinder coordinates, u, v, z the electric Hertz vector, Π_e is (Jones, 1964, p. 85) as follows:

$$\Pi_e(u, v, z) = U(u)V(v)Z(z), \tag{4.A.76}$$

and

$$\frac{\partial^2 Z}{\partial z^2} = m^2 Z, \tag{4.A.77}$$

$$\frac{\partial^2 U}{\partial u^2} + \left[(m^2 + k^2)u^2 - h \right] U = 0, \tag{4.A.78}$$

$$\frac{\partial^2 V}{\partial v^2} + [(m^2 + k^2)v^2 + h]V = 0. \tag{4.A.79}$$

where m and h are separation constants. Both the equations in U and V can be treated similarly. Substitutions of:

$$u = \left[4\left(m^2 + k^2 \right) \right]^{-1/4} X, \text{ and}$$

$$i\left(v + 1/2 \right) = \left(4\left(m^2 + k^2 \right)^{-1/2} \right) X, \tag{4.A.80}$$

where X is a placeholder, gives:

$$\frac{\partial^2 U}{\partial u^2} + \left(\frac{1}{4} X^2 - i\left(v + \frac{1}{2} \right) \right) U = 0. \tag{4.A.81}$$

A further substitution of $X = x\exp\left[(1/4)\pi i \right]$ gives:

$$\frac{\partial^2 U}{\partial u^2} + \left(v + \frac{1}{2} - \frac{1}{4} x^2 \right) U = 0, \tag{4.A.82}$$

which we recognize again as Weber's equation. As before, the solutions are WHWFs:

$$U_n(x) = 2^{-n/2}\exp\left[-x^2/4 \right] H_n\left(x/\sqrt{2} \right) \quad n = 0,1,2,\dots. \tag{4.A.83}$$

The Hertzian vectors, Π_e and Π_m together form a six vector system (Nisbet, 1955), and TE and TM waves in a waveguide can be defined in terms of Π_e and Π_m and shown to be generated by exciter systems that are equivalent respectively to electric and magnetic oscillating dipoles parallel to the direction of propagation (Essex, 1977). The two Hertzian vector approach is related to Whittaker's (1904, 1951) expression of an electromagnetic field in terms of two scalar functions.

The vectors, Π_e and Π_m can be united in a covariant formulation resulting in a skew tensor of rank two (Nisbet, 1955; McCrea, 1957; Chapou Fernández et al., 2009). This Hertz's tensor is defined as follows:

$$\Pi^{0i} = \left(\Pi_e \right)_i, \quad \Pi^{ij} = \varepsilon^{ijk} \left(\Pi_m \right)_k. \tag{4.A.84}$$

where ε^{ijk} is the Levi–Civita permutation symbol. This potential tensor, $\Pi^{\mu\nu}$, $\mu, \nu = 0,1,2,3,$ has space-time components corresponding to the electric Hertz vector, Π_e components, and purely spatial components corresponding to the magnetic Hertz vector, Π_m components. The Hertz tensor transforms according to the gauge chosen.

It should also be noticed that $\Pi^{\mu\nu}$ has the characteristic of the eight-dimensional algebra of octonions (Conway & Smith, 2003; Baez, 2002, 2005; Baez & Huerta, 2011), that have transformation properties different than those of vectors and tensors.

An alternative approach to the treatment of Hertz vectors is the so-called "source scalarization" method, whereby any given distribution of arbitrarily oriented sources is reduced to an equivalent distribution of single-component parallel electric and magnetic sources (Weigelhofer, 2000; Georgieva & Weiglhofer, 2002; Weigelhofer & Georgieva, 2003). The objective is to produce a complete description of the EM field in terms of two so-called "scalar wave" potentials – actually vector potentials. The method has been extended to the case of propagation in stratified gyrotropic media (De Visschere, 2009).

6. Turning from classical mechanics to quantum mechanics, a quantum mechanical derivation commences with the Hamiltonian of a particle as follows:

$$H = \frac{p^2}{2m} + \frac{1}{2}m\omega^2 x^2, \qquad (4.A.85)$$

where x is the position operator, p is the momentum operator: $p = -i\hbar\, \partial/\partial x$ and where the first term is the kinetic energy of the particle, and the second is the potential energy.

The one-dimensional Schrödinger wave equation, inspired by an optical analogy, is as follows:

$$-\frac{\hbar^2}{2m}\frac{d^2\psi}{dx^2} + V(x)\psi = E\psi, \qquad (4.A.86)$$

or

$$-\frac{\hbar^2}{2m}\frac{d^2\psi}{dx^2} + \frac{kx^2\psi}{2} = E\psi. \qquad (4.A.87)$$

This equation can be written in dimensionless form using the following substitutions:

$$\xi = \alpha x;\ \alpha^4 = mk/\hbar^2;\ \lambda = \left(2E/\hbar\right)\left(m/k\right)^{1/2} = \left(2E\right)/\hbar\omega. \qquad (4.A.88)$$

The wave equation in this dimensionless form becomes Weber's equation:

$$\frac{d^2\psi}{d\xi^2} + \left(\lambda - \xi^2\right)\psi = 0. \qquad (4.A.89)$$

Alternatively, by a coordinate transformation:

$$x = \left(\frac{\hbar}{m\omega}\right)^{1/2}\xi \qquad (4.A.90)$$

the wave equation becomes:

$$\left(\frac{d^2}{d\xi^2} + \frac{2E}{\hbar\omega} - \xi^2\right)\psi = 0, \qquad (4.A.91)$$

which is also Weber's equation. Except when $\xi = \infty$, the solutions are as follows:

$$\psi = \exp\left(\pm(1/2)\xi^2\right)^{1/2} y, \tag{4.A.92}$$

for which

$$\left(\frac{d^2}{d\xi^2} - 2\xi\left(\frac{d}{d\xi}\right) + \frac{2E}{\hbar\omega} - 1\right) y = 0, \tag{4.A.93}$$

with solutions for polynomials of degree n

$$-2n + \frac{2E}{\hbar\omega} - 1 = 0, \tag{4.A.94}$$

resulting in eigenvalues:

$$E_n = \left(n + 1/2\right)\hbar\omega. \tag{4.A.95}$$

The solutions for y are as follows:

$$y_n = H_n(\xi) = (-1)^n \exp\left[\xi^2\right]\frac{d^n}{d\xi^n}\exp\left[-\xi^2\right] \tag{4.A.96}$$

where $H_n(\xi)$, $n = 0,1,2,\ldots$ are Hermite polynomials, satisfying:

$$\left(\frac{d^2}{d\xi^2} - 2\xi\frac{d}{d} + 2n\right)H_n(\xi) = 0, \tag{4.A.97}$$

for which $2n = \lambda - 1$ or $\lambda = 2n + 1$. Therefore, substituting, the solutions for ψ are:

$$\psi = \exp\left(\pm\frac{1}{2}\xi^2\right)^{1/2} H_n(\xi), \quad n = 0,1,2,\ldots. \tag{4.A.98}$$

which are WHWFs

7. An alternative quantum mechanical view is to commence with the time-independent one-dimensional Schrödinger equation in the bra-ket (notation for describing quantum states) form:

$$H|\psi\rangle = E|\psi\rangle, \tag{4.A.99}$$

which is seen to be a Weber's equation:

$$\frac{-\hbar^2}{2m}\frac{d^2\psi(x)}{dx^2} + \frac{1}{2}m\omega^2 x^2 \psi(x) = E\psi(x). \tag{4.A.100}$$

which has a general solution:

$$\langle x \,|\, \psi_n \rangle = \frac{1}{\sqrt{2^n n!}} \left(\frac{m\omega}{\pi\hbar} \right)^{1/4} \exp\left(-\frac{m\omega x^2}{2\hbar} \right) H_n\left(\sqrt{\frac{m\omega}{\hbar}} x \right), \quad n = 0, 1, 2, \dots. \qquad (4.A.101)$$

where $H_n(x) = (-1)^n \exp(x^2) \dfrac{d^n}{dx^n} \exp(-x^2)$ are Hermite polynomials; or

$$\psi_n(z) = \left(\frac{\alpha}{\pi} \right)^{1/4} \frac{1}{\sqrt{2^n n!}} H_n(z) \exp\left(-z^2/2 \right), \quad n = 0, 1, 2, \dots \qquad (4.A.102)$$

for $z = \sqrt{\alpha} x$ and $\alpha = m\omega/\hbar$, which is a normalized form of the WHWFs.

The corresponding energy levels are again:

$$E_n = \hbar\omega\left(n + 1/2 \right), \qquad (4.A.103)$$

and the expectation value for the potential energy is

$$\langle V_n \rangle = \int\limits_{-\infty}^{+\infty} \left(\frac{1}{2} \right) \psi_n^* kx^2 \psi_n(x)\, dx = (1/2)k \frac{2n+1}{2\alpha^2} = (1/2)(n+1/2)\hbar\omega = (1/2)E_n. \quad (4.A.104)$$

Therefore,

$$\Delta x \Delta p = \frac{1}{2}(2n+1)\hbar, \; n = 0, 1, 2, \dots \qquad (4.A.105)$$

Substituting $x \to t$, $p \to f$ and $\hbar \to 1$ for the classical case gives the TBPs for the WHWFs:

$$\Delta t \Delta f = \frac{1}{2}(2n+1), \quad n = 0, 1, 2, \dots \qquad (4.A.106)$$

It should be noted that these time-bandwidth products (TBPs) refer to one standard deviation of the signal duration, and one standard deviation of the signal BW, that is, not the 90% or 99% support of signal duration and BW in general use in engineering.

APPENDIX 4.B Relationship of WHWFs to Fractional Calculus

Herrmann (2011, p. 19) states that the integral representation of Hermite polynomials:

$$H_n(x) = e^{x^2} \frac{2^{n+1}}{\sqrt{\pi}} \int\limits_0^\infty e^{-t^2} t^n \cos\left(2xt - \frac{\pi}{2} n \right) dt, n \in \mathbb{N}, \qquad (4.B.1)$$

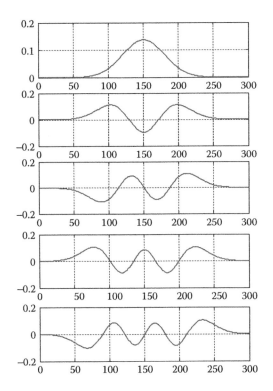

FIGURE 4.B.1
Weber–Hermite (WHWFs) polynomials 0–5. The axes are general: the x-coordinate could be time and the y-axis could be amplitude.

may be extended to a definition of fractional order Hermite polynomials by

$$H_n(x) = e^{x^2} \frac{2^{n+1}}{\sqrt{\pi}} \int_0^\infty e^{-t^2} t^\alpha \cos\left(2xt - \frac{\pi}{2}\alpha\right) dt, \alpha \in \mathbb{C}, \tag{4.B.2}$$

but that it is "an open question" whether orthogonality survives. We show here that orthogonality survives but only in the complex Fourier domain.

This can be seen by taking 6 WH polynomials (0–5)* – the number 6 is arbitrary (Figure 4.B.1), and fractionally FT these polynomials according to:

$$FracFT_a = \int_{-\infty}^{+\infty} K_a(f,t) f(t) dt \tag{4.B.3}$$

---

* There are other appellations for the WHWFs, for example, Hermite–Gaussian functions. I prefer the designation WHWF because (a) Hermite–Gaussian implicates Gaussian in all polynomials, $n = 0, 1, 2, \dots$ instead of just the first for $n = 0$; (b) Weber's equation is more general than Hermite's equation; (c) "Weber–Hermite" follows the *Mathematical Encyclopedia* (Hazewinkel, 2002) usage; and (d) other texts, for example, (Morse & Feshback, 1953, vol. 2, p. 1642; Jones, 1964, p. 86), have used "Weber–Hermite".

where

$$K_a(f,t) = A_a \exp\left[i\pi\left(\cot(\alpha f^2) - 2\csc(\alpha ft) + \cot(\alpha t^2)\right)\right] \qquad (4.B.4)$$

$$A_a = \sqrt{1 - i\cot(\alpha)} \qquad (4.B.5)$$

$$\alpha = a\pi/2 \qquad (4.B.6)$$

In the case $a = 1$, then $\alpha = \pi/2$, and

$$FracFT_1(f) = FT(f) = \int_{-\infty}^{+\infty} \exp[-i2\pi ft] f(t) dt \qquad (4.B.7)$$

The variable, a, ranges from -2 to $+2$. When $a = +1$, the transformation is the conventional forward Fourier transformation; when $a = -1$, the transformation is the conventional inverse Fourier transformation. When $a = 0$, the untransformed time-domain signal is obtained. At other values of a, the ath-order FracFT results. The FracFT spectra of the WHWFs are shown in the following Figure 4.B.2.

If the dot product is taken *between* these six complex signals of Figure 4.B.2 for the 400 lines ($0 < a < 2$) but not for $a = 1$, the result is near zero. Also, if the dot product is taken across any single complex signal ($0 < a < 2$) but not for $a = 1$, the result is near zero.

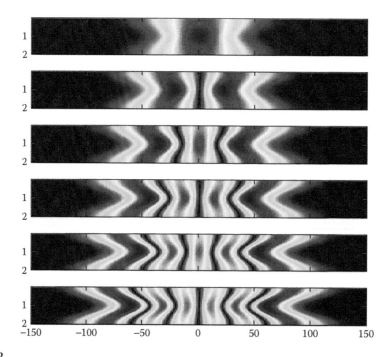

FIGURE 4.B.2
The absolute of 400 fractional Fourier transforms of, top to bottom, WHWF0 - 5, with a ranging from 0, through 1 (conventional Fourier transform) to 2, the 400th. The axes are general: the x-coordinate could be frequency and the y-axis could be amplitude.

However, if the dot product is taken at $a = 1$ for each polynomial, which is the conventional FT, the result is unity. Thus, although orthogonality is lost between fractional WHWFs for $\alpha \in \mathbb{C}$ see Equation 4.B.2, orthogonality is retained for the FRFT of WHWFs, that is, $FRFT_a(H_n(x))$, $n \in \mathbb{N}$, $0 < a < 2$, for $n \in \mathbb{N}$, $a \in \mathbb{N}$, $a \neq 1$ – see Equation 4.B.1.

References

Abe, S. & Sheridan, J.T. (1994) Optical operations on wave functions as the Abelian subgroups of the special affine Fourier transformations. *Opt. Lett.*, 19, pp. 1801–1803.

Abramowitz, M. & Stegun, C.A., (Eds.) (1972) "Parabolic Cylinder Functions". Chapter 19 in *Handbook of Mathematical Functions with Formulas, Graphs and Mathematical Tables*, Dover, New York, pp. 685–700.

Almeida, L.B., (1994) The Fractional Fourier Transform and Time–frequency Representations. *IEEE Trans. Signal Processing*, 42, pp. 3084–3091.

Baez, J.C., (2002) The octonions. *Bull. American Math. Soc.*, 39, pp. 145–205.

Baez, J.C., (2005) Review of: *On quaternions and octonions: Their geometry, arithmetic, and symmetry*, by John H. Conway and Derek K. Smith, A. K. Peters, Ltd., Natick, MA, 2003, *Bull. American Math. Society*, 42, pp. 229–243.

Baez, J.C. & Huerta, J., (2011) The strangest numbers in string theory. *Scientific American*, 304, pp. 60–65.

Barrett, H.H., (1984) The Radon transform and its applications. In *Progress in Optics XXI*, Chapter 3, pp. 217–286, Elsevier, Amsterdam.

Barrett, T.W., (1972) Conservation of Information, *Acustica*, 27, pp. 44–47.

Barrett, T.W., (1973a) Analytic Information Theory, *Acustica*, 29, pp. 65–67.

Barrett, T.W., (1973b) Structural Information Theory, *J. Acoust. Soc. Am.* 54, pp. 1092–1098.

Barrett, T.W., (1996) Active signalling systems. United States Patent 5,486,833 dated January 23rd, 1996.

Barrett, T.W., (2008) Method and application of applying filters to n-dimensional signals and images in signal projection space. United States Patent No 7,426,310, dated Sep. 16, 2008.

Barrett, T.W., (2012) *Resonance and Aspect Matched Adaptive Radar (RAMAR)*, World Scientific, Singapore.

Barrett, T.W., (2012) *Resonance and Aspect Matched Adaptive Radar (RAMAR)*, World Scientific, Singapore.

Bell, A.J. & Sejnowski, T.J., (1995) An information-maximization approach to blind separation and blind convolution. *Vision Research*, 37, pp. 1129–1159.

Bell, M.R. (1988) Information theory and radar: Mutual information and the design and analysis of radar waveforms and systems. Ph.D. dissertation, California Institute of Technology, Pasadena.

Bell, M.R., (1993) Information theory and radar waveform design. *IEEE Trans. Information Theory*, 39, pp. 1578–1597.

Bendat, J.S., (1990) *Nonlinear System Analysis and Identification from Random Data*, (John Wiley, New York).

Bendat, J.S. & Piersol, A.G., (1993) *Engineering Applications of Correlation & Spectral Analysis*, 2nd edition, (John Wiley, New York).

Bendat, J.S. & Piersol, A.G., (2000) *Random Data Analysis and Measurement Procedures*, 3rd Edition, (John Wiley, New York).

Bernardo, L.M. (1996) ABCD matrix formalism of fractional Fourier optics. *Opt. Eng.*, 35, pp. 732–740.

Bienvenu, G. & Kopp, L., (1983) Optimality of high resolution array processing using the eigensystem approach," *IEEE Trans. Acoustics, Speech and Signal Processing*, 31, pp. 1234–1248.

Born, M. & Wolf, E., (1999) *Principles of Optics*, 7th Edition, Cambridge University Press.

Candan, C., Kutay, M.A. & Ozaktas, H.M., (2000) The discrete Fractional Fourier Transform. *IEEE Trans. Signal Processing*, 48, pp. 1329–1337.

Cariolaro, G., Erseghe, T., Kraniauskas, P. & Laurenti, N., (1998) Unified framework for the Fractional Fourier Transform. *IEEE Trans. Signal Processing*, 46, pp. 3206–3219.

Chapou Fernández, J.L., Granados Samaniego, J., Vargas, C.A. & Velázques Arcos, J.M., (2009) Hertz tensor, current potentials and their norm transformations. *Progress in Electromagnetics Research Symposium, Proceedings, Moscow, Russia*, Aug 18–21, pp. 529–534.

Chen, Q., Hunag, N., Riemenschneider, S. & Xu, Y. (2006) A B-spline approach for empirical mode decomposition. *Advances in Computational Mathematics* 24, pp. 171–195.

Choi, H.I. & Williams, W.J., (1989) Improved time–frequency representation of multi-component signals using exponential kernels. *IEEE Trans. Acoust. Speech Signal Processing*, 37, pp. 862–871.

Cichocki, A. & Amari, S., (2002) *Adaptive Blind Signal and Image Processing*, Wiley, New York.

Claasen, T.A.C.M. & Mecklenbräuker, W.E.G., (1980a) The Wigner distribution – A tool for time–frequency signal analysis – Part I: Continuous time signals. *Philips J. Research*, 35, pp. 217–250.

Claasen, T.A.C.M. & Mecklenbräuker, W.E.G., (1980b) The Wigner distribution – A tool for time–frequency signal analysis – Part II: Discrete time signals. *Philips J. Research*, 35, pp. 276–300.

Claasen, T.A.C.M. & Mecklenbräuker, W.E.G., (1980c) The Wigner distribution – A tool for time–frequency signal analysis – Part III: Relations with other time–frequency transformations. *Philips J. Research*, 35, pp. 372–389.

Cohen, L., (1989) Time–frequency Distributions - A Review. *Proc. IEEE*, 77, pp. 941–981.

Cohen, L., (1995) *Time–frequency Analysis*, (Prentice-Hall, Englewood Cliffs, New Jersey).

Conway, J.H. & Smith, D.K., (2003) *On Quaternions and Octonions: Their Geometry, Arithmetic, and Symmetry*, A. K. Peters, Ltd., Natick, MA.

Cover, T.M. & Thomas, J.A., (1991) *Elements of Information Theory*, Wiley, New York.

Datig, M. & Schlurmann, T., (2004) Performance and limitations of the Hilbert-Huang transformation (HHT) with an application to irregular water waves. *Ocean Engineering*, 31, pp. 1783–1834.

Deans, S.R., (1981) Hough transform from the Radon transform. *IEEE Transactions on Pattern Analysis and Machine Intelligence*, 3, pp. 185–188.

Deans, S.R., (1983) *The Radon Transform and Some of Its Applications*, John Wiley, New York.

Deley, G.W., (1970) Waveform Design. Chap. 3, M.I. Skolnik (Ed.) *Radar Handbook*, McGraw-Hill.

De Visschere, P., (2009) Electromagnetic source transformations and scalarization in stratified gyrotropic media. *Progress in Electromagnetics Research*, 18, pp. 165–183.

Dickinson, B.W. & Steiglitz, K., (1982) Eigenvectors and functions of the discrete Fourier transform. *IEEE Trans. Acoust. Speech Signal Process.* ASSP-30, pp. 25–31.

Ding, Z. & Li, Y., (2001) *Blind Equalization and Identification*, Marcel Dekker, New York.

Durka, P., (2007) *Matching Pursuit and Unification in EEG Analysis*, Artech House, Norwood, MA.

Dym, H. & McKean, H.P., (1972) *Fourier Series and Integrals*, Academic Press, New York.

Erseghe, T., Kraniauskas, P. & Cariolaro, G., (1999) Unified fractional Fourier transform and sampling theorem. *IEEE Trans. Signal Processing*, 47, pp. 3419–3423.

Flandrin, P. (1998) *Time–frequency and Time-Scale Analysis*, Academic Press. Volume 10 in the series: *Wavelet Analysis and Applications*.

Flandrin, P. & Gonçalvès, P., (2004) Empirical mode decompositions as data-driven wavelet-like expansions. *Int. J. of Wavelets, Multiresolution and Information Processing*, 2, pp. 477–496.

Flandrin, P., Rilling, G. & Gonçalvès, P., (2003) Empirical mode decomposition as a filter bank. *IEEE Signal Processing Letters*, 10, pp. 1–4.

Flandrin, P., Rilling, G. & Gonçalvès, P., (2004) Empirical mode decomposition as a filter bank. *IEEE Signal Processing Letters*, 11, pp. 112–114.

Frazer, G. & Boashash, B., (1994) Multiple window spectrogram and time–frequency distributions. *Proc. IEEE Int. Conf. Acoustic, Speech, and Signal Processing – ICASSP'94, volume IV*, pp. 293–296.

Gauss, C. F. (1866) Disquisitiones Generales Circa Seriem Infinitam $\left[\frac{\alpha\beta}{1\cdot\gamma}\right]\chi + \left[\frac{\alpha(\alpha+1)\beta(\beta+1)}{1\cdot 2\cdot\gamma(\gamma+1)}\right]\chi^2 + \left[\frac{\alpha(\alpha+1)(\alpha+2)\beta(\beta+1)(\beta+2)}{1\cdot 2\cdot 3\cdot\gamma(\gamma+1)(\gamma+2)}\right]\chi^3 +$ etc. Pars Prior. *Commentationes Societies Regiae Scientiarum Gottingensis Recentiores, Vol. II.* 1812. Reprinted in *Gesammelte Werke, Bd. 3*, pp. 123–163 & 207–229.

Gelfand, I.M., Graev, M.I., & Vilenkin, N. Ya., (1966) *Generalized Functions. Volume 5, Integral Geometry and Representation Theory*. Academic Press, New York.

Georgieva, N.K. & Weiglhofer, W.S., (2002) Electromagnetic vector potentials and the scalarization of sources in a nonhomogeneous medium. *Phys. Rev. E*, 66, 046614–1–8.

Ghavami, M., Michael, L.B. & Kohno, R., (2007) *Ultrawideband Signals and Systems in Communication Engineering*, 2nd Edition, Wiley.

Ginkel, M. van, Hendricks, C.L.L. & Vliet, L.J. van, (2004) A short introduction to the Radon and Hough transforms and how they relate to each other, Number QI-2004–01 in the Quantitative Imaging Group Technical Report Series, Delft University of Technology, Delft, The Netherlands.

Guerci, J.R., (2010) *Cognitive Radar*, Artech House, MA.

Haykin, S. (2006). Cognitive Radar: A Way of the Future, *IEEE Signal Processing Magazine*, 23, pp. 30–40.

Hazewinkel, M., (Ed.) (2002), *Encyclopaedia of Mathematics*, Springer, New York.

Helgason, S., (1999) *The Radon Transform*, 2nd Edition, Birkhäuser, Boston.

Herrmann, R., (2011) *Fractional Calculus: An Introduction for Physicists*. World Scientific.

Hertz, H., (1889) *Ann. d. Physik*, 36, 1.

Hertz, H., (1893) The forces of electric oscillations, treated according to Maxwell's theory. *Wiedemann's Ann*. 36, pp. 1–23, 1889. Reprinted in H. Hertz, *Electric Waves*, Macmillan, 1893, reprinted Dover Publications, 1962.

Hlawatsch, F. & Boudreaux-Bartels, G.F., (1992) Linear and quadratic time–frequency representations. *IEEE Signal Processing Magazine*, 9, pp. 21–67.

Hough, P.V.C., (1962) Method and means for recognizing complex patterns. United States Patent No. 3,069,654 dated 1962.

Huang, N.E. & Attoh-Okine, N.O., (2005) *The Hilbert-Huang Transform in Engineering*, (Taylor & Francis).

Huang, N.E. & Shen, S.S.P., (2005) *Hilbert-Huang Transform and Its Application*, (World Scientific).

Huang, N.E., Shen, Z. & Long, R.S., (1999) A new view of nonlinear waves – the Hilbert spectrum. *Ann. Rev. Fluid Mech.*, 31, pp. 417–457.

Huang, N.E., Shen, Z., Long, S.R., Wu, M.C., Shih, H.H. & Zheng, Q., (1998) The empirical mode decomposition and the Hilbert spectrum for nonlinear and nonstationary time series analysis. *Proc. R. Soc. Lond. A*, 454, pp. 903–995.

Huang, N.E. & Wu, Z.H., (2008) A review on Hilbert-Huang transform: Method and its applications to geophysical studies. *Reviews of Geophysics*, 46, RG2006, doi:10.1029/2007RG000228.

Huang, N.E., Wu, M.L., Long, R.S., Shen, S.S., Qu, W.D., Gloersen, P. & Fan, K.L., (2003) A confidence limit for the empirical mode decomposition and Hilbert spectral analysis. *Proc. Roy. Soc. Lond. A*, 460, pp. 1597–1611.

Hyvärinen, A., (2001) Complexity pursuit: Separating interesting components from time series. *Neural Computation*, 13, pp. 883–898.

Hyvärinen, A., Karhunen, J. & Oja, E., (2001) *Independent Component Analysis*, Wiley, New York.

Illingworth, J. & Kittler, J., (1988) A survey of the Hough transform. *Computer Vision, Graphics and Image Processing*, 44(1), pp. 87–116.

Jackson, J.E., (2003) *A User's Guide to Principle Components*, Wiley, New York.

Janssen, A.J.E.M., (1981) Positivity of weighted Wigner distributions. *SIAM J. Mathematical Analysis*, 12, pp. 752–758.

Janssen, A.J.E.M., (1982) On the locus and spread of pseudo-density functions in the time–frequency plane. *Philips J. Research*, 37, pp. 79–110.

Janssen, A.J.E.M., (1984) Positivity properties of phase-plane distribution functions. *J. Math. Phys.*, 25, pp. 2240–2252.

Jolliffe, I.T., (2002) *Principal Component Analysis*, 2nd edition, Springer, New York.

Jones, D.S., (1964) *The Theory of Electromagnetism*, Pergamon Press, New York.

Kak, A.C., & Slaney, M. (2001) *Principles of Computerized Tomographic Imaging*, Society for Industrial and Applied Mathematics, Philadelphia, PA, 2001.

Landau, H.J. & Pollak, H.O., (1961) Prolate Spheroidal wave functions, Fourier Analysis and Uncertainty – II, *Bell Syst. Tech. J.*, 40, pp. 65–84.

Landau, H.J. & Pollak, H.O., (1962) Prolate spheroidal wavefunctions, Fourier analysis and uncertainty – III: The dimension of the space of essentially time- and band-limited signals", *Bell Syst. Tech. J.*, 41, pp. 1295–1336.

Li, B., Tao, R. & Wang, Y., (2007) New sampling formulae related to linear canonical transform. *Signal Processing*, 87, pp. 983–990.

Li, Y. (2008) Wavelet-fractional Fourier transforms. *Chinese Physics* B, 17, pp. 170–179.

Lohmann, A.W., (1993) Image rotation, Wigner rotation, and the Fourier transform. *J. Opt. Soc. Am.*, A10, pp. 2181–2186.

Lohmann, A.W., Mendlovic, D., Zalevsky, Z. & Dorsch, R.G., (1996) Some important fractional transformation for signal processing. *Opt. Commun.*, 125, pp. 18–20.

Lohmann, A.W. & Soffer, B.H., (1993) Relationship between two transforms: Radon-Wigner and fractional Fourier," in *Annual Meeting, OSA Technical Digest Series (Optical Society of America, Washington, D.C., 1993)*, Vol 16, p. 109.

Mallat, S., (1999) *A Wavelet Tour of Signal Processing*, 2nd Edition, Academic Press, New York.

Mallat, S. & Zhang, Z., (1993) Matching pursuit with time–frequency dictionaries. *IEEE Trans. Signal Processing*, 41, pp. 3397–3415.

Marple, S.L., (1987) *Digital Spectral Analysis with Applications*, Prentice-Hall, Englewood Cliffs, New Jersey.

Martin, W. & Flandrin, P., (1985) Wigner-Ville spectral analysis of nonstationary process. *IEEE Trans. Acoust., Speech, Signal Processing*, 33, pp. 1461–1470.

McBride, A.C. & Kerr, F.H., (1987) On Namias' fractional Fourier transforms. *IMA J. Appl. Math.*, 39, pp. 159–175.

McCrea, (1957) Hertzian electromagnetic potentials. *Proceedings of the Royal Society*, A, 240, pp. 447–457.

Meyer, Y., (1993) *Wavelets: Algorithms & Applications*, Society for Industrial & Applied Mathematics, Philadelphia.

Morse, P.M. and Feshbach, H., (1953) *Methods of Theoretical Physics*, 2 Volumes, McGraw-Hill, NY.

Moshinsky, M. & Quesne, C., (1971) Linear canonical transformations and their unitary representations. *J. Math. Phys.*, 12, pp. 1772–1783.

Moyal, J.E., (1949) Quantum mechanics as a statistical theory. *Proc. Camb. Phil. Soc.*, 45, pp. 99–124.

Namias, V., (1980) The fractional order Fourier transform and its application to quantum mechanics. *J. Inst. Math. Appl.*, 25, pp. 241–265.

Nisbet, A., (1955) Hertzian electromagnetic potentials and associated gauge transformations. *Proceedings of the Royal Society*, A, 231, pp. 250–262.

Onural, L. (1993) Diffraction from a wavelet point of view. *Opt. Lett.*, 18, pp. 846–848.

Ozaktas, H.M., Barshan, B., Mendlovic, D. & Onural, L., (1994) Convolution, filtering and multiplexing in fractional domains and their relation to chirp and wavelet transforms. J. *Opt. Soc. Am.*, A11, pp. 547–559.

Ozaktas, H.M. & Mendlovic, D., (1993) Fourier transforms of fractional order and their optical interpretation. *Optics Communications*, 101, pp. 163–169.

Ozaktas, H.M., Zalevsky, Z. & Kutay, M.A., (2001) *The Fractional Fourier Transform with Applications in Optics and Signal Engineering*, (John Wiley, New York).

Panofky, W.G.H. & Phillips, M., (1962) *Classical Electricity and Magnetism*, 2nd Edition, Addison-Wesley, Reading, Massachusetts.

Papoulis, A., (1977) *Signal Analysis*, McGraw-Hill, New York.

Pei, S.C. & Ding, J.J., (2000) Closed form discrete fractional and affine fractional transforms. *IEEE Trans Signal Processing*, 48, pp. 1338–1353.

Pei, S.C. & Ding, J.J., (2001) Relations between fractional operations and time–frequency distributions, and their applications. *IEEE Trans Signal Processing*, 49, pp. 1638–1655.

Pei, S.C. & Ding, J.J., (2002) Eigenfunctions of linear canonical transform. *IEEE Trans Signal Processing*, 50, pp. 11–26.

Radon, J., (1917) Über die Bestimmung von Funktionen durch ihre Integralwerte längs gewisser Mannigfaltigkeiten. *Berichte Sächsische Akademie der Wissenschaften, Leipzig, Mathematisch-Physikalische Klasse*, 69, pp. 262–277.

Ramm, A.G. & Katsevich, A.I., (1996) *The Radon Transform and Local Tomography*, CRC Press, Boca Raton.

Raveh, I. & Mendlovic, D., (1999) New properties of the Radon Transform of the Cross Wigner/ Ambiguity Distribution Function. *IEEE Trans. Signal Processing*, 47, pp. 2077–2080.

Righi, A. (1901) Sui campi elettromagnetici e particolarmente su quelli creati, da cariche elettriche o da poli magnetici in movimento. *Nuovo Cimento*, 2, pp. 104–121.

Rihaczek, A.W., (1969) *Principles of High-Resolution Radar*, McGraw-Hill, NY.

Rihaczek, A.W. & Hershkowitz, S.J., (1996) *Radar Resolution and Complex-Image Analysis*, Artech House, MA.

Rihaczek, A.W. & Hershkowitz, S.J., (2000) *Theory and Practice of Radar Target Identification*, Artech House, MA.

Saxena, R. & Singh, K., (2005) Fractional Fourier transform: A novel tool for signal processing. *J. Indian Inst. Sci.*, 85, pp. 11–26.

Schmidt, R.O., (1986) Multiple emitter location and signal parameter estimation. IEEE Trans. Antennas & Propagation, AP-34, pp. 276–280.

Shan, P-W. & Li, M., (2010) Nonlinear time-varying spectral analysis: HHT and MODWPT, *Mathematical Problems in Engineering*, Volume 2010, Article ID 618231, doi:10.1155/2010/618231.

Slepian, D., (1964) Prolate spheroidal wave functions, Fourier analysis and uncertainty – IV: Extensions to many dimensions; generalized prolate spheroidal functions, *Bell System Tech. J.*, 43, pp. 3009–3057.

Slepian, D., (1978) Prolate spheroidal wave functions, Fourier analysis and uncertainty – V. The discrete case, *Bell System Technical J.*, vol. 57, pp. 1371–1430.

Slepian, D. & Pollak, H.O., (1961) Prolate speheroidal wave functions, Fourier analysis and uncertainty – I. *Bell System Tech. J.*, 40, pp. 43–64.

Stone, J.V., (2004) *Independent Component Analysis: A Tutorial Introduction*, MIT Press, Cambridge, MA.

Stone, J.V., (2005) Independent Component Analysis. *Encyclopedia of Statistics in Behavioral Sciences*, B.S. Everitt & D.C. Howell, Editors, Volume 2, pp. 907–912, John Wiley.

Thomson, D.J., (1982) Spectrum estimation and harmonic analysis. *Proc. IEEE*, 70, pp. 1055–1096.

Ville, J., (1948) Theorie et applications de la notion de signal analytique. *Cables et Transmission*, 2A, pp. 61–74.

Weber, H., (1869) Über die Integration der partiellen Differentialgleichung: $\partial^2 u/\partial x^2 + \partial^2 u/\partial y^2 + k^2 = 0$, *Math. Ann.*, 1, pp. 1–36.

Weiglhofer, W.S., (2000) Scalar Hertzian potentials for nonhomogeneous uniaxial dielectric-magnetic mediums. *Int. J. Appl. Electrom.*, 11, pp. 131–140.

Weiglhofer, W.S. & Georgieva, N., (2003) Vector potentials and scalarization for nonhomogeneous isotropic mediums. *Electromagnetics*, 23, pp. 387–398.

Whittaker, E.T., (1902) On the functions associated with the parabolic cylinder in harmonic analysis. *Proc. London Math. Soc.*, 35, pp. 417–427.

Whittaker, E.T., (1903) On the partial differential equations of mathematical physics. *Math. Ann.*, 57, pp. 333–355.

Whittaker, E.T., (1904) On an expression of the electromagnetic field due to electrons by means of two scalar potential functions. *Proc. London Math. Soc.*, Series 2, 1, pp. 367–372.

Whittaker, E.T., (1951) *History of the Theories of Aether and Electricity: Volume 1: The Classical Theories; Volume II: The Modern Theories 1900–1926*, New York, Dover Publications.

Whittaker, E.T., & Watson, G.N., (1927) *A Course of Modern Analysis*, 4th Edition, Cambridge University Press.

Wiener, N., (1933) *The Fourier Integral and Certain of Its Applications*, Cambridge U. Press.

Wiener, N., (1958) *Nonlinear Problems in Random Theory*. MIT Press & Wiley.

Wigner, E.P., (1932) On the quantum correction for thermodynamic equilibrium. Phys. Rev., 40, pp. 749–759.

Wood, J.C. & Barry, D.T., (1994a) Tomographic time–frequency analysis and its application toward time-varying filtering and adaptive kernel design for multicomponent linear-FM signals. *IEEE Trans Signal Processing*, 42, pp. 2094–2104.

Wood, J.C. & Barry, D.T., (1994b) Linear signal synthesis using the Radon-Wigner transform. *IEEE Trans Signal Processing*, 42, pp. 2105–2111.

Wood, J.C. & Barry, D.T., (1994c) Radon transformation of time–frequency distributions for analysis of multicomponent signals. *IEEE Trans Signal Processing*, 42, pp. 3166–3177.

Woodward, P.M., (1953) *Probably and Information Theory with Applications to Radar*, Artech, MA, USA.

Woodward, P.M. & Davies, I.L., (1950) A theory of radar information. *Phil. Mag.*, 41, p.1001.

Xie, S., He, Z. & Fu, Y., (2005) A note on Stone's conjecture of blind signal separation. *Neural Computation*, 17, pp. 321–330.

Xu, Y., Haykin, S. & Racine, R.J., (1999) Multiple window time–frequency distribution and coherence of EEG using Slepian sequences. *IEEE Trans. Biomed. Eng.*, 49, pp. 861–866.

Zadeh, L.A., (1950a) Frequency analysis of variable networks. *Proc. I.R.E.*, 38, pp. 291–299.

Zadeh, L.A., (1950b) Correlation functions and power spectra in variable networks. *Proc. I.R.E.*, 38, pp. 1342–1345.

5

Modeling of Ultrawideband (UWB) Impulse Scattering by Aerial and Subsurface Resonant Objects Based on Integral Equation Solving

Oleg I. Sukharevsky, Gennady S. Zalevsky, and Vitaly A. Vasilets

CONTENTS

5.1 Introduction and Objectives

This Chapter examines the use of ultrawideband (UWB) signals (signals with spectrum confined within a very high frequency [VHF] and ultra high frequency [UHF] bands) to enhance the non-cooperative radar identification of objects. Many radar books contain a classic figure showing the enhancement of the radar cross section (RCS) of a perfectly electrically conducting (PEC) sphere, when the signal frequency falls in the Mie resonance region. This condition occurs when the ratio of the circumference to the wavelength lies between 1 and 10. The experimental studies mentioned in Chapter 4 described how the use of radar resonant waveband sounding signals can stimulate secondary radiation of objects. This chapter describes how to predict those reflected fields.

The stealth aircraft and missile threat have encouraged many countries to build VHF radar systems to enhance the detection of these low contrast and relatively electrically small targets. These VHF band radars [1–3] are effective instruments for obtaining radar information about resonant size objects, such as missiles, unmanned aerial vehicles (UAVs), and small airplanes. Additionally, stealth aircraft radar absorbing coatings commonly do not work effectively in this waveband.

Ground-penetrating radars (GPRs) use UWB signals with a spectrum confined within a frequency band of 100…4000 MHz [4–6] because they provide the best ground penetration. These special UWB signals can help to identify subsurface objects, such as different types of mines against a background of interfering items. They can also differentiate between the radar responses from mines located at small depth (some centimeters) and the strong reflection from the ground interface. Mines can also have radar resonant properties like the small aircraft mentioned earlier. Moreover, it is necessary to account for the fact that GPR antenna system operates in the near field zone of the object.

Over the last thirty years, specialists of many countries have demonstrated the possibility of object identification based on their complex natural resonances (CNRs) in UWB radar signals. Many articles and books describe CNRs' features and separation techniques for both airborne [7–12], and subsurface objects in [13–16]. In theory, CNRs do not depend on the object's aspect angle with respect to the signal. This property reduces the number of parameters in an identification algorithm [7–16]. Sounding with UWB signals produces the most significant advantages as compared to signals with smaller fractional bandwidths when searching for resonant-sized objects.

Developing effective algorithms for processing the signal scattered from radar objects requires investigation of their scattering characteristics. Fortunately, we can predict scattering characteristics by computer simulation. The most commonly used simulation methods include the following: (1) asymptotic high-frequency methods (AHFMs), (2) the method of finite-differences time-domain (FDTD), and (3) the method of boundary integral equations (IEs). The AHFMs give good accuracy when predicting the scattering for electrically large objects [1,17–20]. The use of the FDTD method is limited by small-sized objects [21,22]. The method of boundary IEs provides the most accurate and universal method for correctly predicting the scattering characteristics of complex-shaped resonant size objects with a wide range of electrical sizes.

Modern numerical methods based on solving the boundary IEs have an extensive literature, particularly for PEC objects in free space [2,23–38], for PEC objects in material media (subsurface objects) [14,16,39–41], for dielectric and composite objects in free space [23,24,32–34,42–47], and for dielectric and composite subsurface objects [15,16,40,41,48,49]. This chapter briefly analyzes the advantages and disadvantages of the known methods of IE solving.

Section 5.2 specified the considered electromagnetic (EM) problems.

Sections 5.3 and 5.4 will give detailed descriptions of numerical methods developed by the authors for solving the considered specific problems. The algorithms described provide ways to numerically simulate EM scattering by the following:

- Aerial objects which can model the object as a PEC surface in the case of cruise missiles, UAVs, and small-scale airplanes when radiated in the VHF band (Subsections 5.3.1–5.3.3).
- Subsurface PEC (Subsection 5.3.4) and dielectric (Section 5.4) objects, such as mines of different types or other explosive objects buried in the ground.

We have based the numerical methods developed here on the IE solution in the frequency domain.

Section 5.5 presents an algorithm for calculating the UWB impulse responses (high-resolution radar range profiles) of objects.

Sections 5.6 and 5.7 present practical examples of UWB impulse response calculations for cruise missiles and buried antitank and antipersonnel mines.

5.2 The Impulse Scattering Problem Formulation

The models of EM wave scattering by resonant size objects considered in this chapter appear in Figure 5.1. The first model is the three-dimensional scatterer V_2 with surface S in free space V_1 with relative permittivity $\varepsilon_1 = 1$ (Figure 5.1a). The object V_2 has a complex relative permittivity $\varepsilon_2 = \varepsilon_2' + i\varepsilon_2''$ (where i is the imaginary unit). For a PEC object, then $\varepsilon_2'' \to \infty$ and EM field components inside the object V_2 are equal to zero. The subsurface object V_2 placed in dielectric half-space V_3 with relative permittivity $\varepsilon_3 = \varepsilon_3' + i\varepsilon_3''$ is also considered (Figure 5.1b). The plane *xoy* defines the interface between two half-spaces V_1 and V_3.

The relative permittivities ε_2 and ε_3 are functions of frequency. The relative permeability is $\mu = 1$ for all materials and media considered here.

The modeling process starts by sounding the object with an EM field (with electrical $\vec{E}^0$ and magnetic $\vec{\mathcal{H}}^0$ vectors) excited by exterior source. The analysis can put the monostatic

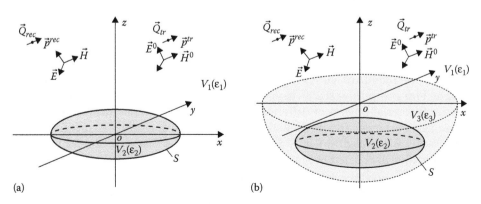

(a)   (b)

FIGURE 5.1
EM wave scattering models for (a) objects in free space and (b) subsurface objects.

or multistatic (general case) transmitting and receiving antennas with centers at points $\vec{Q}_{tr}$, and $\vec{Q}_{rec}$, respectively, either in far-field, or near-field zone with respect to the radar object. Figure 5.1 shows the unit vectors $\vec{p}^{tr}$, $\vec{p}^{rec}$ which respectively indicate the orientation of magnetic vectors of incident and scattered fields.

All EM fields considered in this chapter have time dependence $\exp(-i\omega t)$ (where $\omega = 2\pi f$ is the cyclic frequency, f is a linear frequency, and t is the current time). These EM fields satisfy the Maxwell's equations at any point of space, where permittivity and permeability are continuous functions of spatial coordinates. At the discontinuities of electrophysical parameters, the EM fields satisfy the boundary conditions consistent with the tangent EM field components being continuous at specified surfaces of discontinuity [23,24,29,34]. For PEC objects, the components of EM field are equal to zero inside the object. The tangent component of electric vector at the PEC object surface also equals to zero. Besides the boundary conditions, the EM fields considered in this chapter satisfy the radiation conditions at infinite distance [23,24,29,34].

This chapter will develop the calculation methods for predicting the EM fields ($\vec{E}$, $\vec{H}$) scattered by resonant size objects in homogeneous (free) space, as well as by objects located in dielectric half-space given various parameters of radar sounding including polarization, spatial, and time–frequency parameters of sounding signal, for monostatic and bistatic cases. These methods are based on solution of boundary IEs in frequency domain.

5.3 Calculation Methods for Obtaining the Scattering Characteristics of Aerial and Subsurface Perfectly Electrically Conducting Resonant Objects Based on a Magnetic Field Integral Equation Solution in the Frequency Domain

The boundary IE method stipulates the calculation of current densities at the object's surface. Having calculated the current densities, the resulting scattered EM field can be calculated by means of Kirchhoff type integral [34]. With regard to PEC objects in free space, various methods for finding electric surface current density exist based upon the following: (1) magnetic field integral equation (MFIE) [2,3,23–27,29–37], (2) electric field integral equation (EFIE) [23,24,28,30,33,34,38], and (3) linear combination of these equations called as combined field integral equation (CFIE) [23,28,29,33,34,38]. IEs are usually solved by the method of moments (MoM) [14,25–29,32,34,35,38,39] where the RWG (Rao–Wilton–Glisson) [25,28,29,34] or other [27,32,33,35] functions are used as basis and testing ones.

Analysis of the literature shows the need to develop methods based on the IE solution and designed for computing scattering characteristics of complex-shaped objects. For this, we consider the aerial objects of different electrical size elements including both smooth parts (fuselage, turbines) and parts with surface fractures, edges, or surfaces with small curvature radii (wings and fins of flying vehicles). In this case, one should be concerned with some peculiarities listed below when developing numerical algorithms.

MFIE and EFIE can have ambiguous solutions at some frequencies caused by natural oscillations inside PEC object [23,24,29]. Such solutions are spurious and they are consequence of imperfections in the mathematical methodology. They can be eliminated by means of applying additional condition (commonly used for MFIE solution), according to which the EM field inside the object is assumed equal to zero [24,29,36]. A drawback of such approach is the necessity of generating additional nodes inside the PEC object.

Another approach to eliminating ambiguous solutions is the use of CFIE [23,28,29,33,34,38]. Such an IE solution does not need any increase in IE dimension, but increases the calculation burden, and requires accounting for peculiarities of both the MFIE and EFIE approaches. At the same time, methods based on the CFIE solution do not solve the problems connected with calculation of scattering characteristics of complex-shaped objects containing either relatively smooth parts, or electrically thin parts with edges.

As a rule, spurious solutions mentioned in the two previous paragraphs appear for volume elements such as fuselage or turbine of flying vehicles. For electrically thin elements, the ambiguous solutions are not usually observed.

At the same time, discretization of IE for a surface containing both volume and electrically thin elements leads to the necessity of solving the large ill-conditioned linear systems of equations (LSE), which reduces the solution accuracy.

Iterative methods for solving IE can eliminate the disadvantage mentioned above [24,29,34]. The known iterative methods operate with full matrix of IE kernels at each iteration, so they are very demanding to computer storage space. If the matrix does not fit into computer RAM, then the computation process involves virtual memory characterized by slow access time. As a result, the calculation time considerably increases due to large number of matrix-to-vector products. The known multilevel fast multipole algorithm (MLFMA) [27–30,34] eliminates this advantage. However, its application is effective in case of sufficiently electrically large objects where application of AHFM [17,19,20], devoid of IE method disadvantages, is more preferable.

It should be noted that the scattering by separate elements of the object surface (either volume parts, or electrically thin parts) can be reduced to an effectively solved LSEs.

This section will present an original calculation method for modeling the scattering characteristics of aerial resonant-sized complex-shaped objects with PEC surfaces [37]. The object can include both smooth volume elements and electrically thin parts with edges. The method developed is based upon MFIE solution. Subsection 5.3.1 proposes the algorithm for solving the MFIE. The algorithm stipulates the object surface discretization with intervals depending on curvature radii of surface elements. For parts with smaller curvature radii, the discretization interval is chosen smaller than that for smooth parts. As a result, we avoid excessive increase of MFIE dimension. The proposed method of placing nodes for the current density together with integration algorithm provides a way to calculate the current density on the surface of electrically thin parts with sufficiently high accuracy. The algorithm also presumes elimination of ambiguous solutions caused by natural oscillations inside PEC object using an additional condition which assumes a zero EM field inside the scatterer. The method proposes a way to generate additional nodes inside the object and determines their sufficient number.

To verify the algorithm accuracy in Subsection 5.3.2, we compared the calculated scattering characteristics of simple objects with those obtained in physical experiments as well as with predictions obtained by other numerical methods.

Further, in Subsection 5.3.3, we propose an iterative MFIE solution method for complex-shaped objects [37]. In contrast to known procedures, the iterative method developed here operates at each iteration with matrices of significantly smaller dimension than complete matrix for whole surface. The method stipulates the decomposition of aerial object surface into parts including volume elements with large curvature radii (smooth elements, such as fuselage, turbine), and into the part composed of electrically thin elements comprising surface parts with small curvature radii (wings and fins). This way is more effective when compared to known iterative methods [29], since it reduces the calculation burden. The gain in computation efficiency is even more appreciable when calculations involve the right-hand side vector of large dimension (for instance, great number of discrete sounding aspect angles). Additionally, it is known that MFIE (inhomogeneous Fredholm IE) has better convergence of iterative algorithm than EFIE (homogeneous Fredholm IE) [29,33], which additionally recommends choosing MFIE for calculating the scattering characteristics of resonant size complex-shaped radar objects.

This developed calculation method provides a simulation of scattering characteristics of missiles, UAVs, and small airplanes in VHF band [37].

In Subsection 5.3.4, the MFIE solution method proposed in Section 5.3.1 is generalized for the case of subsurface PEC objects (mines and other explosive objects buried in a ground) [16,40,41].

The proposed methods provide ways to compute the frequency scattering characteristics for aerial and subsurface objects for given specified polarization, spatial, and time-frequency parameters of the sounding signal.

5.3.1 A Method for Solving the Magnetic Field Integral Equation for a Perfectly Electrically Conducting Object in Free Space

We can find the MFIE for electric current density at the surface S of PEC object by applying the Lorentz reciprocity theorem [24,34] to the desired EM field $(\vec{E}, \vec{H})$, and to a field of an auxiliary point magnetic source $(\vec{E}_1^m, \vec{H}_1^m)$ in free space V_1 (Figure 5.1a) [2,3,31,36,37]. We can introduce the equation under consideration as the system of two scalar equations as follows:

$$\begin{cases} \vec{\tau}_2^0 \cdot \vec{J}^e(\vec{Q}_0)(1+\Omega_{s_0}(\vec{Q}_0)) + \dfrac{2}{i\omega}\displaystyle\int\limits_{S\backslash s_0} \vec{E}_1^m\left(\vec{Q}\middle|\vec{Q}_0,\,\vec{\tau}_1^0\right)\cdot\vec{J}^e\left(\vec{Q}\right)ds_Q = 2\vec{\tau}_1^0\cdot\vec{H}^0\left(\vec{Q}_0\right), \\[2mm] -\vec{\tau}_1^0 \cdot \vec{J}^e(\vec{Q}_0)(1+\Omega_{s_0}(\vec{Q}_0)) + \dfrac{2}{i\omega}\displaystyle\int\limits_{S\backslash s_0} \vec{E}_1^m\left(\vec{Q}\middle|\vec{Q}_0,\,\vec{\tau}_2^0\right)\cdot\vec{J}^e\left(\vec{Q}\right)ds_Q = 2\vec{\tau}_2^0\cdot\vec{H}^0\left(\vec{Q}_0\right), \end{cases} \tag{5.1}$$

where:

$\vec{Q},\,\vec{Q}_0 \in S$ are the integration and observation points, respectively;

$\vec{\tau}_1^0$ and $\vec{\tau}_2^0$ are mutually orthogonal unit vectors tangential to the surface S at point $\vec{Q}_0$ to form the right-hand system with interior unit normal $\vec{v}^0$ at this point;

$\vec{J}^e(\vec{Q}) = \vec{n}\times\vec{H}(\vec{Q})$ is the desired surface electric current density;

$\vec{H}^0$ and $\vec{H}$ are magnetic vectors of the initial and complete EM fields, respectively;

$\vec{E}_1^m\left(\vec{Q}\middle|\vec{Q}_0,\,\vec{\tau}_{1(2)}^0\right) = -i\omega(\vec{\tau}_{1(2)}^0\times\vec{\nabla}G(\vec{Q}_0,\,\vec{Q}))$ is the electric vector of the point magnetic source;

$G(\vec{Q}_0,\,\vec{Q}) = G(R) = \exp\left(ik_1R\right)/4\pi R,\ R = \left|\vec{Q}-\vec{Q}_0\right|$;

$k_1 = 2\pi/\lambda_1$ and λ_1 are the wavenumber and the wave length in free space V_1, respectively.

We can carry out the integration in Equations 5.1 over the object's surface excluding the singular point neighborhood s_0 placed inside the sphere with radius $\rho_0 \ll \lambda_1$, and centered at point $\vec{Q}_0$. At the small element s_0, we can assume that $\vec{J}^e(\vec{Q}) \approx \vec{J}^e(\vec{Q}_0)$. In this case, we can approximately analytically calculate the integral over s_0 as [2,3,31,36,37] follows:

$$I_{s_0} \approx \vec{\tau}_{2(1)}^0 \cdot \vec{J}^e\left(\vec{Q}_0\right)\Omega_{s_0}\left(\vec{Q}_0\right), \tag{5.2}$$

where

$$\Omega_{s_0}\left(\vec{Q}_0\right) = \frac{c_{11} + c_{22}}{2ik_1}\left[\left(ik_1\rho_0 - 2\right)\exp\left(ik_1\rho_0\right) + 2\right], \tag{5.3}$$

and

$$c_{1(2)} = \frac{\partial^2 F\left(\vec{Q}(u_1,\ u_2)\right)}{\partial u_{1(2)}^2}\left(\vec{\tau}_{1(2)}^0\right)^2 \Bigg/ \left[2\left|\nabla F\left(\vec{Q}\right)\right|_{\vec{Q}=\vec{Q}_0}\right]. \tag{5.4}$$

Here $\left(u_1,\ u_2\right)$ is the local rectangular coordinate system centered at point $\vec{Q}_0$, with its axes directed along vectors $\vec{\tau}_1^0,\ \vec{\tau}_2^0$, respectively; $F(\vec{Q}(u_1,\ u_2)) = 0$ is the equation of surface S in the vicinity of the point $\vec{Q}_0$.

To eliminate ambiguous solutions of the MFIE caused by natural oscillations inside the PEC object, the system (Equations 5.1) can be supplemented with the equations in which the EM field inside the object is assumed equal to zero [24,29,36,37]. Such additional equations can be represented as [36,37] follows:

$$\begin{cases} \dfrac{1}{i\omega}\displaystyle\int_S \vec{E}_1^m\left(\vec{Q}\big|\vec{Q}_0,\ \vec{\tau}_1^0\right)\cdot\vec{J}^e\left(\vec{Q}\right)ds_Q = \vec{\tau}_1^0\cdot\vec{H}^0\left(\vec{Q}_0\right), \\[4mm] \dfrac{1}{i\omega}\displaystyle\int_S \vec{E}_1^m\left(\vec{Q}\big|\vec{Q}_0,\ \vec{\tau}_1^0\right)\cdot\vec{J}^e\left(\vec{Q}\right)ds_Q = \vec{\tau}_2^0\cdot\vec{H}^0\left(\vec{Q}_0\right), \end{cases} \tag{5.5}$$

where $\vec{Q}_0 \in V_1 \setminus S$.

The desired current density can be determined by solving the overdetermined system of equations (ODSE) formed after the discretization of Equations 5.1 and 5.5. The least squares method provides a common solution method for solving the system of equations. After solving the mentioned system, we can represent the magnetic component of EM field scattered by PEC resonant object [2,3,31,36,37] as follows:

$$i\omega\vec{p}^{rec}\cdot\left(\vec{H}\left(\vec{Q}_{rec}\right) - \vec{H}^0\left(\vec{Q}_{rec}\right)\right) = -\int_S \vec{E}_1^m\left(\vec{Q}\big|\vec{Q}_0,\vec{p}^{rec}\right)\cdot\vec{J}^e\left(\vec{Q}\right)ds_Q, \tag{5.6}$$

where $\vec{Q}_{rec} \notin V_2$ and $\vec{p}^{rec}$ is the unit vector, denoting orientation of magnetic vector of EM field scattered by the PEC object (Figure 5.1).

The algorithm for calculating the PEC object scattering characteristic includes the following four stages:

1. Create a scatterer surface model: This means approximating the object surface by parts of easily parameterized ellipsoids. This approach speeds and unifies the processes of surface model creation and calculation surface integral at the later stages of current density calculation and scattered field modeling. The authors have found that modeling the surface as ellipsoid parts is more efficient when compared to models constructed of surfaces of different types (sphere, cylinder, cone, plate, etc.). Further, we find it preferable for solving most problems concerning calculation of the real radar object scattering characteristics.

   After approximating the surface, we will relate constituent ellipsoid parts to a common rectangular coordinate system and will quantize them. At any point $\vec{Q}$, the parameters of the approximating ellipsoid (its semi-axes, center coordinates, and angles which determine its orientation in the space), the point coordinates, and components of two tangential vectors $\vec{\tau}_1$ and $\vec{\tau}_2$ are specified. For the best use of computer memory, we set the sampling interval for generation of current density nodes proportional to the radii of the surface S curvature. At segments with the smaller curvature radii, the sampling interval also decreases.

2. Form LSE: At this stage, model surface S is split into a set of N elements and assumes a constant current density at each. After introducing the indices of observation $n_0 = \overline{1, N}$ and integration $n = \overline{1, N}$ points, and after representing the desired current density in the form of two tangential components $J_{1(2),n} = \vec{\tau}_{1(2),n} \cdot \vec{J}_n$, the system of scalar equations (5.1) becomes:

$$
\begin{cases}
J^e_{2,n_0} A^{12}_{n_0,n_0} + \displaystyle\sum_{\substack{n=1 \\ (n \neq n_0)}}^{N} \left[J^e_{1,n} A^{11}_{n_0,n} + J^e_{2,n} A^{12}_{n_0,n} \right] = B_{1,n_0}, \\[2em]
J^e_{1,n_0} A^{21}_{n_0,n_0} + \displaystyle\sum_{\substack{n=1 \\ (n \neq n_0)}}^{N} \left[J^e_{1,n} A^{21}_{n_0,n} + J^e_{2,n} A^{22}_{n_0,n} \right] = B_{2,n_0},
\end{cases}
\tag{5.7}
$$

where

$$
A^{pq}_{n_0,n} =
\begin{cases}
(-1)^q + \Omega_{s_0}\left(\vec{Q}_{n_0}\right) + \dfrac{2}{i\omega} \displaystyle\int\limits_{s_{n_0} \backslash s_0} \left(\vec{\tau}_{q,n_0} \cdot \vec{\mathcal{E}}^m_1\left(\vec{Q}\big|\vec{Q}_{n_0}, \vec{\tau}_{p,n_0}\right) \right) ds_Q, & \text{if } n = n_0, \\[2em]
\dfrac{2}{i\omega} \displaystyle\int\limits_{s_n} \left(\vec{\tau}_{q,n} \cdot \vec{\mathcal{E}}^m_1\left(\vec{Q}\big|\vec{Q}_{n_0}, \vec{\tau}_{p,n_0}\right) \right) ds_Q, & \text{if } n \neq n_0;
\end{cases}
\tag{5.8}
$$

$p = 1, 2; q = 1, 2;$

$$
B_{p,n_0} = 2\vec{\tau}_{p,n_0} \cdot \vec{H}^0\left(\vec{Q}_{n_0}\right).
\tag{5.9}
$$

Now we consider the elements of Equation 5.8 in more detail. To single out the singularity and compute elements $A^{pq}_{n_0,n}$, we apply the following algorithm [36,37]. Each elementary area s_{n_0} containing the singular point is divided again into N_0 cells $s_{n_0} = \sum_{g=1}^{N_0} s'_g$ as shown in Figure 5.2. Then, the distances ρ_{0g} between s_{n0} center ($\vec{Q}_{n_0}$) as well as the centers of s'_g elements are determined, and the condition $k_1\rho_{0g} > \gamma_0$ (Figure 5.2) is checked. Those elements, for which condition is held, do not have any singularity and they constitute the part of s_{n0}, over which the function $\vec{\tau}_{q,n_0} \cdot \vec{\mathcal{E}}^m_1\left(\vec{Q}\big|\vec{Q}_{n_0}, \vec{\tau}_{p,n_0}\right)$ is integrated numerically. The elements s'_g, for which $k_0\rho_{0g} \leq \gamma_0$, are excluded from numerical algorithm and the integral over surface

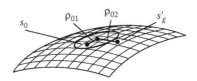

FIGURE 5.2
Integration over surface element containing singular point ($k_1\rho_{01} < \gamma_0, k_1\rho_{02} > \gamma_0$).

they lie at is replaced with the value $\Omega_{s_0}\left(\vec{Q}_{n_0}\right)$ calculated using (5.3). The value $\gamma_0 = 0.01$ was determined experimentally.

If the element of the surface s_0 in Figure 5.2 is rather flat, the value $\Omega_{s_0}\left(\vec{Q}_{n_0}\right)$ tends to zero and can be neglected. Besides, if elements s_n of the surface are small, then $\left|A_{n_0,\,n_0}^{pq}\right| \approx 1$.

Then calculate the integrals in Equation 5.8 by means of the five-point Gauss formula.

3. Eliminate the ambiguous solutions caused by natural oscillations inside the PEC object: The mathematical modeling results for PEC objects show that if the object's three-dimensional electrical sizes a_j $(j = \overline{1,\ 3})$ are less than $0.6\lambda_1$, then, as a rule, MFIE do not produce ambiguous solutions. The LSE shown in Equation 5.7 will solve for the surface electric current density in a matrix format as follows:

$$\mathbf{A}\vec{J}^e = \vec{B}. \tag{5.10}$$

In another case, if $\min(a_j) \geq 0.6\lambda_1$ (in particular, this condition is carried out for fuselage of flying vehicle), then the MFIE can have ambiguous solutions caused by natural oscillations of the PEC object cavity. As it was noted above, for elimination of previously mentioned spurious solutions, we can apply the additional Equations 5.5 represented similarly to Equations 5.7 which produces:

$$\begin{cases} \sum_{n=1}^{N}\left[J_{1,n}^e A_{n01,n}^{11} + J_{2,n}^e A_{n01,n}^{12} \right] = B_{1,n01}, \\ \sum_{n=1}^{N}\left[J_{1,n}^e A_{n01,n}^{21} + J_{2,n}^e A_{n01,n}^{22} \right] = B_{2,n01}, \end{cases} \tag{5.11}$$

where

$$A_{n01,n}^{pq} = \frac{1}{i\omega}\int_{S_n}\left(\vec{\tau}_{q,n01}\cdot\vec{E}_1^m\left(\vec{Q}\big|\vec{Q}_{n01},\ \vec{\tau}_{p,n01}\right)\right)ds_Q, \tag{5.12}$$

$p = 1, 2; q = 1, 2;$

$$B_{p,n01} = \vec{\tau}_{p,n01}\cdot\vec{H}^0\left(\vec{Q}_{n01}\right), \tag{5.13}$$

where $\vec{Q}_{n01}$ are additional points inside the object V_2;

$$n_{01} = \overline{N+1,\ N+1+N_1};$$

N_1 is the number of additional points inside the object.

Integrals such as Equation 5.12, as like those of Equation 5.8 above, are calculated by means of five-point Gauss formula.

Uniting Equations 5.7 and 5.11 forms the ODSE of $(2(N+N_1)\times 2N)$ dimension. To obtain the surface electric current density, then solve the system shown in Equation 5.10 with the use of least-squares method.

4. Calculation of the EM field scattered by the PEC object: Ultimately, we need to model the object scattering characteristics in both the frequency and time domains

(see the discussion of time-and-frequency domain signal processing by Barrett in Chapter 4.) To determine the magnetic component of the EM field scattered by the object at a fixed frequency, substitute the current density (calculated for given sounding conditions) into integral representation shown in Equation 5.6, which, after surface integral sampling, appears as follows:

$$\vec{p}^{rec} \cdot \vec{H}\left(\vec{Q}_{rec}\right) = \left(\vec{p}^{rec} \cdot \vec{H}^0\left(\vec{Q}_{rec}\right)\right) -$$

$$-\frac{1}{i\omega} \sum_{n=1}^{N} \left(\left(\vec{\tau}_{1,n} \cdot \vec{\mathcal{E}}_1^m\left(\vec{Q}_n \middle| \vec{Q}_{rec}, \ \vec{p}^{rec}\right)\right) J_{1,n}^e + \left(\vec{\tau}_{2,n} \cdot \vec{\mathcal{E}}_1^m\left(\vec{Q}_n \middle| \vec{Q}_{rec}, \ \vec{p}^{rec}\right)\right) J_{2,n}^e \right) \Delta s_n. \tag{5.14}$$

To obtain the UWB impulse response, it is necessary to apply the inverse Fourier transform to the determined using the Equation 5.14 frequency function. In particular, for calculations carried out in this work, we used the algorithm described below in Section 5.5.

For calculations using the method developed here, chose the number of nodes N, for calculating the electric current density from condition of guaranteeing the algorithm intrinsic convergence characterized by the value of δ_a. In this chapter, the intrinsic convergence of algorithm, we consider the property of the resulting object RCS value σ_N to tend to the fixed value σ_{N_c} given increasing N:

$$\lim_{N \to \infty} \sigma_N = \sigma_{N_c}, \tag{5.15}$$

and starting from some value $N = N_a$, satisfies the next condition as follows:

$$\frac{\left|\sigma_{N_c} - \sigma_{N_a}\right|}{\sigma_{N_c}} \cdot 100\% \leq \delta_a, \tag{5.16}$$

where σ_{N_a} and σ_{N_c} are the RCS values, calculated for the number of nodes N_a and N_c, respectively, and $N_c > N_a$.

In the calculations carried out in this chapter, we chose the number of nodes N_a to correspond to $\delta_a = 1\%$.

The calculations showed that to guarantee the given value of δ_a, the intervals between the nodes of current density at relatively smooth parts of the object's surface should be set as $r_a = (0.09\ldots0.24)\lambda_1$. In the neighborhoods of edges, or at the parts with small radii of curvature, then the calculations should sample the surface at smaller intervals. If the surface has edges (discontinuities), then we perform a smoothing approximation. For example, we represented the base of cylinder as part of thin ellipsoid with minor semi-axis $a_e = 10^{-5}\lambda_1$ as shown in Figure. 5.3. This approximation realized a smooth junction between the lateral surface and base of cylinder.

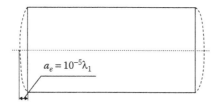

FIGURE 5.3
Example of the smoothing approximation of the cylinder edge.

Near the cylinder edge, we chose the small neighborhood $0.01\lambda_1 < \rho < 0.2\lambda_1$ (where ρ is the distance to the edge). This sets the sampling interval in this neighborhood as $r_a = 0.03\lambda_1$. The sampling interval r_a has the same value at the elements with small radii of curvature and does not generate current density nodes in the neighborhood of edge, where $\rho \leq 0.01\lambda_1$. This approach correctly accounts for the current increase near the edge. Figure 5.4a, shows the cross section in the plane formed by wave vector and $\vec{H}^0$ of calculated electric current density component $J^{e\perp}$ (which is orthogonal to the cylinder axis) at lateral surface of cylinder with radius $a_c = 2.2\lambda_1$ and length $d_c = 0.933\lambda_1$. The next Figure 5.4b shows the orientation of vectors. Figure 5.4a also shows the same current density approximated by $\upsilon_1\rho^{\upsilon_2}$ function. The least squares method determined the coefficients $\upsilon_1 = 0.513$ and $\upsilon_2 = -0.292$.

The current density at the edge from the side of cylinder's base behaves similarly with $\upsilon_1 = 0.554$, $\upsilon_2 = -0.29$. The obtained results matched well with the theory, according to which the current near the wedge with interior angle of $\pi/2$ meets the Meixner condition, and increases proportionally to $\rho^{-1/3}$ [50,51]. It is important to note that the algorithm developed here does not require any additional conditions at the edges as it was done in [51] for example.

To eliminate the ambiguous solutions caused by natural oscillations, we determined the number of necessary additional nodes from the same considerations needed to ensure the value $\delta_a = 1\%$ from a formula similar to Equation 5.16.

On the basis of the numerical results, we determined that we should generate additional points inside the object within the volume limited by analytic continuation of the object surface, but spaced by $\approx 0.1\lambda_1$ from it using the Halton series [52], or placed directly on to the mentioned analytic continuation. The number of entire auxiliary points must constitute 10...15% of the number of points used to calculate the surface electric current density [36,37].

To investigate the accuracy and applicability boundaries of the method developed, we computed the RCS of simple shape PEC objects. The results were compared with the physical experiment data and with the results by other known numerical methods.

5.3.2 Verification of the Calculated Scattering Characteristics of Simple Shape Perfectly Electrically Conducting Objects

To verify the calculated scattering results, we used canonical test objects including the PEC sphere, cylinder, ellipsoid, electrically thin disc, and cube. Our calculation compared the RCS of the objects versus their electrical size or aspect angle for monostatic and multistatic radar measurements.

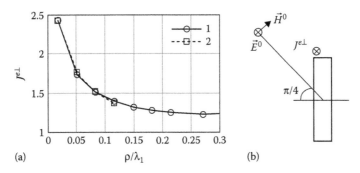

(a)   (b)

FIGURE 5.4
Calculation of the electric current density in the neighborhood of a PEC cylinder's edge. (a) Data obtained (1) by solving the MFIE and (2) by an approximation with the function $\upsilon_1\rho^{\upsilon_2}$ and (b) orientation of electric and magnetic vectors of incident EM field and current density component.

Figure 5.5 shows the orientation and parameters of the objects used in our calculations. The ellipsoid half-axes $a_{e\,x}, a_{e\,y}, a_{e\,z}$ are parallel to the corresponding coordinate axes as shown in Figure 5.5a. We aligned the cylinder axis with the Ox coordinate axis as shown in Figure 5.5b. The monostatic and multistatic RCS in E- and H-planes were determined by similar methods that are described in [23,29,53–55]. In the monostatic radar case, the RCS in E-plane corresponds to the vector orientation of $\left(\vec{H}^0 \,\|\vec{H}\| \, xOy\right)$; in the H-plane RCS, the vector is oriented as $\left(\vec{H}^0 \,\|\vec{H}\| \perp xOy\right)$. In multistatic cases, the reverse situation applies and the RCS in the E-plane corresponds to the vector orientation of $\left(\vec{H}^0 \,\|\vec{H}\| \perp xOy\right)$; and in H-plane, it corresponds to $\left(\vec{H}^0 \,\|\vec{H}\| \, xOy\right)$. The monostatic radar measurement aspect angle β is measured from Ox axis in xOy plane as shown in Figure 5.5.

When comparing the different methods, we analyzed the relative error value as follows:

$$\delta_{met}(f,\beta) = \frac{\left|\sigma_1(f,\beta) - \sigma_2(f,\beta)\right|}{\sigma_2(f,\beta)} \cdot 100\%, \tag{5.17}$$

where $\sigma_1(f,\beta)$ is the test object RCS given fixed frequency and aspect angle, obtained using the method developed here, the latter being based on the MFIE solution;

$\sigma_2(f,\beta)$ is the test object RCS obtained using other numerical methods, or from physical experiments.

The analogous function $\delta_{met}(f,\beta_1)$ is analyzed when considering the bistatic RCS.

Using the method developed here, we computed the monostatic and bistatic RCS σ_s of a PEC sphere of radius a_s. Results are compared with the corresponding RCS obtained by the Mie series [29,53]. In calculations using the Mie series, we retained 50 summands. Figure 5.6 shows the calculation results of the PEC sphere monostatic RCS (normalized to πa_s^2) versus its electrical size $k_1 a_s$ as computed by our method and by the Mie series. Figure 5.7 shows bistatic RCS versus bistatic angle β_1. The RCS is shown in dB ($10\log(\sigma_s/\pi a_s^2)$) for fixed value $k_1 a_s = 2\pi$. The difference between results calculated using our method and the Mie series is not higher than $\sigma_{met} = 1\%$.

Let us now consider the peculiarities of EM wave scattering by a PEC cylinder and ellipsoid of resonant sizes. These scatterers make interesting test objects because their surfaces include both smooth parts and elements with small curvature radii.

The results shown in Figures 5.8 through 5.11 demonstrate the normalized monostatic RCS in dB ($10\log(\sigma_c/\pi a_c^2)$) of PEC cylinders with various radii a_c and heights d_c.

Figure 5.8 shows the RCS of PEC cylinder with $a_c = 0.216\lambda_1$, $d_c = 2.76\lambda_1$ calculated using our method based on solving the MFIE, and experimental data [23]; as well as results calculated using FEKO™ (Suite 5.5) EM simulation software [38]. This software allows the user to select from different numerical methods. In this chapter, we chose the methods based on solving the EFIE and CFIE for the FEKO™ simulation. For the considered cylinder, we did not observe any ambiguous IE solutions during the RCS calculations.

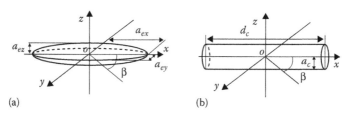

(a)   (b)

FIGURE 5.5
Orientation of scatterers and their parameters: (a) ellipsoid and (b) cylinder.

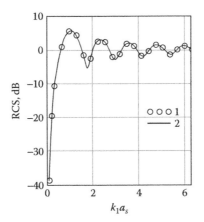

FIGURE 5.6
A comparison of the calculated monostatic RCS of a PEC sphere versus electrical size: (1) Results obtained using the algorithm developed here. (2) Results obtained using the Mie series.

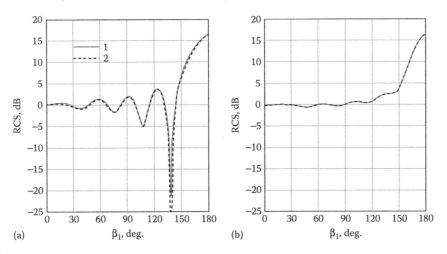

FIGURE 5.7
A comparison of the calculated bistatic RCS of a PEC sphere plotted against the bistatic angle β_1 with $k_1 a_s = 2\pi$ in E-plane (a) and H-plane (b), calculated (1) using the method developed here and (2) using the Mie series.

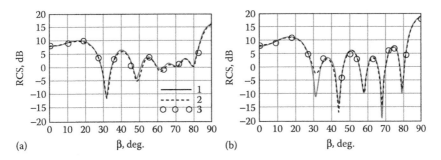

FIGURE 5.8
A comparison of the monostatic RCS of a PEC cylinder with $a_c = 0.216\lambda_1$ and $d_c = 2.76\lambda_1$ versus the aspect angle β in (a) the E-plane and (b) the H-plane. Results were calculated using (1) the developed method (MFIE, LSE); (2) the FEKO™ software (EFIE); and (3) physical experimental data. (From Mittra, R., *Computer Techniques for Electromagnetics*, 1973. With permission.)

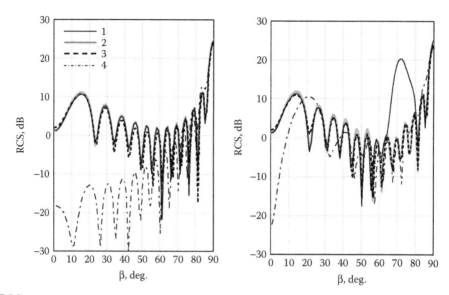

FIGURE 5.9

A comparison of the calculated monostatic cylinder and ellipsoid (Figure 5.5) RCS versus the aspect angle β in (a) the E-plane and (b) the H-plane. Traces (1–3) show PEC cylinder ($a_c = 0.31\lambda_1$, and $= 6.32\lambda_1$) RCS. Traces (4) show ellipsoid ($a_{ex} = 3.16\lambda_1, a_{ey} = 0.2\lambda_1,$ and $a_{ez} = 0.31\lambda_1$) RCS. The results were obtained using the developed method (1) MFIE (LSE); (2, 4) MFIE (ODSE); and (3) the FEKO™ software (CFIE).

Figure 5.9 shows the RCS of a PEC cylinder with $a_c = 0.31\lambda_1$ and $d_c = 6.32\lambda_1$, calculated using the developed method and FEKO™ software. In this case, the ambiguous MFIE solutions led to significant RCS errors for the aspect angles of 60...80 degrees in the H-plane. The curves in this figure show the data obtained by solving LSE and ODSE (the influence of ambiguous solutions was eliminated). In this figure, the RCS ($10 \log(\sigma_e / \pi a_{ez}^2)$) of the triaxial ellipsoid with semi-axes $a_{ex} = 3.16\lambda_1$, $a_{ey} = 0.2\lambda_1$ of Figure 5.5a is also shown for reference.

Figure 5.10 shows the RCS of a PEC cylinder with $a_c = 0.5\lambda_1$ and $d_c = 5\lambda_1$ calculated using the developed method (MFIE, ODSE) and FEKO™ software (CFIE) together with the physical experiment results [54]. The cylinder under consideration has a relatively large electrical size in both radii and height. Therefore, Figure 5.10 also presents the results obtained using the AHFM [17,19,20]. This latter method was developed for calculating the scattering characteristics of radar objects of large electrical sizes. It presumes the resulted scattered EM field presented in the form of sum of fields scattered by smooth parts of the PEC object surface and its edges.

Figure 5.11 shows the RCS of a PEC cylinder, close to the edge of which the current density was investigated in Subsection 5.3.1 and shown in Figure 5.4. The radius of cylinder exceeds by more than twice its height ($a_c = 2.2\lambda_1$ and $d_c = 0.933\lambda_1$), so the edges have a significant influence on the calculated RCS.

As shown in Figures 5.8 through 5.11, the results obtained using the developed method coincide well with the physical measurement results and with the data obtained by other methods. It should be noted that the difference between data obtained by the developed method and FEKO™ software shown in Figures 5.8 through 5.11 does not exceed $\delta_{met} = 1...3\%$ (except some deep notches in scattering diagrams). The calculation results presented in Figure 5.8 coincide well with the physical experiment data. A more appreciable difference between the experimental and calculated data is observed, given the sounding

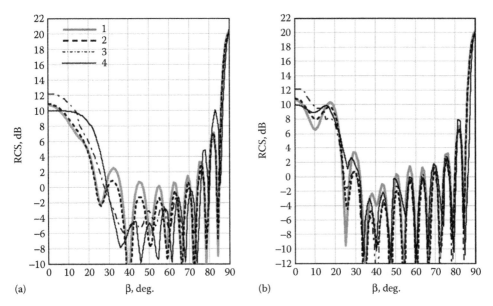

FIGURE 5.10
A comparison of the monostatic RCS of a PEC cylinder ($a_c = 0.5\lambda_1$ and $d_c = 5\lambda_1$) versus the aspect angle β in (a) the *E*-plane and (b) the *H*-plane. Results were calculated using (1) the developed method (MFIE, ODSE); (2) the FEKO™ software (FIE); (3) the AHFM; and (4) the data of physical experiment measurements. (From Ufimtsev, P.Y. *Method of Edge Waves in Physical Theory of Diffraction*, 1962. With permission.)

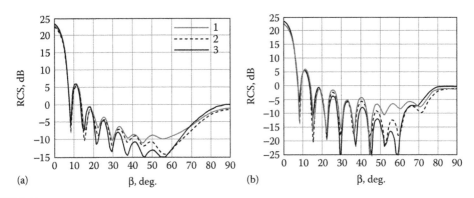

FIGURE 5.11
A comparison of the calculated monostatic RCS of PEC cylinder ($a_c = 2.2\lambda_1$ and $d_c = 0.933\lambda_1$) versus aspect angle β in (a) the *E*-plane and (b) the *H*-plane. Results were calculated using the developed method (1) the MFIE, ODSE; (2) the FEKO™ software (CFIE); and (3) the AHFM.

direction deviating from that perpendicular to the cylinder axis as shown in Figure 5.10. Notice the greater difference in the *E*-plane. At the same time, the qualitative behavior of the calculation results agrees well with experimental data. It is clear that the AHFM provides the best accuracy for such scatterer aspect angles, at which the visible electrical sizes of object are relatively large. In Figure 5.10 ($a_c = 0.5\lambda_1$, $d_c = 5\lambda_1$), these sounding directions correspond to approximately 90 degrees with respect to cylinder axis; and in Figure 5.11, ($a_c = 2.2\lambda_1$, $d_c = 0.933\lambda_1$) such directions correspond to sounding along the cylinder axis.

Figure 5.12 shows the monostatic RCS of an electrically thin PEC disc with a radius $a_d = 0.239\lambda_1$ and thickness $d_d = 0.02\lambda_1$ for the two planes (normalized RCS $10 \log(\sigma_d/\pi a_d^2)$

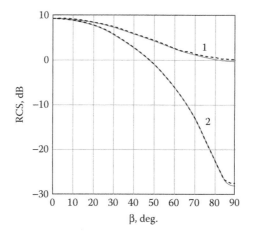

FIGURE 5.12

A comparison of the calculated monostatic RCS of a PEC disk ($a_d = 0.239\lambda_1$ and $d_d = 0.02\lambda_1$) versus the aspect angle β. Trace (1) shows the E-plane and (2) the H-plane. The results were calculated using the developed method (MFIE, LSE, solid lines) and FEKO™ software (EFIE, dashed lines).

is shown in dB). The relative error between data obtained by the developed method and FEKO™ software does not exceed $\delta_{met} = 1\ldots3\%$.

Figure 5.13 compares the RCS of an electrically thin disk approximated by the two-axial ellipsoid ($a_{ey} = a_{ez}$, $a_{ex}/a_{ey} = 0.05$) with the electrical size $k_1 a_{ey}$, as shown in Figure 5.5a calculated by the developed method, and determined for an infinitely thin PEC disc ($a_d = a_{ey} = a_{ez}$) [53]. These plots show the RCS expressed in dB as $10 \log\left(\sigma_e / \pi a_{ey}^2\right)$. The small thickness $d_d = 0.02\lambda_1$, for which the developed method provides good accuracy, demonstrates a way to simulate and predict radar EM scattering by real objects.

Calculating the bistatic RCS of a PEC cube with the edge length $a_{cube} = 3.015\lambda_1$ produced the results shown in Figure 5.14. The plots correspond to the sounding direction perpendicular to the cube face $(\beta = 0)$ and along the Ox axis. The bistatic angle β_1 is measured in the xOy plane. In this case, we normalized the cube RCS to its maximum value corresponding

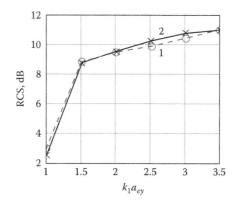

FIGURE 5.13

The calculated monostatic RCS of a two-axial PEC ellipsoid (1) with ratio of minor and major semi-axes of $a_{ex}/a_{ey} = 0.05$ versus its electric size. (2) Rigorous calculations of the RCS of an infinitely thin PEC disk ($a_d = a_{ey} = a_{ez}$). (From King, R. et al., *The Scattering and Diffraction of Waves*, 1959. With permission.)

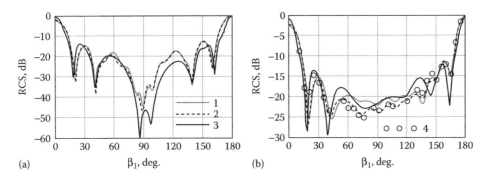

FIGURE 5.14
A comparison of the bistatic RCS of a PEC cube ($a_{cube} = 3.105\lambda_1$) versus the bistatic angle β_1 in (a) the *E*-plane and (b) the *H*-plane. These results were calculated using (1) the developed method (MFIE, ODSE); (2) the FEKO™ software (CFIE); (3) the AHFM; and (4) the physical experiment data. (From Penno, R.P. et al., Scattering from a perfectly conducting cube, *Proceedings of the IEEE*, 77, 815–823, 1989. © 1989 IEEE.)

to $\beta_1 = 180$ degrees. Comparing calculation results and experimental data [55] in *H*-plane shows a good correspondence.

Comparing the results of the RCS computed using the developed method against known data and other methods shows nearly identical results, and validates the developed method. The described methods show how to make precise calculations of the scattering characteristics of smooth objects or scattering surfaces with edges and breaks as well as electrically thin objects. This algorithm has applications for simulating the scattering characteristics of relatively large objects. This method works for radar signals up to the high-frequency region where the asymptotical methods work without the disadvantages of IE methods.

5.3.3 A Method for Calculating the Radar Scattering Characteristics of Complex-Shaped Aerial Perfectly Electrically Conducting Objects Using an Iterative Algorithm

Real aerial radar objects have a wide range of shapes including both relatively smooth elements (e.g., aircraft fuselage sections and turbine housings) and electrically thin parts with edges (wings and stabilizers). As considered above at the beginning of Section 5.3, the application of boundary IE method for calculating the scattering characteristics of complex-shaped objects means solving ill-conditioned and large dimension LSE. These cases use the known iterative methods of IE solution [24,29,34]. But their application requires a large computer memory with long computational times.

This Subsection proposes an iterative method which operates with matrices of significantly smaller dimension than a complete matrix for the whole surface S and works by splitting the surface S into two convenient parts $S = S_1 + S_2$ as follows:

1. For the surface S_1, the method uses a set of surfaces of smooth resonant size elements such as the fuselage and turbines. This allows the existence of ambiguous MFIE solutions caused by the natural oscillations inside the PEC object.

2. For the surface S_2, the method uses a set of electrically thin resonant size surfaces with edges (surface patches with small radii of curvature), such as wings, stabilizers, and fins to model the surface.

This splits the whole surface, we present the system of equations 5.10 as follows [37]:

$$\begin{cases} C_{11}\vec{J}_{s_1}^e + C_{12}\vec{J}_{s_2}^e = \vec{B}_{s_1}, \\ C_{21}\vec{J}_{s_1}^e + C_{22}\vec{J}_{s_2}^e = \vec{B}_{s_2}, \end{cases} \tag{5.18}$$

where $C_{\vartheta\varsigma}$ ($\vartheta = 1, 2$; $\varsigma = 1, 2$) are rectangular submatrices (in the general case), which together constitute the original matrix in 5.10 resulting in the following:

$$\mathbf{A} = \begin{pmatrix} C_{11} & C_{12} \\ C_{21} & C_{22} \end{pmatrix}. \tag{5.19}$$

In this case, $\vec{J}_{s_1}^e$ and $\vec{J}_{s_2}^e$ indicate the current densities at S_1 and S_2, respectively; $\vec{B}_{s_1}$ and $\vec{B}_{s_2}$, in the same manner, correspond to a location of observation point at S_1 and S_2. The elements of matrices $C_{\vartheta\varsigma}$ and the right-hand vectors $\vec{B}_{s_\vartheta}$, as before in Subsection 5.3.1, are calculated using the Equations 5.7 through 5.9 and Equations 5.11 through 5.13.

We can solve the equation system 5.18 through an iterative process [37] as follows:

$$\begin{cases} C_{11}\vec{J}_{s_1}^{e(\xi)} = \vec{B}_{s_1} - C_{12}\vec{J}_{s_2}^{e(\xi-1)}, \\ C_{22}\vec{J}_{s_2}^{e(\xi)} = \vec{B}_{s_2} - C_{21}\vec{J}_{s_1}^{e(\xi-1)}, \end{cases} \tag{5.20}$$

where $\xi = \overline{1, \xi_{st}}$ is the iteration index; ξ_{st} is the number of iterations needed for obtaining the current density steady-state values.

As an initial approximation for the surface current density $\vec{J}_{s_\vartheta}^{e(0)}$, we used the following:

$$\vec{J}_{s_\vartheta}^{e(0)} = \vec{J}_{s_\vartheta}^{e(0)}\left(\vec{Q}_{n_0}\right) = 2\left(\vec{v}_0 \times \vec{H}^0\left(\vec{Q}_{n_0}\right)\right). \tag{5.21}$$

The actual value of ξ_{st} depends on the calculation accuracy requirements and can be determined during the iterative process.

If an aerial object contains the elements, which allow the presence of spurious solutions described above in the introduction to Section 5.3, you can use the least-squares method to solve the corresponding system Equations 5.20 that includes the rectangular matrices $C_{\vartheta\varsigma}$. Otherwise, we can write solution to the system Equations 5.20 as follows:

$$\begin{cases} \vec{J}_{s_1}^{e(\xi)} = C_{11}^{-1}\left(\vec{B}_{s_1} - C_{12}\vec{J}_{s_2}^{e(\xi-1)}\right), \\ \vec{J}_{s_2}^{e(\xi)} = C_{22}^{-1}\left(\vec{B}_{s_2} - C_{21}\vec{J}_{s_1}^{e(\xi-1)}\right). \end{cases} \tag{5.22}$$

We can split the surface S in more than two parts, however this significantly complicates the algorithm.

In this manner, the numerical algorithm based on Equations 5.7 through 5.13 and 5.18 through 5.22 allow us to reduce the problem of determining the electric current density at the complex-shaped resonant size object's surface to solving the equation systems for the parts S_1 and S_2 separately. These latter equation systems have sufficiently smaller dimensions as compared with the full system corresponding to the whole surface S. The interaction between S_1 and S_2 is taken into account during iterations.

As before, the magnetic vector of the EM field scattered by a complex-shaped object can be obtained after substituting the electric current density thus determined into the Equation 5.14.

To analyze the accuracy and convergence of the developed algorithm, we computed the test object RCS using different methods [37]. The test object as shown in Figure 5.15 consisted of two ellipsoids – the first had semi-axes $a_{e\,x1} = 3.16$ m, $a_{e\,y1} = 0.31$ m, and $a_{e\,z1} = 0.3$ m; and the second had $a_{e\,x2} = 0.22$ m, $a_{e\,y2} = 3$ m, and $a_{e\,z2} = 0.02$ m. The method assumes that a monochromatic EM wave with wavelength $\lambda_1 = 1$ m illuminated the object. Figure 5.16 shows the test object monostatic RCS versus azimuth aspect angle β measured off the xOy plane. The position $\beta = 0$ corresponds to illumination along the Ox axis, when given the elevation angle $\varepsilon = 3°$ degrees as shown in Figure 5.15. The results corresponding to the horizontal ($\vec{H}^0$ and $\vec{H}$ are oriented in the plane orthogonal to the wing plane xOy and contained the illumination direction vector) and the vertical ($\vec{H}^0$ and $\vec{H}$ are parallel to the wing plane) polarizations. In the iterative algorithm, we chose S_1 as the larger ellipsoid's surface, and the surface of electrically thin ellipsoid corresponded to S_2. The number of iterations performed was $\xi_{st} = 15$. Figure 5.16 compares the calculation results obtained by solving the full system (without splitting the S, black solid lines) with those obtained using the iterative algorithm of MFIE solving (gray lines), and by the FEKO™ software, in which CFIE was used (black dash lines). The data shown in Figure 5.16 demonstrate that the RCS calculation results for test object are considered to coincide well for all three methods.

Figure 5.17 illustrates the convergence of results obtained by the developed iterative method (circles are for horizontal polarization, squares are for vertical polarization). The ordinate axis shows the value of maximum error as follows:

$$\delta_\xi = \max \left| \frac{\sigma_\xi(\beta) - \sigma_{st}(\beta)}{\sigma_{st}(\beta)} \right|, \%, \tag{5.23}$$

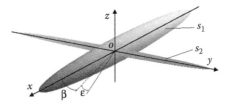

FIGURE 5.15
The test object.

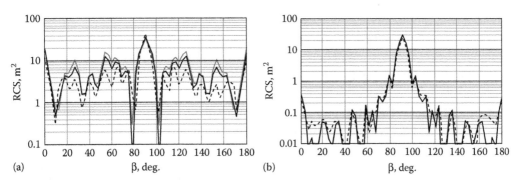

FIGURE 5.16
Test object monostatic RCS calculation results for (a) horizontal and (b) vertical polarization.

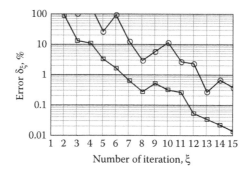

FIGURE 5.17
Results (circles are for horizontal polarization, squares are for vertical polarization) showing the convergence of the developed iterative algorithm as the number of iterations increases.

where $\sigma_\xi(\beta)$ is the object RCS corresponding to ξ iterations, given fixed aspect angle ε and $\sigma_{st}(\beta)$ is the steady-state RCS of the object, which is the value averaged over three latest iterations as follows:

$$\sigma_{st}(\beta) = \frac{1}{3} \sum_{\xi=\xi_{st}-3}^{\xi_{st}} \sigma_\xi(\beta). \tag{5.24}$$

As shown in Figure 5.17, the iterative algorithm developed here converges quite rapidly. Let us note that the error of the RCS calculation using the developed method for getting the MFIE solution does not exceed $\delta_a = 1...3\%$ as determined by Equation 5.16 given that number of the current nodes goes up to $N = 955$. In the case when FEKO™ software is used (when CFIE is applied), such accuracy is only achieved when the number of nodes goes as high as $N = 14300$. The similar accuracy is obtained in the frequency range of $100...400$ MHz ($\lambda_1 = 3...0.75$ m). Having analyzed the results of this simulation, we can state that for the resonant-sized objects considered (such as missiles, UAVs, and small-scale airplanes in VHF band), the convergence of iterative algorithm characterized by the error of $\delta_\xi \leq (1...3)\%$ is reached given $\xi = \xi_{st} \approx 25$.

The example calculations used a relatively simple test object. For the more complex scatterers, there can be situations when the calculation algorithm based on solving the full system of equations would not provide results with such accuracy. The iterative algorithm also produces more stable results. At the same time, the developed algorithm does not require large computer memory and reduces the computational burden.

The method proposed here is more effective for modeling the scattering characteristics of such complex-shaped objects of resonant sizes as missiles, UAVs, and small-scale airplanes in the VHF band [37].

Further, in Section 5.6, we will demonstrate and discuss the developed methods to UWB radar high-resolution range profiles of cruise missiles.

5.3.4 The Magnetic Field Integral Equation for a Subsurface Perfectly Electrically Conducting Resonant Object

This section presents the EM scattering model of a subsurface PEC object as shown in Figure 5.1b. We can get the MFIE for the electric current density at the surface of the PEC object V_2 buried in the ground by applying the Lorentz reciprocity theorem [24,34] in regions V_1 and V_3 to the desired EM field $(\vec{E}, \vec{H})$ and to the field of the auxiliary point magnetic source $\left(\vec{\mathcal{E}}_{1,3}^m, \vec{\mathcal{H}}_{1,3}^m \right)$ considering the existence of the interface between the half-spaces V_1 and V_3, [16,40,41]. This yields a system of scalar equations as follows:

$$\begin{cases} \vec{\tau}_2^0 \cdot \vec{J}^e \left(\vec{Q}_0 \right) \left(1 + \Omega_{s_0} \left(\vec{Q}_0 \right) \right) + \dfrac{2}{i\omega} \int\limits_{S \backslash s_0} \vec{\mathcal{E}}_{1,3}^m \left(\vec{Q} \middle| \vec{Q}_0, \, \vec{\tau}_1^0 \right) \cdot \vec{J}^e \left(\vec{Q} \right) ds_Q = 2 \vec{\tau}_1^0 \cdot \vec{H}^0 \left(\vec{Q}_0 \right), \\ -\vec{\tau}_1^0 \cdot \vec{J}^e \left(\vec{Q}_0 \right) \left(1 + \Omega_{s_0} \left(\vec{Q}_0 \right) \right) + \dfrac{2}{i\omega} \int\limits_{S \backslash s_0} \vec{\mathcal{E}}_{1,3}^m \left(\vec{Q} \middle| \vec{Q}_0, \, \vec{\tau}_2^0 \right) \cdot \vec{J}^e \left(\vec{Q} \right) ds_Q = 2 \vec{\tau}_2^0 \cdot \vec{H}^0 \left(\vec{Q}_0 \right), \end{cases} \tag{5.25}$$

where $\vec{Q}_0$, $\vec{Q} \in S$; and $\vec{H}^0$ is the magnetic vector of initial field, which also takes into account the presence of interface between V_1 and V_3.

In Equation 5.3 that defines the term $\Omega_{s_0} \left(\vec{Q}_0 \right)$, we should substitute k_1 with $k_3 = k_1 \sqrt{\varepsilon_3}$ (the wavenumber in half-space V_3).

The equations for $\vec{\mathcal{E}}_{1,3}^m$ are rather cumbersome and not shown in this chapter. The derivation of the required formulae is explicitly described in the well-known classical literature [56,57].

We can calculate the magnetic vector of the field scattered by the subsurface PEC object using the integral representation [16,40,41] as follows:

$$i\omega \vec{p}^{rec} \cdot \left(\vec{H} \left(\vec{Q}_{rec} \right) - \vec{H}^0 \left(\vec{Q}_{rec} \right) \right) = - \int\limits_S \vec{\mathcal{E}}_{1,3}^m \left(\vec{Q} \middle| \vec{Q}_0, \vec{p}^{rec} \right) \cdot \vec{J}^e \left(\vec{Q} \right) ds_Q, \tag{5.26}$$

where $\vec{Q}_{rec} \in V_1$ as shown in Figure 5.1b.

Note that Equations 5.25 and 5.26 have the same form as Equations 5.1 and 5.6 used above for computing the scattering in free space. Therefore, to obtain the scattering characteristics of subsurface PEC object, we can apply the numerical algorithm described in Subsection 5.3.1. To account for the media interface, we can replace the function $\vec{\mathcal{E}}_1^m$ corresponding to the free space with $\vec{\mathcal{E}}_{1,3}^m$, which accounts for the interface between the two half-spaces. We can choose the function $\vec{H}^0$ in the same way.

Later, Section 5.7 will demonstrate the results of the UWB impulse response calculation for metallic mines buried in the ground obtained using the developed algorithm.

5.4 Calculation Methods for Obtaining the Scattering Characteristics of Aerial and Subsurface Dielectric Resonant Objects Based on an Integral Equation Solution in the Frequency Domain

Several different ways exist to find the IEs for solving the problems associated with EM scattering by dielectric objects. The peculiarities of different type boundary IEs have been analyzed in [24,42]. According to some researchers, only those boundary IEs have a unique solution at all frequencies that involve both electric and magnetic equivalent currents; and that satisfy the boundary conditions for both electric and magnetic tangential vectors. The system of boundary IEs of the Müller type [16,24,32,33,40–42,44,47] (system of inhomogeneous Fredholm IEs of the second kind) can be considered as such equations.

The boundary IEs regarding dielectric objects as well as PEC objects are solved using the MoM and its modifications [16,24,32,33,40–42,44,47].

However, an analysis of the literature cited above [32–34,42–49] indicates that the known solving methods have some limitations and demand further development.

Subsection 5.4.1 will briefly discuss the method for solving the Müller type IE system with regard to the equivalent electric and magnetic current densities at the surface of the homogeneous dielectric three-dimensional scatterer in free space. The kernels of the system are the differences of EM fields of auxiliary electric and magnetic point sources. The algorithm for finding the solution for the considered IE system then becomes the basis of the method for calculating the scattering characteristics of free space homogeneous dielectric scatterers.

Subsection 5.4.2 will compare the RCS calculation results for simple dielectric objects obtained using different methods.

Subsection 5.4.3 will generalize the method for solving the IE system that was developed for an object in free space on a case of subsurface dielectric object [16,40,41].

5.4.1 A Method for Solving the System of Integral Equations for a Dielectric Object in Free Space

Let us consider the problem of EM scattering by a dielectric homogeneous three-dimensional object V_2 with relative permittivity $\varepsilon_2 = \varepsilon_2' + i\varepsilon_2''$ in free space V_1 with $\varepsilon_1 = 1$ shown in Figure 5.1a.

We start by introducing the EM fields of auxiliary point sources (dipoles) $\left(\vec{\mathcal{E}}_\alpha^{e(m)}\left(\vec{Q} \middle| \vec{Q}_0, \vec{\tau}^0\right), \vec{\mathcal{H}}_\alpha^{e(m)}\left(\vec{Q} \middle| \vec{Q}_0, \vec{\tau}^0\right) \right)$ shown here as the EM field (at point $\vec{Q}$) of a point electric (magnetic) source (located at point $\vec{Q}_0$ with unit vector-moment $\vec{\tau}^0$). We take as given that the relative permittivities of the object V_2 and the surrounding surface V_1 coincide, $\varepsilon_1 = \varepsilon_2 = \varepsilon_\alpha \ (\alpha = 1, 2)$.

To find the system of IE for the equivalent electrical and magnetic current densities at the object's surface, we can apply the Lorentz reciprocity theorem [24,34] in the regions V_1 and V_2 to the desired EM field $\left(\vec{E}, \vec{H}\right)$ and, consequently, to the field of the electric and magnetic point sources described above, which gives the following:

$$
\begin{cases}
\vec{\tau}_2^0 \cdot \vec{J}^m\left(\vec{Q}_0\right)(\varepsilon_2 + \varepsilon_1) - 2\varepsilon_2 \vec{\tau}_1^0 \cdot \vec{E}^0\left(\vec{Q}_0\right) = \dfrac{2}{i\omega} \int\limits_{S \backslash s_0} \left(\Delta \vec{\mathcal{H}}^e\left(\vec{\tau}_1^0\right) \cdot \vec{J}^m\left(\vec{Q}\right) + \varepsilon_0^{-1} \Delta \vec{\mathcal{D}}^e\left(\vec{\tau}_1^0\right) \cdot \vec{J}^e\left(\vec{Q}\right) \right) ds_Q, \\[2em]
-\vec{\tau}_1^0 \cdot \vec{J}^m\left(\vec{Q}_0\right)(\varepsilon_2 + \varepsilon_1) - 2\varepsilon_2 \vec{\tau}_2^0 \cdot \vec{E}^0\left(\vec{Q}_0\right) = \dfrac{2}{i\omega} \int\limits_{S \backslash s_0} \left(\Delta \vec{\mathcal{H}}^e\left(\vec{\tau}_2^0\right) \cdot \vec{J}^m\left(\vec{Q}\right) + \varepsilon_0^{-1} \Delta \vec{\mathcal{D}}^e\left(\vec{\tau}_2^0\right) \cdot \vec{J}^e\left(\vec{Q}\right) \right) ds_Q, \\[2em]
\vec{\tau}_2^0 \cdot \vec{J}^e\left(\vec{Q}_0\right) - \vec{\tau}_1^0 \cdot \vec{H}^0\left(\vec{Q}_0\right) = -\dfrac{1}{i\omega} \int\limits_{S \backslash s_0} \left(\Delta \vec{\mathcal{H}}^m\left(\vec{\tau}_1^0\right) \cdot \vec{J}^m\left(\vec{Q}\right) + \Delta \vec{\mathcal{E}}^m\left(\vec{\tau}_1^0\right) \cdot \vec{J}^e\left(\vec{Q}\right) \right) ds_Q, \\[2em]
-\vec{\tau}_1^0 \cdot \vec{J}^e\left(\vec{Q}_0\right) - \vec{\tau}_2^0 \cdot \vec{H}^0\left(\vec{Q}_0\right) = -\dfrac{1}{i\omega} \int\limits_{S \backslash s_0} \left(\Delta \vec{\mathcal{H}}^m\left(\vec{\tau}_2^0\right) \cdot \vec{J}^m\left(\vec{Q}\right) + \Delta \vec{\mathcal{E}}^m\left(\vec{\tau}_2^0\right) \cdot \vec{J}^e\left(\vec{Q}\right) \right) ds_Q,
\end{cases}
$$

$$(5.27)$$

where $\vec{Q}_0$, $\vec{Q} \in S$ are the observation and integration points, respectively;

$\vec{\tau}_1^0$ and $\vec{\tau}_2^0$ are mutually orthogonal unit vectors tangent to S at point $\vec{Q}_0$; these vectors form the right-hand system with the unit vector $\vec{v}^0$, which is the external normal to S at this point;

$\vec{J}^e(\vec{Q}) = \vec{n} \times \vec{H}(\vec{Q})$ and $\vec{J}^m(\vec{Q}) = \vec{n} \times E(\vec{Q})$ are the densities of equivalent electric and magnetic currents, respectively ($\vec{v}$ is the normal at integration point); and $(\vec{E}^0, \vec{H}^0)$ is the initial EM field;

$$\Delta \vec{\mathcal{H}}^e(\vec{\tau}^0) = \varepsilon_2 \vec{\mathcal{H}}_2^e(\vec{Q}|\vec{Q}_0, \vec{\tau}^0) - \varepsilon_1 \vec{\mathcal{H}}_1^e(\vec{Q}|\vec{Q}_0, \vec{\tau}^0),$$ (5.28)

$$\Delta \vec{\mathcal{D}}^e(\vec{\tau}^0) = \vec{\mathcal{D}}_2^e(\vec{Q}|\vec{Q}_0, \vec{\tau}^0) - \vec{\mathcal{D}}_1^e(\vec{Q}|\vec{Q}_0, \vec{\tau}^0),$$ (5.29)

$$\vec{\mathcal{D}}_{1(2)}^e(\vec{Q}|\vec{Q}_0, \vec{\tau}^0) = \varepsilon_0 \varepsilon_{1(2)} \vec{\mathcal{E}}_{1(2)}^e(\vec{Q}|\vec{Q}_0, \vec{\tau}^0),$$ (5.30)

$$\Delta \vec{\mathcal{H}}^m(\vec{\tau}^0) = \vec{\mathcal{H}}_2^m(\vec{Q}|\vec{Q}_0, \vec{\tau}^0) - \vec{\mathcal{H}}_1^m(\vec{Q}|\vec{Q}_0, \vec{\tau}^0),$$ (5.31)

$$\Delta \vec{\mathcal{E}}^m(\vec{\tau}^0) = \vec{\mathcal{E}}_2^m(\vec{Q}|\vec{Q}_0, \vec{\tau}^0) - \vec{\mathcal{E}}_1^m(\vec{Q}|\vec{Q}_0, \vec{\tau}^0).$$ (5.32)

The components of EM fields of auxiliary dipoles in homogeneous space V_α, which corresponds to ε_α, can be represented as follows [29,34]:

$$\vec{\mathcal{E}}_\alpha^e(\vec{Q}|\vec{Q}_0, \vec{\tau}^0) = (\varepsilon_0 \varepsilon_\alpha)^{-1} g_{\alpha 1}(\vec{Q}_0, \vec{Q}),$$ (5.33)

$$\vec{\mathcal{H}}_\alpha^e(\vec{Q}|\vec{Q}_0, \vec{\tau}^0) = g_{\alpha 2}(\vec{Q}_0, \vec{Q}),$$ (5.34)

$$\vec{\mathcal{E}}_\alpha^m(\vec{Q}|\vec{Q}_0, \vec{\tau}^0) = -g_{\alpha 2}(\vec{Q}_0, \vec{Q}),$$ (5.35)

$$\vec{\mathcal{H}}_\alpha^m(\vec{Q}|\vec{Q}_0, \vec{\tau}^0) = \mu_0^{-1} g_{\alpha 1}(\vec{Q}_0, \vec{Q}),$$ (5.36)

where:

$$g_{\alpha 1}(\vec{Q}_0, \vec{Q}) = \left[\vec{\tau}^0 k_\alpha^2 G_\alpha(\vec{Q}_0, \vec{Q}) + \vec{\nabla}\left(\vec{\tau}^0 \cdot \vec{\nabla} G_\alpha(\vec{Q}_0, \vec{Q}) \right) \right],$$ (5.37)

$$g_{\alpha 2}(\vec{Q}_0, \vec{Q}) = i\omega\left(\vec{\tau}^0 \times \vec{\nabla} G_\alpha(\vec{Q}_0, \vec{Q}) \right),$$ (5.38)

$$G_\alpha(\vec{Q}_0, \vec{Q}) = G_\alpha(R) = \frac{\exp(ik_\alpha R)}{4\pi R};$$ (5.39)

$k_\alpha = \omega\sqrt{\varepsilon_0 \varepsilon_\alpha \mu_0} = 2\pi/\lambda_\alpha$ is the wavenumber in medium V_α with ε_α;
λ_α is the wavelength in V_α;
ε_0 and μ_0 are the permittivity and permeability of free space.

The system of Equations 5.27 represents system of the Fredholm inhomogeneous IEs of the second kind (of Müller type [24,42]). The kernels of the equations are singular when $\vec{Q}_0$ coincides with $\vec{Q}$. However, as shown in [40,41], the boundary integrals in Equations 5.27 possess the finite values given that $\vec{Q}_0$ belongs to S, and this system can be solved numerically.

For this purpose, we can approximate the dielectric object's surface with parts of three-axial ellipsoids, which, in their turn, are split into facets in the same way as it was in case of PEC object. As a result, the surface S gets represented by a set of N electrically small elemental portions s_n ($n = \overline{1, N}$), the surface current densities $\vec{J}_n^e, \vec{J}_n^m$ at which are assumed to be constant, so that they can be put outside the integral sign. As a result, we can write the first equation of the system of Equations (5.27) as follows:

$$\vec{\tau}_{2n0}^0 \cdot \vec{J}_{n_0}^m (\varepsilon_2 + \varepsilon_1) - 2\varepsilon_2 \vec{\tau}_{1,n_0}^0 \cdot \vec{E}_{n_0}^0 = \frac{2}{i\omega} \sum_{n=1}^{N} \left(\vec{J}_n^m \int_{s_n} \Delta \vec{\mathcal{H}}^e \left(\vec{\tau}_{1n_0}^0 \right) ds_Q + \varepsilon_0^{-1} \vec{J}_n^e \int_{s_n} \Delta \vec{\mathcal{D}}^e \left(\vec{\tau}_{1n_0}^0 \right) ds_Q \right), \quad (5.40)$$

where the indices $n_0 = \overline{1, N}$, $n = \overline{1, N}$, show the current observation and integration points, respectively. Similar to Equation 5.40, the rest of the system of Equations 5.27 is represented in discrete form.

We can compute the surface integrals over elements s_n using the algorithm similar to that developed for solving the MFIE, given the PEC object and described in Subsection 5.3.1. The surface element s_{n_0} that contains a singularity is, in its turn, represented by a set of N_0 surface parts $s_{n_0} = \sum_{g=1}^{N_0} s_g'$ as shown in Figure 5.2. The surface parts s_g' that lie outside the small neighborhood s_0 of singular point, for which the condition of $k_3 \rho_{0\,g} > \gamma_0$ is held true (where $\rho_{0\,g}$ is the distance between the centers of s_{n_0} ($\vec{Q}_{n_0}$) and s_g'; the value of $\gamma_0 = 0.01$ was found experimentally during computations, given simple shape objects), constitute the s_{n_0}, and the integral over this surface being computed numerically using the five-point quadrature Gauss formula. The integral over small neighborhood of singularity s_0 that includes the elements s_g' corresponding to the condition of $k_3 \rho_{0\,g} \leq \gamma_0$ is assumed to be zero.

Integrals over surface parts s_n do not contain any singularities and they are also computed by the five-point Gauss formula.

Having thus found the densities of equivalent currents, we can compute the components of EM field scattered by dielectric object as follows:

$$\begin{cases} i\omega \vec{p}^{rec} \cdot \left(\vec{E} \left(\vec{Q}_{rec} \right) - \vec{E}^0 \left(\vec{Q}_{rec} \right) \right) = \int_S \left(\Delta \vec{\mathcal{H}}^e \left(\vec{p}^{rec} \right) \cdot \vec{J}^m \left(\vec{Q} \right) + \varepsilon_0^{-1} \Delta \vec{\mathcal{D}}^e \left(\vec{p}^{rec} \right) \cdot \vec{J}^e \left(\vec{Q} \right) \right) ds_Q, \\ \\ -i\omega \vec{p}^{rec} \cdot \left(\vec{H} \left(\vec{Q}_{rec} \right) - \vec{H}^0 \left(\vec{Q}_{rec} \right) \right) = \int_S \left(\Delta \vec{\mathcal{H}}^m \left(\vec{p}^{rec} \right) \cdot \vec{J}^m \left(\vec{Q} \right) + \Delta \vec{\mathcal{E}}^m \left(\vec{p}^{rec} \right) \cdot \vec{J}^e \left(\vec{Q} \right) \right) ds_Q, \end{cases} \quad (5.41)$$

where $\vec{Q}_{rec} \notin V_2$ as shown in Figure 5.1a.

Based on Equations 5.27 through 5.41, we designed the numerical algorithm that includes the first, second, and fourth stages of the algorithm developed for the case of PEC object described in Subsection 5.3.1. As we have mentioned above, the IE system of the Müller type as in Equation 5.27 has unique solutions at all frequencies. Therefore, the third stage is not needed and can be eliminated.

Below in Subsection 5.4.2, we compare the results of our RCS computation for dielectric objects with the data obtained using the FEKO™ software.

5.4.2 Verification of the Calculated Scattering Characteristics of Simple Shape Dielectric Objects

We chose two dielectric scatterers as the test objects: (1) the electrically thin ellipsoid of Figure 5.5a with relative permittivity $\varepsilon_2 = 10 + i2$ and with semi-axes $a_{e\,x} = 0.9\lambda_1, a_{e\,y} = 0.2\lambda_1$, and $a_{e\,z} = 0.02\lambda_1$ and (2) and the cylinder of Figure 5.5b with $\varepsilon_2 = 2.2 + i0.1$, radius $a_c = 0.216\lambda_1$

and height $d_c = 2.76\lambda_1$. We calculated each test object monostatic RCS as a function of the aspect angle β of Figure 5.5 in the E-plane and H-plane determined by the same order as in Subsection 5.3.2.

We calculated the test object RCS using the previously developed method based on a system of Müller type IEs 5.27 solution, and using the FEKO™ software. Results of ellipsoid and cylinder monostatic RCS are presented in Figures 5.18 (10 $\log(\sigma_e/\pi a_{e\,z}^2)$) and 5.19 (10 $\log(\sigma_c/\pi a_c^2)$), respectively.

A comparison of RCS data from two methods shows in both cases how the relative error value δ_{met} determination using Equation 5.17 does not exceed 1...3 % with the exception

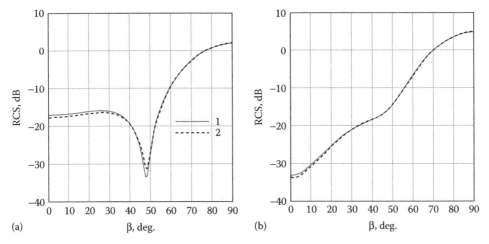

(a)   (b)

FIGURE 5.18
A comparison of the calculated monostatic RCS of the dielectric ellipsoid ($\varepsilon_2 = 10 + i2$, $a_{e\,x} = 0.5\lambda_1$, $a_{e\,y} = 0.02\lambda_1$, and $a_{e\,z} = 0.2\lambda_1$) in (a) the E-plane and (b) the H-plane. Results were calculated against the aspect angle β using (1) the developed method, and (2) the FEKO™ software.

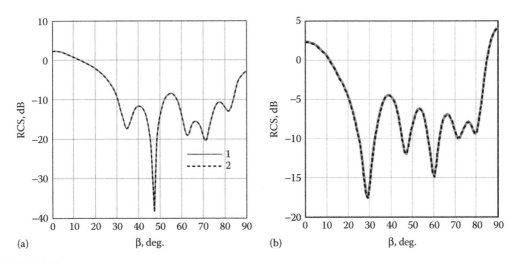

(a)   (b)

FIGURE 5.19
A comparison of the calculated RCS of the dielectric cylinder ($\varepsilon_2 = 2.2 + i0.1$, $a_c = 0.216\lambda_1$, and $d_c = 2.76\lambda_1$) in (a) the E-plane and (b) the H-plane. Results were calculated against the angle β using (1) the developed method, and (2) the FEKO™ software.

of electrically thin ellipsoid RCS, corresponding to $\beta = 0...10$ degrees in both polarization planes and to $45...50$ degrees in *E*-plane where the error δ_{met} does not exceed 8%.

5.4.3 System of Integral Equations for a Subsurface Dielectric Object

We extended our method to determining the RCS of a dielectric object V_2, located in the dielectric half-space V_3 as shown in Figure 5.1b. To obtain the necessary IEs, we can apply the Lorentz reciprocity theorem. For this purpose, we introduce the next set of auxiliary EM fields as follows:

$\left(\vec{E}_{1,2}^{e(m)} \left(\vec{Q} \middle| \vec{Q}_0, \vec{\tau}^0 \right), \ \vec{\mathcal{H}}_{1,2}^{e(m)} \left(\vec{Q} \middle| \vec{Q}_0, \vec{\tau}^0 \right) \right)$ is the EM field of the auxiliary point electric (magnetic) source which takes into account the presence of the boundary between dielectric half-spaces V_1 and V_3 (Figure 5.1b), given that relative permittivity ε_3 of half-space V_3 surrounding the object V_2 is equal to ε_2;

$\left(\vec{E}_{1,3}^{e(m)} \left(\vec{Q} \middle| \vec{Q}_0, \vec{\tau}^0 \right), \ \vec{\mathcal{H}}_{1,3}^{e(m)} \left(\vec{Q} \middle| \vec{Q}_0, \vec{\tau}^0 \right) \right)$ is the EM field of the auxiliary point electric (magnetic) source which takes into account the presence of the boundary between dielectric half-spaces V_1 and V_3, given that object V_2 relative permittivity ε_2 is equal to ε_3 of half-space V_3.

Applying the Lorentz reciprocity theorem to the desired EM field $\left(\vec{E}, \ \vec{H} \right)$ and to the auxiliary fields introduced above in regions $V_1 \cup V_3$ and V_2 produces the system of IEs similar to the system shown in Equation 5.27 for a dielectric object in a homogeneous space as follows:

$$\left\{ \begin{aligned} &\vec{\tau}_2^0 \cdot \vec{J}^m \left(\vec{Q}_0 \right) (\varepsilon_2 + \varepsilon_3) - 2\varepsilon_2 \vec{\tau}_1^0 \cdot \vec{E}^0 \left(\vec{Q}_0 \right) = \frac{2}{i\omega} \int\limits_{S \backslash s_0} \left(\Delta \vec{\mathcal{H}}_{2,3}^e \left(\vec{\tau}_1^0 \right) \cdot \vec{J}^m \left(\vec{Q} \right) + \varepsilon_0^{-1} \Delta \vec{\mathcal{D}}_{2,3}^e \left(\vec{\tau}_1^0 \right) \cdot \vec{J}^e \left(\vec{Q} \right) \right) ds_Q, \\[2ex] &-\vec{\tau}_1^0 \cdot \vec{J}^m \left(\vec{Q}_0 \right) (\varepsilon_2 + \varepsilon_3) - 2\varepsilon_2 \vec{\tau}_2^0 \cdot \vec{E}^0 \left(\vec{Q}_0 \right) = \frac{2}{i\omega} \int\limits_{S \backslash s_0} \left(\Delta \vec{\mathcal{H}}_{2,3}^e \left(\vec{\tau}_2^0 \right) \cdot \vec{J}^m \left(\vec{Q} \right) + \varepsilon_0^{-1} \Delta \vec{\mathcal{D}}_{2,3}^e \left(\vec{\tau}_2^0 \right) \cdot \vec{J}^e \left(\vec{Q} \right) \right) ds_Q, \\[3ex] &\vec{\tau}_2^0 \cdot \vec{J}^e \left(\vec{Q}_0 \right) - \vec{\tau}_1^0 \cdot \vec{H}^0 \left(\vec{Q}_0 \right) = -\frac{1}{i\omega} \int\limits_{S \backslash s_0} \left(\Delta \vec{\mathcal{H}}_{2,3}^m \left(\vec{\tau}_1^0 \right) \cdot \vec{J}^m \left(\vec{Q} \right) + \Delta \vec{E}_{2,3}^m \left(\vec{\tau}_1^0 \right) \cdot \vec{J}^e \left(\vec{Q} \right) \right) ds_Q, \\[2ex] &-\vec{\tau}_1^0 \cdot \vec{J}^e \left(\vec{Q}_0 \right) - \vec{\tau}_2^0 \cdot \vec{H}^0 \left(\vec{Q}_0 \right) = -\frac{1}{i\omega} \int\limits_{S \backslash s_0} \left(\Delta \vec{\mathcal{H}}_{2,3}^m \left(\vec{\tau}_2^0 \right) \cdot \vec{J}^m \left(\vec{Q} \right) + \Delta \vec{E}_{2,3}^m \left(\vec{\tau}_2^0 \right) \cdot \vec{J}^e \left(\vec{Q} \right) \right) ds_Q, \end{aligned} \right.$$

$$(5.42)$$

where $\vec{Q}_0, \ \vec{Q} \in S$;

$\left(\vec{E}^0, \ \vec{H}^0 \right)$ is the initial EM field, taking into account the presence of boundary between the half-spaces V_1 and V_3;

$$\Delta \vec{\mathcal{H}}_{2,3}^e \left(\vec{\tau}^0 \right) = \varepsilon_2 \vec{\mathcal{H}}_{1,2}^e \left(\vec{Q} \middle| \vec{Q}_0, \vec{\tau}^0 \right) - \varepsilon_3 \vec{\mathcal{H}}_{1,3}^e \left(\vec{Q} \middle| \vec{Q}_0, \vec{\tau}^0 \right), \tag{5.43}$$

$$\Delta \vec{\mathcal{D}}_{2,3}^e \left(\vec{\tau}^0 \right) = \vec{\mathcal{D}}_{1,2}^e \left(\vec{Q} \middle| \vec{Q}_0, \vec{\tau}^0 \right) - \vec{\mathcal{D}}_{1,3}^e \left(\vec{Q} \middle| \vec{Q}_0, \vec{\tau}^0 \right), \tag{5.44}$$

$$\vec{\mathcal{D}}_{1,2(3)}^e \left(\vec{Q} \middle| \vec{Q}_0, \vec{\tau}^0 \right) = \varepsilon_0 \varepsilon_{2(3)} \vec{E}_{1,2(3)}^e \left(\vec{Q} \middle| \vec{Q}_0, \vec{\tau}^0 \right), \tag{5.45}$$

$$\Delta \vec{\mathcal{H}}_{2,3}^m \left(\vec{\tau}^0 \right) = \vec{\mathcal{H}}_{1,2}^m \left(\vec{Q} \middle| \vec{Q}_0, \vec{\tau}^0 \right) - \vec{\mathcal{H}}_{1,3}^m \left(\vec{Q} \middle| \vec{Q}_0, \vec{\tau}^0 \right), \tag{5.46}$$

$$\Delta \vec{E}_{2,3}^m \left(\vec{\tau}^0 \right) = \vec{E}_{1,2}^m \left(\vec{Q} \middle| \vec{Q}_0, \vec{\tau}^0 \right) - \vec{E}_{1,3}^m \left(\vec{Q} \middle| \vec{Q}_0, \vec{\tau}^0 \right). \tag{5.47}$$

After finding the equivalent electric $\vec{J}^e$ and magnetic $\vec{J}^m$ current densities, we can calculate the EM field scattered by subsurface dielectric object in half-space V_1 of Figure 5.1b using the following representation:

$$
\begin{cases}
i\omega \vec{p}^{rec} \cdot \left(\vec{E}\left(\vec{Q}_{rec} \right) - \vec{E}^0 \left(\vec{Q}_{rec} \right) \right) = \int\limits_S \left(\Delta \vec{\mathcal{H}}^e_{2,3} \left(\vec{p}^{rec} \right) \cdot \vec{J}^m \left(\vec{Q} \right) + \varepsilon_0^{-1} \Delta \vec{\mathcal{D}}^e_{2,3} \left(\vec{p}^{rec} \right) \cdot \vec{J}^e \left(\vec{Q} \right) \right) ds_Q, \\
\\
-i\omega \vec{p}^{rec} \cdot \left(\vec{H}\left(\vec{Q}_{rec} \right) - \vec{H}^0 \left(\vec{Q}_{rec} \right) \right) = \int\limits_S \left(\Delta \vec{\mathcal{H}}^m_{2,3} \left(\vec{p}^{rec} \right) \cdot \vec{J}^m \left(\vec{Q} \right) + \Delta \vec{\mathcal{E}}^m_{2,3} \left(\vec{p}^{rec} \right) \cdot \vec{J}^e \left(\vec{Q} \right) \right) ds_Q,
\end{cases}
\tag{5.48}
$$

where $\vec{Q}_{rec} \in V_1$.

As mentioned above, the derivation of the EM field components takes into account the presence of dielectric half-space boundary. $\vec{\mathcal{E}}^{e(m)}_{1,2}$, $\vec{\mathcal{H}}^{e(m)}_{1,2}$, $\vec{\mathcal{E}}^{e(m)}_{1,2}$, $\vec{\mathcal{H}}^{e(m)}_{1,2}$ are described in the references [56,57].

The numerical algorithm developed from Equations 5.42 through 5.48 allows calculating frequency responses of dielectric subsurface objects. These could have great value in detecting dielectric mines buried in the ground.

5.5 Method for Computing the UWB Impulse Responses for Resonance-Sized Objects

We have designed our methods for computing the scattering characteristics of resonance-sized radar objects based on solving the IEs in the frequency domain. Equations 5.6, 5.26, together with 5.41 and 5.48 provide a way to compute the complex frequency response function at a fixed point of reception $\vec{Q}_{rec}$. We can relate this function to the magnetic field scattered by the radar object in the following manner:

$$
G_{rec}\left(f \right) \equiv \vec{p}^{rec} \cdot \vec{H}\left(\vec{Q}_{rec}, f \right) \equiv \vec{p}^{rec} \cdot \vec{H}\left(\vec{Q}_{rec} \right).
\tag{5.49}
$$

The sounding signal spectrum $G_{ss}\left(f \right)$ is taken into account while setting the functions describing the incident field as follows:

$$
\vec{H}^0 \left(\vec{Q}_0 \right) = G_{ss}\left(f \right) \vec{H}^0_{01}\left(\vec{Q}_0, f \right),
\tag{5.50}
$$

$$
\vec{E}^0 \left(\vec{Q}_0 \right) \equiv G_{ss}\left(f \right) \vec{E}^0_{01}\left(\vec{Q}_0, f \right),
\tag{5.51}
$$

where $\vec{H}^0_{01}\left(\vec{Q}_0, f \right)$ and $\vec{E}^0_{01}\left(\vec{Q}_0, f \right)$ are the vector functions of magnetic and electric fields of the source of illumination signal within the frequency range of interest with sounding signal amplitude spectrum density $\left| G_{ss}\left(f \right) \right| = 1$ and its phase spectrum $\phi_{ss}\left(f \right) = \arg\left(G_{ss}\left(f \right) \right) = 0$.

We can determine the time domain response as the inverse Fourier transform of $G_{rec}\left(f \right)$. However, the integrand can happen to be fast oscillating. To compute the time response $U(t)$ at fixed reception point $\vec{Q}_{rec}$, we need to apply the quadrature formula that uses a piecewise linear approximation of the signal spectrum $G_{rec1}\left(\omega \right) = G_{rec}\left(f \right)$ between its values at discrete frequencies $\omega_m = 2\pi f_m$ ($m = \overline{1, M-1}$, M is the number of discrete frequencies, at which the spectrum is defined) as follows:

$$G_{rec1}(\omega) = a_{1m}\omega + a_{2m}, \quad [\omega_m \le \omega \le \omega_{m+1}]. \tag{5.52}$$

Accounting for Equation 5.52, we can write the inverse Fourier transform of the frequency spectrum as follows:

$$U(t) = \frac{1}{2\pi} \int_{\omega_H}^{\omega_k} G_{rec1}(\omega) \exp(-i\omega t) d\omega \approx \frac{1}{2\pi} \sum_{m=1}^{M-1} \int_{\omega_m}^{\omega_{m+1}} (a_{1m}\omega + a_{2m}) \exp(-i\omega t) d\omega. \tag{5.53}$$

Given the previous expression, we can analytically compute the integral of Equation 5.53 over the section $[\omega_m, \omega_{m+1}]$ as follows:

$$\int_{\omega_m}^{\omega_{m+1}} (a_{1m}\omega + a_{2m}) \exp(-i\omega t) d\omega =$$

$$= F_m(t) = \begin{cases} 0,5a_{1m}\left(\omega_{m+1}^2 - \omega_m^2\right) + a_{1m}\left(\omega_{m+1} - \omega_m\right), \quad \text{given} \quad t = 0, \\[2mm] \dfrac{a_{1m}}{t}\left[\exp(i\omega_{m+1}t)(i\omega_{m+1}t - 1) - \exp(i\omega_m t)(i\omega_m t - 1)\right] + \\[2mm] \quad + \dfrac{a_{1m}}{it}\left[\exp(i\omega_{m+1}t) - \exp(i\omega_m t)\right], \quad \text{given} \quad t \ne 0. \end{cases} \tag{5.54}$$

Accounting for Equation 5.54, we can write the expression for computing the impulse response of radar object in compact form as follows:

$$U(t) \approx \frac{1}{2\pi} \sum_{m=1}^{M-1} F_m(t). \tag{5.55}$$

To compute the unknown coefficients a_{1m} and a_{2m} in (5.54), we will write the system as follows:

$$G_{rec1}(\omega_m) = G_{rec1m} = a_{1m}\omega_m + a_{2m}, \tag{5.56}$$

$$G_{rec1}(\omega_{m+1}) = G_{rec1m+1} = a_{1m}\omega_{m+1} + a_{2m}. \tag{5.57}$$

Subtracting Equation 5.57 from 5.56, we can solve a_{1m} and a_{2m} as follows:

$$a_{1m} = \frac{G_{rec1m+1} - G_{rec1m}}{\omega_{m+1} - \omega_m}, \tag{5.58}$$

$$a_{2m} = \frac{G_{rec1m}\omega_{m+1} - G_{rec1m+1}\omega_m}{\omega_{m+1} - \omega_m}. \tag{5.59}$$

So, the sum of Equation 5.55 together with additional expressions in Equations 5.54, 5.58 and 5.59 provides us with an apparatus for computing the impulse responses of actual radar objects by the discrete values' frequency responses G_{rec1m} in Equation 5.49 including the case of radar sounding with a UWB signal. To do so, we need to account for the sounding signal spectrum according to Equations 5.50 and 5.51 in the expressions for the incident field by entering the Equations 5.1, 5.6, 5.25 through 5.27, 5.41, 5.42 and 5.48.

While developing computer simulations, we found that in order to obtain reliable impulse responses of resonance-sized objects of complex shape in the VHF band using the transform of Equation 5.55, one should first receive the discrete values of object's frequency response spaced by not more than 5...10 MHz.

In this work, we used the formula shown in Equation 5.55 to compute the high-resolution range profiles of a cruise missile, the impulse responses of antitank and antipersonnel mines buried in a ground when illuminated by UWB signals. We will describe these responses in Sections 5.6 and 5.7.

5.6 Computed UWB High-Resolution Range Profiles of Complex-Shaped Aerial Resonant Objects

This subsection represents the results of calculated cruise missile high-resolution range profiles for VHF band UWB sounding signals corresponding to VHF and UHF bands. As mentioned earlier, the aerial object surface modeling algorithm approximates the object as multiple ellipsoid parts. Section 5.3 showed how to model the object surface as ellipsoids. Figure 5.20 shows the surface models of the Taurus KEPD 350 and AGM-86C cruise missiles. The models consist of 26 and 13 ellipsoid parts, respectively. We considered the surfaces of missiles as PEC. The calculations did not take rotor modulation into account.

First we calculated the cruise missile frequency responses with iterative algorithm described in Subsection 5.3.3. Then we obtained the range profiles using the inverse Fourier transform based on a piecewise-linear approximation of object frequency responses described in Section 5.5.

The cruise missile range profiles correspond to a sounding signal with a uniform amplitude-frequency spectrum and zero phase-frequency spectrum. Calculation results have been obtained for bandwidths $\Delta f = 300$ MHz ($f_{min} = 100$ MHz, $f_{max} = 400$ MHz; $\lambda_{1\,max} = 3$ m, $\lambda_{1\,min} = 0.75$ m) and $\Delta f = 100$ MHz ($f_{min} = 200$ MHz, $f_{max} = 300$ MHz; $\lambda_{1\,max} = 1.5$ m, $\lambda_{1\,min} = 1$ m). The calculations used a 10-MHz frequency sampling interval. Frequency responses were weighted with a Hamming function. In this case, the range resolution was $\delta r = 0.71$ m for $\Delta f = 300$ MHz and $\delta r = 2.13$ m for $\Delta f = 100$ MHz. Our figures show the range profiles obtained for a monostatic radar configuration. Results correspond to the horizontal ($\vec{E}$ is parallel to wing plane), and to the vertical ($\vec{E}$ is placed in plane containing the illumination direction vector and perpendicular to wing plane) polarizations. The elevation angle is $\varepsilon = 3°$ which shows a typical illumination from lower half-space corresponding to a ground based radar tracking the missile. The range profiles are considered for three azimuth aspects $\beta = 0°$ (front illumination), $\beta = 90°$ (side illumination), and $\beta = 60°$. Figure 5.21 shows the conditions for calculating the range profiles corresponding missile orientation

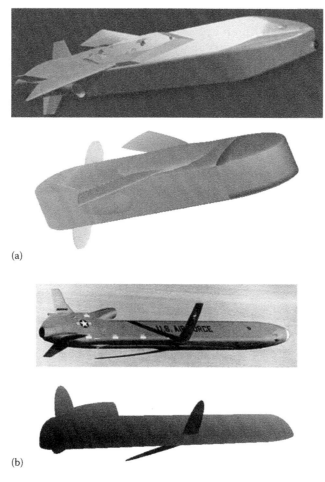

(a)

(b)

FIGURE 5.20
Cruise missiles and their surface models. (a) Taurus KEPD 350 German/Swedish Air Launched Cruise Missile and (b) AGM-86C American Air Launched Cruise Missile (ALCM).

in horizontal plane. Figures 5.22 through 5.25 show the calculated range profiles. The local maximums corresponding to missile construction elements are marked by numbers for easy interpretation. The cruise missile range profile amplitudes for given frequency bands are normalized to maximum value of impulse response for aspect angle $\beta = 90°$ and horizontal polarization. In Figures 5.22 through 5.25, the distance r lies along horizontal axis and positive values correspond to moving off from illumination point.

To evaluate the effect of our calculation method, we need to examine the range profiles determined with range resolution $\delta r = 0.71$ m shown as the black lines in Figures 5.22 through 5.25.

The VHF band missile range profiles have well defined local maximums which correspond to structural components of the missiles. The range profiles of Taurus KEPD 350 shown in Figures 5.22 and 5.23 vary depending on the aspect angle as follows:

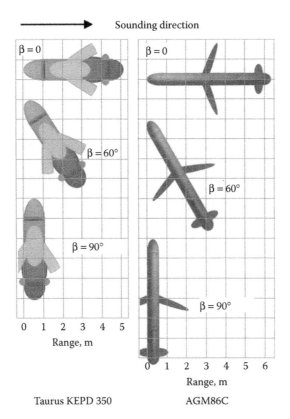

FIGURE 5.21
Cruise missile aspect angle conditions used for determining the range profiles.

1. Aspect angle β = 0° contains the following maximums corresponding to: (1) the fore body; (2) the leading edge assemblies; (3) the wing trailing edges; and (4) the fins. The responses of the horizontally oriented wings have a substantially smaller amplitude in the vertical polarization.

2. Aspect angle β = 60° for horizontal polarization shows maximums observed at: (1) the fore body; (2) the leading edge assembly (illuminated at a right angle as shown in Figure 5.21); (3) the response conditioned by interaction between fuselage and far wing (along illumination direction); (4) the rear fuselage; and (5) the response from leading edge assembly is not observable for the vertical polarization at the same aspect angle. It is necessary to note that at aspect angle β = 60° for both polarizations, the returns show a response induced by "creeping" wave which rounds fuselage rear.

3. For aspect angle β = 90°, the range profiles of the Taurus KEPD 350 contain: (1) a practically undistorted response from fuselage and (2) a "creeping" wave response caused by the high-resolution pulse traveling around the fuselage and returning with a delay. The amplitude of second maximum is considerably larger for vertical polarization explained by prevalent currents which flow athwart the fuselage axis. Responses conditioned by "creeping" waves are well-known phenomenon consistent with resonant size object radar sounding. These responses contain information about object geometrical dimensions [7–12].

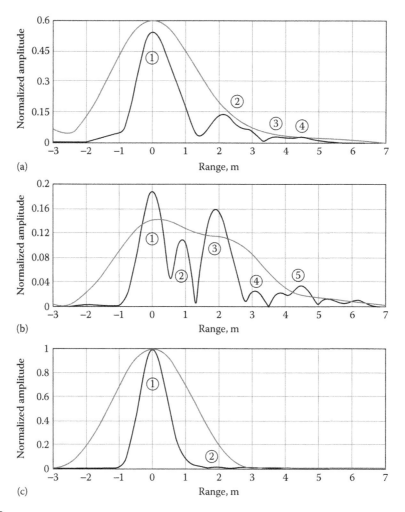

FIGURE 5.22
Calculated high-resolution range profiles of the Taurus KEPD 350 cruise missile for a UWB radar sounding on horizontal polarization. Radar profiles correspond to aspect angles (a) $\beta = 0°$, (b) $\beta = 60°$, and (c) $\beta = 90°$. The black and gray lines show range resolution $\delta r = 0.71$ m and $\delta r = 2.13$ m, respectively.

Figures 5.24 and 5.25 show AGM-86C ALCM range profiles:

1. For the front aspect angle $\beta = 0°$ of Figure 5.21 for both polarizations, the responses have the following local maximums corresponded to: (1) fore body (double response), (2) and (3) wings, (4) the turbine air inlet, and (5) fins.

2. For the aspect angle $\beta = 60°$, the AGM-86C range profiles have significant differences for orthogonal polarizations. The range profiles for both polarizations include: (1) responses representing superposition of fore body and near wing (along the sounding direction) reflections; (2) the result of interactions between fuselage central part, wings, and air inlet of the turbine; (3) the response of rear fuselage with the fins; and (4) like the Taurus KEPD 350, the AGM-86C range

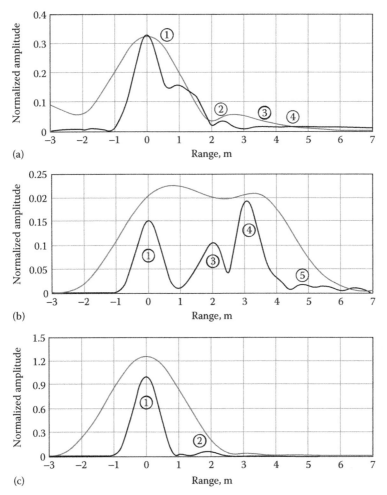

FIGURE 5.23
Calculated high-resolution range profiles of the Taurus KEPD 350 cruise missile for a UWB radar sounding on vertical polarization. Radar profiles correspond to aspect angles (a) $\beta = 0°$, (b) $\beta = 60°$, and (c) $\beta = 90°$. The black and gray lines show range resolution $\delta r = 0.71$ m and $\delta r = 2.13$ m, respectively.

profiles for $\beta = 60°$ contain responses induced by "creeping" waves which bend around the fuselage rear.

3. For aspect angle $\beta = 90°$, the range profiles for both polarizations contain three maximums: (1) a relatively weak response near the wing; (2) a practically undistorted response from the fuselage; and (3) a response induced by "creeping" wave which comes around the fuselage.

Note that the orthogonal polarizations of the cruise missile radar profiles have significant differences which occur near the resonant band. The cruise missile range profiles in UHF and super high frequency (SHF) bands practically coincide with orthogonal polarizations in majority of aspect angles due to large electrical sizes of missile main construction elements including the smooth parts and linear edges [20].

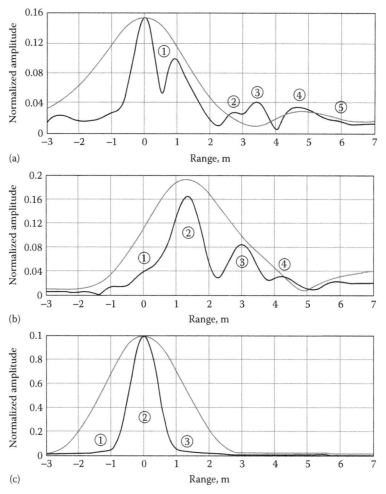

FIGURE 5.24
Calculated high-resolution range profiles of the AGM-86C cruise missile for a UWB radar sounding on horizontal polarization. Radar profiles correspond to aspect angles (a) $\beta = 0°$, (b) $\beta = 60°$, and (c) $\beta = 90°$. The black and gray lines show range resolution $\delta r = 0.71$ m and $\delta r = 2.13$ m, respectively.

Range profiles of the considered objects in the case of larger range resolution $\delta r = 2.13$ m are shown in the gray lines of Figures 5.22 through 5.25. Using a larger range resolution signal only allows determining the object length in the radial direction. This effect appears in the range profiles for aspect angles different from side illumination when the longitudinal dimension is larger than $\delta r = 2.13$ m as shown in the azimuth aspect angles of 0° and 60° in Figures 5.22 through 5.25.

An analysis of the simulation results suggests the following practical conclusions for range profiles of cruise missiles as resonant size objects. These show the following features:

1. The range profiles can contain responses conditioned by complex interactions between separate construction elements (particularly, range profiles for aspect angle $\beta = 60°$).

2. The range profiles of cruise missiles in VHF band have significant differences for horizontal and vertical polarization. Based on these results, horizontal polarization signals produce better results for this class of object.

3. As expected, the range profiles of the considered cruise missiles depend on the object aspect angle. Such phenomenon would complicate the algorithm of aerial object non-cooperative identification using only range profiles.

4. We can get an important advantage using sounding signals with the resonant wavelength band of the target. The effects observed in Figures 5.22 through 5.25 demonstrate the existence of "creeping" waves or CNRs associated with the object dimensions. Properties and methods of CNR's extraction are described in [7–12]. CNRs do not depend on the object aspect angle, and they can serve as identification signs for reducing the number of identification algorithm parameters [7–16].

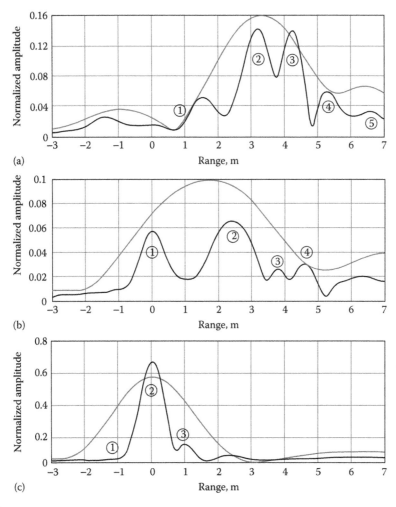

FIGURE 5.25
Calculated high-resolution range profiles of the AGM86C cruise missile for a UWB radar sounding on vertical polarization. Radar profiles correspond to aspect angles (a) $\beta = 0°$, (b) $\beta = 60°$, and (c) $\beta = 90°$. The black and gray lines show range resolution $\delta r = 0.71$ m and $\delta r = 2.13$ m, respectively.

Our results show that using VHF band signals provide the best results for cruise missile detection and identification. The calculation results show how the RCS median values of cruise missiles in VHF band can exceed the analog values in UHF and SHF bands by two orders for azimuth aspect angles from 0 to 45 deg. We can also suggest increasing the reflected signal intensity which leads to longer detection ranges for this class of aerial objects. It also improves the accuracy of target coordinate estimation.

5.7 Calculated UWB Impulse Responses of Buried Land Mines

This section presents the results of numerical simulations of the impulse responses of M15 and MK7 metal antitank mines and the DM11 antipersonnel plastic mine. Our impulse response calculations used the constructed mine surface models shown in Figure 5.26. Table 5.1 gives the sizes and properties of the mines considered.

Our calculations examined the buried mine responses when illuminated by a UWB signal with a uniform amplitude-frequency spectrum and zero phase-frequency spectrum in the band from 100 to 1000 MHz. Our signals had a 50-MHz spectral line interval. The mine frequency responses were weighted with Hamming window function.

Our simulations used a bistatic GPR with small electrical dimension antennas (magnetic dipoles with vector moments $\vec{p}^{tr}$ and $\vec{p}^{rec}$, respectively, oriented along the ox axis, as shown in Figure 5.1b). We selected a 0.5 m horizontal distance between the antennas and height above the ground interface. The antenna system assumed complete isolation between the transmit and receive elements. Our configuration put the centers of transmitting and receiving dipoles located at points with coordinates $x_{tr} = 20$ cm $y_{tr} = 0$, and $x_{rec} = -15.36$ cm, $y_{rec} = 35.36$ cm, respectively. Coordinates $x = y = 0$ corresponded to mine center at all considered cases. We assumed subsurface objects buried at the depth of $h_{ob} = 6$ cm.

Our calculations assumed the mines were surrounded by two types of soil with various densities ρ, and moisture contents W: gray loam having $\rho = 1.2 \, \text{g/cm}^3$, $W = 10\%$, and brown loam with parameters $\rho = 1.2 \, \text{g/cm}^3$, $W = 10\%$ [58].

Figure 5.27 shows the UWB impulse responses of the M15, the MK7, and the DM11 buried in gray loam and brown loam as indicated.

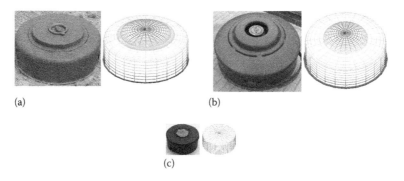

(a)   (b)

(c)

FIGURE 5.26
Mines and models used to compute UWB impulse responses. (a) The M15 antitank metal mine, (b) the MK7 antitank metal mine, and (c) the DM-11 antipersonnel plastic mine.

TABLE 5.1

Summary of Mines

Mine	Type	Diameter (mm) d_{mine}	Height (mm) h_{mine}	Material
M15	Antitank	334	124	Metal (PEC)
MK7	Antitank	325	127	Metal (PEC)
DM-11	Antipersonnel	81	37	Plastic ($\varepsilon_2 = 2.2$)

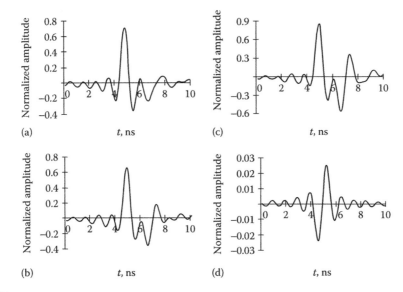

FIGURE 5.27
Computed UWB radar impulse responses of buried mines. (a) M15 in gray loam, (b) MK7 in gray loam, (c) MK7 in brown loam, and (d) DM-11 in gray loam.

Figure 5.27 shows the observed effects inherent in radar sounding of resonant size objects. We applied the numerical methods considered in Subsection 5.3.4 and 5.4.3 to develop the mine identification algorithm against a background of interfering items for the mines CNRs [16]. The calculation results confirm that the CNRs of mines do not depend on a relative positioning of the GPR antenna system and the mine. However, implementation of the algorithm requires performing measurements of the soil electrophysical characteristics.

5.8 Conclusions

This chapter presented newly developed numerical methods for simulating UWB scattering characteristics of metal and dielectric resonant sizes radar targets. These methods can help to predict the VHF band scattering characteristics of aerial radar targets, such as missiles, UAV, and small-scale airplanes which could appear in future wars. The methods can also simulate the scattering characteristics of the objects located in dielectric media such as different types of mines including nonmetallic ones. Predicting UWB responses of dielectric objects has potential applications in nondestructive testing and medical imaging for detecting masses of different dielectric properties such as material defects and effects of disease on organs.

Developed methods are based on IE solution in frequency domain and provide an easily implemented approach using multiple single-frequency signals. Sections 5.3 and 5.4 described numerical methods which have some advantages for the scattering simulation characteristics of air and subsurface resonant size radar objects. To determine the accuracy of these methods, we compared the calculated RCS of simple objects with the results found in physical experiments and results calculated by other numerical methods.

The developed methods have applications for estimating the scattering characteristics of air and subsurface objects of resonant sizes in the spatial-frequency and spatial-time domains. The offered algorithms simulate the monostatic and bistatic signals reflected by the objects when given the sounding signal polarization, spatial and time–frequency parameters. They have potential applications in developing signal-processing methods for predicting radar responses and identification methods based on signal content analysis.

References

1 Knott, E.F., Shaeffer, J.F., and Tuley, M.T. *Radar Cross Section*, 2nd edition. SciTech Publishing, Inc., Raleigh, NC. 2004.

2 Zalevsky, G.S. and Sukharevsky, O.I. Numerical method of resonance-size air object scattering characteristic calculation based on integral equation solving, In *Proceedings of the 12th International Conference on Mathematical Methods in Electromagnetic Theory, MMET'08*, June 29 –July 2, 2008, Odessa, Ukraine, pp. 334–336.

3 Sukharevsky, O.I., Zalevsky, G.S., Nechitaylo, S.V., and Sukharevsky, I.O., *Simulation of Scattering Characteristics of Aerial Resonant-Size Objects in the VHF Band*, Radioelectronics and Communications Systems, 2010, Vol. 53, No. 4, pp. 213–218.

4 Taylor, J.D. et al. *Ultrawideband Radar: Applications and Design*, CRC Press, Taylor & Francis Group. Boca Raton, FL, 2012.

5 Grinev, A.Y. et al. *Questions of Subsurface Radar*, Radiotechnika. Moscow. 2005 (in Russian).

6 Jol, H.M. et al. *Ground Penetrating Radar Theory and Applications*, Elsevier, Amsterdam, 2008.

7 Gaunaurd, G.C., Uberall, H., and Nagl, A. Complex-frequency poles and creeping-wave transients in electromagnetic-wave scattering, *Proceedings of the IEEE*, Vol. 31, No 1, pp. 172–174, 1983.

8 Baum, C., Rothwell, E., Chen, K., and Nyquist, D. The singularity expansion method and its application to target identification, *Proceedings of the IEEE*, Vol. 79, No 10, 1481–1492, 1991.

9 Mooney, J.E., Ding, Z., and Riggs, L.S. Robust target identification in white Gaussian noise for ultra wide-band radar systems, *IEEE Transactions on Antennas and Propagations*, Vol. 46, No 12, pp. 1817–1823, 1998.

10 Mooney, J.E., Ding, Z., and Riggs, L.S. Performance Analysis of a CLRT automated Target discrimination scheme, *IEEE Transactions on Antennas and Propagations*, Vol. 49, No. 12, pp. 1827–1835, 2001.

11 In-Sik C., Joon-Hu L., Hyo-Tae K., and Rothwell, E.J. Natural frequency extraction using late-time evolutionary programming-based CLEAN, *IEEE Transactions on Antennas and Propagations*, Vol. 51, No. 12, pp. 3285–3292, 2003.

12 Chen, W.C and Shuley, N.V.Z. Robust Target Identification Using a Modified Generalized Likelihood Ratio Test, *IEEE Transactions on Antennas and Propagations*, Vol. 62, No 1, pp. 264–273, 2014.

13 Chan, L.C., Moffat, D.L. and Peters, L. Jr. A Characterization of Subsurface Radar Targets, *Proceedings of the IEEE*, Vol. 67, No 7, pp. 991–1000, 1979.

14 Vitebsky, S. and Carin, L. Resonances of Perfectly Conducting Wires and Bodies of Revolution Buried in a Lossy, Dispersive Half-Space, *IEEE Transactions on Antennas and Propagations*, Vol. 44, No 12, pp. 1575–1583, 1996.

15 Geng, N., Jackson, D.R., and Carin, L. On the Resonances of a Dielectric BOR Buried in a Dispersive Layered Medium, *IEEE Transactions on Antennas and Propagation*, Vol. 47, No 8, pp. 1305–1313, 1999.

16 Zalevsky, G.S., Muzychenko, A.V., and Sukharevsky, O.I. Method of radar detection and identification of metal and dielectric objects with resonant sizes located in dielectric medium, *Radioelectronics and Communications Systems*, Vol. 55, No 9, pp. 393–404, 2012.

17 Shirman, Ya.D. et. al. *Computer Simulation of Aerial Target Radar Scattering Recognition, Detection and Tracking*, Artech House, Norwood, M.A. 2002.

18 Ufimtsev, P.Y. *Theory of Edge Diffraction in Electromagnetics*, Tech Science Press. Encino, CA. 2003.

19 Sukharevsky, O.I, Gorelyshev, S.A., and Vasilets, V.A. UWB Pulse Backscattering from Objects Located near Uniform Half-Space. In Taylor, J.D. et al. *Ultrawideband Radar: Applications and Design*. CRC Press Taylor & Francis Group. Boca Raton, FL, pp. 253–284, 2012.

20 Sukharevsky, O.I. et al. *Electromagnetic Wave Scattering by Aerial and Ground Radar Objects*, CRC Press Taylor & Francis Group. Boca Raton, FL, 2014.

21 Çağatay U., Gonca Ç., Mustafa Ç., and Levent S. Radar Cross Section (RCS) Modeling and Simulation, Part 1: A Tutorial Review of Definitions, Strategies, and Canonical Examples, *IEEE Antennas and Propagations Magazine*, Vol. 50, No 1, pp. 115–126, 2008.

22 Gonca Ç., Mustafa Ç., and Levent S. Radar cross section (RCS) Modeling and Simulation, Part 2: A Novel FDTD-Based RCS Prediction Virtual Tool for the Resonance Regime, *IEEE Antennas and Propagations Magazine*, Vol. 50, No 2, pp. 81–94, 2008.

23 Mittra, R. *Computer Techniques for Electromagnetics*, Pergamon Press. Oxford. 1973.

24 Vasilyev, Ye.N. *Exitation of a Body of Revolution*, Radio I svyaz. Moscow. 1987 (in Russian).

25 Rius, J.M., Úbeda, E., and Parron, J. On the Testing of the Magnetic Field Integral Equation With RWG Basis Functions in Method of Moments, *IEEE Transactions on Antennas and Propagation*, Vol. 49, No 11, pp. 1550–1553, 2001.

26 Levent, G. and Özgür, E. Singularity of the Magnetic Field Integral Equation and its extraction, *IEEE Antennas and Wireless Propagation Letters*, Vol. 4, pp. 229–232, 2005.

27 Özgür E. and Levent G. Linear-Linear Basis Functions for MLFMA Solutions of Magnetic Field and Combined Field Integral Equations, *IEEE Transactions on Antennas and Propagation*, Vol. 55, No 4, pp. 1103–1110, 2007.

28 Eibert, T.F. Some scattering results computed by surface-integral-equation and hybrid finite-element – boundary-integral techniques, accelerated by the multilevel fast multipole method, *IEEE Antennas and Propagation Magazine*, Vol. 49, No 2, pp. 61–69, 2007.

29 Gibson, W.C. *The Method of Moments in Electromagnetics*. Chapman & Hall, Taylor & Francis Group. Boca Raton, FL, 2008.

30 Ylä-Oijala, P., Taskinen, M., and Järvenpää, S. Advanced Surface Integral Equation Methods in Computational Electromagnetics, *Proceedings of International Conference on Electromagnetics in Advanced Applications, ICEAA '09*, September 14–18, 2009, Torino, Italy, pp. 369–372.

31 Zalevsky, G.S., Nechitaylo, S.V, Sukharevsky, O.I., and Sukharevsky, I.O. EM Wave Scattering by Perfectly Conducting Disk of Finite Thickness, *Proceedings of the 13th International Conference on Mathematical Methods in Electromagnetic Theory, MMET'10*, September 6–8, 2010, Kyiv, Ukraine, 1 CD-ROM.

32 Ubeda, E., Tamayo, J.M., and Rius, J.M. Taylor-Orthogonal Basis Functions for the Discretization in Method of Moments of Second Kind Integral Equations in the Scattering Analysis of Perfectly Conducting or Dielectric Objects, *Progress in Electromagnetics Research*. Vol. 119, pp. 85–105, 2011.

33 Su Y., Jian-Ming J., and Zaiping N. Improving the Accuracy of the Second-Kind Fredholm Integral Equations by Using the Buffa-Christiansen Functions, *IEEE Transactions on Antenna and Propagations*, Vol. 59, No 4, pp. 1299–1310, 2011.

34 Volakis, J.L. and Sertel, K. *Integral Equation Methods for Electromagnetics,* SciTech Publishing, Inc. Raleigh, NC, 2012.

35 Ubeda, E., Tamayo, J.M., Rius, J.M., and Heldring, A. Stable Discretization of the Electric-Magnetic Field Integral Equation With the Taylor-Orthogonal Basis Functions, *IEEE Transactions on Antennas and Propagation,* Vol. 61, No 3, pp. 1484–1490, 2013.

36 Zalevsky, G.S. and Sukharevsky, O.I. Secondary Emission Characteristics of Resonant Perfectly Conducting Objects of Simple Shape, *Proceedings of International Conference on Antenna Theory and Techniques, ICATT'13,* September 16–20, 2013, Odessa, Ukraine, pp. 145–147.

37 Zalevsky, G.S. and Sukharevsky, O.I. Calculation of Scattering Characteristics of Aerial Radar Objects of Resonant Sizes Based on Iterative Algorithm, *Radioelectronics and Communications Systems,* Vol. 57, No 6, pp. 13–25, 2014.

38 FEKO™ Comprehensive Electromagnetic Solutions. The Complete Antenna Design and Placement Solution. Online: http://www.feko.info.

39 Vitebsky, S., Sturgess, K., and Carin, L. Short-Pulse Plane-Wave Scattering from Buried Perfectly Conducting Bodies of Revolution, *IEEE Transactions on Antennas and Propagation,* Vol. 44, No 2, pp. 143–151, 1996.

40 Sukharevsky, O.I. and Zalevsky, G.S. EM Wave Scattering by Resonance-Size Buried Objects, *Telecommunication and RadioEngineering,* Vol. 51, No 9, pp. 80–87, 1997.

41 Sirenko, Yu.K., Sukharevsky, I.V., Sukharevsky, O.I., and Yashina, N.P. *Fundamental and Applied Problems of the Electromagnetic Wave Scattering Theory,* Krok, Kharkov. 2000 (in Russian).

42 Harrington, R.F. Boundary Integral Formulations for Homogeneous Material Bodies, *Journal of Electromagnetic Waves and Applications,* Vol. 3, No 1, pp. 1–15, 1989.

43 Ylä-Oijala, P. and Taskinen, M. Application of Combined Field Integral Equation for Electromagnetic Scattering by Dielectric and Composite Objects, *IEEE Transactions on Antennas and Propagation,* Vol. 53, No 3, pp. 1168–1173, 2005.

44 Ylä-Oijala, P. and Taskinen, M. Well-Conditioned Müller Formulation for Electromagnetic Scattering by Dielectric Objects, *IEEE Transactions on Antennas and Propagation,* Vol. 53, No 10, pp. 3316–3323, 2005.

45 Ylä-Oijala, P., Taskinen, M., and Sarvas, J. Surface Integral Equation Method for General Composite Metallic and Dielectric Structures with Junctions, *Progress in Electromagnetics Research,* Vol. 52, pp. 81–108, 2005.

46 Lu, C.C. and Zeng, Z.Y. Scattering and Radiation Modeling Using Hybrid Integral Approach and Mixed Mesh Element Discretization, *Progress in Electromagnetic Research Symposium, Proceedings,* August 22–26, 2005, Hangzhou, China, pp. 70–73.

47 Ylä-Oijala, P., Taskinen, M., and Järvenpää, S. Analysis of Surface Integral Equations in Electromagnetic Scattering and Radiation Problems. *Engineering Analysis with Boundary Elements,* Vol. 32, pp. 196–209, 2008.

48 Geng, N. and Carin, L. Wide-Band Electromagnetic Scattering from a Dielectric BOR Buried in a Layered Lossy Dispersive Medium, *IEEE Transactions on Antennas and Propagation.* Vol. 47, No 4, pp. 610–619, 1999.

49 Kucharsky, A.A. Electromagnetic Scattering by Inhomogeneous Dielectric Bodies of Revolution Embedded Within Stratified Media, *IEEE Transactions on Antennas and Propagation,* Vol. 50, No 3, pp. 405–407, 2002.

50 Meixner, J. The Behavior of Electromagnetic Fields at Edges, *IEEE Transactions on Antennas and Propagation,* Vol. 20, No 4, pp. 442–446, 1972.

51 Stevens, N. and Martens, L. An Efficient Method to Calculate Surface Currents on a PEC Cylinder with Flat end Caps, *Radio Science.* Vol. 38, No 1, 2003 [Electron resource]:http://www.onlinelibrary.wiley.com/doi/10.1029/2002RS002768/pdf.

52 Halton, J.H. On the Efficiency of Certain Quasi-Random Sequences of Points in Evaluating Multi-Dimensional Integrals. *Numerische Mathematik.* Vol. 2, No 2, pp. 84–90, 1960.

53 King, R. and Tai Tsun Wu. *The Scattering and Diffraction of Waves,* Harvard University Press, Cambridge, 1959.

54 Ufimtsev, P.Y. *Method of Edge Waves in Physical Theory of Diffraction*, Sov. Radio. Moscow. 1962 (in Russian).

55 Penno, R.P., Thiele, G.A. and Pasala, K.M. Scattering from a perfectly conducting cube, *Proceedings of the IEEE*. Vol. 77, No 5, pp. 815–823, 1989.

56 Sommerfeld, A. *Partial Differential Equations in Physics (Lectures on Theoretical Physics Vol. VI)*, Academic Press, 1964.

57 Felsen, L.B. and Marcuvitz, N. *Radiation and Scattering of Waves*, Prentice-Hall, Inc. Englewood Cliffs, New Jersey, 1973.

58 Hipp, J.E. Soil Electromagnetic Parameters as Functions of Frequency, Soil Density, and Soil Moisture. *Proceedings of the IEEE*. Vol. 62, No 1, pp. 98–103, 1974.

6

Nondestructive Testing of Aeronautics Composite Structures Using Ultrawideband Radars

Edison Cristofani, Fabian Friederich, Marijke Vandewal, and Joachim Jonuscheit

CONTENTS

6.1 Introduction to Extremely High-Frequency Microwave Ultrawideband Radars for the Nondestructive Testing of Aeronautical Structures

This chapter reveals some of the most relevant capabilities of ultrawideband (UWB) radar sensors applied to nondestructive testing (NDT). It will show the value, strengths, and limitations of UWB NDT technology imposed by technology or physical limits. The value of UWB radar sensors in the extremely high frequency (EHF) microwave band, that is, from 10 gigahertz (GHz) to 300 GHz [1], still needs practical demonstrations for the majority of the existing and prospective NDT applications. Most of those applications have been sufficiently covered after decades of using well-known NDT techniques operating in different frequency bands. However, recent developments in hardware have made possible a new generation of affordable and compact EHF UWB sensors as a technology that is worth considering in the NDT world. Our demonstration will show UWB NDT radar as a possible technology of choice for certain dielectric materials such as aramid or fiberglass composites.

Unfortunately, UWB radar cannot perform in-depth inspection of carbon parts due to their conductive nature, although surface inspection still remains possible. In the recent years, several groups have started investigating the possibilities of EHF UWB imagery, given the sustained growth of air transport and the extremely high safety standards imposed to older as well as cutting-edge aircraft. Moreover, Boeing's newest model – the Boeing 787 Dreamliner – or the prospective Airbus models foresee drastic reductions in fuel consumption because composite materials account for at least 50% of the aircraft's total structural weight [2,3]. The fiberglass composite materials are used in very critical points in an aircraft and can deteriorate due to impacts and mechanical stress. Some of these parts, as shown in Figure 6.1, are the nose radome, leading flaps, wing-to-body fairings, or the vertical stabilizers. The industry needs a need for fast, noninvasive, in-depth inspection of these critical parts, and EHF UWB radars as an NDT merit serious consideration.

The typical industry NDT approach works on selected or all parts of an aircraft on a regular basis in order to meet safety standards and maximize the material lifespans.

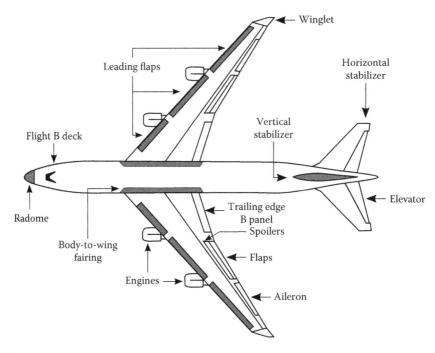

FIGURE 6.1
A typical modern commercial airplane uses both metal and composite structures. The dark areas show typical parts made of fiberglass composites or laminates which EHF UWB radar sensors could potentially inspect.

Commercial constraints emphasize reducing aircraft downtime during maintenance while keeping cost-effectiveness and safety standards whenever possible. The reduction of aircraft downtime is not always possible depending on the material to be inspected and the chosen NDT technique. The most widely used NDT techniques include infrared thermography [4–6], X-ray radiography, and three-dimensional (3D) computed tomography [7,8], high-frequency microwave systems (approximately 30 GHz) [9–14], or ultrasonic testing [15,16]. These techniques can evaluate defects within fiberglass composite materials and reveal surface, subsurface, and any other in-depth information, but have important restrictions on their operation. For example:

- Infrared thermography: in-situ, low-cost inspection lacks suitability for thick composite materials.
- X-ray radiography and 3D computed tomography: provide very high spatial resolution with very expensive sensors. These present major health risks due to the ionizing X-ray radiation, which requires confined testing areas and increased aircraft downtime.
- High-frequency microwave systems: provide in-situ, low-cost inspection requiring near-field measurements. These yield exploitable spatial resolutions but have very limited depth inspection capabilities.
- Ultrasonic testing: requires direct material contact to perform the inspection. The typically used immersion baths and coupling gels imply disassembling parts whenever possible. This adds to aircraft downtime.

When compared to X-ray based techniques, this chapter will show how EHF UWB radar systems have the most attractive capabilities of hazard-free operation by using nonionizing radiation [17,18]. Furthermore, EHF UWB radar systems have the remarkable ability to penetrate most nonmetal, nonpolarized materials, such as aramids, or fiberglass composites, and other innovative structures. Such a high material penetration allows in-depth or 3D NDT imagery. Thanks to the outstanding bandwidth of modern UWB sensors, their capability to acquire in-depth information is a reality and can be systematically performed by using frequency-modulated continuous-wave (FMCW) systems and optic components. This lets the user to focus radiated beams into very thin beam spots onto or within the inspected material [19,20]. It can also use unfocused beams and focus them digitally using image processing software. Section 6.2 will describe these two approaches to FMCW UWB radar NDT along with the capabilities of the three (100, 150, and 300 GHz) UWB radars used in materials testing. Section 6.3 will show examples of measurements performed on fiberglass aeronautical structures to detect typical defects. This will help the reader understand how EHF UWB radars can make valuable contributions to the NDT industry. Section 6.4 will introduce the potential value of using a complementary semi-automatic image processing algorithm during aircraft materials' inspection. It will show several examples of how to simplify the operator's task using a visual representation of detected defects. Section 6.5 will present an innovative technique which increases both spatial and depth resolutions in UWB imagery by means of fusing data from two UWB radar sensors operating in adjacent frequency bands. Finally, Section 6.6 will summarize the capabilities and restrictions of UWB radars applied to NDT and suggest future trends.

6.2 UWB Radar NDT Applications

6.2.1 A Survey of Existing NDT Techniques

Any new inspection NDT technology must produce comparable or superior results compared to well established and trusted techniques. This chapter will compare the results of EHF UWB radar inspection with currently established NDT techniques with particular emphasis on infrared thermography, X-ray radiography, and ultrasound techniques. The underlying work focuses on volume inspection techniques which only require access from one side of the object under test. This presents a great advantage compared to transmission mode X-ray radiography which requires access to the front and backside of the test sample. However, due to its superior image resolution, the X-ray technique serves as a good reference for comparisons with regard to defect detections.

Specifically, X-ray radiography can detect inhomogeneous material distributions caused by absorption and scattering effects within the respective sample regions. The resulting X-ray images come from discontinuities in materials, varying thicknesses, and defects, such as cracks or foreign object inclusions.

To present a complete comparison, we performed ultrasonic transmission measurements. Ultrasound requires critical coupling of the acoustic waves to the structure because variations of the ultrasonic propagation will influence the measured intensity distribution. Nevertheless, ultrasonic technology can provide reflection geometry measurements and can investigate objects only accessible from one side. One method of choice is the pulse echo technique, in which the acoustic back-reflection of the ultrasound pulse on a boundary layer within the sample is detected by the ultrasound transceiver. The amplitude and runtime of the signal and, consequently, the distance of the reflection can be determined.

Infrared thermography inspection of an object relies on the analysis of the temperature distribution across the test sample. In this case, the sample requires careful heating in advance of the measurements. Since a suitable infrared camera makes fast noncontact measurements and generates fast results as images, it makes an attractive technique.

All the previously described X-ray, ultrasonic, and thermography NDT techniques require either adequate sample preparation, sample contact (directly or by a coupling medium), and radiation protection. However, they cannot provide depth information for a better defect localization. EHF UWB radars do not inherently have any of these shortcomings.

6.2.2 NDT with FMCW UWB Radar Sensors

UWB FMCW radar has great potential as a NDT measurement technique. This proposed measurement technique uses a FMCW radar system which uses a voltage-controlled oscillator (VCO) as signal source. The designer can program the VCO output frequency as a linearly swept saw-tooth ramp sequence. A data acquisition unit (DAQ) works in combination with a low-pass filter to provide a continuous sweep frequency signal from the sampled digital output signal of the DAQ. As shown in Figure 6.2, the VCO drives both an active multiplier chain and a harmonic mixing receiver unit connected to a directional coupler.

The coupler guides the up-converted VCO signal from the multiplier chain into a circular horn antenna for free-space emission and directs the received signal by the same horn antenna to the harmonic mixing receiver. The VCO signal which goes into the harmonic mixer serves as local oscillator (LO) signal for the superposition of the received backscattered radiation from a target in front of the antenna. This results in a beat signal, the beat

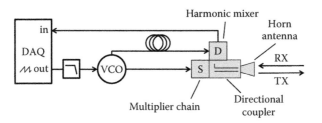

FIGURE 6.2
The FMCW UWB radar system for NDT. A single-horn antenna transmits the signal from the multiplier chain (S) and feeds the received signal back into the directional coupler for detection (D).

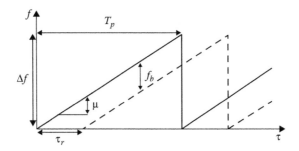

FIGURE 6.3
The FMCW radar output showing the transmitted (solid line) and received signal (dashed line) frequency versus elapsed time.

frequency f_b of which is linearly proportional to the time delay τ_r between the transmitted (solid) and received (dashed) signals, as shown in Figure 6.3.

Consequently, the relationship between the time delay and the beat frequency is given by Equation 6.1 as follows:

$$f_b = \frac{\tau_r \Delta f}{T_p} = \tau_r \mu, \tag{6.1}$$

where:
$\mu = \Delta f / T_p$ indicates the chirp rate
Δf indicates the modulation bandwidth
T_p indicates the time period of the frequency sweep.

Since the time delay τ_r is proportional to the optical path difference of the mixed signals, Equation 6.2 describes the distance d of a target as follows:

$$d = \frac{c_0}{2n_v} \tau_r = \frac{c_0}{2n_v} \cdot \frac{f_b T_p}{\Delta f} = \frac{c_0}{2n_v} \cdot \frac{f_b}{\mu}, \tag{6.2}$$

where the relationship between the speed of light in the vacuum c_0 and the frequency-dependent refractive index n_v of the propagated medium is considered. The received FMCW signal can be expressed as Equation 6.3 and a delayed version by τ_r of the transmitted signal can be expressed as Equation 6.4 as follows:

$$s_t(\tau) = \exp\left(j2\pi\left[\phi_0 + f_c\tau + \frac{\mu}{2}\tau^2 \right] \right), \tag{6.3}$$

$$s_r(\tau) = s_t(\tau - \tau_r) = \exp\left(j2\pi\left[\phi_0 + f_c(\tau - \tau_r) + \frac{\mu}{2}(\tau - \tau_r)^2 \right]\right), \tag{6.4}$$

where:
ϕ_0 denotes the initial phase
τ denotes the time since the start of the FMCW cycle
f_c denotes the sensor's carrier frequency.

The beat signal presents a much lower bandwidth, obtained after homodyne detection, as described by Equation 6.5. Equation 6.6 gives a general expression for the beat signal as follows:

$$s_b(\tau) = s_t(\tau) \cdot s_r^*(\tau) \propto \tag{6.5}$$

$$\propto \exp\left(j2\pi f_c\tau_r \right)\cdot\exp\left(j2\pi f_b\tau \right)\cdot\exp\left(-j\pi f_b\tau_r \right),$$

$$s_b(\tau) \cong \exp\left(-j\frac{4\pi d}{\lambda} \right)\cdot\exp\left(-j2\pi f_b\tau \right). \tag{6.6}$$

The three terms in Equation 6.5 from left to right indicate the cross-range phase history, the range chirp, and the so-called residual video phase. Neglecting the video phase, because of the small size compared to the other two terms [21], results in Equation 6.6. The resolution of the beat signal frequency Δf_{b_min} is equal to $1/T_p$. Consequently, the range measurement resolution Δd_{min} does not depend on the operating frequencies of the measurement devices and has a limited set by the modulation bandwidth Δf, as described by Equation 6.7 as follows:

$$\rho_r = \Delta d_{min} = \frac{c_0}{2n_v}\cdot\frac{\Delta f_{b\_min}T_p}{\Delta f} = \frac{c_0}{2n_v\Delta f}. \tag{6.7}$$

The DAQ records and discretizes the analog output signal of the receiver module, while a delay line for the LO signal can shift the beat frequencies of the targets of interest to the proper operating range of the system components. Fourier analysis of the received mix of beat signals from multiple targets in different depths can determine the distance of each target as well as the amplitudes of the received signals. Since the measurement system acquires a real-valued band-limited signal, the fast fourier transform (FFT) can provide the required complex discrete-time analytical signal [22] for further processing of the coherent measurements. Afterwards, the signal processing applies a Hamming window to compensate the errors introduced by the band-limited signal conversion. In order to maintain the required linearity for the frequency sweep, the nonlinearity of the VCO is characterized and compensated by an adjusted VCO driving voltage. While this approach allows for static corrections of a given setting, the application of a resampling algorithm on the acquired data in combination with a reference loop for parallel observation of the VCO signal can compensate for the dynamic changes of the VCO behavior [23]. Since the static correction method preserves already sufficient results for the underlying objective, this requires no additional development efforts to implement supplementary correction techniques.

Prior to measuring a sample, the system's noise m_{noise} is determined by deflecting the emitted signal so that no measureable back-scattered signal portions can be detected by the transceiver. Afterwards, a reference measurement m_{ref} is performed on a system-integrated direct back-reflector located at the center of the system's working distance. This gives a flat amplitude response and a fixed phase center of the frequency sweep for future processing. Hence, the system calibrates the received signal from a tested sample data m_{ucal} as follows:

$$m_{cal} = \frac{m_{ucal} - m_{noise}}{m_{ref} - m_{noise}}. \tag{6.8}$$

In order to use the system for volume inspection, the measurement unit combines with a two-axis translation unit to perform two-dimensional (2D) raster scans of the sample. This allows the acquisition of 3D image data, since each measurement point of the scanned sample area contains depth information. Figure 6.4 shows samples scanned in a vertical plane.

In the EHF microwave band, the achievable image resolution increases with higher center frequencies, whereas in return, the transmittance of the object typically decreases. The transparency of the sample strongly depends on its nature and for this reason, the NDT radar has three different measurement units operating at different frequencies in parallel to acquire three images with depth information.

Additionally, the radar system can operate complementary synchronized receiver units to perform transmission measurements, however, the present work does not use these. The specifications of the used transceiver units are described in Table 6.1.

In addition to a quasi-optical configuration, which comprises a set of Teflon lenses to focus the emitted radiation onto or into the sample, a lens-less synthetic aperture radar (SAR) configuration can record the image data. While the latter requires additional

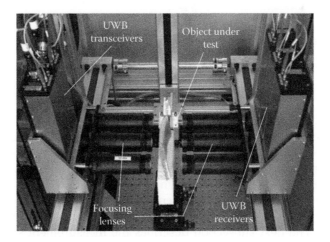

FIGURE 6.4
Photo of the FMCW UWB imaging system, operating at EHF microwave frequencies.

TABLE 6.1

Specifications of the Three FMCW UWB Measurement Units in the Focused Configuration

FMCW UWB Unit	100 GHz	150 GHz	300 GHz
Frequency band	70–110 GHz	110–170 GHz	230–320 GHz
Bandwidth	40 GHz	60 GHz	90 GHz
Output power	2 mW	100 μW	60 μW

FIGURE 6.5
The quasi-optical system configuration can sharpen UWB beam propagation. The finest spatial resolution occurs at the beam waist, but degrades at any other depth.

post-processing steps for image reconstruction, the quasi-optical approach directly acquires the image line by line. The following subsections describe both configurations and comments on their advantages and disadvantages.

6.2.2.1 The Quasi-Optical FMCW UWB Radar for NDT

The quasi-optical configuration can obtain a high-resolution image with strong signal amplitude from object features within the focal plane by focusing the emitted radiation. Images remain focused within a depth of focus b as shown in Figure 6.5, where the beam radius w_0 or beam waist is minimal. Outside this depth of focus, the beam-waist effect will produce unfocused or blurry imagery.

Both the depth of focus and the lateral resolution w_0 depend on the focal length f', the center wavelength λ_c, and the illuminated lens diameter D, which is described as follows:

$$b = \frac{4\lambda_c f'^2}{\pi D^2}. \tag{6.9}$$

Whereas the diameter of the beam waist also gives a measure for the lateral resolution, found by another approved approximation as follows:

$$w_0 \approx 1,22 \cdot \frac{\lambda_c f'}{D}. \tag{6.10}$$

Consequently, the designer can choose a larger focal length to increase the depth of focus for a roughly homogenous inspection of thicker samples. The longer focal length has the disadvantage of worse spatial resolution. For this reason, the scanner system operates with a lens configuration giving a 50-mm, 100-mm, or 200-mm focal length, depending on the requirements of the sample. The results shown in this chapter used a focal length of 50 mm, as the depth of focus is sufficient for the given samples and a high resolution is preferable for defect detections. A SAR approach makes it possible to overcome the limitations of the lens configuration. The next subsection describes the measurement set-up and the SAR configuration signal processing.

6.2.2.2 UWB NDT Sensors Using Synthetic Aperture Radar

By proper design and signal processing, UWB sensors can perform either out-of-focus or SAR measurements. The sensor can radiate energy using its real aperture or a specific wide-beam antenna. These measurements using the latter do not suffer from the beam-waist effect and can provide a constant but coarser spatial resolution. In the case of focused EHF UWB radar systems, the size of each pixel in the image is diffraction-limited [24] and spatial resolution degrades for ranges outside the beam spot. Thanks to the much larger illuminated area and the measurement overlapping (provided that the scanning step size satisfies a minimum spatial sampling to avoid spatial aliasing), then the applied SAR algorithms can focus the wide-beam measurements [25]. By coherently integrating the echoes received during the measurement, the sensor synthetically creates a much larger aperture and can provide a spatial resolution as fine as half the sensor's real physical aperture or $L_a/2$. The following expression approximates the cross-range or spatial resolution:

$$\rho_s = \frac{L_a}{2} \cong \frac{c_0/n_v}{2 f_c \sin\theta_a},$$
(6.11)

where θ_a is the opening angle of the FMCW UWB sensor. The technical specifications of the FMCW UWB sensor in the wide-beam configuration are listed in Table 6.2.

Performing this signal integration in SAR data processing can significantly improve SNR values. Compared to focused FMCW UWB radar, the SAR has a simpler design because it no longer needs beam-focusing lenses or a complex calibration. The SAR relaxes the scanning step, which means generating potentially less acquired data and faster scanning times. Figure 6.6 shows a typical monostatic NDT measurement radar, which illuminates a material by a wide-beam sensor on a platform performing 2D raster scans.

TABLE 6.2

Specifications of the Two FMCW UWB Measurement Units in the SAR Configuration (*Theoretical Values)

Measurement Unit	100	150
Frequency band	70–110 GHz	110–170 GHz
Bandwidth	40 GHz	60 GHz
Depth resolution* ρ_r (in vacuum, $n_v = 1$)	3.75 mm	2.5 mm
Aperture opening angle (−3 dB)	16°	16°
Wide-beam spatial resolution* ρ_s (in vacuum, $n_v = 1$)	6 mm	4 mm

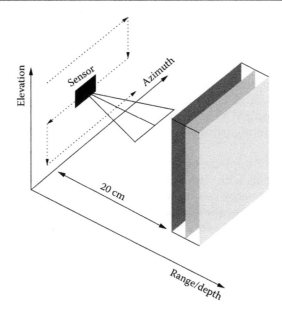

FIGURE 6.6
Schematic representation of a 2D raster scan of a material using the wide-beam FMCW UWB approach.

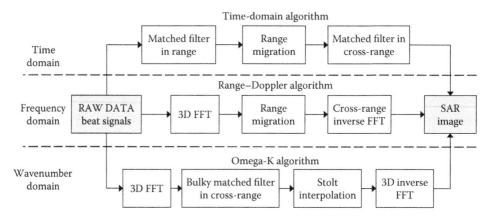

FIGURE 6.7
Conceptual block diagram describing the typical steps performed in three SAR algorithms operating in different domains.

The evolution of the range history for a given point scatterer follows a hyperbolic range function $d(t) = \sqrt{d_{min}^2 + V_{plat}^2 t^2}$, where d_{min} is the minimum sensor-to-scatterer distance, V_{plat} is the platform's speed, and t is the slow time or the time cross-range (azimuth or elevation dimensions) also referring to the platform's position. Due to the hyperbolic nature of the collected data, processing software must apply range migration to compensate for the curvature of $d(t)$ to allow accurate focusing of the collected data. Applying one of the many well-known and widely used SAR algorithms can achieve further range and cross-range energy compaction.

Figure 6.7 shows three of the most well-known SAR focusing algorithms. These include *time domain*, *frequency domain* (range–Doppler), and *wavenumber domain* (omega-K). The frequency domain (second) and wavenumber domain (third) types have remarkably efficient computational loads, but only offer an approximate solution.

The time-domain algorithm. This signal-processing algorithm delivers an optimum SAR focusing solution. However, it has an extremely heavy computational load due to the use of time domain matched filters. This makes the time-domain algorithm an improbable choice for NDT applications, where azimuth and elevation data are obtained and require coherent processing. Current processing capabilities (a 64-bit Intel® Quad processor at 2.4 GHz running MATLAB® on Linux) suggest. In contrast, frequency-domain or wavenumber-domain algorithms can generate a sufficiently accurate SAR image in less than two minutes.

As shown in Figure 6.7 (top), the basic steps to focus SAR raw data include matched filtering the raw beat signals with a model-generated or reference-transmitted signal in the time domain so that the energy is compressed in range. The process then performs a range migration to the range-compressed data in elevation and azimuth to correct the curved SAR data. Finally, the process sequentially applies a time-domain matched filtering in azimuth and elevation.

The range–Doppler algorithm. Algorithms in the frequency domain, such as the range–Doppler algorithm and its modified versions [21], are computationally efficient because they perform the most intensive SAR signal processing in the frequency domain [26–29], thanks to the extremely optimized implementations of the direct and inverse FFT. While approximations in the frequency domain processing apply, these methods produce perfectly useful results in the NDT field.

The range–Doppler algorithm use several main stages which includes the following:

1. Application of a three-dimensional FFT to the time-domain raw data to transform it to Doppler domain data. A positive aspect of FMCW systems is that the data do not need to be range-compressed in the frequency domain and inverse-transformed as in pulsed radar systems, hence reducing the computational load.

2. Subsequently, range migration and range interpolation compensate the hyperbolic curvature of the 3D data.

3. Cross-range matched filtering is efficiently performed in the horizontal and vertical Doppler domains, compacting the energy initially spread along the cross-range evolution for every target in the scene.

4. Finally, it applies an inverse FFT to the Doppler domains to produce a 3D SAR image back in the range-time-time domain or a full spatial representation. The range–Doppler offers a good trade-off between computational load and image reconstruction. This makes it a good candidate for applying SAR in a FMCW UWB NDT radar system.

The Omega-K Algorithm

The omega-K algorithm operates in the wavenumber domain [29]. As in the range–Doppler algorithm, an FFT in range on the raw data produces a very efficient range compression that can be used as a starting point for further processing which goes as follows [27,29,30,31]:

1. A 3D FFT transforms the data into the full frequency domain which produces range, azimuth, and elevation wavenumber frequencies, or f_r, f_u and f_v, respectively.

2. Applying a matched filter center initially at a reference range achieves azimuth and elevation compression. This compression will only focus targets located at the reference range. Applying a change of variables in the wavenumber domain, known as Stolt mapping or interpolation, focuses the remaining ranges [29].

3. The resulting phase evolution is transformed from quadratic to linear, as described by Equations 6.12 and 6.13, which enables an optimum focusing for all ranges.

4. Finally, applying an inverse Fourier transform to the 3D data retrieves the 3D SAR image.

$$\sqrt{\left(f_0+f_r\right)^2-\frac{c^2 f_{u,v}^2}{4V_r^2}} \approx \left(f_0+f_r\right)-\frac{c^2 f_{u,v}^2}{8V_r^2\left(f_0+f_r\right)^2}, \tag{6.12}$$

$$\sqrt{\left(f_0+f_r\right)^2-\frac{c^2 f_{u,v}^2}{4V_r^2}} = f_0+f_r'. \tag{6.13}$$

6.2.3 Discussion and Conclusions about SAR Algorithms for UWB Radar NDT

As discussed, the time-domain approach for SAR imagery is extremely time-consuming due to the intensive use of time-domain matched filters. In an industry-oriented application, SAR algorithms operating in the frequency or wavenumber domain are the most likely choice. Given the very similar performance of range–Doppler and omega-K in terms of imagery reconstruction and time consumption, the more widespread range–Doppler algorithm is chosen for presenting results in this chapter.

6.3 Materials Testing and Comparison of UWB Radar NDT Results

6.3.1 Description of the Tested Composite Materials

Examining several fiberglass test samples similar to ones used in aeronautical structures provided a quantitative assessment of EHF UWB radar NDT techniques. These samples reproduce examples of the most typical defects due to misprocess during material production. Examples of defects include paper or polyethylene sheets used for storing and separating the different sample plies and which can remain glued to the fiberglass sheet during production; Teflon or air gaps to simulate delaminations between plies in the structure; and foreign material intrusions such as water vapor. Figures 6.8 through 6.10 show the three main types of structures tested. Each structure consists of multi-ply skins made of fiberglass and epoxy resin. The structures consisted of fiberglass laminates, reinforced sandwich structures, and fiberglass-reinforced plastic C-sandwich structures. Each material could develop defects which would materially weaken the structural member and require easy detection NDT with UWB radar.

The UWB radar trials evaluated show seven different fiberglass-reinforced plastics used in this chapter and their particularities tested in using UWB radar methods. Table 6.3 summarizes the results.

6.3.2 Results of Focused FMCW UWB Radar Tests on Selected Samples

The FMCW radar described earlier measured the samples described in Table 6.3. The radar measured each piece from both sides using reflection geometry at three different frequency bands using the quasi-optical configuration described earlier. Each test sets the surface perpendicular to the transceiver beam's propagation axis. To achieve the highest spatial resolution, samples were placed in the focus of the 50-mm focusing lens. Figure 6.11 shows the effects of increasing the sensor's operation frequency and bandwidth in the

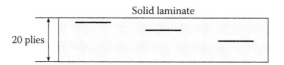

FIGURE 6.8
Example of a 20 fiberglass ply solid laminate with man-made defects between plies 3 and 4, at quarter and at half depth.

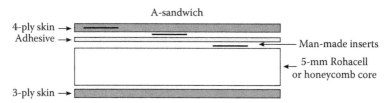

FIGURE 6.9
Fiberglass-reinforced plastic A-sandwich structures made of a 4-ply anterior skin, an adhesive layer, a 5-mm ROHACELL® foam or Nomex™ honeycomb (HC) core, and 3-ply posterior skin. Man-made defects can be located inside the anterior skin, between the anterior skin and the adhesive, or between the adhesive and the core. Small amounts of water injected in the inner skin of these structures reproduced vapor residuals during manufacturing.

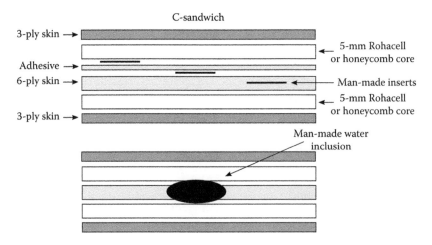

FIGURE 6.10
Fiberglass-reinforced plastic C-sandwich structures made of a 3-ply anterior skin, a 5-mm ROHACELL® or Nomex™ HC core, an adhesive layer, a 6-ply inner skin, a second core, and a 3-ply posterior skin. Man-made defects can be located between the first core and the adhesive, between the adhesive and the 6-ply skin, and inside the 6-ply skin.

front views. The three examples taken from the FMCW UWB sensor at 100, 150, and 300 GHz show different feature separation and localization capabilities on the fiberglass solid laminate structure. Although the 150-GHz and 300-GHz sensors produce finer imagery, the impact of lower transmitted power and lower material penetration is clearly visible.

Despite the possible negative effects of noisier images, the chart in Figure 6.12 shows a robust behavior in fiberglass solid laminates, for both front- and back-sides and at all frequencies.

While every underlying FMCW UWB measurement of the described laminate sample provides very positive results and allows an easy identification of nearly all inserts, more complex structures such as the ROHACELL® foam or the Nomex™ HC sandwich structures involve stronger signal absorbance and distracting scattering effects, which made defect detection more difficult as illustrated in Figure 6.13.

Figure 6.13 shows the front- and back-side images of sample GFRP_03, an A-sandwich with Nomex™ HC core acquired by the 150-GHz sensor. Despite the distracting structure of the Nomex™ HC core in the background of the front-side image, the inserts of the sample can still be satisfactorily identified, whereas the identification throughout the core is more challenging. Consequently, the 100-GHz images of sample GFRP_13, a C-sandwich with ROHACELL® foam core in Figure 6.14 show similar results between front- and back-side measurements.

In Figure 6.15, the comparison charts of the A-sandwiches with Nomex™ HC and ROHACELL® foam core show a significant difference between the two cores. While the sensors perform better on the Nomex™ HC sample, the difference between front- and back-sides of the ROHACELL® foam core sandwich is much lower. This can be explained by the homogenous background of the foam structure in connection with high reflections at its boundary layers, which lead to a poor signal contrast between core and inserts.

Figure 6.15 also reveals the limited penetration capabilities at higher center frequencies, as the results of measurements by the 300-GHz sensor demonstrate. Usually it requires a

TABLE 6.3

All the Materials Tested Using UWB Radar Including the Type of Structure, Materials, Dimensions, and Defects Are Found

Reference	GFRP_01	GFRP_02	GFRP_03
Picture			
Structure	Solid laminate	A-sandwich and ROHACELL® core	A-sandwich and Nomex™ HC core
Sample h × w × d (mm)	340 × 200 × 5	340 × 200 × 8	340 × 200 × 8
Type and size of defects	Inserts from 6 × 6 to 25 × 25 mm	Inserts from 6 × 6 to 25 × 25 mm	Inserts from 6 × 6 to 25 × 25 mm

Reference	GFRP_13	GFRP_14	GFRP_18	GFRP_20
Picture				
Structure	C-sandwich and ROHACELL® core	C-sandwich and Nomex™ HC core	C-sandwich and Nomex™ HC core	C-sandwich and Nomex™ HC core
Sample h × w × d (mm)	340 × 200 × 15	340 × 200 × 15	150 × 100 × 15	150 × 100 × 15
Type and size of defects	Inserts from 6 × 6 to 25 × 25 mm	Inserts from 6 × 6 to 25 × 25 mm	Water inclusion (0.2 ml of water)	Water inclusion (0.75 ml of water)

trade-off between resolution and penetration capabilities for best results, as the comparison charts of C-sandwiches in Figure 6.16 indicate.

In case of the investigated C-sandwiches, the difference between the core types is smaller than with the A-sandwiches, since the distractions created by the Nomex™ HC cores increased leading to a similar signal contrast.

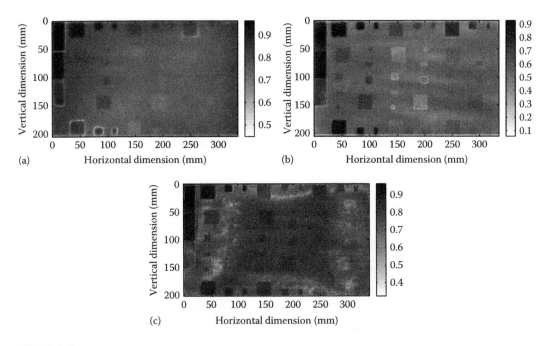

FIGURE 6.11
FMCW UWB images of the solid laminate sample with inserts measured at (a) 100 GHz, (b) 150 GHz, and (c) 300 GHz.

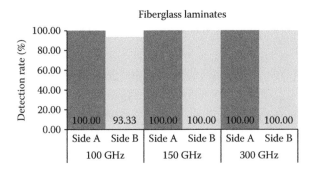

FIGURE 6.12
Comparison chart of the three FMCW UWB sensors defect detection rates on fiberglass solid laminates for both front- and back-sides.

The tests included FMCW UWB radar measurements to investigate the water inclusions in C-sandwiches with Nomex™ HC cores. Water vapor confined within the composite's layers during the manufacturing process can weaken the integrity of the material. Water inclusions create a remarkable contrast by exploiting the almost isotropic energy dispersion of water in the EHF microwave band, as shown in Figures 6.17 and 6.18.

Despite the small volume of water used in sample GFRP_18 (0.2 ml), foreign inclusion detection was possible using the 100-GHz and 150-GHz UWB radars. Likewise, the 0.75-ml water inclusion in GFRP_20 presents an exceptional contrast with respect to the C-sandwich structure. As shown in Figure 6.19, the 100-GHz and 150-GHz UWB sensors completely identified the water inclusions. The 300-GHz sensor missed the water

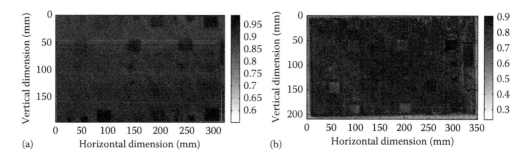

FIGURE 6.13
150-GHz sensor images of the sample GFRP_03 (A-sandwich with Nomex™ HC core): (a) front-side and (b) back-side.

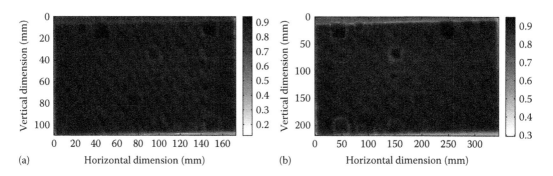

FIGURE 6.14
100-GHz sensor images sample GFRP_13 (C-sandwich with ROHACELL® core): (a) front-side and (b) back-side.

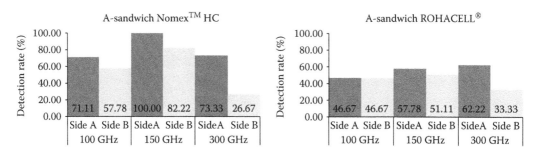

FIGURE 6.15
Comparison of defect detection rates of the three FMCW UWB sensors on A-sandwiches with ROHACELL® foam and Nomex™ HC cores, for both front- and back-sides.

inclusions due to the lower power, reduced penetration capabilities, and lower reflectivity contrast at that frequency band.

The defect detection capabilities of the discussed FMCW UWB systems have been extensively compared in Table 6.4 to the NDT methods briefly described in Section 2 as follows: X-ray radiography, ultrasound, and infrared thermography, as well as manual inspection. Measurements on the same test samples were performed using each technique in the comparison. The table evaluates each parameter according to the literature using a range from

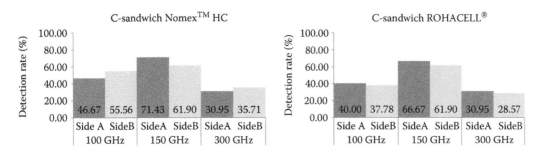

FIGURE 6.16
Comparison the detection rates of the three FMCW UWB sensors on A-sandwich samples with ROHACELL® foam and Nomex™ HC cores, for both front- and back-sides.

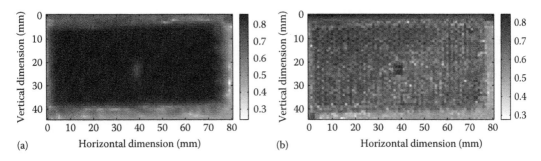

FIGURE 6.17
Radar images sample GFRP_18 (C-sandwich with Nomex™ HC core and a 0.2-ml water inclusion). (a) 100 GHz and (b) 150 GHz.

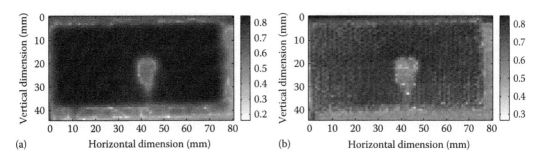

FIGURE 6.18
Radar images of the sample GFRP_20 (C-sandwich with Nomex™ HC core and a 0.75-ml water inclusion) (a) 100 GHz and (b) 150 GHz.

not suitable/low (− −) to suitable/high (+ +). FMCW UWB radar shows an overall performance comparable to that of other conventional techniques.

As demonstrated in Table 6.4, all techniques except X-ray radiography were performed successfully. This is in part due to poor detection of inserts when using radiography on solid laminates. This can be explained by the quasi-homogeneous solid structure of the sample, which provides a good transparency to EHF microwave radiation as well as a reasonable thermal conductivity and a good acoustic coupling efficiency throughout the

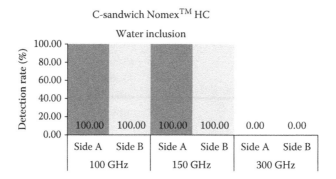

FIGURE 6.19

Comparison of the detection rates of the three FMCW UWB sensors on C-sandwiches with Nomex™ HC cores, for both front- and back-sides.

TABLE 6.4

Comparison of Objective and Subjective Performance Benefits of FMCW UWB Radar Images With Conventional NDT Methods

Technique	Material penetration	See-through capability	Portable	Cost	Health risk	Technique maturity	Overall performance
FMCW UWB radar	+ +	+ +	+	−	+ +	+	+ +
Manual inspection	− −	− −	+ +	+ +	−	+ +	−
X-ray	− −	−	− −	− −	− −	+ +	− −
Ultrasound	+ +	+ +	−	−	+ +	+ +	+
Thermography	+ +	+ +	+	−	+ +	+ +	+ +

sample. In contrast, the index of refraction of this sample structure and its inserts is almost the same in the X-ray band as in air ($n \approx 1$), leading to a very poor image contrast of object features and, consequently, to a low defect detection rate. The situation is different for structures with Nomex™ HC core, since the scattering effect of the Nomex™ HC structure has a positive influence on the contrast of the radiography images. However, the inserts of the A-sandwich sample located between solid laminate plies on the front-side of the core provided a suitable thermal behavior for thermography measurements. The FMCW UWB measurements outperform all other methods in reflection mode for measurements through the back-side of the sample. The test demonstrated the obvious advantage of EHF UWB radars for the C-sandwich measurements. EHF UWB radar performed convincingly, whereas all other applied techniques either suffered from the thermal or acoustic properties of the cores or from the very poor refractive index throughout the whole sample structure.

6.3.3 Results of Synthetic Aperture FMCW UWB Radar Imaging of Selected Samples

This section shows the results of imaging selected samples measured in wide-beam FMCW UWB radar mode and illustrates the possibilities that these measurements can offer as well as their physical limitations. Given the lack of beam-waist effects in wide-beam measurements – and since tests showed irrelevant differences between front and back measurements – only front sides were measured and investigated. The coarser spatial or

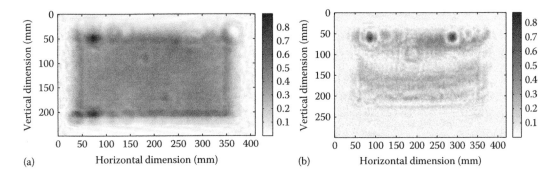

FIGURE 6.20
SAR images of sample GFRP_13 (C-sandwich with ROHACELL® core and inserts) (a) 100 GHz and (b) 150 GHz. The images show several defects present with very high contrasts. However, the poor spatial resolution of the SAR approach did not show shapes and approximate sizes.

cross-range resolution intrinsic to this technique makes imaging of medium and small size defects difficult or even unattainable. For this reason, an equitable comparison between focused and wide-beam FMCW UWB radar measurements is not possible. Instead, examples are given which can be extended to very thick fiberglass laminate or structure materials.

Figure 6.20 shows two SAR images of a C-sandwich structure with ROHACELL® (reference GFRP_13) measured at 100 GHz and 150 GHz. The initial inspections showed several defects present with very high contrasts. However, the poor spatial resolution of the SAR approach did not show shapes and approximate sizes. The trained eye could see several other less-bright defects when exploring other depths. Despite the smallest defects in the sample present dimensions (6 × 6 mm) roughly matching the sensors' spatial resolution, traces of diffraction effects can be observed and therefore, a defect can be guessed. Teflon defects present a much brighter behavior, which is consistent when increasing the operation frequency, and these defects are typically trivial to locate.

Contrary to what can be observed in focused measurements, the honeycomb pattern present in sample GFRP_14 (a C-sandwich structure with Nomex™ HC core and inserts) does not interfere the visual interpretation of the images, but generates a disturbing phenomenon which masks most of the defects in 100-GHz measurements. The slightly finer spatial and depth resolutions of the 150-GHz FMCW UWB radars allow a better reconstruction and positioning of certain defects, as illustrated in Figure 6.21.

Water inclusions may act as shapeless or spherical bodies dispersing the wide-angle plane wave fronts in countless arbitrary directions, producing low reflections from the volume with the water inclusion. This assumption implies that this type of artifacts will appear as low reflectivity regions inside the composite material. Figure 6.22 shows a great contrast in reflectivity caused by the water inclusion for both the 100-GHz and 150-GHz SAR measurements, enabling unequivocal foreign inclusion detections.

Despite the modest results provided by SAR UWB radar for the specific inspected samples, the great potential for inspection of very thick materials remains untouched. Although materials in the order of several centimeters thick are not common or even foreseen in the aeronautics field, other existing or prospective fields requiring in-situ and fast maintenance of fiberglass composites may find SAR UWB radar a desirable option to consider.

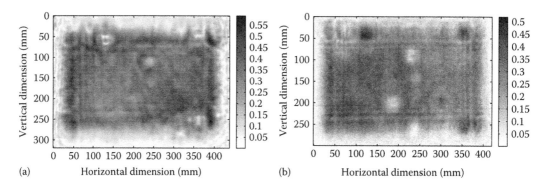

FIGURE 6.21
SAR images of sample GFRP_14 (C-sandwich with Nomex™ HC core and inserts): (a) 100 GHz and (b) 150 GHz.

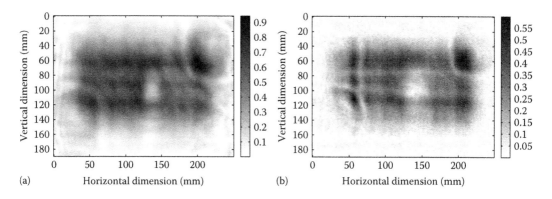

FIGURE 6.22
SAR images of sample GFRP_20 (C-sandwich with Nomex™ HC core and a 0.75-ml water inclusion) demonstrate the ability to detect water volumes in the material: (a) 100 GHz and (b) 150 GHz.

6.4 Semiautomatic Image Processing for EHF UWB NDT

6.4.1 Introduction and Objectives

The complex task of manually evaluating 3D EHF UWB imagery requires an experienced operator, which means human ability or physical fatigue can limit the reliability of the results. In certain situations, only a human operator can interpret the rich texture and patterns in large volumetric imagery. As an example, inhomogeneous and nonuniform fiber structures cause a fluctuating signal background, which makes the signal strength over a certain depth cross section to vary even when no defect is present. In this case, only an experienced operator and visual inspection can give reliable results. In other cases, image processing algorithms can reduce the operator's work for unequivocally detectable targets. The human operator can then focus their expertise on areas of interest with possible defects. Moreover, a statistical analysis of extracted defects can help to identify the type of defect, material, shape, or volume. Figure 6.23 shows a semi-automatic, that is,

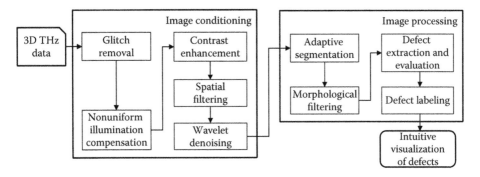

FIGURE 6.23
A semi-automatic image processing algorithm block diagram with two stages. Image conditioning used to ensure maximum performance in further processing stages. Image processing for the semi-automatic detection of defects.

operator-dependent, image processing algorithm implemented to enhance reliability and reproducibility in the EHF UWB imagery defect identification process. The algorithm has two major blocks: image conditioning and processing.

6.4.2 Preparing NDT Radar Image Data

6.4.2.1 NDT Radar Image Conditioning

In order to improve the performance of the semi-automatic detection, the algorithm can apply proper data conditioning to the acquired 3D data to avoid or reduce undesired effects and phenomena. These data conditioning steps include the following:

- Glitch removal: spurious glitches (outliers in an image) of various unknown sources are scarce, but may occur in current EHF radar measurements. Simple local detection methods can find such glitches to avoid unnecessary false alarms.

- Nonuniform illumination compensation: sample misalignment or nonflat samples produce gradient-like illumination patterns when measured. The shape of these patterns is a priori unknown, although strategies assume that nonuniform illumination is a nonstationary process and therefore use this assumption to remove the effects [32].

- Contrast enhancement: applying image histogram equalization can enhance the imagery contrast to maximize possible small, low-contrast defect feature extractions.

- Spatial filtering: some structures include Nomex™ HC cores which are clearly visible to sensors with fine spatial resolutions as shown in Figure 6.24a. This can cause the algorithm to fail in detecting clear defects due to the omnipresent honeycomb pattern. Given that this pattern is periodic, spatial filtering provides a powerful option to reduce these undesired effects.

- Figure 6.24b shows the 2D Fourier transform of the front-view of Nomex™ HC-cored sample in Figure 6.24a, in which most of the energy is concentrated around the lower frequencies. The bright points in the frequency domain indicate the presence of the honeycomb structure and can be removed by applying 2D spatial or frequency filtering. Figure 6.24c shows the filtered image with no traces of the honeycomb structure.

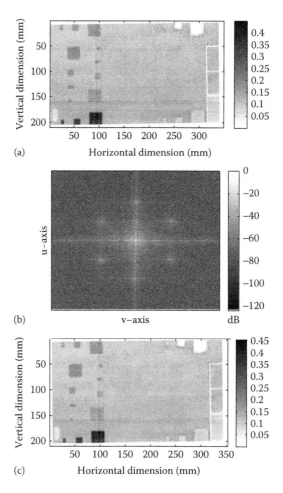

FIGURE 6.24
Example of spatially filtering UWB radar images. (a) The bright points in the frequency domain image indicate the presence of the honeycomb structure, (b) a 2D FFT of the image shows the energy concentrated in the lower frequencies and (c) the spatially filtered image presents no traces of the honeycomb structure.

- Wavelet denoising: FMCW UWB systems can produce a sufficiently high signal-to-noise ratio (SNR) imagery for many NDT purposes, however, some cases may require further noise reduction. In that regard, the de-noising process must not alter the structure and intensity values of the features found in the data. Wavelets-based denoising performs successfully when compared to spatial or frequency domain filtering. The resulting denoised images will less likely present false alarms in the semi-automatic defect detection.

6.4.2.2 NDT Radar Image Processing

The previous section showed how to preprocess 3D data and prepare it for further processing. Given the 3D nature of the data, the next steps can be performed in a 2D (range or cross-range data cuts) or 3D fashion. However, applying the 2D approach sequentially on cross-range data cuts will yield much lower computational complexity and very similar results. The core of the image processing detection includes the following:

- Image segmentation by adaptive thresholding: given the knowledge of the histogram of each imagery cut obtained after contrast enhancement, thresholding is optimized to segment regions with similar pixel values. A plausible assumption is that such regions describe defects or structures within the sample presenting similar reflectivity properties.

- Morphological filtering: applies successive morphological operators [32] to images resulting from adaptive segmentation. The process chooses structural elements comparable in shape and size to the expected defects in the sample, and then filters out all segmented regions and clutters not resembling the structural elements. Considered shapes for structural elements include the following: circular for impacts, square or rectangular for foreign inclusions, long and thin structures for cracks or disbonds, or disk shapes for water inclusions. At this point, the process becomes semi-automatic and needs human intervention.

- Defect extraction: The process extracts pixel values of the resulting candidates from the FMCW UWB data and evaluates several exclusion conditions, such as the appearance of a given defect in several cross-range cuts (actual defects show certain persistence in depth, false alarms due to the processing may not); and comparison of statistical pixel-level descriptors of the candidate; and those of the immediate surrounding area (detection likelihood is based on this local comparison).

- Defect labeling: In the last step, the process registers the remaining defects in the 3D resulting image and puts them into unique classes. A bounding cube describes each defect (therefore, storing the position and volume), statistical descriptors, and detection likelihood figures. The process can further interpret the results based on size, shape, or material; the latter assumes that defects of a same material will present very similar statistical descriptors.

6.4.3 Image Processing Applied to UWB Radar Measurements

This section will show how to apply semi-automatic image processing to selected samples including several phenomena in the focused and SAR configurations using FMCW UWB radars. Figure 6.25 shows the image processing results for the backside of the solid laminate GFRP_01 at 100 GHz. The left image shows a detection likelihood of the detected defects. Several of the smallest defects are difficult to spot during visual inspection and only appear at certain depths, which may confuse the operator. The image processing algorithm takes into account defect perseverance in depth, and the more a given defect appears at different ranges, the higher the likelihood of being positively detected. Obviously, in an automatically considered situation, such small defects may vary in size or shape when inspecting several depths. Those defects not appearing in the detection image were not possible to label as clear positive detections. Therefore, the semi-automatic possibilities of the image processing algorithm empower the operator to modify the detection sensitivity, at the cost of increasing the risk of false alarms.

Figure 6.26 shows an example of a trade-off between high sensitivity and false detections. On the one hand, the original focused UWB image shows several large defects easily located with the naked eye. On the other hand, the operator can select a high sensitivity in the image processing algorithm to locate those smaller defects appearing at certain ranges

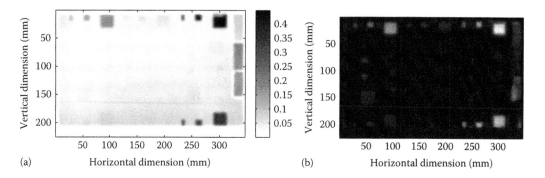

FIGURE 6.25
Example of semi-automatic image processing for UWB NDT radar defect detection. (a) The focused image of the backside of sample GFRP_01 (solid laminate with inserts) measured at 100 GHz and (b) processed image showing the detection likelihood generated by the image processing algorithm.

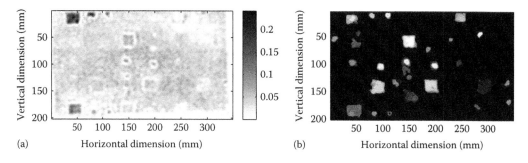

FIGURE 6.26
Example of a sensitivity and false alarm trade-off in automatic detection applied to 150-GHz radar measurements. (a) The focused image of the front-side of sample GFRP_14 (C-sandwich with Nomex™ HC core and inserts) and (b) the detection likelihood generated by the image processing algorithm.

or masked by the honeycomb pattern in the core. The process triggers several false alarms linked to the repetition of the honeycomb pattern at different ranges. In this case, the image processing acts as a complementary tool to the operator's experience.

In Figure 6.27, the focused UWB measurements perfectly detect a 0.75-ml water inclusion with a maximum likelihood, given the high contrast between water and the sandwich structure.

The image processing can reveal defects in SAR measurements, which are not clear after visual inspection. SAR images may present very complex patterns, intensity fluctuations, or diffraction phenomena (halo-like shapes). The image processing integrates all this information to simplify the scene to the maximum, as shown in Figure 6.28. The sample GFRP_13 yields defect detections bypassed by the naked eye and generally associated with low received power or low defect reflectivity.

Finally, SAR imagery can detect water inclusions as shown in Figure 6.29. Visual inspection of the SAR image reveals areas where water could be present in several depths,

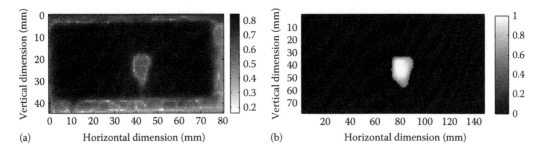

FIGURE 6.27
Example of a NDT 100-GHz radar water inclusion detected by image processing. (a) Focused image of the front-side of sample GFRP_20 (C-sandwich with Nomex™ HC core and a 0.75-ml water inclusion and (b) the detection likelihood generated by the image processing algorithm.

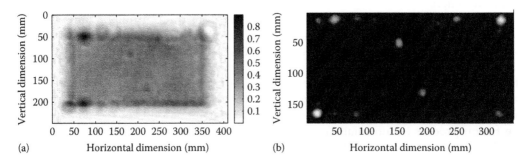

FIGURE 6.28
NDT 100-GHz SAR defects invisible to the naked eye. (a) SAR image of the front-side of sample GFRP_13 (C-sandwich with ROHACELL® core and inserts) measured at 100 GHz and (b) the defect detection likelihood generated by the image processing algorithm.

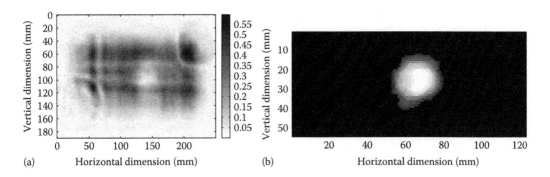

FIGURE 6.29
A 0.75-ml water inclusion in a sample GFRP_20 sample of a C-sandwich with Nomex™ HC core. (a) SAR image of the front-side and (b) the detection likelihood generated by the image processing algorithm.

although no clear decision can be made regarding its location and size. The detection likelihood figure shows a well-shaped structure, resulting from the superposition of water inclusion detections at several depths.

6.5 NDT Data Fusion Techniques

The imaging results of the EHF UWB radar measurements presented in Section 6.3 demonstrated the great benefits of the higher spatial resolution of the 300-GHz measurement sensor for defect detection. The lower transmittance of objects in the 300-GHz frequency band limits this class of sensor when compared to units with other operating frequencies. However, in addition to the better lateral resolution of the 300-GHz unit, the superior depth resolution by the 90-GHz modulation bandwidth allows a much better separation of image features from distracting signal information in adjoining depths for enhanced defect detection.

Since the proposed measurement system can operate three different transceivers in parallel, a data fusion approach could provide better depth resolution by merging the acquired data from the measurement heads. A common calibration procedure, a mandatory flat sensor amplitude response, and common fixed phase center ensure that the three units will provide consistent spatial data. Furthermore, the requirement of a joint frequency axis with uniform step size should cover the frequency range without interruption. If a gap occurs within the sequence of frequency points of the data-sets, then the fusion process cannot mathematically derive the missing data from the surrounding data. Nevertheless, a simple filling algorithm can cover small gaps without creating strong artifacts. In the case of larger gaps, such as between the 150-GHz and 300-GHz measurement units (gap size of nearly 60 GHz), applying adapted filling algorithms based on certain assumptions about the sample can eventually give satisfactory results. Generally speaking, the following sections will focus on how to merge the data from the 100-GHz and 150-GHz unit to create a data-set with a 100-GHz bandwidth resulting in a depth resolution that competes with the 300-GHz heads at the good transparency of EHF microwave radiation.

As already mentioned, merging the obtained data-sets correctly requires a correct joint frequency axis with equal frequency step size. However, Table 6.5 indicates a different step size and granularity of the given measurement heads, which leads to different limits in the time-of-flight information represented by the FFT of the obtained data-sets, as shown in Figure 6.30.

Since the Fourier transformed data at high positive or negative times are close to zero, the waveforms may be cut or extended by zero-padding and then retransformed by inverse

TABLE 6.5

Frequency Parameters of the 100-GHz and 150-GHz Sensors

Head	100 GHz	150 GHz
Frequency range	70.02–110.86 GHz	115.85–174.53 GHz
Frequency points	900	875
Frequency step size	45.38 MHz	67.06 MHz

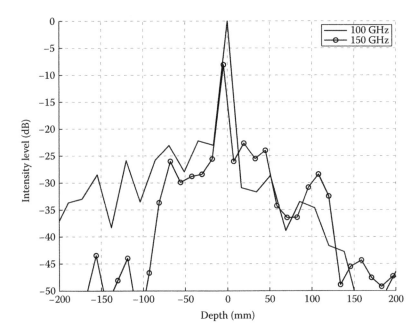

FIGURE 6.30
Fourier transforms of the obtained data-sets from the measurements of a 3-mm thick polyethylene sample by the 100-GHz and 150-GHz sensors.

Fourier transformation. This results in new data-sets with the same shape as before, but now they have an equal frequency step size.

Before merging the data-sets, the frequency gap between them needs filling to preserve frequency information throughout the whole frequency range without interruption. This means that in the present case, a frequency gap of 5.0 GHz between the 100-GHz and 150-GHz head has to be filled. Some obvious ideas such as zero filling between the highest value from the 100-GHz head and the lowest value from the 150-GHz head creates strong artifacts. On the contrary, a simple heuristic stretch approach leads to adequate results. Hereby, all measured values from the upper 2.5 GHz of the 100-GHz head used for two frequency steps rather than one, this creates 2.5 GHz of additional data used for filling up the lower half of the gap. Stretching the lower 2.5 GHz of the 150-GHz head down in the same way fills the upper half of the gap. In addition to results of the zero filling and described stretch approach, Figure 6.31 shows an even better result by a stretch approach with an overlapping frequency region.

Within this slightly more sophisticated approach, the data gets stretched by 3.33 GHz rather than 2.5 GHz (two thirds of the gap), which creates a data overlap in the central third of the gap. In this overlap region, the position of closest approach between the respective data of both heads becomes determined and the arithmetic mean of these two values is derived. At higher frequencies, the 150-GHz data are selected and at lower frequencies, the 100-GHz data are selected. The stretching method with frequency overlap usually makes an improvement because data with high-phase errors are discarded and the discontinuity at the contact point of both frequency ranges is minimized.

Figure 6.32 allows a closer look at the resulting depth resolution in comparison to the single-head data for 100, 150, and 300 GHz. The graphs show the Fourier-transformed

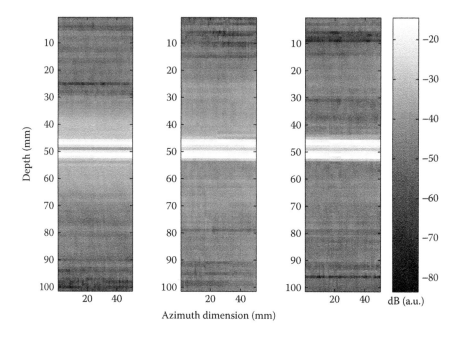

FIGURE 6.31
Comparison between zero filling, stretch method, and stretch method with overlap for filling the gap between data-sets to be fused.

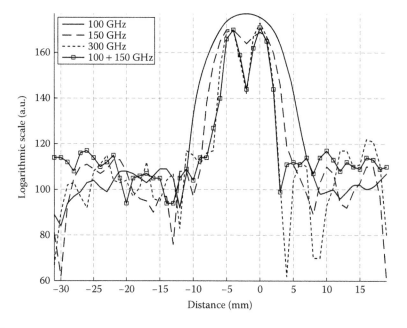

FIGURE 6.32
Hamming-filtered Fourier transform of the acquired data by the measurement of a 3-mm thick polyethylene sample.

and Hamming-filtered information of the acquired data from a 3-mm thick polyethylene sample in a fixed XY-position. Appling a window function eliminates the strong effects of the discontinuous data at the upper and lower band edges on the Fourier transform. This corresponds to a bandwidth of 150 GHz which is achieved by zero padding the Hamming-filtered data on the frequency scale.

As can be seen in Figure 6.32, the 100-GHz head cannot resolve the two surfaces which are 3 mm apart corresponding to an optical thickness of 4.5 mm. The 150-GHz head just resolves the surfaces but would not resolve a thinner sample of 2 mm. The combined data have much thinner line width than the 150-GHz data. Combining data helps to clearly resolve the surfaces and could also resolve the surfaces of much thinner steps. The resolution is similar to the data taken with the 300-GHz head. The image representation of this comparison is given in Figure 6.33.

Due to the remaining nonlinearity of the frequency sweep, phase errors particularly at the beginning of the sweep can result in considerable errors in the fused data-set. Also the applied stretch method to fill the frequency gaps can lead to a worse signal contrast between object features, especially if the gap is large. Therefore, the proposed method should only be used to fill small gaps, which are eventually smaller than in the presented case. It has to be taken into account that the discussed data fusing approach requires the refractive index and hence, the optical thickness of the material to be almost equal throughout the relevant frequency range. A recently published more sophisticated approach [34], which is similar to the one shown can eventually be adapted for this application.

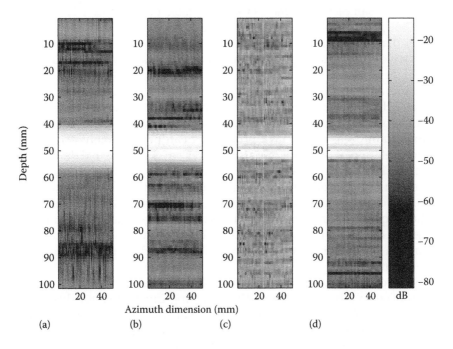

FIGURE 6.33

Comparison of depth images of a 3-mm thick polyethylene sample. The images show the measurements by the (a) 100 GHz, (b) 150 GHz, (c) the fused data (100 GHz + 150 GHz), and (d) 300-GHz sensor, respectively.

6.6 Conclusions

The contact-free capabilities to noninvasively penetrate nonconductive materials, while providing millimeter resolutions have caused an increasing industrial interest in new measurement systems operating in the EHF microwave frequency band.

The UWB radar system presented compares favorably with well-established NDT methods such as X-ray radiography, infrared thermography, and ultrasound for defect detection in glassfiber-reinforced plastic composite structures. In this comparison, the FMCW UWB radar system shows its strengths especially on sandwich structures with ROHACELL® foam and Nomex™ HC cores, proving to be the only considered technique providing the capability to determine the inserts of the inspected test panels throughout the cores.

Furthermore, the FMCW radar technology allows in-depth measurements from one side of the sample, which does not require access to the other side of the sample for defect detection. However, the results reveal the necessity of finding a trade-off between resolution (higher operating frequency) and penetration depth (lower operating frequency), as the absorption of the sample materials increases with the sensor operating frequency. Together with the detection of inserts in composite sample structures, the presented UWB radar system has also successfully determined water inclusions in sandwich structures.

This chapter presented two different measurement approaches, such as a quasi-optical focused and a wide-beam (SAR) configuration. Although the quasi-optical approach provides good lateral resolutions, the optics and the beam-waist effect limit the in-depth capabilities. As a complementary method, wide-beam measurements can especially help to investigate thicker samples since the sensor's available output power limits the in-depth inspection. The spatial resolution remains constant for all depths, although being significantly coarser than that of the quasi-optical approach.

Two innovative techniques were introduced in this chapter: a semi-automatic image processing algorithm and a data fusion technique for combining two UWB sensors. The semi-automatic image processing can work as a complementary tool to visual inspection of EHF UWB imagery. It exploits the in-depth data generated by the volumetric measurements obtained by the performed 2D raster scans in an efficient and fast way. In certain occasions, image processing can be extremely useful since it can ease the interpretation of complex or noisy imagery and even reveal defects invisible to the naked eye. The data fusion technique combined two UWB sensors in adjacent frequency bands in order to obtain a much larger bandwidth. The combination of bandwidths uses several signal processing approaches, provided that the frequency gap between the frequency bands is not too large. This larger bandwidth implies a great improvement in the depth resolution and the ability of the system to locate defects and other features in depth, comparable to that of the 300-GHz sensor but keeping a better material penetration.

In the recent years, many different active EHF microwave system concepts have been developed to overcome the lack of real-time imaging capabilities for different potential applications [35,36]. The current commercially available high-performance computing technology allows the realization of completely new system designs for industrial and even public environments and gives future UWB radar systems a great potential in the field of NDT [37,38].

Acknowledgments

The research leading to these results has received funding from the European Union Seventh Framework Programme FP7/2007-2013 under grant agreement no. 266320, "The DOTNAC project" (http://www.dotnac-project.eu).

Acronyms List

2D: Two-dimensional
3D: Three-dimensional
DAQ: Data acquisition unit
EHF: Extremely high frequency
FFT: Fast Fourier transform
FMCW: Frequency-modulated continuous-wave
GHz: Gigahertz
HC: Honeycomb
LO: Local oscillator
NDT: Nondestructive testing
SAR: Synthetic aperture radar
VCO: Voltage-controlled oscillator

References

1. IEEE Standard Letter Designations for Radar-Frequency Bands," IEEE Std 521–2002 (Revision of IEEE Std 521–1984), 2003. doi: 10.1109/IEEESTD.2003.94224.
2. Boeing, "AERO magazine" <http://www.boeing.com/commercial/aeromagazine/articles/qtr_4_06/AERO_Q406.pdf> 4th Qtr. 2006 (Accessed 14 October 2014)
3. Chady, T., AIRBUS VERSUS BOEING—COMPOSITE MATERIALS: The sky's the limit..., *LE MAURICIEN*, <http://www.lemauricien.com/article/airbus-versus-boeing-composite-materials-sky-s limit>, 6 September 2014 (Accessed October 2014).
4. Swiderski, W., Nondestructive Testing of Honeycomb Type Composites by an Infrared Thermography Method, *Proc. IV Conferencia Panamericana de END*, Buenos Aires, Octubre (2007).
5. Durrani, T.S., Rauf A., Boyle K., Lotti, F. and Baronti S., Thermal imaging techniques for the non destructive inspection of composite materials in real-time, *Acoustics, Speech, and Signal Processing, IEEE International Conference on ICASSP '87.*, vol.12, no., pp.598, 601, Apr 1987. doi: 10.1109/ICASSP.1987.1169621.
6. Shepard, S.M., Flash thermography of aerospace composites, *Proceedings of the 4th Pan American Conference for NDT*, Buenos Aires, Argentina, October 2007.
7. Blom A.F. and Gradin P.A., Radiography, Chapter 1 in *Non-Destructive Testing of Fibre-Reinforced Plastics Composites*, vol. 1, J. Summerscales, ed., Elsevier Applied Science, New York, NY, 1987.
8. Krumm, M., Kasperl S. and Franz, M., Reducing non-linear artifacts of multi-material objects in industrial 3D computed tomography, *NDT and E International*, 41(4), 242–251, 2008.

9. Hochschild, R., Applications of Microwaves in Nondestructive Testing, *NDT*, Vol. 21, No. 2, March-April 1963, pp. 115–120.

10. Case, J.T., Ghasr, M.T. and Zoughi, R., Optimum Two-Dimensional Uniform Spatial Sampling for Microwave SAR-Based NDE Imaging Systems, *IEEE Transactions on Instrumentation and Measurement*, Vol.60, no.12, pp. 3806,3815, Dec. 2011. doi: 10.1109/TIM.2011.2169177.

11. Hatfield, S.F., Hillstrom, M.A., Schultz, T.M., Werckmann, T.M., Ghasr, M.T. and Donnell, K.M., UWB microwave imaging array for nondestructive testing applications, *Instrumentation and Measurement Technology Conference (I2MTC), 2013 IEEE International*, pp. 1502,1506, 6–9 May 2013. doi: 10.1109/I2MTC.2013.6555664.

12. Bahr, A.J., Experimental Techniques in Microwave ND. *Review of Progress in Quantitative Nondestructive*, Vol.14, pp. 593–600, 1995.

13. Busse, G., NDT Online Workshop 2011 - Microwave NDT System for Industrial Composite Applications, <www.ndt.net/article/CompNDT2011/papers/8_Meier.pdf>, 30th April 2011, (Accessed August 2014).

14. Zoughi, R., *Microwave Non-Destructive Testing and Evaluation*. The Netherlands: Kluwer, 2000.

15. Siqueira, M.H.S., Gatts, C.E.N., da Silva, R.R. and Rebello, J.M.A., The use of ultrasonic guided waves and wavelets analysis, *Ultrasonics*, 41(10), 785–797 (2004).

16. Salazar, A., Vergara, L. and Llinares, R., Learning Material Defect Patterns by Separating Mixtures of independent Component Analyzers from NDT Sonic Signals, *Mechanical Systems and Signal Processing*, 24(6), 1870–1886 (2010).

17. Roth, D.J., Seebo, J.P., Trinh, L.B., Walker, J.L. and Aldrin, J.C. Signal Processing Approaches for Terahertz Data Obtained from Inspection of the Shuttle External Tank Thermal Protection System Foam, In *Proc. Quantitative Non-Destructive Evaluation* 2006.

18. Tanabe, Y., Oyama, K., Nakajima K., Shinozaki, K. and Nishiuch, Y., Sub-terahertz imaging of defects in building blocks, *NDT & E International*, 42(1), 28–33 (2009).

19. Am Weg, C., von Spiegel, W., Henneberger, R., Zimmermann, R., Loeffler, T. and Roskos, H.G. (2009). Fast active THz cameras with ranging capabilities. *Journal of Infrared, Millimeter, and Terahertz Waves*, 30(12), 1281–1296.

20. Keil, A., Hoyer, T., Peuser, J., Quast, H. and Loeffler, T., (2011, October). All-electronic 3D THz synthetic reconstruction imaging system. In *IEEE 36th International Conference on Infrared, Millimeter and Terahertz Waves (IRMMW-THz)*, 2011 (pp. 1–2).

21. de Wit J.J.M., Meta, A. and Hoogeboom, P., Modified range-Doppler processing for FM-CW synthetic aperture radar, *IEEE Geosci. Remote Sens. Lett.*, vol. 3, no. 1, pp. 83–87, Jan. 2006.

22. Marple, S. Lawrence Jr., Computing the Discrete-Time Analytic Signal via FFT, *IEEE Transactions on Signal Processing*, vol. 47, no. 9, September 2009.

23. Vossiek, M., v. Kerssenbrock, T. and Heide, P., Signal processing methods for millimetrewave FMCW radar with high distance and Doppler resolution," in *27th European Microwave Conference*, Jerusalem, 1997, pp. 1127–1132.

24. Redo-Sanchez, A., Karpowicz, N6., Xu, J. and Zhang, X. Damage and defect inspection with terahertz waves, *International Workshop on Ultrasonic and Advanced Methods for Nondestructive Testing and Material Characteristics*, pp. 67–78, 2006.

25. Cumming, I.G. and Wong, F.H. *Digital Processing of Synthetic Aperture Radar Data*, Artech House, Boston, 2005.

26. Bamler, R., A Systematic Comparison of SAR Focusing Algorithms. In *Proceedings of the IEEE International Geoscience and Remote Sensing Symposium IGARSS*, pp. 1005–1009, June 1991.

27. Bamler, R., A Comparison of Range-Doppler and Wavenumber Domain SAR Focusing Algorithms, *IEEE Transactions on Geoscience and Remote Sensing* 1992.

28. Cristofani, E., Brook, A. and Vandewal, M., 3-D synthetic aperture processing on high-frequency wide-beam microwave systems. *Proc. SPIE 8361, Radar Sensor Technology XVI, 83610E (May 1, 2012)*; doi:10.1117/12.919409.

29. Cumming, I.G., Neo Y.L. and Wong, F.H., Interpretations of the omega-K algorithm and comparisons with other algorithms, *IGARSS 2003 2003 IEEE International Geoscience and Remote Sensing Symposium Proceedings* IEEE Cat No03CH37477, vol. 00, no. 1, pp. 1455–1458, 2003.

30. Stolt, R.H. Migration by Fourier transform, *Geophysics*, vol. 43, no. 1, p. 23, 1978.
31. Cristofani, E., Vandewal, M., Matheis, C. and Jonuscheit, J., In-depth high-resolution SAR imaging using Omega-k applied to FMCW systems, *Radar Conference (RADAR), 2012 IEEE*, vol., no., pp. 0725,0730, 7–11 May 2012 doi: 10.1109/RADAR.2012.6212233
32. Withagen, P.J., Schutte, K. and Groen, F.C.A., Global Intensity Correction in Dynamic Scenes. *International Journal of Computer Vision*, (86)1:33–47, 2010.
33. Pitas, I. and Venetsanopoulos, A. *Nonlinear Digital Filters: Principles and Applications*, Kluwer Academic Publishers, Boston, MA, 1990.
34. Tian, J., Sun, J., Wang, G., Wang, Y. and Tan, W. (2013). Multiband radar signal coherent fusion processing with IAA and apFFT. *Signal Processing Letters, IEEE*, 20(5), 463–466.
35. Friederich, F., Von Spiegel, W., Bauer, M., Meng, F., Thomson, M.D., Boppel, S. and Roskos, H.G. (2011). THz active imaging systems with real-time capabilities. *Terahertz Science and Technology, IEEE Transactions on*, Vol. 1 (1), 183–200.
36. Kahl, M., Keil, A., Peuser, J., Loeffler, T., Paetzold, M., Kolb, A. and Bolívar, P. H., (2012, May). Stand-off real-time synthetic imaging at mm-wave frequencies. In *Proc. SPIE 8362, Passive and Active Millimeter-Wave Imaging XV, 836208*. doi:10.1117/12.919104.
37. Baccouche, B., Keil, A., Kahl, M., Haring Bolívar, P., Löffler, T., Jonuscheit, J. and Friederich, F. (2015). A sparse array based sub-terahertz imaging system for volume inspection, *European Microwave Conference* (EuMC), Paris, 2015, pp. 438–441. doi: 10.1109/EuMC.2015.7345794.
38. Ahmed, S.S., Schiessl, A. and Schmidt, L. A novel fully electronic active real-time imager based on a planar multistatic sparse array. *Microwave Theory and Techniques, IEEE Transactions*, Vol. 59, No.12 (2011): pp. 3567–3576.

7

Modeling of UWB Radar Signals for Bioradiolocation

Lanbo Liu

CONTENTS

7.1 Introduction

7.1.1 Abstract

This chapter describes the approach of using the finite-difference time-domain (FDTD) numerical simulation approach and synthetic computational experiments to investigate the effectiveness of the ultrawideband (UWB) radar technique for human vital sign detection, known as bioradiolocation. The direct use of bioradiolocation can provide contactless monitoring of human vital signs in security surveillance, biomedical engineering applications, and search and rescue of victims under collapsed building debris caused by catastrophic earthquakes, along with many other applications. The first section summarizes the major features of human vital signs pertinent to UWB radar detection. Next, it discusses the approach to generating of human vital signs for numerical simulation. Then, it gives a brief summary of the FDTD numerical simulation technique. In a sequence from simple to more complicated, it then demonstrates the effectiveness of this modeling approach in three following sections, with each of the three describing a typical scenario in vital sign detection. The first example is the through-wall detection of a single person. It describes the detailed approach for how to embed the model of a living human subject behind a concrete wall into the computational domain. Besides vital sign detection, it also briefly discusses the possibility of imaging the existence of a human being with MIMO interferometric radar in this section. The second example is a search and rescue scenario model at an earthquake disaster site. The model consists of two human beings with different characteristics of vital signs, that is with different cardiorespiration features, posed in different positions, and buried at different depths in the debris. This model of the collapsed building was developed based on a real situation from an earthquake disaster site. Analysis of the synthetic data indicates that the UWB impulse radar can identify and separate the human subjects' vital sign for a radar record as short as 20 sec. The third example is a scenario of line-of-sight, noncontact detection for the vital signs from three human subjects simultaneously for monitoring purpose. All of the simulation results have been verified with physical experiments using impulse UWB radar with real human subjects in presence.

In each of the three examples, advanced signal processing of source separation and signal processing using empirical mode decomposition (EMD) were conducted to identify and locate the human subjects. The FDTD numerical simulation results show that UWB radar in bioradiolocation is a promising technique for the purposes of through-wall detection and search and rescue of living victims at disaster sites, along with other practical purposes.

7.1.2 Introduction to Bioradiolocation Radar

Ultrawideband (UWB) radar signals can be used in many applications, including biomedical imaging, geophysical imaging, vehicular radar, communication, and so on. Examples of imaging applications include ground-penetrating radar (GPR), through-the-wall imaging to detect the location or movements of objects, security surveillance, search and rescue, and medical systems. Vehicular radar systems are commonly used for collision avoidance and roadside assistance. UWB communication systems are useful for high data rate transmission in harsh propagation environments, such as indoor applications with dense multipath channels, for consumer electronics, and for covert operations.

Bioradiolocation is a method for detection and diagnostic monitoring of humans, even behind obstacles, by means of radar (Bugaev et al., 2004). This technique has significant meanings on several applications such as contactless medical measurements, remote psychoemotional state estimation and battlefield or debris evacuation, and so on (Bugaev et al., 2004; Ivashov et al., 2004). This is the additional advantage when using UWB radar for bioradiolocation, because UWB communication links can send raw radar return data to other users. This could allow many users access to real-time (or near-real-time) radar return information for related searches. For example, multiple radar systems could search an area in both monostatic and bistatic or multistatic modes to increase the probability of finding all victims quickly.

To serve as a tool for testing the software algorithms in bioradiolocation, this chapter presents an approach to generate synthetic data with finite-difference time-domain (FDTD) numerical simulation methods to reproduce a realistic field scenario and produce a synthetic data set that can be used in vital sign detection algorithm verification. One of the most severe disastrous results of a major earthquake is the loss of lives in collapsed buildings. Great efforts have been devoted worldwide to search and rescue for survivors underneath collapsed building debris. A timely search and rescue is the key component of hazard mitigation after a major earthquake. UWB radar technique bears potential to assist this kind of efforts. For example, one type of UWB impulse radar system known as a "life detector" has been put in test for search and rescue at the Wenchuan earthquake hyper-center region in southwestern China in May 2008 (NSFC news 2008; Cist, 2009; Lv et al., 2010). A multifrequency continuous wave (CW) radar life detection system based on Doppler frequency shift was also developed and tested in Europe (Bimpas et al., 2004).

This chapter is organized as follows. Section 7.2 gives a brief description of the major features of human vital sign that are critical to be transformed into a numerical model that can be used in vital sign (breathing and heartbeat) UWB radar detection. Section 7.3 describes how to generate the numerical model of a living human's chest to get the cardio-respiratory motion. Section 7.4 describes the finite-difference time-domain (FDTD) method with the living human subject embedded into structure model. Section 7.5 describes the numerical simulation of an impulse radar signal propagation and the generation of the synthetic data for a single human subject. Section 7.6 deals with the vital sign analysis, with emphasis on the application of empirical mode decomposition method for a model with two victims buried under collapsed building debris. Section 7.7 describes the analysis of the numerically generated data for line-of-sight detection of the respiratory signals from three subjects with the verification by a physical experiment under laboratory conditions. Section 7.8 summarizes the main findings and gives the concluding remarks.

7.2 Fundamental Observations to Direct Bioradiolocation Signal Modeling

In recent years, bioradiolocation has been widely applied to detect cardiorespiratory signals (Boric-Lubecke et al., 2009; Bugaev et al., 2004; Ivashov et al., 2004; Yarovoy et al., 2006; Attiya et al., 2004; Immoreev and Samkov, 2003; Zaikov et al., 2008; Sisma et al., 2008; Yarovoy et al., 2007). Its application can be achieved using continuous wave signals (Bugaev et al., 2004; Ivashov et al., 2004), Doppler radar (Boric-Lubecke et al., 2009), or UWB techniques (Yarovoy et al., 2006; Attiya et al., 2004; Immoreev and Samkov, 2003; Zaikov et al., 2008; Sisma et al., 2008; Yarovoy et al., 2007; Narayanan, 2008). Fundamentally, bioradiolocation with UWB radar relies on the modulation of a reflected radar signal by the movements of human beings. In a brief summary, the signature of a human body movement can be generated by the following (Bugaev et al., 2004):

1. Breathing with a frequency range between 0.2 and 0.5 Hz, and the resulting thorax and chest movement amplitude can reach 0.5–1.5 cm

2. Heartbeat with a frequency range between 0.8 and 2.5 Hz, and the chest motion amplitude can reach 2–3 mm

3. Articulation or movement of the vocal apparatus (lips, tongue, larynx)

4. Movements of other body parts

Since the cardiorespiratory movement exists constantly even if a person stays quiet and motionless, detection of the cardiorespiratory signatures represents the primary task in UWB radar live detection. This is the so-called vital sign detection.

Compared with a CW radar, UWB impulse radar radiates a very short duration pulse and thus intrinsically possesses a wide spectral band, resulting in small spatial resolution and low energy consumption at the same time (Yarovoy et al., 2006). Prior to the studies on detecting cardiorespiration signatures, some studies (Attiya et al., 2004; Immoreev and Samkov, 2003) have shown that UWB impulse radar system is capable of detecting human beings by identifying the impedance contrast between the human skin and the air. Recent research (Yarovoy et al., 2006) also shows the efficiency of UWB impulse radar systems for capturing human breathing motion even when the subject is otherwise motionless. Based on these studies, Zaikov et al., (2008) reported an experiment using UWB radar for the detection of trapped people. These key advantages led a number of researchers (Yarovoy et al., 2006; Zaikov et al., 2008; Sisma et al., 2008; Yarovoy et al., 2007) to turn their attention to the application of UWB impulse radar in through-wall life detection by searching the cardiorespiration signals from a live person.

Thus, in this chapter, the FDTD simulations use an impulse source for generating the radar radiation wave field. For the most frequently encountered applications in the field for UWB radar vital sign detections, I simulated three case studies for generating synthetic datasets, in an order of simple to complex. The three cases are as follows:

1. Through-wall vital signs detection for mono-subject
2. Vital signs detection of two subjects at an earthquake disaster site
3. Remote measurement on cardiorespiratory movement of three patients

For all three cases, the simulation results are analyzed with the classic fast Fourier transform (FFT) and the other more advanced time–frequency analysis empirical mode decomposition (EMD, Huang and Wu, 2008) to extract the vital sign signals. To meet the objective of detecting human vital sign in an adverse environment, a set of given technical requirements such as bandwidth, power budget, dynamic range, and stability can be uniformly attributed to different radar systems for their functionality assessment (Saha and Williams, 1992). For the two commonly used UWB systems, that is the time-domain impulse radar and the step frequency continuous wave (SFCW), the impulse radar is moderately fast for data acquisition but might have a problem with its linear dynamic range. A way to get around this bottleneck might be division of the total operational downrange in subranges and operation within these subranges. On the other hand, while the stepped-frequency continuous wave technique possesses a very good total power budget and dynamic range, it also suffers from a relatively slow data acquisition. This means the SFCW system needs a much longer time than the impulse system measurement time, even with a fast frequency sweep rate (Saha and Williams, 1992). The FDTD numerical simulation technique can flexibly handle different parameter inputs; thus, it can be a great tool to assist UWB bioradiolocation system design. Nevertheless, since the radar signals acquired by SFCW algorithm will be eventually converted to impulse source propagation the FDTD simulation in this chapter focuses on the impulse source propagation cases.

7.3 Generation of Human Vital Signs for Numerical Simulation

Modeling human cardiorespiratory signals involves two time scales: (1) the UWB radar signal travel time at the scale of nanoseconds (ns); (2) the vital signal recording time at tens to hundreds seconds. It is noteworthy to point out that the two time scales differ with each other for nine orders of magnitude. For the detection of cardiorespiratory signals with a frequency band of 0.2–2 Hz, a sampling interval of 0.05–0.1 sec (10–20 Hz) is fine enough. Thus, a UWB radar signal propagating at a time scale of nanosecond can be safely treated as an infinitesimal instant. Using this philosophy, we treat the dynamic chest of a living human as a group of static snapshots.

Based on a slice of a magnetic resonance image (MRI) of a generic human chest, a quasi-static cardiorespiration model was constructed with 45 instants (or statuses) standing for a total recording time of 9 sec, as shown in Figure 7.1. With a 0.2 sec time interval

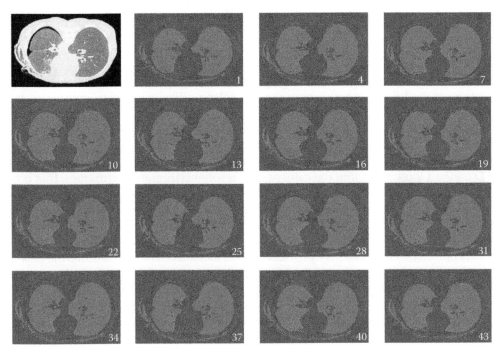

FIGURE 7.1
Example of 15 snapshots at an interval of every 3 from a 45-status cycles of human vital activities showing 2 respiratory periods and 11 heartbeats based on an MRI image of the cross section of a generic human chest (the upper left grayscale image).

between two adjacent statuses; this two-dimensional (2D) model contains two cycles of breath (lung movement) and 11 cycles of heartbeat (heart deformation). Consequently, this model gives a breath frequency of 0.22 Hz and a heartbeat frequency of 1.22 Hz. The motion amplitude of breath and heartbeat are set to be 5–15 mm and 2–3 mm, respectively. Figure 7.1 shows a group of 15 statuses from every 3 snapshots out of the total of 45 statuses for conducting the finite-difference time-domain (FDTD) simulation.

The model shown in Figure 7.1 is only one example. Actually, the cardiorespiration model is designed to be flexible for simulating human subjects with different heartbeat and breathing frequencies to accommodate different vital statuses, as shown in the following sections. Models of living human subjects at different locations will be embedded into the static structure model for conducting the FDTD simulation of the buried living human beings in a variety of realistic situations.

7.4 Finite-Difference Time-Domain Modeling Techniques

Among the different numerical simulation methods, the finite-difference time-domain (FDTD) modeling technique (Yee, 1966; Taflove and Hagness, 2000; Liu and Arcone, 2003; Liu et al., 2011) is a powerful tool to simulate and comprehend UWB radar field survey data. The staggered grid, two-dimensional (2D) FDTD method (Yee, 1966) in conjunction with the perfectly matched layer (PML) (Berenger, 1994) as the absorptive boundary condition (ABC) to truncate the computation domain for suppressing unwanted artificial reflections is an efficient and robust algorithm. It has been widely used in electromagnetic wave propagation simulations (Liu and Arcone, 2003; Liu et al., 2011).

The FDTD method is arguably the simplest, both conceptually and in terms of implementation, of the full-wave techniques used to solve problems in electromagnetics. It can accurately tackle a wide range of problems. However, as with all numerical methods, it does have its share of artifacts and the accuracy is contingent upon the implementation. The FDTD method can solve complicated problems, but it is generally computationally expensive. Solutions may require a large amount of memory and computation time. The FDTD method loosely fits into the category of "resonance region" techniques, that is ones in which the characteristic dimensions of the domain of interest are somewhere on the order of a wavelength in size. If an object is very small compared with a wavelength, quasi-static approximations generally provide more efficient solutions. Alternatively, if the wavelength is exceedingly small compared with the physical features of interest, ray-based methods or other techniques may provide a much more efficient way to solve the problem.

To approximate the derivatives in the Maxwell equations that describe the EM wave propagation with finite-differences, we used a staggered difference algorithm in a 2D space domain as proposed by Yee (1966). Figure 7.2 shows the staggered grid system for the transverse magnetic (TM) mode with the y-component of the electric field being vertical and in alignment

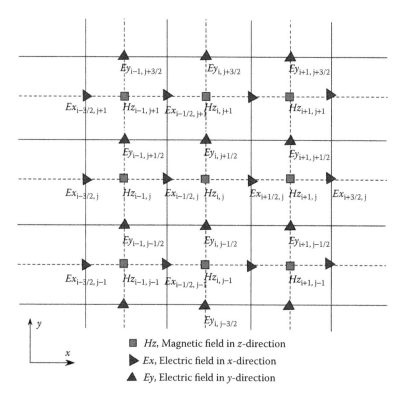

FIGURE 7.2
A sketch of 2D staggered grid system in x–y plane for simulating TM mode radar wave propagation with the FDTD method. The magnetic field is in z-direction (perpendicular to the x–y plane) and the electric fields (Ex and Ey) are in the x–y plane.

with the borehole axis and the *x*-component of the electric field being horizontal and parallel to the surface as in Figure 7.2. The marching in time domain is also staggered between the computations of the electric field *E* and the magnetic field *H*. Yee's second-order staggered grid algorithm remains the most economical and robust way to carry out FDTD computations (Taflove and Hagness, 2000). For using the second-order staggered grid algorithm, a minimum of 20 grids per shortest wavelength is needed to reach the numerical stability.

Berenger's (1994) perfectly matched layer (PML) technique was adopted for the absorption boundary condition (ABC). We have achieved highly effective suppression of the unwanted reflections from domain boundaries. Numerical experiments demonstrated that use of an eight-layer PML boundary condition could reduce the reflected error 30–40 dB over any previously proposed absorption boundary conditions (Yuan et al., 1997). Though any radar survey in real world is 3D in nature, a 2D FDTD model is a good compromise that can catch the major propagation characteristics and possesses the advantage of fast and efficient numerical computation.

7.5 UWB Radar Detection of a Single Person behind a Wall

This section describes the situation of detection of a single person behind the wall. First of all, let us examine the most fundamental phenomena that UWB radar can register for the vital sign from a living human subject.

Figure 7.3 shows a piece of real radar signal data acquired in the laboratory for a human subject in normal breathing, breath holding, and speaking status. The UWB impulse radar

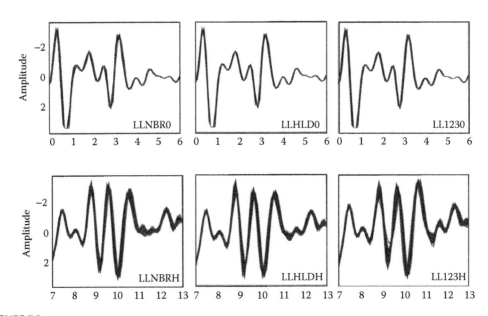

FIGURE 7.3
The dataset of the A-scan traces superimposed together for a human subject in conditions of normal breath (241 traces, left column), breath holding (170 traces, middle column), and speaking (254 traces, right column). The top row shows the early time cases for a time window from 0 to 6 ns. The bottom row shows the same case for a time window from 7 to 13 ns.

was placed on one side of a concrete block wall with a thickness of 20 cm formed by cinder blocks (no reinforcing steel bars). The UWB impulse radar antenna was coupled closely to the wall at the same height of the subject's chest, while on the other side, the subject was standing upright at the distance of 1 m from the wall. The GPR system collected 1 recording trace per 0.1 sec, with 8 stacking to minimize the random error. The subject was asked to stand as quietly and still as possible to reduce whole body motion. Figure 7.3 shows the simple superposition of all the recorded traces one on another for all the radar profiles for the 3 statuses of one human subject. This is the most intuitive and direct way to present the features of the cardiorespiratory signature in time domain. The top row of Figure 7.3 shows the superimposed traces of the early time corresponding to the direct transmitter–receiver coupling and the reflection from the wall, while the bottom row shows the reflection from the subject.

Since the physical ambient environment should have no time-dependent variation, the traces of early time are basically repeating of one another in the time window. In contrast, for the later time windows in the bottom rows of Figure 7.3, an apparent variation in the radar reflections can be found for the human subject, which makes the superimposed traces appear "thicker" or blurred. Obviously, in contrast to the response to the "static" wall, the changes in behavior in the recording traces are associated with the presence of the living human subject. Further examination for the "normal breath" status (the lower left panel labeled LLNBRH of Figure 7.3) revealed that in spite of the time–axis variation in the entire reflection portion being relatively low, there is a relatively large variation in a short-time window between 11 and 12 ns in travel time (sample number 110–120) that deserves further investigation for this human subject.

Now, we describe the construction of the numerical model, which can mimic the situation of a living human subject behind the wall. In this model, the dynamic cardiorespiration model described in Section 7.3 is embedded into the 2D computational domain to mimic the through-wall setup as shown in Figure 7.4. The geometry is very similar to the layout of a laboratory experiment. The total number of grid is 2000×1000, with a grid size of 1 mm $\times$ 1 mm to occupy a 2 m $\times$ 1 m 2D area. Using the generic values of dielectric constants for air and dry wall, as well as the published values for human tissues (Saha and Williams, 1992; Gabriel, et al., 1996; Bronzino, 1999; Egot-Lemaire, et al., 2009), the complete list of the dielectric constants for different materials at 1 GHz is shown in Table 7.1.

In the FDTD simulation, the radar transmitted a UWB impulse with a central frequency of 1 GHz from the source placed on the right side of the wall, while a human chest model as shown in Figure 7.1 was placed on the other side. A receiving point was placed 10 cm away from the source. The total recording length was set to be 16 ns, identical to that of the physical experiment, with a total of 8000 time steps at the sampling interval of 0.002 ns. Because the time elapsed from the instant the radar wave begins radiating to the time a reflection arrives back from the human body is about eight orders of magnitude shorter than the time scale of human's cardiorespiration period (1–5 sec), it is a reasonable approximation to treat each of the 45 statuses as a quasi-static snapshot as if the chest movement is "frozen" at each moment. Therefore, the 45 simulations were repeated to form a profile in recording time that contains 256 traces at an interval of 0.2 sec, as shown in Figure 7.5.

After background removal, it is clear that the chest displacement caused by cardiorespiration mainly appears after 9.8 ns (sample number 490). The simulation results will guide us as a reference for searching the cardiorespiration signals in our laboratory experiments data. A quick spectral analysis of the synthetic record in Figure 7.5 using the fast Fourier

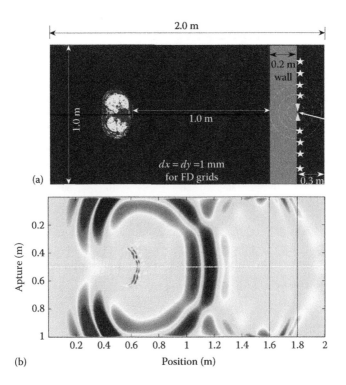

FIGURE 7.4
Experimental and simulation geometry. (a) Layout of the FDTD computational domain for the case of a single human subject behind the wall, where the grid size is chosen to be 1 × 1 mm. The location of dynamic cardiorespiration model from Figure 7.1 wall and UWB impulse radar (location of Tx/Rx) are set up to be as similar as the real experiment. The perfect matched layer (PML) plays a function of absorbing outgoing waves and reduces reflection and (b) a snapshot of radar wave field at 7 ns after firing the source impulse. The trace at the bottom is the wave field profile along the central array denoted by white crosses.

TABLE 7.1

Relative Dielectric Constant Used in the Through-wall Model

Air	Dry Wall	Skin	Heart	Lung	Bone
1	4	38	60	35	40

transform (FFT) indicates that, as expected, the dominant frequency for the respiration is 0.22 Hz in Figure 7.6a, and the heartbeat is at 1.22 Hz in Figure 7.6b, in exact agreement with the prescribed 45-status cardiorespiration model shown in Figure 7.1. The harmonic spectral peaks f_m between the peak of breathing f_b and the peak of heartbeat f_h are associated with intermodulation of the frequency of breath and heartbeat at a family of frequencies of $f_m = lf_h \pm kf_b$ with $k, l = 0, 1, 2, 3, \ldots$

As another approach for signal extraction, the Hilbert–Huang transform (HHT) time–frequency analysis described in Chapter 4 can be used to characterize the cardiorespiratory radar data. The HHT, a novel approach pioneered by Huang (Huang and Wu, 2008), is an adaptive time–frequency analysis tool particularly suitable for analyzing the nonlinear and nonstationary signals. The analysis procedure consists of two techniques in conjunction: the empirical mode decomposition (EMD) and the Hilbert spectral analysis (HSA). The

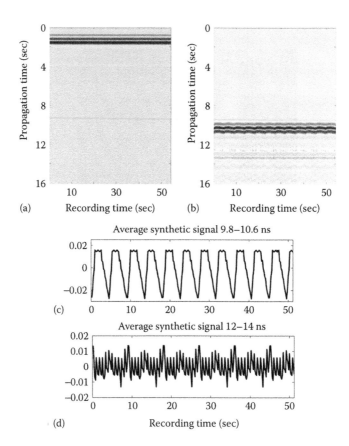

FIGURE 7.5
The simulation result based on the synthetic model. The left panel (a) shows the original record in terms of the sampler number, with the sampling interval of 0.02 ns, and the trace number with a recording interval of 0.2 sec, (b) after background removal, the reflections from the human body can be clearly seen at the travel time of 9.8 ns, corresponding to the sample number of 490, (c) the signal of respiration is the dominant part at 9.8–10.3 ns, and (d) after 12 ns, the heartbeat about 1.2 Hz can be seen.

EMD treats the signal as a collection of many coexisting simpler oscillatory modes called the intrinsic mode functions (IMFs). Each IMF has to satisfy the following conditions:

1. The number of extremes must be equal to, or at most differ by only one from, the number of zero-crossings, which is similar to the traditional narrowband requirements for the stationary Gaussian process.

2. At any point, the mean value of the upper and lower envelopes defined by the local maxima and minima, respectively, is zero. These upper and lower envelopes are determined by some interpolation algorithm such as cubic splines. This condition ensures phase function for getting the bias-free instantaneous frequency.

The EMD method separates those IMFs from the original signal one by one by means of an algorithm called shifting process until the residue appears to be monotonic (Huang and Wu, 2008). Thus, the original signal is written as the sum of all IMFs. This decomposition can be simply achieved in the time domain, and proved to be powerful of extracting physically meaningful features from extreme noisy background (Huang and Wu, 2008,

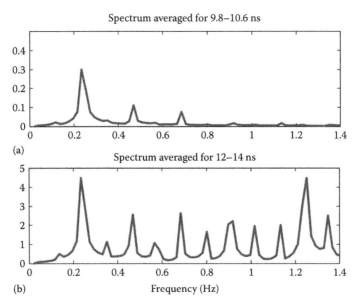

FIGURE 7.6
The frequency spectra for the time sequences in Figure 7.5.c (a) and Figure 7.5.d (b), respectively. The signal of human respiration is the dominant part for reflections at 9.8–10.3 ns (a). For the record at 12–14 ns, the spectrum of the heartbeat at 1.2 Hz is about the same height as that of respiration at 0.2 Hz.

Narayanan, 2008). Next, the IMFs of each radar profile can be transformed from the recording-time domain to frequency domain using the Hilbert transform-based spectral analysis (HSA), a technique based on instantaneous parameters obtained from classic Hilbert transform, and proved to be a better analysis tool than the conventional FFT for nonlinear, nonstationary signals (Huang and Wu, 2008). The principle and procedure of HSA has been comprehensively reviewed in (Huang and Wu, 2008). Based on this observation, the HSA was used to determine the frequency spectra of certain IMF, which mainly contain respiration feature. The result of HHT analysis for the synthetic simulation and the observed radar signals are shown in Figure 7.7. The respiration is clearly identified around 0.22 Hz and 9.5–11 ns along travel time direction. In addition, a slight energy concentration can be found around 1.2 Hz and 12 ns in travel time, which is likely associated with heartbeat but far from being a convincing evidence of the ability to detect heartbeat by this particular UWB radar system. These two features in general coincide with the spectral information using FFT shown in Figure 7.6, with the addition of information on the travel time location of the cardiorespiratory energy.

The HHT analysis is also applied to the observed radar signals shown in Figure 7.3 with the resulting HSA time–frequency plot as shown in Figure 7.8. The features of respiration seem to be complex, with a concentrated distribution in both frequency and travel time directions. Since the subject stands at a location of 1 m from the wall during the experiment, strong reflected energy appears around 12 ns in all conditions. It also clearly indicates that the frequency for normal breathing is around 0.35 Hz.

For the case of breath holding, the energy level is clearly much lower, but some energy still exists, which may result from the subject's whole body movement or involuntary motions rather than respiration. For the case of the subject speaking, the time–frequency distribution of the energy level of the subject's respiration has no noticeable increase, which implies that he spoke in a relatively calm manner during the experiment.

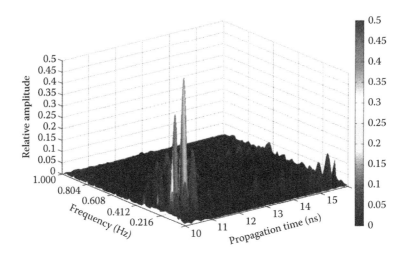

FIGURE 7.7
Result of time–frequency analysis for the result of synthetic simulation using HHT. The motion of human respiration is clearly identified as 0.22 Hz around 10 ns in the travel time direction. The feature of the heartbeat around 1.2 Hz is relative weak, but can still visible around the travel time of 12.5 ns.

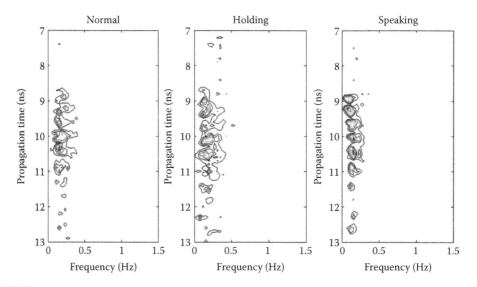

FIGURE 7.8
The HSA time–frequency analysis results in terms of power (amplitude squared) of UWB data shown in Figure 7.3 for the three conditions: normal breathing, breath holding, and speaking.

7.6 The Model of Collapsed Building with Two Living Victims

One of the most disastrous impacts of a major earthquake is loss of lives due to collapsed buildings. Great worldwide efforts have developed new search and rescue methods for survivors trapped underneath collapsed building debris. Search and rescue in a timely manner is the key component of earthquake hazard mitigation after a major earthquake. The UWB radar technique has great potential to assist in victim location. The numerical model of UWB radar detection of two victims buried underneath collapsed building debris is discussed in this section. Similar as done in the last section, this approach uses the two human cardiorespiratory models of Figure 7.1 with different vital sign features and embeds them into a larger model of the digitized collapsed building debris. This model came from a real collapsed building at the hypocenter of Yushu earthquake, which struck on April 14, 2010, in the northeast edge of the Tibet plateau. Collapsed buildings in this disaster claimed more than 2,000 lives. Figure 7.9 shows the model of the collapsed building with two buried human beings.

The human model on the bottom center (Human1) is placed facing down, with a heartbeat frequency of 1.25 Hz and breath frequency of 0.12 Hz. The human model (Human2) on the left side is facing right, with a heartbeat frequency of 1.67 Hz and breath frequency of 0.35 Hz. The total number of grid of this collapsed building is 3791 × 1991, with a uniform grid size of 1 mm × 1 mm to occupy a 2D area of 3.77 m horizontally and 1.97 m in vertical direction. For the collapsed building pieces, I assigned the dielectric constant of 6.25. The electric conductivity and dielectric constants of human tissues are modeled in more details based on published data (Saha and Williams, 1992; Bronzino, 1999; Egot et al., 2009), and presented in Table 7.1.

In the FDTD simulation for detecting the buried live human beings, the UWB impulse source with the central frequency of 1 GHz was placed on the surface and transmitting, while the human beings were buried in the depth of the debris as in Figure 7.9. For detection and imaging purposes, the total recording length was set to be 67 ns with a total of 33,500 time steps at a sampling interval of 0.002 ns. An array of 37 receiving points, with 10 cm spacing, was also placed on the surface. For the simulation purposes, the source can be placed at any lateral position of the surface. Figure 7.10 shows a series of snapshots of the FDTD-generated wave fields when the impulse radar source is placed at the upper-left corner on the surface of the model as shown in Figure 7.9.

From Figure 7.10, it is clear that the incident radar wave interacts with the building debris and produces a very complicated scattered field. Furthermore, it also interacts with the buried human beings with relatively strong reflection, due to the large contrast of human body and the ambient media, as seen from Human1 at the center from the snapshot at $t = 16$ ns

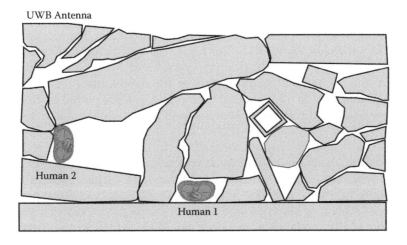

FIGURE 7.9
Model of two human subjects buried in a collapsed building at an earthquake hazard site with a size of 3.74 m × 1.97 m. (Reprinted from *Ad Hoc Networks*, Liu, L. et al., Numerical simulation of UWB impulse radar vital sign detection at an earthquake disaster site, 34–41, Copyright 2014, with permission from Elsevier.)

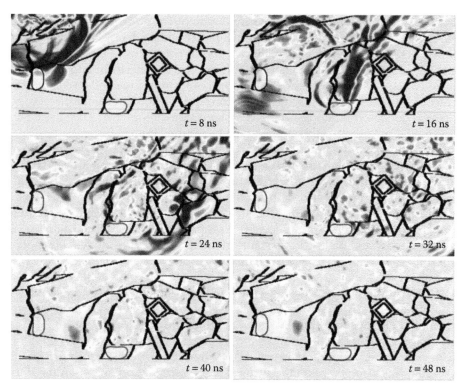

FIGURE 7.10
Simulated EM wave field snapshots at elapsed time of 8, 16, 24, 32, 40, and 48 ns. The source is a TE-polarized dipole at the center of the model located above the surface. (Reprinted from *Ad Hoc Networks*, Liu, L. et al., Numerical simulation of UWB impulse radar vital sign detection at an earthquake disaster site, 34–41, Copyright 2014, with permission from Elsevier.)

(the upper right panel), as well as from Human2 on the left from the snapshot at $t = 24$ ns (the middle left panel). The wavelength has become shorter inside human bodies, in response to much slower radar wave propagation velocity inside the human tissues.

The most intuitive way to present the features of the cardiorespiratory signature is in time domain as shown in Figure 7.11. The piece of synthetic data shown in Figure 7.11 is collected from the pair of fixed transmitting source and receiving point (10 cm apart) placed directly above Human2 on the surface of the debris, the Tx/Rx in Figure 7.9, for continuous illumination into the collapsed building for a given period of time, for example 20 sec. Thus, we have a collection of 200 time traces if the time interval between two contiguous traces is 0.1 sec.

Since the essential physical environment of the building debris should have no time-dependent variation, the traces of early time are basically repeating, except when there is activity of living human subjects which may result in time-varying signatures. Figure 7.11 has easily observed periodic variation with a period of 2.8 s (about seven breathing cycles along the axis of recording time) around 15 ns in propagation time. This should be the sign of breathing of Human2 with a frequency of 0.35 Hz, which is placed about 1 m below the surface that needs 16.6 ns for the radar wave to reach the subject and reflected back to the receiver. It is also obvious to see the breath of Human1 appearing at the propagation time around 36 ns as the periodic variation with a period of 8.3 s (less than three breathing cycles). Also, the registration of heartbeat signal from Human2 can also be marginally

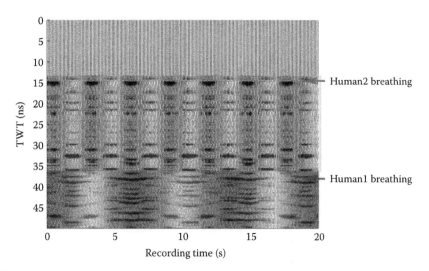

FIGURE 7.11
A collection of common-offset time traces over 20 s in recording time (the horizontal axis) and 50 ns in the two-way travel (TWT) time (the vertical axis). (Reprinted from *Ad Hoc Networks,* Liu, L. et al., Numerical simulation of UWB impulse radar vital sign detection at an earthquake disaster site, 34–41, Copyright 2014, with permission from Elsevier.)

recognized around 22 ns in propagation time evidenced by the shorter period amplitude oscillation superimposed on the longer period larger amplitude signal of Human2's breathing. This feature is reconfirmed in the frequency domain via the Fourier transform of the time-domain expression as shown in the time–frequency analysis using empirical mode decomposition (EMD), also known as the Hilbert–Huang transform (HHT, Huang and Wu, 2008), as shown in Figure 7.12.

As shown in Figure 7.12, besides the clear expression of the breathing signals for Human2 with a frequency about 0.35 Hz starting at propagation time around 15 ns, as well as the breathing signals for Human1 with a frequency about 0.12 Hz starting at propagation time of around 36 ns as previously shown in the time domain record in Figure 7.11, we can also barely notice vital signs from heartbeat of Human2 at 1.67 Hz around 22 ns in propagation time. However, it is nearly impossible to pick up the heartbeat for Human1, due to the greater depth under the debris, and possibly because of laying faced down with much less chest motion as seen from the rear. Nevertheless, the other newly emerged weak signal detection technologies may shed more light to help solving this problem, such as getting the material transfer functions using time reversal (Wua et al., 2010), and also possibly the blind source separation algorithms (Sawada et al., 2004).

To verify the validity of the numerical simulation for this more complicated situation, we have tested the vital sign separation algorithm using observed data obtained by a controlled case study carried out inside the laboratory. A ground-penetrating radar system with an impulse source at 1 GHz was used. The radar antenna was placed against a cinder block wall about 1.5 m from the ground. The height is about the same level of peoples' chest. The plan view of the experimental setup is shown in Figure 7.13. Two human subjects (S1 and S2 in Figure 7.13) stand still in the opposite side of the wall at different positions with different face directions, similar to the earthquake debris model shown in Figure 7.9.

In the experiment, we first acquired the background baseline without the presence of human subjects. Then, the two human subjects were positioned in the location as shown

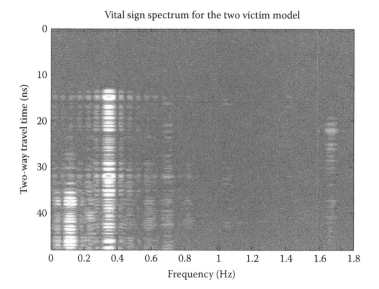

FIGURE 7.12
The time-spectral analysis using empirical mode decomposition (EMD) for the vital signs generated by the two buried humans in the model shown in Figure 7.9. (Reprinted from *Ad Hoc Networks*, Liu, L. et al., Numerical simulation of UWB impulse radar vital sign detection at an earthquake disaster site, 34–41, Copyright 2014, with permission from Elsevier.)

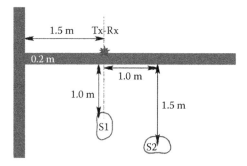

FIGURE 7.13
Laboratory experiment setup for using a 1-GHz impulse radar placed behind the wall with two human subjects on the opposite side, in a position similar to the case in Figure 7.9. (Reprinted from *Ad Hoc Networks*, Liu, L. et al., Numerical simulation of UWB impulse radar vital sign detection at an earthquake disaster site, 34–41, Copyright 2014, with permission from Elsevier.)

in Figure 7.13. After the elimination of the strong direct coupling and reflection from the opposite side of the wall, the time-domain records of the laboratory experiments were obtained as shown in Figure 7.14. The total recording time is about 70 s. The top panel is a background check without human subjects present; the bottom panel is the record with human subjects. The reflections from the two human subjects start to show up at about 8 ns (for S1) and 13 ns (for S2), with noticeable amplitude undulations.

Using the same time–frequency analysis approach with the EMD described in the last section, the results are shown in Figure 7.15, for background (left panel), with human subjects present (middle), and the difference of the two situations (right panel). The signatures of breathing from the two subjects are clearly seen. For the subject close to the antenna (S1),

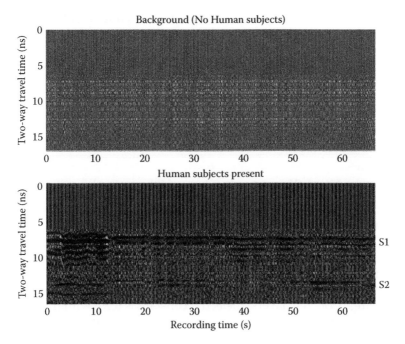

FIGURE 7.14
The raw time-domain laboratory test data using a 1-GHz GPR system. The upper panel is the time domain record without human subjects present. The bottom panel is the time-domain records with the two human subjects positioned at the location shown in Figure 7.13. (Reprinted from *Ad Hoc Networks*, Liu, L. et al., Numerical simulation of UWB impulse radar vital sign detection at an earthquake disaster site, 34–41, Copyright 2014, with permission from Elsevier.)

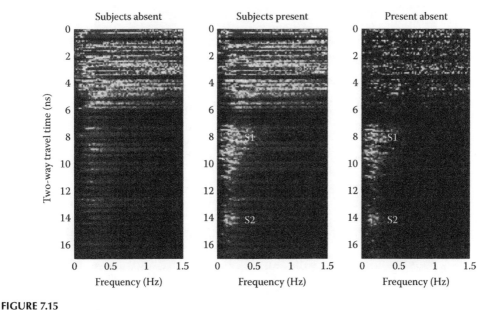

FIGURE 7.15
The time–frequency analysis of the vital signs in the radar records generated by two human subjects positioned on the opposite side of the wall with the UWB radar. (Reprinted from *Ad Hoc Networks*, Liu, L. et al., Numerical simulation of UWB impulse radar vital sign detection at an earthquake disaster site, 34–41, Copyright 2014, with permission from Elsevier.)

the breathing is shown between 8 and 10 ns in propagation time, and for the subject farther away from the antenna (S2), the breathing is shown between 13 and 15 ns in propagation time, with much weaker signal power. Both subjects have a breathing frequency of about 0.2 Hz. The analysis cannot catch heartbeat information for both subjects. Nevertheless, for the purpose of vital sign detection, using breath alone is enough to classify a person as alive.

7.7 Line-of-Sight Vital Sign Monitoring of Three Patients

As the most complicated case of these three examples, this section describes the FDTD numerical simulation for a scenario of simultaneously monitoring the vital signs from three patients in a line-of-sight (LOS) fashion. Figure 7.16 shows the monitoring setup. For this case, I intended to simulate the application of the UWB impulse radar on contactless medical measurement. As Figure 7.16 shows, this simulation is applied in a model space with the total number of grids of 4200 × 3200, with a grid size of 1 mm × 1 mm to occupy a 4.2 m × 3.2 m area. Concrete layers (dielectric constant is 6.25) with the thickness of 1 m surrounded the model space to mimic the walls, ceiling, and floor of a real room. Three human subjects with different heartbeat and respiration frequencies shown in Figure 7.16 lie in beds (dielectric constant is also 6.25) and facing up.

In the FDTD computation, an impulse source at a central frequency of 1 GHz together with single receiver is placed at the location shown in Figure 7.16. The total recording length was set to be 21.1 ns with a total number of 10,550 time steps at a sampling interval

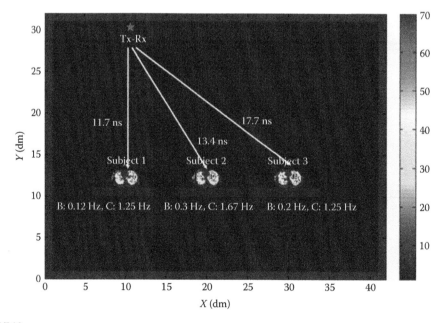

FIGURE 7.16
Model of line-of-sight monitoring of three patients in a 4.2 m × 3.2 m room. The breathing and heartbeat parameters for each patient are marked below the bed for each person. The two-way travel time to the radar antennas for each patient is also marked.

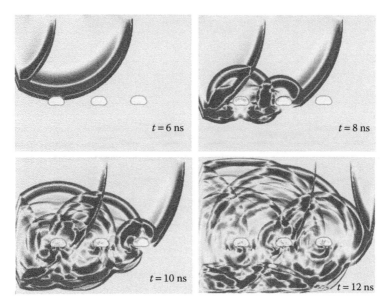

FIGURE 7.17
Simulated EM wave field snapshots at elapsed time of 6, 8, 10, and 12 ns. The source is a TE-polarized dipole located at a position in the model domain shown in Figure 7.16.

of 0.002 ns. Figure 7.17 shows a series of snapshots of the FDTD-generated wave fields when the impulse radar source is placed at the upper left quarter of the model. From Figure 7.17, it is clear that the incident radar wave first interacts with the left wall of the building (t = 6 ns) and produces a strong reflection field. Furthermore, it starts to interact with the human subjects and produces relatively strong reflection (t = 8, 10 ns), due to the large contrast of human body and the ambient media (air). Also, it clearly shows that the wavelength becomes shorter inside human bodies, in response to much slower radar wave propagation velocity inside the human tissues. The wave field becomes very complex after it interacts with all the three subjects, and some of the wavefronts start to reach the receiver antenna on the upper-left corner.

Figure 7.18 shows the background-removed synthetic dataset collected from the tri-patient model of Figure 7.16. The traces included are organized in the same way as displayed in the Figure 7.11. Periodical change can be observed from 12 to 21 ns. The classic spectral analysis using FFT clearly showed the signature for all the respiration signals, and the heartbeat signature can be barely identified, as shown in Figure 7.19.

The analysis of the synthetic data is shown in Figure 7.20. Their observed frequencies in Figure 7.19 after FFT are applied horizontally along a different propagation time. It is clear to observe the signal of ~0.1 Hz starting from ~10 ns, ~0.2 Hz starting from ~16 ns, and ~0.3 Hz starting from 12 ns. They are coinciding with the respiration frequencies of models Subject 1, Subject 3, and Subject 2, respectively. Due to the distance variance in each human model to the receiver, the respiration frequencies appear at different propagation time of the record. Besides respiration, the frequency component of ~1.25 Hz and ~1.25 Hz observed from Figure 7.19 coincides with heartbeat of Subjects 1&3 and Subject 2, respectively. These can be more clearly observed after averaging the spectrum from 10.2 ns to 20 ns. As Figure 7.19 shows, both respiration and heartbeat of three human bodies can be observed with some effort.

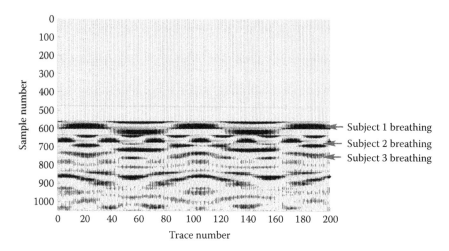

FIGURE 7.18
A collection of common-offset time traces over 20 s in recording time (the horizontal axis) and 21 ns in propagation time (the vertical axis). The signature of breathing for all the subjects can be clearly identified.

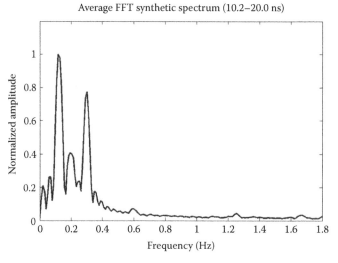

FIGURE 7.19
The spectral analysis result of the vital signal simulation results shown in Figure 7.18.

As shown in Figure 7.20, the breathing signals for all three subjects are clearly displayed at the right position, in agreement with the parameters estimated and marked in Figure 7.16. However, it is barely possible to pick up the heartbeats at 1.25 Hz and 1.67 Hz.

Again, to verify the validity of the numerical simulation for this more complicated situation, we have tested the vital sign detection algorithm using observed data obtained by a controlled case study carried out inside the laboratory. This study used the same ground-penetrating radar system as shown for the two subject case. The radar antenna was placed against a cinder block wall (thickness 20.32 cm, i.e., 8″) about 1.5 m from the ground. The height is about the same level of peoples' chest. The plan view of the experiment setup is shown in Figure 7.21. Three human subjects (SM, WZ, and ZL in Figure 7.21) stand still in

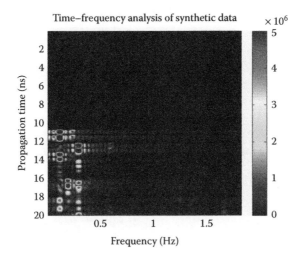

FIGURE 7.20
The time-spectral analysis using empirical mode decomposition for the vital signs generated by the three human subjects in the model shown in Figure 7.16.

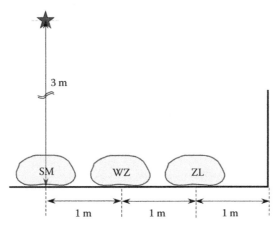

FIGURE 7.21
Physical measurement setup of line-of-sight monitoring using 1-GHz GPR for three patients in a laboratory.

the opposite side of the wall at different positions with all faced toward the GPR antenna, similar to the tri-patient monitoring model shown in Figure 7.16.

For the laboratory physical experiment, the radar recording length is 30 ns for each trace. More than 600 traces are recorded at an interval of 0.1 s. The original radar data for the case of without human beings present (marked as "Background Test," the top panel in Figure 7.22) and with people present (marked as "Regular test," the bottom panel in Figure 7.22) are shown in Figure 7.22.

Using the same time–frequency analysis approach with EMD described in previous sections gives the results shown in Figure 7.23, for the background (left panel), with human subjects present (middle), and the difference of the two situations (right panel). The signatures of breathing from the three subjects can be identified. For the subject close to the antenna (SM), the breathing is shown around 20 ns in propagation time, and for the subject farther away from the antenna (ZL), the breathing is shown around 23 ns in

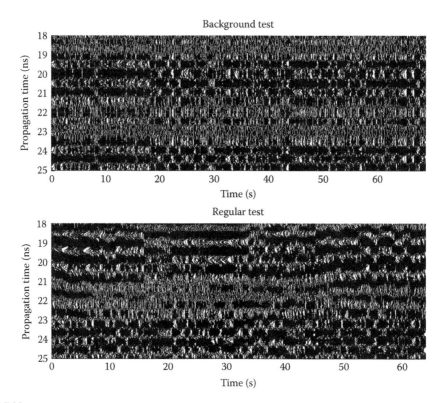

FIGURE 7.22
The raw time-domain laboratory test data using a 1-GHz GPR system. The upper panel is the time domain record without human subjects present. The bottom panel is the time-domain records with the two human subjects positioned at the location shown in Figure 7.21.

propagation time, with weaker signal power. All of the three subjects have a breathing frequency of about 0.25 Hz. The analysis cannot catch heartbeat information for all three subjects. Nevertheless, for the purpose of vital sign detection, using breath alone might be enough to classify a person as alive.

7.8 Summary and Conclusions

This chapter presented an approach to numerically simulating UWB radar responses to the vital signs of buried human beings. This approach can serve as a synthetic data generator for producing data that can be supplementary to acquire field data in different physical scenarios (Ivashov et al., 2004; Zhuge et al., 2007; Lazaro et al., 2010; Suksmono et al., 2010). Laboratory tests verified the vital signatures presented in the simulated data. The work in the near future for the community should continuously validate this approach in comparison with physical measurements under more realistic conditions closer to a real earthquake hazard site. Robust, fast detection algorithm for on-site characterization and classification is the key for the success of UWB radar life search and rescue in earthquake hazard mitigation.

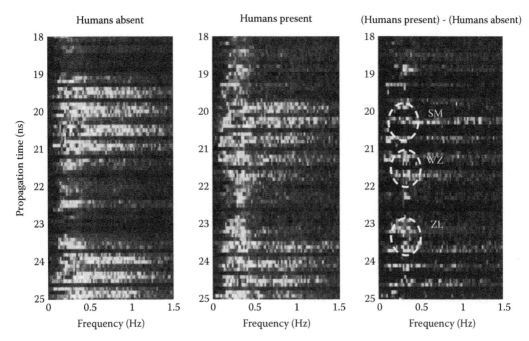

FIGURE 7.23
Model of two human subjects buried in a collapsed building at an earthquake hazard site with a size of 3.74 m x 1.97 m.

In conclusion, the identification of vital signals from multiple human subjects using single radar is a reachable goal. The demonstrations showed the value of empirical model decomposition using Hilbert–Huang transform (HHT) as also covered in Chapter 4. Blind Source Separation (BSS) for multiple subject cases is possible. Passive and active time reversal (TR) for ambient Green's function for facilitating source separation can aid in UWB materials penetrating signal analysis.

The techniques of this chapter can have potential applications in advancing materials penetrating radar technology for multiple applications.

References

Anishchenko, L. N., Bugaev, A. S., Ivashov, S. I. and Vasiliev, I. A. 2009. Application of Bioradiolocation for Estimation of the Laboratory Animals' Movement Activity, *Proceedings of the Progress in Electromagnetics Research Symposium, PIERS 2009, Moscow, Russia, 18–21 August.*

Attiya, A. M., Bayram, A., Safaai-Jazi, A. and Riad, S. M. 2004. UWB applications for through-wall detection, IEEE, *Antennas Propag. Soc. Int. Symp.*, 3: 3079–3082, June.

Berenger, J. P., 1994. A perfectly matched layer for the absorption of electromagnetic waves, *J. Comput. Phys.* 114: 185–200.

Bimpas, M., Paraskevopoulos, N., Nikellis, K., Economou, D. and Uzunoglu, N. 2004. Development of a three band radar system for detecting trapped alive humans under building ruins, *Progress in Electromagnetics Research* 49: pp. 161–188.

Boric-Lubecke, O., Lubecke, V., Mostafanezhad, M., Isar, P., B-K, Massagram, W. and Jokanovic, B. 2009. Doppler radar architectures and signal processing for heart rate extraction, *Microwave Review* 12. pp. 12–17.

Bronzino, J. D. 1999, *The Biomedical Engineering Handbook, 2nd ed.* CRC Press, Boca Raton, FL.

Bugaev, A. S., Chapursky, V. V., Ivashov, S. I. Razevig, V. V., Sheyko. A. P. and Vasilyev, I. A. 2004. Through Wall Sensing of Human Breathing and Heart Beating by Monochromatic Radar, *Proceedings of Tenth International Conference on Ground Penetrating Radar*, Vol. 1. pp. 291–294, June.

Bugaev, A. S., Vasilyev, I. A., Ivashov, S. I., Parashin, V. B., Sergeev, I. K., Sheyko, A. P. and Schukin, S. I. 2004. Remote control of heart and respiratory human system by radar, *Biomedical Technologies and Radioelectronics*. No. 10, pp. 24–31.

Bugaev, A. S., Vasilyev, I. A., Ivashov, S. I., Razevig, V. V. and Sheyko, A. P. 2003. Detection and remote control of human beings behind obstacles with radar using, *Radiotechnique*, No 7, pp. 42–47.

Cist, D. B. 2009. Non-destructive evaluation after destruction: using ground penetrating radar for search and rescue, *NDTCE'09, Non-Destructive Testing in Civil Engineering*, Nantes, France, June 30–July 3.

Egot-Lemaire, S., Pijanka, J., SulÂ'e-Suso, J., and Semenov, S. 2009. Dielectric spectroscopy of normal and malignant human lung cells at ultra-high frequencies. *Phys. Med. Biol.* 54(8): 2341–2357.

Gabriel, C., Gabriel, S. and Corthout, E. 1996. The dielectric properties of biological tissues: I. Literature survey, Phys. Med. Biol. 41: 2231–2249.

Huang, N., and Wu, Z. 2008. A review of Hilbert-Huang transform: Method and its applications to geophysical studies. *Reviews in Geophysics* 46: RG2006.

Immoreev, I. J., and Samkov, S. V. 2003. Ultra-wideband radar for remote detection and measurement of parameters of the moving objects on small range, *Ultra Wideband and Ultra Short Impulse Signals*. Sevastopol, Ukraine. pp. 214–216,.

Ivashov, S. I., Razevig, V. V., Sheyko A. P. and Vasilyev, I. A. 2004, Detection of human breathing and heartbeat by remote radar, *Progress in Electromagnetic Research Symposium*, 663–666, Pisa, Italy.

Lazaro, A., Girbau, D., and Villarino, R. 2010. Analysis of vital signs monitoring using an IR-UWB radar. *Progress in Electromagnetics Research* 100: 265–284.

Li, J., Liu, L., Zeng, Z. and Liu, F. 2014. Advanced signal processing for vital sign extraction with applications in UWB radar detection of trapped victims in complex environment. *IEEE Journal of Selected Topics in Applied Earth Observations and Remote Sensing*. 7(3): 783–791.

Liu, L., and Arcone, S. 2003. Numerical simulation of the wave-guide effect of the near-surface thin layer on radar wave propagation. *Journal of Environmental & Engineering Geophysics* 8(2): 133–141.

Liu, L., and Liu, S. 2014. Remote detection of human vital sign with stepped-frequency continuous wave radar. *IEEE Journal of Selected Topics in Applied Earth Observations and Remote Sensing*. 7(3): 775–782.

Liu, L., Liu, S., Liu, Z. and Barrowes, B. 2011. Comparison of two UWB radar techniques for detection of human cardio-respiratory signals. *Proceedings of the 25th International Symposium on Microwave and Optics Technologies (ISMOT 2011)*, (ISBN: 978–80–01–04887–0), pp. 299–302, June 20–23, Prague, Czech Republic, EU.

Liu, L., Liu, Z. and Barrowes, B. 2011. Through-wall bio-radiolocation with UWB impulse radar: observation, simulation and signal extraction. *IEEE Journal of Selected Topics in Applied Earth Observations and Remote Sensing* 4(3): 791–798.

Liu, L., Mehl, R., Wang, W., and Chen, Q. 2015. Applications of the Hilbert-Huang Transform for Microtremor Data Analysis Enhancement. *Journal of Earth Science* 26(6): 799–806.

Liu, L., Liu, Z., Xie, H., Barrowes, B. and Bagtzoglou, A. 2014. Numerical simulation of UWB impulse radar vital sign detection at an earthquake disaster site. *Ad Hoc Networks* 13: 34–41.

Liu, Z., Liu, L. and Barrowes, B. 2010. The application of the Hilbert-Huang Transform in through-wall life detection with UWB impulse radar. *Progress In Electromagnetics Research Symposium (PIERS 2010)*, Cambridge, MA, USA, July, 2010, PIERS ONLINE, Vol. 6, No. 7, 695–699.

Lv, H., Lu, G. H., Jing, X. J. and Wang, J. Q. 2010. A new ultra-wideband radar for detecting survivors buried under earthquake rubble. *Microwave and Optical Technology Letters*. 52(11): 2621–2624.

Narayanan, R. M. 2008. Through-wall radar imaging using UWB noise waveforms. *Journal of the Franklin Institute.* 345: 659–678.

生命探测雷达"在汶川地震救灾中发挥作用, http://www.nsfc.gov.cn/publish/portal0/tab88/info3036.htm

Saha, S., and Williams, P. A. 1992. Electric and dielectric properties of wet human cortical bone as a function of frequency. *IEEE Trans. Biomed. Eng.* 39: 1298–1304.

Sawada, H., Mukai, R., Araki, S. and Makino, S. 2004. A robust and precise method for solving the permutation problem of frequency-domain blind source separation. *IEEE Trans. Speech & Audio Processing.* 12(5): 530–538.

Sisma, O., Gaugue, A., Liebe, C. and Ogier, J.-M. 2008. UWB radar: vision through a wall. *Telecommun Syst.* 38(1–2): 53–59.

Suksmono, A. B., Bharata, E., Lestari, A. A., Yarovoy, A. G. and Ligthart, L. P. 2010. Compressive Stepped-Frequency Continuous-Wave Ground-Penetrating Radar. *IEEE Geoscience and Remote Sensing Letters* 7: 665–669.

Taflove, A., and Hagness, S. C. *Computational Electromagnetics,* Artech House, Boston, MA, 2000.

Wua, B.-H., Too, G.-P. and Lee, S. 2010. Audio signal separation via a combination procedure of time-reversal and deconvolution process. *Mechanical Systems and Signal Processing.* 24: 1431–1443.

Yarovoy, A. G., Ligthart, L. P., Matuzas J. and Evitas, B. 2006. UWB radar for human being detection. *IEEE Aerosp. Electron. Syst. Mag.* 21: pp. 10–13.

Yarovoy, A. G., Zhuge, X., Savelyev, T. G. and Ligthart, L. P. 2007. Comparison of UWB Technologies for Human Being Detection with Radar. *Proc. of the 4th European Radar Conference,* Munich, Germany, pp. 295–298.

Yee, K. S. 1966. Numerical solution of initial boundary value problems involving Maxwell's equations in isotropic media, *IEEE Trans. Antennas Propag.* 14: 302–307.

Zaikov, E., Sachs, J., Aftanas, M. and Rovnakova, J. 2008. Detection of trapped people by UWB radar. *German Microwave Conf., Hamburg, Germany.*

Zhuge, X., Savelyev, T. G., and Yarovoy, A.G. 2007. Assessment of electromagnetic requirements for UWB through-wall Radar. *International Conference on Electromagnetics in Advanced Applications.* pp. 923–926.

8

Bioradiolocation as a Technique for Remote Monitoring of Vital Signs

Lesya Anishchenko, Timothy Bechtel, Sergey Ivashov, Maksim Alekhin, Alexander Tataraidze, and Igor Vasiliev

CONTENTS

8.1 Introduction

8.1.1 Chapter Overview

This chapter summarizes the results of investigations into the applications of radar in medicine. Bioradars provide a wide range of possibilities for remote and noncontact monitoring of the psycho-emotional state and physiological condition of many macro-organisms. In particular, this chapter provides information on the technical characteristics of bioradars designed at Bauman Moscow State Technical University (BMSTU), Russia, in international collaboration with Franklin & Marshall College, Lancaster, PA, and on experiments using these radars. The results of experiments demonstrate that bioradars of BioRASCAN type may be used for simultaneous remote measurements of respiration and cardiac rhythm parameters. In addition, bioradar-assisted experiments for detection of various sleep disorders are described. The results prove that bioradiolocation allows accurate diagnosis and estimation of obstructive sleep apnea severity, and could be used in place of polysomnography, which is considered the current standard medical evaluation method, but requires direct patient contact.

8.1.2 Bioradiolocation Fundamentals and Applications

Bioradiolocation (BRL) is a modern sensing technique that provides the capability to detect and monitor persons remotely (even when behind optically opaque obstacles) without applying any contact sensors. It is based on reflected radar signal modulation by oscillatory movements of human limbs and organs. Electromagnetic waves reflected from the human body contain specific biometric modulations, which are not present when the waves interact with motionless objects. The main contributors to these signals are heartbeat; contractions of vessels; gross movements of limbs; and the reciprocal movements of the chest wall and abdomen associated with breathing. Thus, a patient's or subject's physical activity and medical (as well as psychological) state determine the characteristics of these signal fluctuations.

Bioradiolocation has a variety of potential application areas in military, law enforcement, medicine, and so on. A detailed list of these areas is given as follows.

- *Antagonist and hostage localization inside buildings during counter-terrorism operations*: There are a few commercially available bioradars for this application; however, these devices have several limitations, which have not yet been overcome, the most serious of which is high attenuation of bioradar signal when propagating through damp or reinforced concrete walls (Greneker 1997, Droitcour 2001).

- *Disaster response*: Bioradars are used for detection of survivors beneath debris after natural or technological disasters (Chen 2000, Bimpas 2004). The most serious challenge lies in multipath propagation of bioradar signal in rubble, which may contain reflectors at all angles. In addition, reduction in false alarms currently requires cessation of all dismantling activities during the period of scanning with bioradar. Similar techniques may be applied during fire emergency searches (Chen 1986). Chapter 7 discusses signal design for biolocation systems.

- *Battlefield or disaster zone triage*: The application of bioradar could help in remotely distinguishing between wounded and dead casualties and thus establish the order of evacuation and treatment priority (Boric-Lubecke 2008).

- *Transportation safety and security*: Examination of transport containers for detecting stowaways, fugitives, or espionage agents at border crossings or other transit terminals (Greneker 1997).

- *Remote diagnostics of psycho-emotional state*: Screening for undue agitation or anxiety during latent or open screening in criminal investigations or at checkpoints (Greneker 1997).

- *Remote speech detection*: Microwave "eavesdropping" to detect speech and speech patterns if not actual transcription of dialogue (Holzrichter 1998).

- *Contactless registration of respiration and cardiac rhythm*: Important for, e.g., burn patients and others for whom contact sensors cannot be applied (Greneker 1997, Droitcour 2001, Staderini 2002).

- *Sleep medicine*: Bioradars can be used to monitor respiration and heartbeat patterns for diagnosis of sleep apnea syndrome (Kelly 2012). In the case of newborns, it is possible to use this method for detection of imminent Sudden Infant Death Syndrome (Li 2010), allowing real-time, immediate, and potentially life-saving response.

- *Estimation of vessel elasticity*: The radar pulse velocity in vessel tissue can reveal patients predisposed to cardiovascular disease (Immoreev 2010).

- *Smart homes*: It is possible to use bioradars to monitoring movements and activities of elderly or invalid subjects in their homes (Li 2010).

- *Tumor tracking*: The size and condition of tumors can be characterized, for example during radiation therapy (Li 2010).

- *Laboratory animal locomotor activity monitoring*: Just as for smart homes, the living quarters of valuable lab animals can be monitored remotely without disturbing them (Anishchenko 2009).

In practice, there are several challenges in achieving reliable registration of respiration and heart rate parameters using bioradar. Among them are: clutter caused by surrounding objects and multipath propagation, artifacts from gross body movements with amplitude much bigger than the desired signals, and problems with isolation and discrimination of respiration and heartbeat signals during prolonged monitoring. These challenges require development of adaptive algorithms for effective extraction of the informative components of bioradar signals, as well as implementation of numerical procedures aimed at improving the accuracy and stability of estimates of physiological parameters such as respiration and heart rate. For example, by applying low-cut filters to reject all near-zero frequencies from the reflected bioradar signal, it is possible to suppress clutter caused by local inanimate objects (Bugaev 2005).

Among the signals suitable for detecting the characteristic signatures of living objects are continuously modulated or nonmodulated microwave signals at frequencies ranging from hundreds of MegaHertz (MHz) to tens of GigaHertz (GHz); narrowband, wideband, or ultrawideband (UWB) signals with appropriate center frequencies; and "noise" pulse signals that have no clearly defined carrier frequency (Immoreev 2010, Li 2012, Wang 2013, Otsu 2011, Ivashov 2004).

The material in this chapter is structured as follows: following the introduction in Section 8.1, Section 8.2 reviews the theory of vital signs registration by means of bioradar using the example of a nonmodulated probing signal. This section will also present two types of processing for separation of co-registered respiratory and heart beat signals. Section 8.3 contains information on methods for verification of the vital signs data recorded by bioradar. Section 8.4 reviews the results of various biomedical experiments using bioradar.

8.2 Bioradiolocation Signal-Processing Methods for Life Signs Monitoring

The presence of biometric information in reflected microwave signals is related to the contraction of the heart, blood vessels, lungs, and other internal organs, as well as movements of the skin surface due to respiration and cardiac pulse. These phenomena are periodic with frequency ranges of approximately 0.8–2.5 Hz for pulse and 0.2–0.5 Hz for respiration. A reflected microwave signal that contains such biometric information is referred to as a biometric radar signal. The useful component of biometric information (related to physiological signatures) is recorded in the parameters of modulation of the biometric radar signal in the time domain as well as in their spectra in the frequency domain. As a result of cyclical recurrence of phenomena related to respiration and pulse, there are corresponding components in the spectrum of a biometric radar signal. The frequencies and amplitudes of these radar signal components roughly match the frequencies and intensities of the breathing and heart beat phenomena. As described below, the theoretical estimation of the spectrum of a reflected signal output from the microwave receiver is of special interest in many applied studies.

8.2.1 Monochromatic Bioradar Signal Processing

8.2.1.1 Theoretical Model of Monochromatic Bioradar Signal

The simplest microwave device capable of recording biometric signatures is a radar with a nonmodulated probing signal described by

$$\dot{u}_0(t) = U_0 e^{j\omega_0 t} \tag{8.1}$$

Dominant reflections arise at the air–skin interface. The effective area of reflection and therefore area of relevant fluctuations of the skin is located within the limits of one Fresnel zone, with radius r_0. The radius of the effective area of reflection can be written as

$$r(t) = r_0 + \Delta r(t), \tag{8.2}$$

where $\Delta r(t)$ is characteristic of the fluctuations of the skin. Based on Equation 8.2, the useful signal recorded by the radar will be registered with attenuation factor q and phase shift $\varphi(t) = -2kr(t)$:

$$\dot{u}_c(t) = q U_0 e^{j\omega_0 t - j\varphi_0 - j2k\Delta r(t)}, \tag{8.3}$$

where:
  $k = 2p/\lambda$ is wavenumber
  λ is radiated signal wavelength
  $\varphi_0 = 2kr_0$

Usually, at the input of the receiver, along with the useful reflected signal, the so-called penetrating signal of the transmitter also registers:

$$\dot{u}_{inp}(t) = q_p U_0 e^{j\omega_0 t - j\varphi_p} + q U_0 e^{j\omega_0 t - j\varphi_0 - j2k\Delta r(t)} \tag{8.4}$$

where q_p and φ_p are the attenuation factor and phase of the penetrating signal.

Further, and without reducing generality, it is possible to accept zero phase for the penetrating signal, or $\varphi_p = 0$. We can represent fluctuations of the skin due to breathing and heart beat by a biharmonic function:

$$\Delta r(t) = \Delta_1 \sin(\omega_1 t) + \Delta_2 \sin(\omega_2 t + \varphi_2) \tag{8.5}$$

where:
$\omega_1 = 2\pi f_1$
$\omega_2 = 2\pi f_2$
$f_1, f_2, \Delta_1, \Delta_2$ are frequencies and amplitudes of breathing and heart beating
φ_2 is a constant phase

In the microwave radar spectrum, it is possible to apply either of the two types of receiver. There may be a coherent receiver with two quadrature phase detectors, or an amplitude receiver. However, to overcome the loss of the basic frequencies in a received signal spectrum, one should use the principles of quadrature coherent signal processing (Bugaev 2004).

8.2.1.2 Basic Experiments on Noncontact Cardiorespiratory Parameters Registration Using a Monochromatic Bioradar

At BMSTU, the BRL technique has been investigated since 2003. During the first series of experiments, a modified ground-penetrating radar (GPR) of RASCAN type with operating frequency of 1.6 GHz was employed. The experiments addressed radar sounding of human cardiorespiratory parameters through a brick wall, and demonstrated that the task of remote diagnostics of human cardiorespiratory parameters using continuous-wave subsurface radar (Bugaev 2004) is technically feasible. A sketch of the experiment is shown in Figure 8.1.

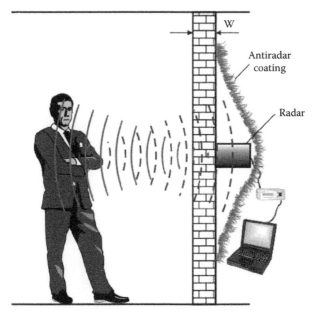

FIGURE 8.1
Sketch of the experiment on through-wall sensing of human cardiorespiratory parameters with a modified ground-penetrating radar ($w = 10$ cm).

The examinee stood behind a 10 cm-thick wall and about 1 m from the wall. The radar antenna was fastened directly at the wall surface. To decrease the interference from reflections of the back lobes of the transmitted signal, the antenna and the part of the wall were veiled by an antiradar (nonreflective) coating with dimensions of 2×2 m.

The reflected radar signals were detected and recorded in the computer memory through an interface module. In Figures 8.2 and 8.3, the pulse record and corresponding signal spectrum for the examinee during a period of breath holding are presented. In Figure 8.2, the breath was held approximately 30 seconds. In Figure 8.3, the hold was about 1 minute. It is clear that with the increase in breath hold time, the amplitude and rate of examinee heartbeat activity are also increased as a result of oxygen deprivation. The results of simultaneous recording of heartbeat and breathing are given in Figure 8.4. The amplitude of breathing oscillations considerably surpasses heart beating vibrations, so a composite response is clearly visible.

The results obtained in the experiments are similar in many respects to the signals registered by time-domain impulse radars in free space (Liu 2012). However, the use of monochromatic wave radars simplifies the experimental installation and subsequent data processing. The experiments on radar sounding of human heartbeat and breathing through a 10-cm-thick brick wall demonstrated that remote diagnostics of human cardiorespiratory parameters using a continuous-wave subsurface radar of RASCAN type is technically feasible.

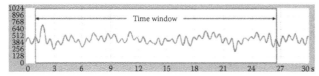
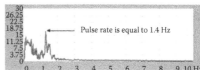

FIGURE 8.2
BRL signal and its spectrum for 30-second breath hold.

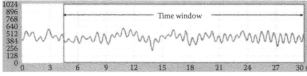
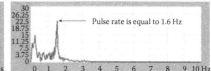

FIGURE 8.3
BRL signal and its spectrum for 60-second breath hold.

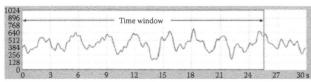
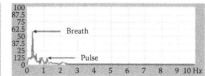

FIGURE 8.4
BRL signal and its spectrum for simultaneous heartbeat and breathing registration.

8.2.2 Comparison of the Two Bioradiolocation Signal-Processing Techniques for Noncontact Monitoring of Cardiorespiratory Parameters

Another possible type of bioradar is multifrequency, which allows the estimation of the range to the subject as well as the cardiorespiratory parameters. Using data from this type of bioradar, a feasibility study of life signs detection and characterization using a multifrequency radar system was performed (Soldovieri 2012). The recordings were processed using two different data-processing approaches, with the performance characteristics for each compared in terms of the accuracy of the frequency characterization for respiration and heartbeat.

The BioRASCAN multifrequency bioradar designed at Remote Sensing Laboratory of BMSTU has a quadrature receiver designed to perform remote monitoring of gross human movements, as well as breathing and heartbeat. This radar operates by transmitting and receiving a continuous-wave field in the frequency range from 3.6 GHz to 4.0 GHz. Based on the dielectric constant of air in this waveband, this provides a spatial resolution of approximately 0.5 cm.

The first data processing approach to be considered was previously presented by D'Urso (2009) and was intended to provide frequency analysis of biometric signals by maximizing the scalar product of the Fourier transform of two signals, one representing the measured signal modulated by reflector displacement and the other derived from a theoretical electromagnetic model. The second approach attempted to provide information not only on the frequency of the life signs signals, but also at recovering information on the range of the subject by establishing a range–frequency matrix. Separation of respiration and heartbeat signals was made by the application of rejection filtration to corresponding line of range–frequency matrix (Bugaev 2005, Anishchenko 2008).

Figure 8.5 shows the experiment in which the subject was located in front of the radar system. In particular, the experiment was carried out with a 20-year-old male subject in good health and fitness, and a professional skier. The distance between the antennas and subject was 1 m. The experiment was divided into two stages. During the first stage, monitoring of breathing and pulse parameters at steady state was performed for about 5 minutes. At the second stage, a breath-holding test was performed. It gave a rough index

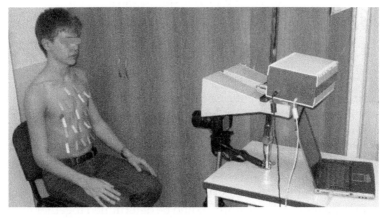

FIGURE 8.5
Photo of the experimental setup.

of cardiopulmonary reserve by measuring the length of time that the subject could hold breath. This is a widely known test in medicine, which is used for estimating physical fitness in the training of pilots, submariners, and divers.

8.2.2.1 The First Bioradiolocation Signal-Processing Approach

As described above, the goal of this test is to detect vital signs (breathing and heartbeat) and to determine their frequency for the case of a human being in free space. This is a simplified model described by Equation 8.5 without the last term corresponding to heartbeat. Computation of the field reflected by the oscillating target exploited quasi-stationarity following the method of Soldovieri (2012).

In real conditions, received signal contains clutter (u_{clut}):

$$u_R = U_0 \exp(-2jk(r_0 + \Delta_1 \sin[\omega_1 t])) + u_{clut} \tag{8.6}$$

Thus, the problem at hand is stated as how to estimate the frequency ω_1 starting from the knowledge of the reflected field measured over a finite time interval $[0,T]$.

The proposed reconstruction procedure included the following steps. First was the removal/mitigation of the static clutter, that is, the u_{clut} term in Equation 8.6. The ideal clutter removal strategy would be based on the difference between the actual signal and the one when no vital signs are present (background signal). Since such a background measurement is not available, an alternative strategy is necessary. In this algorithm, the static clutter removal is carried out by the following steps: First, we compute the mean value u_{mean} of the signal over the interval domain; then, we subtract u_{mean} from the measured one $u_R(t)$ to achieve $\tilde{u}_R(t) = u_R(t) - u_{mean}$. The subsequent processing is then performed on $\tilde{u}_R(t)$. Next, Fourier transformation is performed on the resulting signal to compute the function $G(\omega_1)$ in Doppler domain. The Fourier transform of the model signal $\exp(-2jk\Delta_1 \sin(\omega_1 t))$ is computed as

$$u_{model}(\omega_1) = \int_0^T \exp(-2jk\Delta_1 \sin(\omega_1 t)) \exp(-j\omega_1 t) dt =$$

$$= \int_0^T \sum_{n=-\infty}^{\infty} J_{-n}(2k\Delta_1) \exp(jn\omega_1 t) \exp(-j\omega_1 t) dt = \tag{8.7}$$

$$= \sum_{n=-\infty}^{\infty} J_{-n}(2k\Delta_1) \mathrm{sinc}\left[\frac{T}{2}(\omega_1 - n\omega_1)\right] \exp\left(-j(\omega_1 - n\omega_1)\frac{T}{2}\right)$$

where we exploit the well-known Fourier expansion of the term $\exp(-2jk\Delta_1 \sin(\omega_1 t))$, and $J_n(\bullet)$ denotes the Bessel function of first kind and nth order. Therefore, the Fourier transform $u_{model}(\omega_1)$ is made up of a train of sinc functions centred at $n\omega_1$.

Finally, the unknown Doppler frequency ω_1 is determined as the quantity that maximizes the scalar product between the modulus of the measured Fourier transform $|G(\omega_1)|^2$ and the modulus of the Fourier transform of the model signal $|u_{model}(\omega_1)|^2$.

It is worth noting that in the above-outlined procedure, the maximum displacement is still unknown. In principle, such a quantity could be determined together with the Doppler frequency to maximize the scalar product. However, in the case at hand, in order to make the determination procedure fast enough for realistic conditions, we assume an estimate of the maximum displacement as $\Delta_1 = 0.5$ cm for the breathing and 1 mm for heartbeat.

8.2.2.2 The Second Bioradiolocation Signal-Processing Approach

The second data-processing approach is designed to gain information not only about the frequency behavior of life signs, but also about the range of the investigated subject (Bugaev 2005, Anishchenko). The steps in this procedure are as follows.

The first step is to build a range–frequency matrix (Bugaev 2005); this matrix contains all possible signal reflections including ones from motionless objects, located in different range cells. These objects are the cause of static clutter. The range–frequency matrix resulting from the suppression of the zero or nearly zero frequencies is given in the upper panel of Figure 8.6.

The separation between the breathing and heartbeat signals is carried out next using rejection of the frequency components corresponding to breathing in the range–frequency matrix, and the result is shown in the lower panel of Figure 8.6.

Reconstruction of breathing and heartbeat signals is carried out by applying an inverse Fourier transform to the matrix row corresponding to the distance to the examinee (1.1 m) and evaluating its phase. Signals obtained in this fashion corresponding to range–frequency matrices from Figure 8.6 are shown in Figure 8.7, which clearly show good performance of this approach in separating breathing and heartbeat signals.

8.2.2.3 Bioradiolocation Signal Reconstruction Results

After separating the signals for breathing and pulse, the reconstruction of these signals is necessary. It can be made at any of the probing frequencies. Figures 8.8 and 8.9 compare the reconstructed signals for respiration and pulse at the 3.6 GHz probing frequency.

For better performance, the recorded bioradar signal was divided into 19 time windows, each consisting of 1024 time samples (for a time interval length of 16.3 seconds). For each of these intervals, the two data-processing approaches described above were applied.

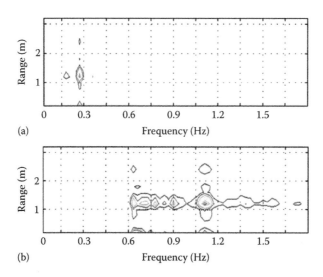

FIGURE 8.6
Range–frequency matrix for the examinee at 1.1 m range: (a) before breathing harmonics rejection and (b) after breathing harmonics rejection.

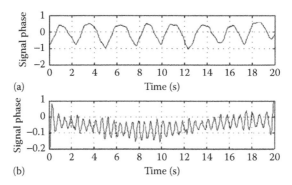

FIGURE 8.7
Reconstructed breathing (a) and heartbeat (b) signals of the subject corresponding to the range–frequency matrices from Figure 8.6 (the signal phase is normalized with respect to its maximum).

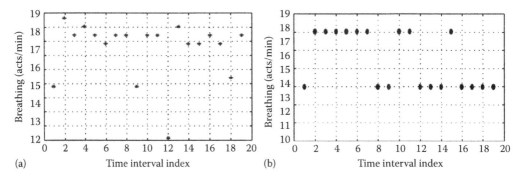

FIGURE 8.8
Comparison between the data processing approaches for respiration monitoring. (a) first data processing approach and (b) second data processing approach.

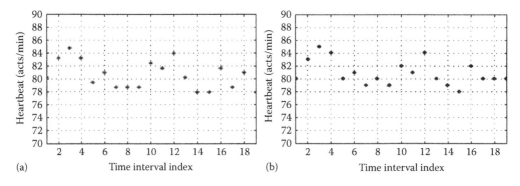

FIGURE 8.9
Comparison between the data processing approaches for pulse monitoring. (a) first data processing approach and (b) second data processing approach.

In both the respiration and pulse recordings, a good agreement is observed between the results for the two data-processing approaches; in particular, an almost uniform breathing behavior is observed with a frequency of 18 beats/min apart from few time intervals.

It can be seen that the average pulse frequency is about 80 beats per minute. In addition, we can note a correlation between the time behavior of the breathing and the one for heart beat. Specifically, when the breathing frequency decreases, the heartbeat frequency also decreases.

8.3 Verification of Bioradiolocation with Standard Contact Methods

For effective application of any new method in medicine, measured vital signs parameters must be verified using current best-practice or gold-standard methods. For heart beat and breathing frequency estimation, the gold-standard methods are ECG and respiratory impedance plethysmography (RIP), respectively (Konno and Mead, 1967). This section presents the verification testing for BioRASCAN measurements.

8.3.1 Verification of *Bioradiolocation* and ECG

Comparative experiments for BioRASCAN and ECG methods were carried out to confirm that the bioradar can be used for accurate heart rate monitoring as shown in Figure 8.10.

The test population of 52 adult examinees consisted of 23 males and 29 females with a mean age 20 ± 1 years (mean $\pm$ SD) participated in the verification experiments. During the experiment, each subject sat in a relaxed pose in front of the bioradar at a distance of 1 m from antennas. For each subject, bioradar and ECG signals were recorded three times for a duration of 1 minute per record. Comparison of the measured heart beat frequency values for both methods were compared, yielding a good agreement with a confidence level of $p = 0.95$. Thus, the feasibility of BRL for simultaneous measurements of breathing and heart rate parameters was demonstrated at a statistically significant level.

8.3.2 Verification of Long Records of Bioradiolocation Signals with Respiratory Plethysmography

Previously, experiments to verify respiratory patterns using BRL were made in idealized conditions with short records, and motionless subjects facing the bioradar atennas (Massagram 2011, Droitcour 2009, Vasu 2011, Alekhin 2013). In this section, we present

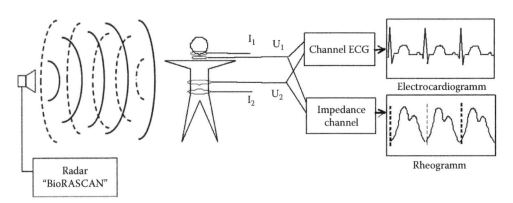

FIGURE 8.10
Sketch of the experiment to verify BioRASCAN measurements against conventional measurements.

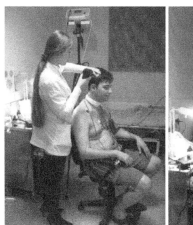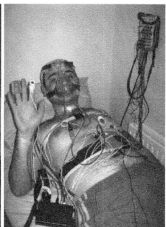

FIGURE 8.11
Photos of the experiment comparing simultaneous registration of BRL (radar) and conventional polysomnography (PSG) data.

experimental results obtained in close-to-real conditions (during a full night of sleep) during which examinees could change their position and move their limbs freely.

Five subjects without sleep breathing disorders underwent sleep study at the Sleep Laboratory of the Federal Almazov Medical Research Centre (Saint Petersburg, Russia). During the experiment, BRL and RIP signals were registered simultaneously throughout a night of sleep as shown in Figure 8.11. Full-night polysomnography (PSG) using an Embla N7000 by Embla Systems LLC, Ontario, Canada, was performed, including registration of respiratory movements by RIP. BRL monitoring was done using a BioRASCAN with operating frequency band 3.6–4.0 GHz.

Detection of inhalation peaks based on BRL and RIP signals was performed. A data-processing algorithm was used for the RIP signal inspiratory peaks detection, consisting of the following steps: summation (half of sum of the abdominal and thoracic RIP signals was used in the subsequent analysis); filtration (3rd-order Butterworth filter operating in 0.2–1.0 Hz frequency range was applied); and detection (inspiratory peaks were detected by an algorithm to find local maxima). The BRL signal shown in Figure 8.12 compared with the RIP signal had the following peculiarities: subjects' movements make more artifacts; there are strong amplitude changes resulting from the subjects' full body displacement; phase shifts result from flip-over signals, making inspiratory peaks downward.

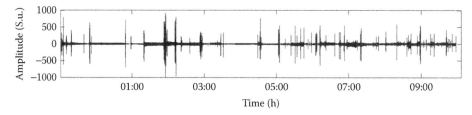

FIGURE 8.12
Original BRL signal recorded prior to signal-processing stage.

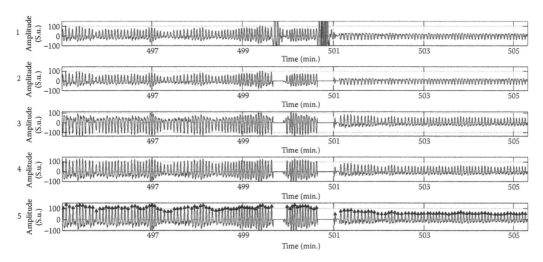

FIGURE 8.13
Results of the inspiratory peak detection algorithm steps: 1 – filtered signal; 2 – signal after movement artifact rejection; 3 – combined signal comprising the best interartifacts parts of signals; 4 – signal after phase shift rejection; 5 – inspiratory peak detection.

The results of BRL signal processing at different steps in the inspiratory peak detection algorithm are given in Figure 8.13.

For the comparison/verification, the start of the BRL and RIP signals were peak-to-peak synchronized. BRL and RIP signals were truncated to the sleep period, that is parts of the signals recorded during waking before and after sleep intervals were deleted. Intervals of RIP signal corresponding to intervals of BRL signal marked as artifacts were also rejected. Inspiratory peak counts per minute (respiratory rhythm) were calculated for the RIP and BRL signals shown in Figure 8.14. Further, Pearson's correlation coefficient was calculated between respiratory rhythms obtained by RIP and BRL methods for each subject.

Although RIP and BRL signals were initially peak-to-peak synchronized, they subsequently desynchronized as shown in Figure 8.15. Since desynchronization is not more than a half of respiratory cycle, this is not a problem, and is apparently due to the absence of continuous or periodic time synchronization between the BioRASCAN and Embla devices.

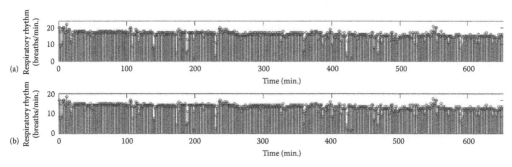

FIGURE 8.14
Comparison of respiratory rhythm estimation by BRL (a) and RIP (b) signals.

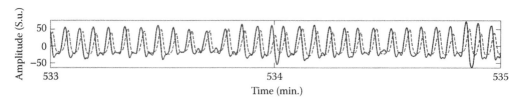

FIGURE 8.15
BRL and RIP signal desynchronization (the solid plot is the BRL signal; the dashed line is the RIP signal).

The cross-correlation coefficient for BRL and RIP results of 0.97 indicates a very strong correlation. The artifact time mean value of 6.8% shows high sensitivity of BRL to movements. Although a small number of subjects were tested, these results proved that bioradars may be used for accurate noncontact monitoring of breathing patterns during prolonged periods, for example in sleep studies.

8.4 Experimental Study on Prospective Applications of Bioradiolocation Technology in Biomedical Practice

This section reviews the bioradar-assisted experiments, which were carried out at BMSTU since 2006. Two bioradars BioRASCAN-4 and BioRASCAN-14 operating in different frequency ranges were used for conducting the experiments. Table 1 presents their technical characteristics (Table 8.1).

8.4.1 Automated *Estimation* of *Sleep Quality* during *Prolonged Isolation*

One of the most promising areas of bioradar application in medicine is somnology (the scientific study of sleep). In 2009, BMSTU conducted the first experiment to investigate the possibility of bioradar usage for noncontact sleep quality estimation. Bioradar signals were recorded during an entire night of sleep for the examinee. Six overnight records were made for a heathly adult male (aged 20 years).

Processing of the recorded signals was performed in stages using MATLAB®. First to eliminate baseline drift, the bioradar signal was filtered using the built-in MATLAB

TABLE 8.1

Technical Characteristics of BioRASCAN Radars

	BioRASCAN-4	BioRASCAN-14
Number of frequencies	16	
Operating frequency band, GHz	3.6–4.0	13.6–14.0
RF output, mW	<3	
Gain constant, dB	20	
Detecting signals band, Hz	0.03–5.00	0.03–10.00
Dynamic range of the detecting signals, dB	60	
Size of antenna block, mm	$370 \times 150 \times 150$	$120 \times 50 \times 50$
Sensitivity, mm	1.0	0.1

Butterworth digital filter with cutoff frequency of 0.05 Hz (filter order of 8). Then, intervals of movement activity were detected. It is obvious that the level of the received signal, which corresponds to calm breathing and whole body movement, differs by more than a factor of 10 in amplitude. However, the main problem in detecting movement artifacts is the fact that the examinee may turn from one side to other during sleep. In this case, the distance between antenna and examinee and the scattering cross section of the target may change. As a result, the level of the received bioradar signal may also vary significantly before and after movement artifact as shown in Figure 8.16a.

That is why it is not enough to use only signal amplitude parameters for the detection of movement episodes. However, episodes during which movement signal artifacts are present contain higher frequency components (greater than 1 Hz) than during times of regular breathing (0.1 to 0.6 Hz). These spectral differences were used in the algorithm for movement artifact detection.

Respiration frequency is estimated only for intervals free from movement artifacts and shown in Figure 8.17. The mean values for breathing intervals were calculated for every 30 seconds as it is usually done while processing somnology data (Rechtschaffen 1968).

Due to the successful results of the studies described above, bioradar experiments were included in the scientific program of the international research project MARS-500 (simulation of prolonged isolation during a manned flight to Mars), which was led by the Institute for Biomedical Problems Russian Academy of Science from June 2010 to November 2011.

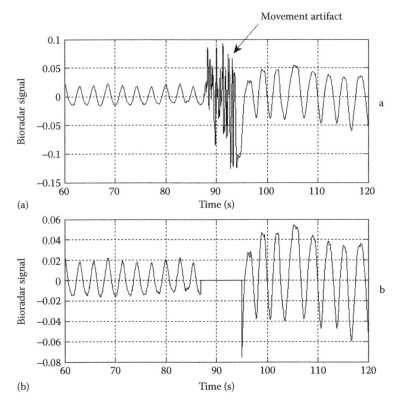

FIGURE 8.16
Bioradar signal before (a) and after (b) movement artifact extraction.

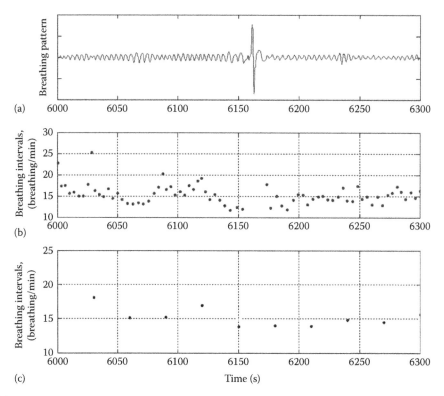

FIGURE 8.17
Example bioradiolocation signals from trials on patients with and without diagnosed sleep disorders. (a) patient with Central Sleep Apnea (CSA), (b) Obstructive Sleep Apnea (OSA), and (c) Normal Calm Sleeping (NCS).

Due to the prolonged isolation, an ethical committee approval and informed consent from all MARS-500 crewmembers was obtained before the start of the experiment. The crew was trained to conduct their own bioradar experiments. During the project, seven series of bioradar experiments were conducted for each of six crew members. As it turned out, it was more convenient to use hourly averages of parameters to record their dynamic ranges during a full night of sleep. The results of the experimental data recorded for one of the MARS-500 crew members are given in Figure 8.18.

It is known that the breathing pattern and movement activity dynamics are characteristic of individuals, and do not usually change greatly from night to night. If changes take place, it may indicate that the examinee suffers from some kind of stress during day time. Using the proposed algorithm, it is possible to monitor breathing and movement activity pattern and thus detect a sleep disturbance caused by daytime stress. In Figure 8.18, the changes in breathing frequency dynamics along the duration of the project for one of the crewmembers are presented. It is clearly seen that for the first half of the experiment (from June 2010 through January 2011) after falling asleep, the respiration frequency of the examinee decreased, but during the last hour of sleep, breathing frequency became higher. However, for the second half of the experiment (from January 2011 through July 2011), the breathing pattern during sleep changed, perhaps indicating stress caused by prolonged isolation.

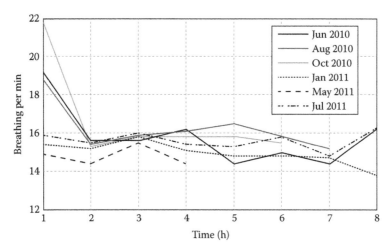

FIGURE 8.18
Respiration frequency dynamics during sleep for one crewmember of MARS-500.

Processing of this experimental data has revealed the individual characteristics of different stages of sleep for the crew members. Some of them have a longer period of falling asleep and more restless sleep, while others, on the contrary, fall asleep faster and have more calm and regular breathing thereafter. Duration of sleep for each subject during the project changed individually, for four out of six crew members, during the first three series of experiments, a greater than 10% decrease in sleep duration was registered. In the second half of the experiment, significant changes in monitored parameters (respiratory rate and the duration of movement artifacts during sleep) did not occur, which indicates a good tolerance of the crew to the conditions of prolonged isolation. Breathing sleep disorders were not detected for any of the crew members.

8.4.2 Noncontact Screening of Sleep Apnea Syndrome Using Bioradiolocation

One of the priority applications of BRL in sleep medicine is detection of sleep disordered breathing (SDB). The objective of a study conducted in 2012 was to estimate the diagnostic ability of BRL for noncontact screening of SDB in adults in comparison with the gold-standard full-night polysomnography (PSG) method.

The seven test subjects included four males and three females, aged 43–62 years, with body mass index or BMI in the range 21.6–57.7. They had a range of severities of obstructive sleep apnea syndrome (OSAS): 4 severe; 1 moderate; 1 mild; 1 normal. The PSG records were collected with an Embla N7000 system in the sleep laboratory of the Almazov Federal Heart, Blood and Endocrinology Centre, while simultaneously monitoring with a BioRASCAN system. Subsequently, PSG records were analyzed by a certified specialist and the verification of corresponding BRL signals was performed manually by a trained operator.

Algorithms applying wavelet transform (WT) and neural network (NN) were previously used for recognition of breathing patterns of BRL signals during noncontact estimation of sleep apnea syndrome (SAS) severity (Alekhin 2013).

The proposed algorithm consists of two main steps. In the first step, WT is applied to extract informative feature. Initially, a general class of wavelets is defined, then a set of wavelet bases with ordinal indexes for wavelet families from the general class is formed,

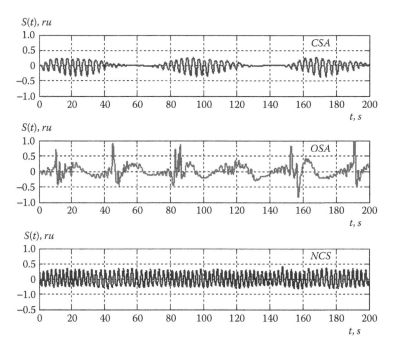

FIGURE 8.19
Bioradiolocation signals for the recognition of clinically verified breathing patterns of BRL signals corresponding to the following syndromes: obstructive sleep apnea (OSA); central sleep apnea (CSA); and normal calm sleeping (NCS) without SDB.

and then the optimal level of wavelet decomposition is determined. In the second step after preliminary estimation of number of NNW hidden layers, the best NNW learning algorithm is applied.

This proposed approach was tested for recognition of clinically verified breathing patterns of BRL signals corresponding to the following syndromes: obstructive sleep apnea (OSA); central sleep apnea (CSA); normal calm sleeping (NCS) without SDB (Figure 8.19).

The analysis of PSG records revealed in total 2700 episodes of SDB: 1279 incidents of OSA; 106 of CSA; 495 for mixed sleep apnea (MSA); and 820 hypopneas (HYPA). The result of verification of BRL signal patterns for SDB in comparison with PSG was as follows: 1955 true positives; 745 false positives; 868 false negatives. Thus, BioRASCAN system displayed a sensitivity of 69% and an accuracy of 72% in noncontact screening for SDB. These results should be considered clinically significant since, in each case, the estimate of apnea–hypopnea index (AHI) for the BRL method overlaps with the same range for the OSAS severity scale using the PSG method.

8.4.3 Human *Physiological Psycho-Emotional State Monitoring* and *Professional Testing*

Another possible area of BRL application is monitoring of the human psycho-emotional state. At BMSTU, we investigated this in experiments involving an internal stress factor. An example of a recorded BRL signal for this kind of test is given in Figure 8.20 for a ringing mobile phone as a stress factor. Note that while the phone was ringing, the amplitude of chest movements caused by breathing became two times lower than without the ringing.

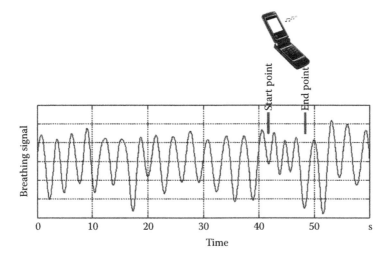

FIGURE 8.20
Received bioradar signal for the additional stress factor experiment. The ringing mobile telephone produces a distinct change in the respiration signal.

As for breathing frequency, its value slightly increased. That is, when the phone rings, the subjects breathing becomes shallow and rapid.

To imitate a stress factor with a longer duration, the standard mental load test was used. During the 5-minute test, the examinee was asked to solve simple mathematical problems. The test sample included 52 subjects (25 males and 27 females, aged 19–21 years). In this case, respiration and heart beat frequency parameters remained almost the same, but their variability did not. A histogram for vital sign frequencies monitored by BRL may be used as a convenient way to represent changes caused by mental load. Histograms of the heart beat intervals before and after the testing for one of the examinees are shown in Figure 8.21. During mental loading, heart rate increased from 1.2 to 1.5 Hz, and pulse interval variability decreased (standard deviation went from 0.25 to 0.06 sec).

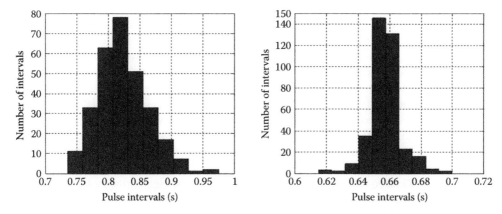

FIGURE 8.21
Histograms of heart beat intervals before and after the standard mental test for one of the examinees.

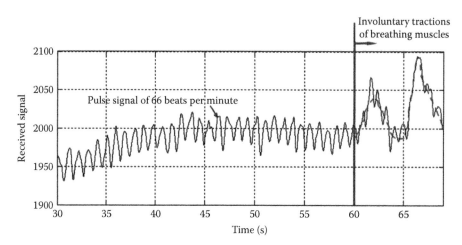

FIGURE 8.22
Received BRL signal for a breath-holding test.

The experimental data analysis showed that performing mental loading leads to a statistically significant change in heart rate (a confidence level of $p = 0.80$), but the changes in respiratory rate were not statistically significant.

The BRL method can be used during the widely known Shtange's and Hench's breath holding tests for estimating cardiorespiratory fitness. These are used in the selection of pilots, submariners, and divers. An example of a recorded bioradar signal for this kind of test is presented in Figure 8.22. After holding the breath for 1 minute , involuntary contraction of respiratory muscles took place because of oxygen starvation. However, examinee continued holding his breath even after this phenomenon. The problem is that the correct duration of this test for the examinee should be estimated without the period when such involuntary contraction occurs as shown on the right side of Figure 8.22. Thus, it was proved that usage of BRL in this type of test would sufficiently improve its accuracy.

8.4.4 Estimation of Small Laboratory Animal Activity by Means of Bioradar

BRL can measure both humans and animals for pharmacology and zoo-psychology studies for developing new medicines or conducting behavioral tests (Kropveld 1993, Anishchenko 2009).

Currently tests of drugs and toxic substances use invasive methods to measure variations in physiological parameters of laboratory animals. The researcher in this case subjectively estimates the animal's locomotor activity visually and has no definitive measure of activity. It is possible to reduce the workload on the researcher by applying automated methods to estimate locomotor activity using specially designed video-tracking systems. The main drawback of this is that it cannot be used for animals in optically opaque mazes.

There are devices based on different physical principles, which can be used for automated monitoring of motor activity parameters of laboratory animals. Some of them use pressure sensors mounted in the cage floor, which allows estimation of the animal movement in the cage (Bederman 1972). Others use light sources and optical sensors integrated in the cage walls (Hideo 1997). Also, for similar purposes, electromagnetic radiation has been proposed (Salmons 1969).

The main drawback of all these devices is the manufacturing complexity of the cage in which the animal is placed during the experiment. Furthermore, these cages are designed for a certain type of laboratory animal for which specific morphometric features need to be considered. That is why, generally, the locomotor activity of the animals is visually estimated by a researcher.

Devices based on the method of BRL are free from the drawbacks listed above. They measure a phase shift of the radar signal reflected from the biological subject, which is caused by body surface displacements. This method does not require design and construction of special cages, and may be used with the plastic containers in which animals are usually kept, or in optically opaque (but electromagnetically transparent) mazes. Another advantage of bioradar sensors in solving the above problems is the possibility of direct automated evaluation of the animals' activity for prolonged periods, as proven by the experiments conducted at BMSTU in 2009. A sketch and photo of the experimental set up are given in Figure 8.23a and b, respectively. During the experiments, the animal (a four-month-old Wistar rat) was placed into a box of $70 \times 70 \times 70$ cm size with dielectric walls. Transmitting and receiving antennas of the radar were directed into the box as shown in Figure 8.23b.

The signal reflected from the animal was recorded for further processing. The distance between the antenna block and carton was approximately 1 m. This short distance was necessary due to the relatively small scattering cross section of the animal. The video signal was also recorded using a simple web camera placed over the box. Behavior and movement activity of the animal during the experiment, as recorded by the camera, was used for the identification of different types of a rat locomotor activity in the BRL signals. It is known that power flux density near radar-receiving antennas declines inversely with the fourth power of the range between antennas and an object, making the power of the reflected signal greatly dependent on the distance between the antenna block and animal. Because of this, an accurate estimation of the rat's movement is challenging. A corner reflector was used to make the power of the reflected signal indifferent to the location of the animal inside the box. It was formed by covering two walls and the floor of the box with metallic film.

Figure 8.24 presents a fragment of the BRL signal in which intervals of inactivity and activity may be easily distinguished by reflected signal amplitude.

If the intensity of the animal's movement activity, as well as different types of activity, needs to be determined, then spectral analysis can be used. Figure 8.25 presents the spectra for BRL signals corresponding to different states or activities of the animal.

The spectra differ greatly in both magnitude and form. Thus, it is possible to distinguish grooming from steady state, sleep, or active movements of the animal using spectral analysis of BRL signals.

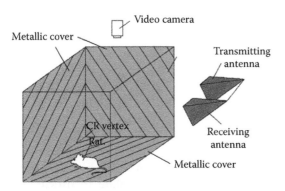

FIGURE 8.23
Schematic and photo of the animal activity experiment set up.

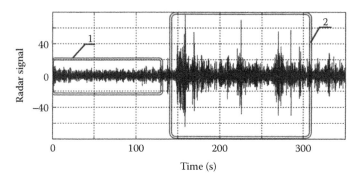

FIGURE 8.24
Radar signal reflected from an animal (1 – steady state, 2 – physical activity).

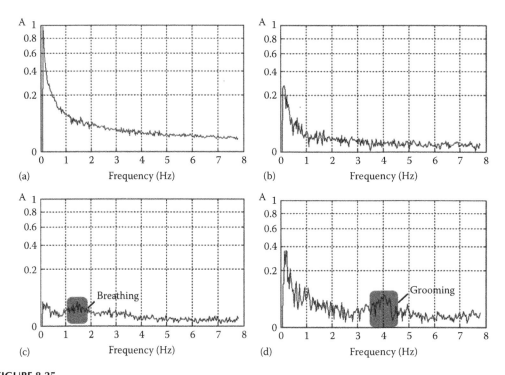

FIGURE 8.25
The results of the animal activity experiments showing the distinct spectra of different activities: (a) Active movements, (b) Steady state, (c) Sleeping, and (d) Grooming.

8.5 Conclusion

This chapter presented the BRL method for remotely monitoring human vital signs and animal activities. It discussed the technical characteristics of the bioradar system created at the Remote Sensing Laboratory of Bauman Moscow State Technical University and the results of experiments using variants of this system. In experiments where bioradar was used simultaneously with standard (contact) methods and video for monitoring respiratory and heart rate parameters, the BRL performed as reliably and effectively, which makes accurate remote noncontact monitoring possible. One set of experiments showed how the BRL is suitable for remote measurement and estimation of human psycho-emotional state. The remote measurements showed the adaptive capabilities of the organism (including tolerance to oxygen starvation) during cardiorespiratory fitness testing. Moreover, BRL can measure and estimate sleep quality and apnea syndrome severity, where it showed a good agreement with gold-standard full night PSG testing.

BRL is a relatively new technology, which now is making its first attempts to enter the remote medical monitoring technology field. Experiments show that it has tremendous possibilities for this.

Editor's Note

The American Food and Drug Administration (FDA) had approved at least one UWB radar for remote patient monitoring. The SensioTec™ Virtual Medical Assistant® (VMA) uses a UWB radar and motion sensors built into a pad which can fit under a hospital bed mattress. The VMA connects to a network of wireless connections to present patient data at remote locations (http://sensiotec.com).

References

Alekhin, M., Anishchenko, L., Tataraidze, A. et al. 2013. "Selection of Wavelet Transform and Neural Network Parameters for Classification of Breathing Patterns of Bio-radiolocation Signals" *Biomedical Informatics and Technology*, T.D. Pham et al. (Eds.): ACBIT 2013, CCIS 404, pp. 175–178. Springer, Heidelberg (2013). http://link.springer.com/chapter/10.1007/978-3-642-54121-6_15

Alekhin, M., Anishchenko, L., Tataraidze, A. et al. 2013. "Comparison of Bioradiolocation and Respiratory Plethysmography Signals in Time and Frequency Domains on the Base of Cross-Correlation and Spectral Analysis," *International Journal of Antennas and Propagation*, 6 pages.

Anishchenko, L.N., Bugaev, A.S., Ivashov, S.I. et al. 2009. "Application of Bioradiolocation for Estimation of the Laboratory Animals' Movement Activity," *PIERS Online*, 5(6), 551–554.

Anishchenko, L.N. and Parashin V. B. 2008. "Design and application of the method for biolocation data processing," *Proceedings of the 4th Russian-Bavarian Conference on Biomedical Engineering at Moscow Institute of Electronic Technology (Technical University)*, pp. 289–294.

Bederman, S. and Lankford L. 1972. *Apparatus for measuring the activity of laboratory animals*, US Pattent 3633011.

Bimpas, M., Paraskevopoulos, N., Nikellis, K. et al. 2004. "Development of a three band radar system for detecting trapped alive humans under building ruins," *Progress In Electromagnetic Research*, PIER 49, 161–188.

Boric-Lubecke, O., Lin, J., Park, B-K. et al. 2008. "Battlefield triage life signs detection technique," Proc. SPIE Defence and Security Symposium, vol. 6947 — Radar Sensor Technology XII, no. 69470J, 10 pages.

Bugaev, A.S., Chapursky, V.V., Ivashov, S.I. et al. 2004. "Through Wall Sensing of Human Breathing and Heart Beating by Monochromatic Radar," *Proceedings of the Tenth International Conference on Ground Penetrating Radar, GPR'2004*, Delft, The Netherlands, vol. 1, 291–294.

Bugaev, A.S., Chapursky, V.V. and Ivashov, S.I. 2005. "Mathematical Simulation of Remote detection of Human Breathing and Heartbeat by Multifrequency Radar on the Background of Local Objects Reflections," *Proc. IEEE International Radar Conference Record*, Arlington, Virginia, USA, 6 pages.

Chen, K.M., Huang, Y., Shang, J. et al. 2000. "Microwave Life Detection Systems for Searching Human Subjects Under Earthquake Rubble or Behind Barrier," *IEEE Trans. Biomed. Eng.*, 27(1), 105–114.

Chen, K.M., Misra, D., Wang, H. et al. 1986. "An X-band microwave life-detection system," *IEEE Trans Biomed Eng 33*, 697–702.

Droitcour, A., Lubecke, V., Lin, J., Boric-Lubecke, O. 2001. "A Microwave Radio for Doppler Radar Sensing of Vital Signs," *Proc IEEE MTT-S Int. Microw. Symp.*, 175–178.

Droitcour, A.D., Seto, T.B., Park, B.K. et al. 2009. "Non-contact respiratory rate measurement validation for hospitalized patients," *Proc. of Annual International Conference IEEE Engineering in Medicine and Biology Society*, 4812–4815.

D'Urso, M., Leone, G. and Soldovieri, F. 2009. "A simple strategy for life signs detection via an X-band experimental set-up," *Progress in Electromagnetics Research C*, vol. 9, 119–129.

Greneker, E.F. 1997. "Radar Sensing of Heartbeat and Respiration at a Distance with Applications of the Technology," *Proc. IEE RADAR-97*, 150–153.

Hideo M. 1997. *Method and apparatus for measuring motion amount of laboratory animal*, US Patent 5608209.

Holzrichter, J.F., Burnett, G.C., Ng, L.C. et al. 1998. "Speech Articulator Measurements Using Low Power EM-Wave Sensors," *J. Acoust. Soc. Am.*, 103(1), 622–625.

Immoreev, I.Y. 2010. "Practical Applications of UWB Technology," *IEEE Aerospace and Electronic Systems Magazine*, 25(2), 36–42.

Ivashov, S.I., Razevig, V.V., Sheyko, A.P. et al. 2004. "Detection of Human Breathing and Heartbeat by Remote Radar," *Progress in Electromagnetics Research Symposium (PIERS 2004)*, Pisa, Italy, 663–666.

Kelly, J.M., Strecker, R.E. and Bianchi, M.T. 2012. "Recent Developments in Home Sleep-Monitoring Devices," *ISRN Neurology*, vol. 2012, Article ID 768794, 10 pages.

Konno, K. and Mead, J. 1967. "Measurement of the separate volume changes of rib cage and abdomen during breathing," *J. Appl Physiol 22*: 407–422.

Kropveld, D. and Chamuleau, R. 1993. "Doppler radar devise as a useful tool to quantify the liveliness of the experimental animal," Med. & Biol Eng. & *Comput*. 31, 340–342.

Li, C. and Lin, J. 2014. "Microwave noncontact motion sensing and analysis," Wiley, New Jersey, 157–185.

Li, J., Liu, L., Zeng, Z. et al. 2012. "Simulation and signal processing of UWB radar for human detection in complex environment," *Proc. of 14th Ground Penetrating Radar (GPR) Conference*, 209–213.

Liu, L., Liu, Z. and Barrowes, B. 2012. "Through-wall bio-radiolocation with UWB impulse radar: observation, simulation and signal extraction," *IEEE Journal on Selected Topics in Applied Earth Observations and Remote Sensing*, vol. 4, no. 4, 791–798.

Massagram, W., Lubecke, V.M., and Boric-Lubecke, O. 2011. "Feasibility assessment of Doppler radar long-term physiological measurements," *Proc. of Annual International Conference IEEE Engineering in Medicine and Biology Society*, 1544–1547.

Otsu, M., Nakamura, R. and, Kajiwara, A. 2011. "Remote respiration monitoring sensor using stepped-FM," *Sensors Applications Symposium (SAS), 2011 IEEE*, 155–158.

Rechtschaffen, A. and Kales, A. 1968. *A Manual of Standardized Terminology: Techniques and Scoring System for Sleep Stages of Human Subjects*. Los Angeles: UCLA Brain Information Service/Brain Research Institute.

Salmons, T. 1969. *Activity detector*, US Patent 3439358.

Soldovieri, F., Catapano, I., Crocco, L. et al. 2012. "A Feasibility Study for Life Signs Monitoring via a Continuous-Wave Radar," *International Journal of Antennas and Propagation*, vol. 2012, Article ID 420178, 5 pages.

Staderini, E.M. 2002. "UWB Radars in Medicine," *IEEE AESS Systems Magazine,* 17, 13–18.

Vasu, V., Heneghan, C., Sezer, S. et al. 2011. "Contact-free Estimation of Respiration Rates during Sleep," *Proc. of 22nd IET Irish Signals and Systems Conference.*

Wang, F-K., Horng, S-H., Peng, K.C. et.al. 2013. "Detection of Concealed Individuals Based on Their Vital Signs by Using a See-Through-Wall Imaging System With a Self-Injection-Locked Radar," *Microwave Theory and Techniques, IEEE Transactions on,* vol. 61, no. 1, 696–704.

9

Noise Radar Techniques and Progress

Ram M. Narayanan

CONTENTS

9.1 Introduction

Although noise is usually considered a nuisance signal contaminating the desired signal in most applications, many researchers have investigated ways to use noise as a remote-sensing tool for a variety of applications (Gupta, 1975). The exploitation of noise signals abounds in areas as diverse as biomedical engineering, circuit theory, communication systems, computers, electroacoustics, geosciences, instrumentation, physical electronics, and reliability engineering. Researchers have used noise as a broadband random signal, as a test signal, as a probe in microscopic investigations, and so on. Some examples of the applications of noise relevant to radar include source detection and location through noise measurements, self-directional microwave communications, electronic countermeasures, measurement of antenna characteristics, impulse response measurements of systems, characterization of nonlinear systems, measurement of linearity and intermodulation in a communications channel, measurement of circuit and device parameters, and prediction of system reliability. This leads us to consider noise as a probing signal for radar applications.

Traditional radar systems have generally employed short pulses of high-frequency waveforms or linear frequency-modulated (LFM) chirp waveforms to perform target detection and obtain target range information. By transmitting either very short pulses in the case of a pulsed radar or a very wide bandwidth signal in the case of an LFM radar, good range resolutions can be achieved to separate small targets in range or to identify unique scattering patterns of range-extended targets. Radar system designers have explored other types of broadband waveforms to achieve high-range resolutions with additional benefits such as immunity from detection, interference, and exploitation. Ultrawideband (UWB) noise waveforms provide these advantages.

Noise radar refers to techniques and applications that use random noise (narrowband or broadband) as the probing transmit waveform and then perform correlation processing, or dual spectral processing, of radar returns for target detection and imaging. This chapter reviews the history of noise radar, presents the fundamental concepts pertaining to noise radar operation, sketches its development over the past 60 years in applications involving target detection, characterization, imaging, and tracking, and discusses novel signal-processing concepts that take advantage of recent developments in hardware realization and implementation.

9.2 A Brief History of Noise Radar

From a historical perspective, the use of noise signals traces back to the very first radar-like experiments performed in 1904. In 1904, Christian Huelsmeyer used noise pulses in the "telemobiloscope," the radar precursor, which used a monostatic configuration (Huelsmeyer, 1904). Alexander S. Popov also used noise pulses in his experiments on ship detection in a bistatic configuration (actually using an early radio communication system). In both cases, a discharge device played the role of a pulse-noise transmitter and a coherer worked as the detector to receive the noise pulses. However, since the processing of the noise radar returns was not coherent, the system could only detect target presence, but could not estimate range or determine Doppler shifts.

The concept of noise radar with coherent reception of radar returns starts as early as the 1950s. Bourret (1957) published the first paper on range-measuring radar based on noise signals using a double differentiation circuit between the delay line and the correlator to enhance detection when using a Gaussian-shaped transmitted power spectral density (PSD). However, concerns were raised that Bourret's scheme was not practically realizable (Hochstadt, 1958) and that noise would mask the detection peak (Turin, 1958).

Horton is generally regarded as the first to describe in detail the design and performance of a continuous wave (CW) noise radar, which used frequency modulation along with the so-called anticorrelation method for signal processing (Horton, 1959). Horton recognized that one way to eliminate range and Doppler ambiguities was to use random noise as the

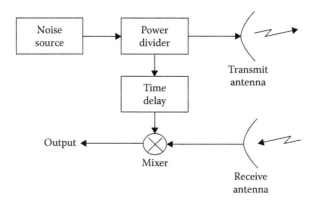

FIGURE 9.1
Simplified architecture of a homodyne correlation noise radar.

modulating function and perform range determination by cross-correlating the return signal with a time-delayed replica of the transmit waveform, as shown in Figure 9.1. In his seminal paper, he derived the fundamental concepts and proposed several implementations, which formed the foundation for noise radar systems developed ever since.

Further papers on this subject were published in the 1960s and 1970s, which may be characterized as a period of initial studies and performance analyses by a handful of researchers. A noise radar system using pseudo-random modulation and correlation processing was demonstrated successfully (Craig, Fishbein, and Rittenbach, 1962). Time-correlation as well as frequency-correlation noise radars were analyzed wherein it was asserted that continuous transmission of true wideband noise would eliminate all ambiguities (Grant, Cooper, and Kamal, 1963). Carpentier (1968) proposed a narrowband noise radar design based on frequency upconversion of a baseband noise signal to a higher frequency, which was transmitted, received, and downconverted back to baseband, and then cross-correlated with a time-delayed replica of the original baseband signal. He demonstrated the improved signal-to-noise ratio (SNR) for this approach, which was given by the time–bandwidth product. Later, he also presented a detailed comparison of noise radar and conventional radar (Carpentier, 1970), and concluded that a randomly phase-modulated sine wave signal was preferred over a thermally generated Gaussian noise signal.

Poirier (1968) proposed a spectrum analyzer (SA) radar concept wherein the power spectrum of the reflection of quasi-monochromatic radiation from an array of discontinuities in a long transmission line was analyzed using the theory of partial coherence. Under certain conditions, it was determined that this spectrum was modulated and that the analysis of the modulation could potentially yield information on the range and radar cross section (RCS) magnitude of a distant reflector. Because of this, the use of power spectrum analysis as a radar technique was proposed and expressions for range resolution and useful range were obtained. The results of some preliminary experimental measurements were also presented. This was seen as an alternate approach to noise radar, obviating the use of the delay line for correlation processing. In this approach shown in Figure 9.2, the reflected and reference signals are first summed up and then the signal frequency is shifted to the intermediate frequency range and transmitted directly to spectrum analyzer. If the propagation time of the noise signal from the reflector and back substantially exceeded the correlation decay time (generally the case for far targets), then the received spectral power density was found to be periodically modulated in frequency

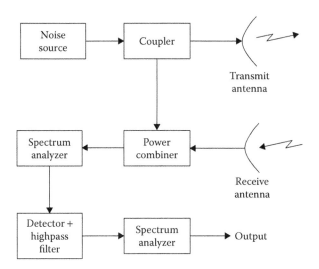

FIGURE 9.2
Simplified architecture of a double spectral processing noise radar.

with the period inversely proportional to the round-trip delay to the target. As a result, one spectral line per target was obtained at the output of the spectrum analyzer, whose location at the frequency axis could provide target range. Dukat (1969) developed a hardware implementation of this concept at X-Band.

A relatively unknown but significant Dutch paper described a noise radar employing delay correlation for range determination and frequency scanning for angle determination (Smit and Kneefel, 1971). Noise radar systems were initially explored for remote sensing of oceans (Chadwick and Cooper, 1971), tornados (Chadwick and Cooper, 1972), and clouds (Krehbiel and Brook, 1979). Furgason et al. (1975) extended the technique to ultrasonic systems operating at 4.8 MHz with improved signal-to-noise ratio (SNR) for detecting wires within water. A low-power noise radar using IMPATT diode noise generators, instead of high-power magnetrons or klystrons, for compact implementation and short-range applications was demonstrated (Forrest and Meeson, 1976). Digital correlation to overcome the slow speed of analog correlators was proposed by Forrest and Price (1978), in which the swept analog delay line was replaced by a series of shift registers, each cell representing a fixed unit of delay corresponding to the appropriate clock period. Since each cell output could be accessed simultaneously, the system could use parallel range processing with obvious advantages in processing speed.

While narrowband noise radars have been proposed and refined over the past 50 years, the concept of ultrawideband (UWB) random noise radar has seen significant development more recently. In contrast to conventional radar, the UWB noise radar transmits a noise or noise-like waveform having a fractional bandwidth of greater than 25%. UWB radars achieve very high-range resolutions, suitable for imaging, owing to their large bandwidths. A significant advancement in the 1990s was the development of a fully coherent UWB noise radar employing heterodyne correlation in which the radar return signal was correlated with a time-delayed and frequency-shifted replica of the transmit signal, as shown in Figure 9.3. This scheme ensured that the correlator output always appeared at the offset (i.e., shift) frequency (instead of at DC), thereby preserving the phase information contained in the return signal. The approach was validated using theoretical analysis

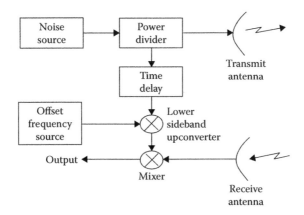

FIGURE 9.3
Simplified architecture of a heterodyne correlation noise radar.

(Narayanan et al., 1998), simulations (Narayanan, Xu, and Rhoades, 1994), and measurements on buried targets (Narayanan et al., 1995).

9.3 Recent Developments in Noise Radar

9.3.1 Noise Radar Modeling and Implementation Using Correlation Detection

A block diagram of a basic correlation-type noise radar system was shown in Figure 9.1. This is called a homodyne correlation noise radar since the correlation process downconverts the received signal to DC for a stationary target and to the Doppler frequency for a moving target. In this scheme, the reflected signal is correlated with a time-delayed replica of the transmitted signal. If the roundtrip time delay to the target matches the internal time delay, a peak is observed in the correlator whose magnitude is proportional to the target reflectivity. The internal time delay then yields the range to the target. The mixer followed by a low-pass filter acts as the correlator.

Let us assume that the transmit signal $v_t(t)$ is given by

$$v_t(t) = a(t)\cos\{[\omega_0 + \delta\omega]t\} \tag{9.1}$$

where:
$a(t)$ is the Rayleigh distributed amplitude that describes amplitude fluctuations
$\delta\omega(t)$ is the uniformly distributed frequency that describes frequency fluctuations over the $\pm\Delta\omega$ range, that is $[-\Delta\omega \le \delta\omega \le +\Delta\omega]$

Assuming that the random variables $a(t)$ and $\delta\omega(t)$ are uncorrelated, we can show that the average power in the signal is $<a^2(t)>/2R_0$, where $<\cdot>$ denotes time average and R_0 is the system impedance. The center frequency f_0 and the bandwidth B are, therefore, given by $f_0 = \omega_0/2\pi$ and $B = \Delta\omega/\pi$, respectively.

If the target complex reflectivity $\bar{\Gamma}$ is given by $\Gamma\exp(j\Theta)$, where Θ is the phase of the target's complex reflection coefficient, then the reflected signal $v_r(t)$ is given by

$$v_r(t) = \Gamma a(t-t_0)\cos\{[\omega_0 + \delta\omega](t-t_0) + \Theta\} \tag{9.2}$$

where t_0 is the roundtrip time to the target, given by $2R/c$, where R is the range to the target, c is the speed of light. The time-delayed transmit replica $v_d(t)$ is given by

$$v_d(t) = a(t - t_d)\cos\{[\omega_0 + \delta\omega](t - t_d)\} \tag{9.3}$$

where t_d is the internal delay.

The cross-correlation of $v_r(t)$ and $v_d(t)$ yields a nonzero output only if the two delays match, that is when $t_0 = t_d$. Since the low-pass filter following the mixer discards the sum frequency signal at and around $2\omega_0$, the low-pass-filtered mixer output $v_0(t)$ is given by

$$v_0(t) = \Gamma a^2(t)\cos\Theta. \tag{9.4}$$

Its average value $< v_0(t) >$ can be derived to be $2R_0P_t\Gamma\cos\Theta$, where P_t is the transmitted power. The average power in the received signal P_0 is then given by

$$P_0 = \frac{< v_0^2(t) >}{2R_0} = 2R_0P_t^2\Gamma^2\cos^2\Theta. \tag{9.5}$$

We note that the received power is proportional to the product $\Gamma^2\cos^2\Theta$, and can actually be zero if Θ is close to 90°. However, in a realistic case, the UWB reflectivity of the target is a function of frequency, that is $\Gamma = \Gamma(\omega)$ and $\Theta = \Theta(\omega)$; therefore, the frequency-averaged product will yield a nonzero value for the average received power. We also note that one cannot isolate magnitude or phase of the target reflectivity at the receiver output. Since the homodyne correlation noise radar downconverts directly to DC, it loses important phase information of the returned signal.

However, there is a way to inject phase coherence in noise radar using a frequency offset to the time-delayed transmit replica, as shown in the heterodyne correlation noise radar of Figure 9.3. This heterodyne correlation scheme cross-correlates the reflected signal from the target with a time-delayed and frequency-offset replica of the transmitted waveform.

Assume that the transmit signal $v_t(t)$ is given by Equation 9.1, such that the target complex reflectivity is $\bar{\Gamma} = \Gamma\exp(j\Theta)$, and t_0 is the roundtrip time to the target. The reflected signal $v_r(t)$ is again given by Equation 9.2. The time-delayed and frequency-offset transmit replica $v_d(t)$ is now given by

$$v_d(t, \omega') = a(t - t_d)\cos\{[\omega_0 + \delta\omega - \omega'](t - t_d)\} \tag{9.6}$$

where:
t_d is the internal delay
ω' is the frequency offset (usually $0.1\omega_0 \leq \omega' \leq 0.2\omega_0$).

As before, the cross-correlation of $v_r(t)$ and $v_d(t, \omega')$ yields a nonzero output only if the two delays match, that is when $t_0 = t_d$. The low-pass-filtered mixer output $v_0(t, \omega')$ is given by

$$v_0(t, \omega') = \Gamma a^2(t)\cos[\omega't + \Theta]. \tag{9.7}$$

which describes a signal centered exactly at ω' (at all times) instead of DC, as in the homodyne correlation noise radar.

If this output is connected to an in-phase/quadrature (I/Q) detector (not shown) fed by the offset frequency generator, we can obtain the in-phase output v_{0I} and quadrature output v_{0Q} given by

$$v_{0I} = \Gamma a^2(t)\cos\Theta, \tag{9.8}$$

$$v_{0Q} = \Gamma a^2(t)\sin\Theta. \tag{9.9}$$

The magnitude and phase angle of the target reflectivity can now be obtained using

$$\Gamma = \sqrt{v_{0I}^2 + v_{0Q}^2},$$ (9.10)

and

$$\Theta = \tan^{-1} \frac{v_{0Q}}{v_{0I}}.$$ (9.11)

Since the output of correlator is always at the offset frequency ω', we observe that the UWB transmit waveform, despite its wide bandwidth, collapses to a single frequency in the receiver. Thus, we can shrink the detection bandwidth at correlator output to enhance the signal-to-noise ratio (SNR). Doppler, if any, will modulate the correlator output and can be extracted from the I/Q detector.

The range resolution for such a cross-correlation radar, ΔR_{CCR}, is given by

$$\Delta R_{CCR} = c/2B.$$ (9.12)

9.3.2 Noise radar Modeling and Implementation Using Double Spectral Processing

To understand double spectrum processing, consider the system shown in Figure 9.2 in which a radar illuminates the target using a Gaussian noise signal with a Gaussian spectrum. A portion of the transmitted signal is combined with the signal reflected from the target and observed on a spectrum analyzer. If the sum signal spectrum analyzer output is detected and passed through a high-pass filter, the original sum spectrum will be converted to a new time function. A second spectrum analyzer can then determine the frequency components present in the new time function. Let the illuminating signal spectrum $\varphi_t(\omega)$ of Gaussian shape be given by

$$\varphi_t(\omega) = \frac{1}{\sigma\sqrt{2\pi}} \exp\left[-\frac{1}{2}\left(\frac{\omega_0 - \omega}{\sigma}\right)^2\right]$$ (9.13)

where:
ω_0 is the center frequency
σ is the variance of the illuminating spectrum.

In this case, the bandwidth B can be defined as the frequency span between the $\pm\sigma$ points around the center frequency as $B = \sigma/\pi$.

For a single target with the reference signal adjusted to be equal to the target return signal, the sum spectrum $\varphi_s(\omega)$ is given by

$$\varphi_s(\omega) = \frac{1}{\sigma\sqrt{2\pi}} \exp\left[-\frac{1}{2}\left(\frac{\omega_0 - \omega}{\sigma}\right)^2\right]\left\{1 + \cos\frac{2\omega R}{c}\right\}$$ (9.14)

where:
R is the range to the target
c is the speed of light.

The time function expression for the received voltage after high-pass filtering can be derived as

$$v_r(t) = \frac{k}{\sigma\sqrt{2\pi}} \exp\left[-\frac{1}{2}S_s\left(\frac{t_0 - t}{\sigma}\right)^2\right]\cos\frac{4\pi R S_s t}{c}$$ (9.15)

where:
t_0 is the roundtrip time to the target, given by $2R/c$
S_s is the sweep speed of the sum signal spectrum analyzer
k is some system related constant

The spectrum of Equation 9.15, which is the product of two time functions, is the convolution of the spectra of the component time functions and results in a spectrum with a variance σ' centered at the modulating frequency $\omega_m = 4\pi RS_s/c$.

The variance σ' is a function only of the variance σ of the illuminating signal and can be derived as

$$\sigma' = \frac{S_s}{\sigma}. \tag{9.16}$$

The range resolution for the double spectral processing radar, ΔR_{DSP}, is given by

$$\Delta R_{DSP} = c/\pi B. \tag{9.17}$$

9.3.3 Ambiguity Function Characterization

The generalized ambiguity function of a waveform is defined from the cross-correlation between the waveform and its time-scaled and delayed version, where the time scaling is related to the relative velocity of the target and its range. As a result of the randomness of the noise radar transmit signals, its ambiguity function fluctuates between different realizations and one needs to express the average ambiguity function over several realizations.

Dawood and Narayanan (2003) investigated the combined range and range rate resolution characteristics of a UWB noise radar using the generalized wideband ambiguity function. In this case, the ambiguity function was defined as the expected or the average value of the response of the correlator matched to a target at a desired range moving at a desired range rate to the return signal from the target at a different range moving at a different range rate. As in the narrowband random noise waveform case, range and range rate resolutions could be controlled independently, the former being inversely related to the transmit bandwidth, while the latter is inversely related to the bandwidth of the integrating filter. Examples of the average ambiguity function for various combinations of bandwidth and integration time are shown in Figure 9.4. For a UWB transmit signal, the compression or stretch due to the range rate on the envelope of the return signal cannot be ignored, necessitating the development of the wideband ambiguity function. It was also shown that, in general, UWB waveforms are not suitable for accurate range rate estimation due to the extended Doppler-spread parameter, that is the product of the transmit bandwidth and the target range rate, unless the correlator was matched in the delay rate as well.

A full characterization of the ambiguity function includes its statistical distribution, but often second-order statistics are sufficient. The ambiguity function can be described as a sum of its average and a noise component. The average value of the ambiguity function includes in the integrand a time-scaled version of the autocorrelation function of the random waveform, while its variance describes the noise floor of the sidelobes. Particularly useful in characterizing the ambiguity function of a random waveform is the second moment statistics describing the sum of the squared average and the variance of the noise

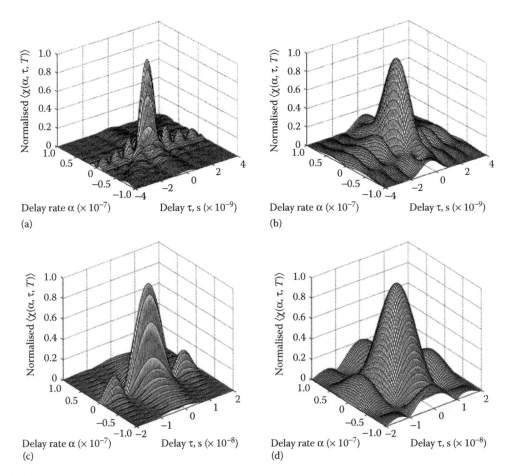

FIGURE 9.4
Average ambiguity function for noise radar having a band-limited rectangular power spectral density (Dawood and Narayanan, 2003). (a) 1-GHz bandwidth and 50-ms integration time, (b) 1-GHz bandwidth and 10-ms integration time, (c) 100-MHz bandwidth and 50-ms integration time, and (d) 100-MHz bandwidth and 10-ms integration time. Note the independent control of the main lobes in the delay (i.e., range) and the delay rate (i.e., velocity) axes as the bandwidth and the integration time are changed. (Reproduced by permission of the Institution of Engineering & Technology. Dawood, M., and Narayanan, R.M., *IEE Proceedings on Radar, Sonar, and Navigation*, 150(5), pp 379–386, 2003.)

floor. Analytical formulations were developed as examples for Gaussian waveforms, phase-modulation by continuous Gaussian noise, and randomized step frequency radar (Axelsson, 2006). Corresponding plots are shown in Figure 9.5.

9.3.4 Phase Noise Characterization in Noise Radar

The Doppler visibility, that is the ability to extract Doppler information over the inherent clutter spectra, is constrained by system parameters, especially the phase noise generated by microwave components. Li and Narayanan (2006) proposed a phase noise model for the heterodyne mixer as applicable for UWB random noise radar and for the local oscillator (LO) in the time domain. The Doppler spectra were simulated by including phase noise contamination effects and compared with their previous experimental results. A genetic

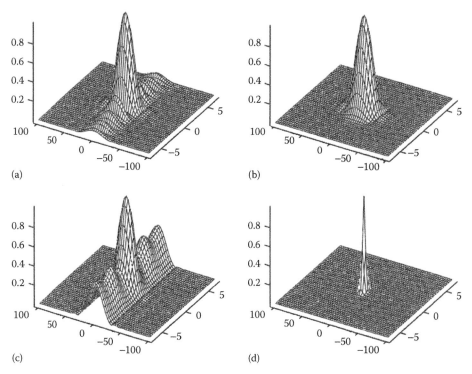

FIGURE 9.5
Squared ambiguity function for at 10-GHz frequency for velocity ranging from –100 to +100 m/s and range varying from –7 to +7 m with a window function $w(t) = \sin^2\left(\pi t/T\right)$ in order to suppress Doppler sidelobes (Axelsson, 2006). (a) Rectangular noise spectrum with 100-MHz bandwidth and 1-ms integration time, (b) Gaussian noise spectrum with 29-MHz 3-dB bandwidth and 1-ms integration time, (c) random phase modulation of 1 rad² phase angle variance of rectangular power spectrum with 100-MHz bandwidth and 1-ms integration time, and (d) random phase modulation of 9 rad² phase angle variance of rectangular power spectrum with 100-MHz bandwidth and 5-ms integration time. Note the highly suppressed sidelobes and the narrow correlation peak for case, and (d). (Permission of SPIE, Axelsson, S.R.J. *SPIE Conference on SAR Image Analysis, Modeling and Techniques.* © SPIE 2006.)

algorithm (GA) optimization routine was applied to synthesize the effects of a variety of parameter combinations to derive a suitable empirical formula for estimating the Doppler visibility in dB. It turns out that the Doppler visibility of UWB random noise radar depends primarily on the following parameters: (1) the LO drive level of the receiver heterodyne mixer, (2) the saturation current in the receiver heterodyne mixer, (3) the bandwidth of the transmit noise source, and (4) the target velocity. Doppler visibility curves generated from this formula matched the simulation results very well over the applicable parameter range within 1 dB, as shown in Figure 9.6. Such a model could be used to quickly estimate the Doppler visibility of random UWB noise radars for tradeoff analysis.

Both correlation processing and spectral interferometry require coherence, so the transmitters and receivers must be driven by highly stable oscillators as phase noise introduced into the system degrades the Doppler resolution of the radar. Morabito et al. (2008) presented a technique to estimate the phase noise process through naturally collected radar data and fit stochastic models to the noise process. Phase noise models can be used in parametric correction techniques, which can have drastically reduced computational

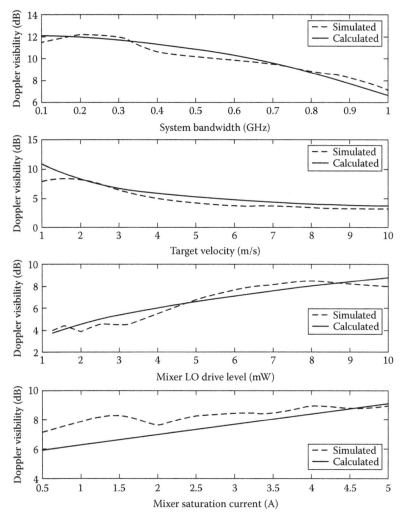

FIGURE 9.6
Comparison between simulated results and empirical model for Doppler visibility as a function of the major parameters impacting the Doppler visibility in a noise radar. (© 2006 IEEE. Reprinted, with permission, from Li and Narayanan, *IEEE Trans Aerospace and Electronic Systems*, 43, 2006. © 2006 IEEE.)

complexity when compared with nonparametric versions. Both autoregressive (AR) and Wiener processes appeared to model the phase noise of a candidate coherent distributed passive radar system quite well compared with the autoregressive integrated moving average (ARIMA) and autoregressive fractionally integrated moving average (ARFIMA) models.

9.3.5 Receiver Operating Characteristics (ROC) and Detection Performance

The performance of a radar system is characterized and evaluated using its receiver operating characteristics, which is a plot of the probability of detection (P_d) as a function of the probability of false alarm (P_f or P_{fa}). Usually, P_f is fixed and signal-processing algorithms are developed to maximize P_d at the given P_f.

The performance of noise radar, assuming a point target, was investigated from a statistical point of view by developing the theoretical basis for the system's receiver operating characteristics (ROC) (Dawood and Narayanan, 2001). Explicit analytical expressions for the joint probability density function (pdf) of the in-phase (I) and quadrature (Q) components of the receiver output were derived under the assumption that the input signals were partially correlated Gaussian processes. The pdf and the complementary cumulative distribution function (cdf) for the envelope of the receiver output were also derived. These expressions were used to relate P_d to P_f for different numbers of integrated samples. Since P_d was found to be primarily dependent upon the correlation between the received and the delayed replicas of the transmitted waveform, plots were presented to demonstrate the effect of this correlation on target detection performance, shown in Figure 9.7.

Mogyla (2012) computed the probability and detection characteristics of coherent detection of reflected signals with the fully known parameters when using stochastic or noise radar signals. It was shown that detection characteristics of deterministic signals represent a theoretical limit to which the detection characteristics of stochastic signals converge. Analytical expression for probability density of decision statistics for the cases of presence at detector's input of the reflected signal only, interference only, and both signal and interference was obtained. The dependence of P_f on the threshold ratio and dependence of P_d on the signal-to-interference ratio (SIR) were computed for different values of bandwidth–duration product of stochastic signal.

9.3.6 Target Detection Considerations in Noise Radar

In a noise radar, the shape of the power spectrum and the number of independent samples during the time of measurement influence the sidelobe suppression in range. As in conventional Doppler radar, the range resolution depends on the signal bandwidth and the Doppler resolution on the time of measurement. If the bandwidth/carrier ratio is not very small, the Doppler resolution could also be limited by the power spectrum width of the transmitted noise.

Axelsson (2003) showed how to reduce this degradation in the correlation procedure using a reference Doppler shift close to that expected from the target injected. The use of a binary or low-bit analog-to-digital converter (ADC) improved the signal-processing rate and reduced costs. In particular, if there are two signals with different Doppler shifts in the same range cell, then false targets may be generated in Doppler for binary or low-bit ADC implementations. Adding an uncorrelated noise signal to the received one before ADC achieved an efficient suppression of the false targets. A comparison of the above approaches is shown in Figure 9.8.

Kwon, Narayanan, and Rangaswamy (2013) presented a target detection method using total correlation based on mutual information theory for noise radar systems, which enabled detection of multiple targets at low SNR regimes. The proposed method utilized the largest eigenvalue of the sample covariance matrix to extract information from the transmitted signal replica, and demonstrated to perform better than the conventional total correlation detector. To avoid ambiguous target detection, they computed the thresholds to guarantee the detection performance with the same number of receiving antenna elements from the appropriate eigenvalue distributions based on random matrix theory. Simulations showed that the proposed detection method could be used in

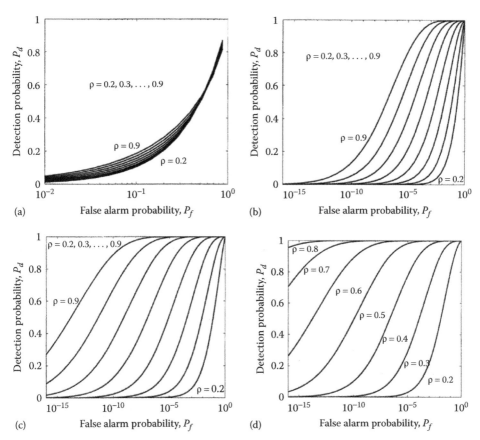

FIGURE 9.7
Receiver operating characteristics for a noise radar as a function of the correlation coefficient ρ between the transmitted and received signals for various values for the number of samples integrated N. (a) N = 1, (b) N = 25, (c) N = 50, and (d) N = 100. Note that the detection probability for a given false alarm probability increases as the correlation coefficient and the number of samples integrated increase. The correlation coefficient can degrade significantly due to dispersion in the medium between the radar and the target, which will result in lowered detection probability. (© 2001 IEEE. Reprinted, with permission, from Dawood, M., and Narayanan, R.M., *IEEE Trans Aerospace and Electronic Systems*, 37, 586–594, 2001.)

intermediate and low SNR environments, and that the thresholds achieved exact target detection. A comparison of the detection performance of the three schemes is shown in Figure 9.9.

9.3.7 Compressive Sensing

Herman and Strohmer (2009) were the first to propose the use of noise transmissions to implement compressive sensing (CS) in radar. They conjectured that by discretizing the time–frequency plane into an $N \times N$ grid, and by assuming that the number of targets K was small (i.e., $K \ll N^2$), a sufficiently "incoherent" pulse could be transmitted and compressive sensing techniques could be employed to reconstruct the target scene. While they explored the use of deterministic Alltop sequences in their analysis, they stated

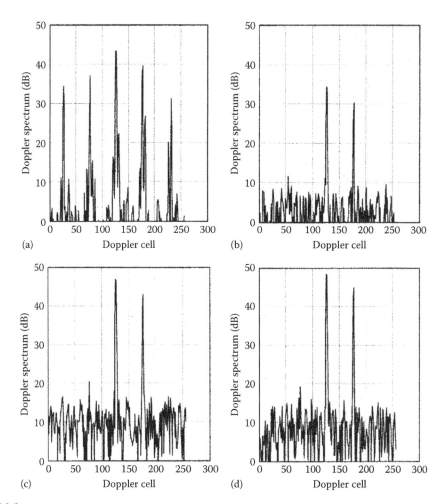

FIGURE 9.8

Simulated Doppler spectrum for two point targets moving with different velocities through the same range cell. (a) Binary ADC with no noise added, (b) binary ADC with extra noise added before ADC (SNR = −10 dB), (c) two-bit ADC for SNR = −6 dB, and (d) three-bit ADC for SNR = −3 dB. A cos²-shaped Hanning window was used to modulate the time samples before implementing the Doppler FFT. Note that the addition of extra noise before the ADC in the receiver channel greatly suppresses the sidelobes in (b) compared with (a). (© 2003 IEEE. Reprinted, with permission, from Axelsson, S.R.J., *IEEE Trans Geoscience and Remote Sensing*, 41, 2703–2720, 2003.)

that transmitting white noise, such as a random Gaussian signal or a constant-envelope random-phase signal, would yield a similar outcome. Since the compressed sensing radar does not use a matched filter at the receiver, this directly impacts Analog-to-digital converter (ADC), and has the potential to reduce the overall data rate and to simplify hardware design.

Shastry, Narayanan, and Rangaswamy (2010) developed the theory behind stochastic waveform-based compressive imaging. Then, they showed how using stochastic waveforms for radar imaging made it possible to estimate target parameters and detect targets by sampling at a rate considerably slower than the Nyquist rate using compressive sensing (CS) algorithms. Thus, it was theoretically possible to increase the bandwidth

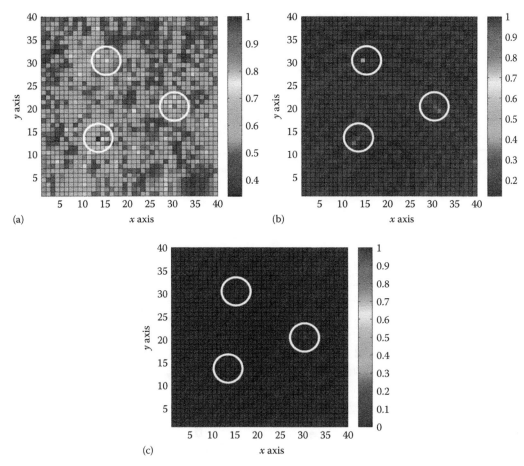

FIGURE 9.9
Detection performance of three targets at an SNR of −10 dB using different methods. (a) Total correlation (TC) approach, (b) modified total correlation (MTC) approach, and (c) modified total correlation (MTC) approach using a threshold. In approach (b), only the largest eigenvalue of the covariance matrix of the receive signal is used. The threshold in (c) is based on random matrix theory. (© 2013 IEEE. Reprinted, with permission, from, Kwon, Y. et al., *IEEE Trans Aerospace and Electronic Systems,* 49(2), 1251–1262, 2013.)

(and hence the spatial resolution) of an UWB radar system using stochastic waveforms, without significant additions to the data acquisition system. Further, there was virtually no degradation in the performance of a UWB stochastic waveform radar system employing compressive sampling. Simulation results showed how to achieve the performance guarantees provided by theoretical results in realistic scenarios.

Jiang et al. (2010) combined CS techniques with a random noise SAR and proposed the concept of random noise SAR based on CS. Their paper presented the radar system block diagram and the detailed analysis of the data processing procedure. The theoretical analysis showed that the sensing matrix of the random noise SAR exhibited good restricted isometry property (RIP). Simulation results showed that the CS-based random noise SAR could reconstruct target images effectively using far fewer samples, as shown in Figure 9.10.

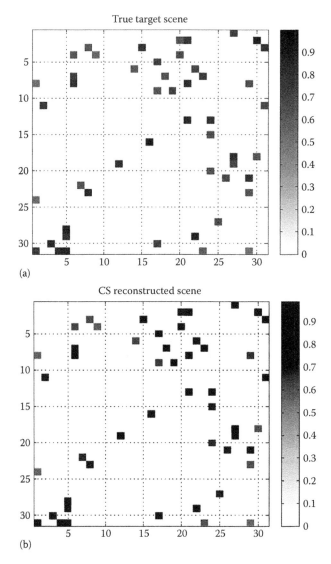

FIGURE 9.10
Simulation results of target recovery using compressed sensing technique in a noise SAR. 10% of the samples are randomly selected and the SNR is assumed to be 20 dB. (a) True target scene and (b) compressively recovered target scene. (© 2010 IEEE. Reprinted, with permission, from Jiang, H. et al., *Proc IEEE International Geoscience and Remote Sensing Symposium*, Honolulu, HI, July 2010, pp. 4624–4627.)

9.4 Novel Noise Radar Applications

9.4.1 Overview

Noise radars have been developed and implemented successfully in a variety of applications within the domain of traditional radar systems. In most cases, they have achieved equivalent performance. Some example applications include through-wall radar, retroreflective or retrodirective radar, Doppler estimation, moving target detection, interferometric and monopulse radar, multiple-input multiple-output (MIMO) radar, bistatic and multistatic radar, millimeter-wave radar, synthetic aperture radar (SAR) and inverse synthetic aperture radar (ISAR), tomographic applications, RF tag implementation, phased array radar, radar networking, and multifunctional systems. This section describes some of these applications.

9.4.2 Doppler Estimation for Noise Signals

Narayanan and Dawood (2000) developed the theoretical foundations for Doppler estimation using UWB noise radar. Although the instantaneous Doppler frequency was random (owing to the wideband nature of the transmit waveform), the averaged Doppler signal converged to the value corresponding to the center frequency of the wideband transmitted signal. Simulation studies and laboratory measurements using a microwave delay line showed that it was possible to estimate the Doppler frequency from targets with linear as well as rotational motion. Field measurements using a photonic delay line demonstrated the success of this technique at a range of about 200 m at target speeds of up to 9 m/s. Analysis showed that the accuracy with which the Doppler frequency can be estimated depended not only on the phase performance of various components within the system, but also upon the random nature and bandwidth (BW) of the transmit waveform and the characteristics of unsteady target motion. Figure 9.11 shows examples of measured Doppler response using a 1–2-GHz noise radar.

Mogyla, Lukin, and Shyian (2002) discussed the practical realizations for the measurement of range and range-rate (i.e., velocity) vector using noise waveforms. They showed that an output SNR of relay-type correlation receivers employing a stepped digital delay line was degraded by about 2 dB in comparison with that of the optimal receiver using a continuous analog delay line. Their presented scheme to measure the velocity vector involved the use of a cross-correlation discriminator utilizing a quadrature detector to yield the sign of the Doppler signal. Figure 9.12 shows the measured Doppler shifts of approaching and receding targets using a noise radar with an I/Q detector.

9.4.3 Moving Target Detection

Stephan and Loele (2000) presented practical results concerning the detection of fixed and moving targets using an X-Band noise radar with a bandwidth of 1.3 GHz. The delayed transmit replica reference was modulated by a signal at 21.4 MHz prior to cross-correlation, thereby accomplishing heterodyne correlation which permitted the detection of target-induced Doppler. Using this approach, the system could detect moving targets and extract the Doppler velocity.

In 2008, Sachs discussed the use of M-sequence pseudo-noise radar for the detection of moving or trapped humans hidden by obstacles (Sachs et al. 2008). The system employed a single transmit and two receive antennas. Moving target localization came from trilateration, which required knowledge of the length of the edges of a triangle built from two reference points (i.e., the receive antennas) and the target. High-pass filtering of the data rejected strong stationary clutter signals and retained Doppler signatures arising from human movement. For detection of buried humans, the repetitive breathing signal was captured using narrowband filtering, which suppressed stationary clutter from rubble. In addition, a sliding average with a short window was performed over the transformed signal in the propagation time direction since the signal flanks consisted of several similarly

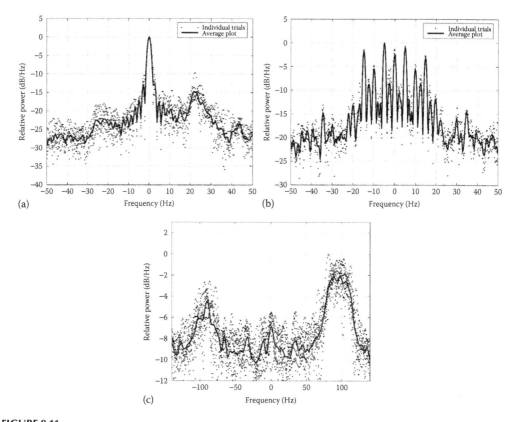

FIGURE 9.11

Measured Doppler spectra using a 1–2-GHz noise radar. (a) Linear motion at short range for a target speed of 2.3 m/s, (b) rotational motion for a target rotation of 75 rpm, and (c) linear motion at moderate range (~200 m) for a target speed of 9 m/s. The solid curve represents the average of the individual realizations shown as dots. Note the broadening of the spectra due to the widebandwidth of the transmit signal. (© 2000 IEEE. Reprinted, with permission, from Narayanan, R.M. and Dawood, M., *IEEE Trans on Antennas and Propagation*, 48(6), 868–878, 2000.)

modulated samples. Figure 9.13 shows the results of tracking human movement behind an obstacle based on a trilateration approach using three antennas.

Chapursky et al. (2004) presented a theoretical analysis of the SNR of a noise waveform transmitting a Gaussian correlation function and demonstrated the identification of slowly moving objects behind a wall made of 13-cm-thick dry plaster sheets using a 1–4-GHz broadband noise radar employing correlation processing and phase-sensitive intersurvey (i.e., background) subtraction. Wang, Narayanan, and Zhou (2009) presented an approach using subtraction of successive frames of the cross-correlation signals between each received element signal of an antenna array and the transmitted signal in order to isolate moving targets in heavy clutter. The back-projection algorithm subsequently obtained images of moving targets. Different models based on the finite-difference time-domain (FDTD) algorithm were set up to simulate different through-wall scenarios of moving targets. The simulation results showed this approach suppressed the heavy clutter and greatly enhanced the signal-to-clutter ratio (SCR). Multiple moving targets could bev detected, localized, and tracked for any random movement. Figure 9.14 shows simulation results for the simultaneous tracking of two humans in a scene behind a wall.

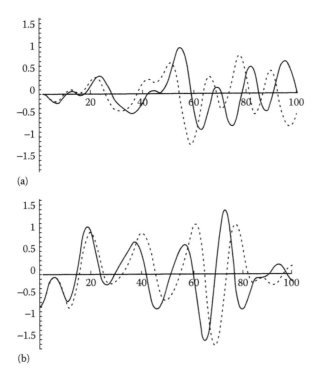

FIGURE 9.12
Time series of the radiated (solid line) and reflected (broken line) signals (Mogyla, Lukin, and Shyian, 2002). (a) Approaching target showing positive Doppler shift and (b) receding target showing negative Doppler shift. (© 2002, BEGELL HOUSE INC., with permission from BEGELL HOUSE INC.)

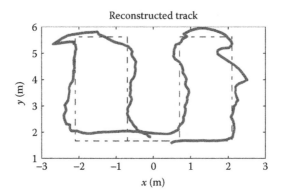

FIGURE 9.13
Tracking of human movement behind a dry wall using an M-sequence PN radar operating over 8–2500 MHz (Sachs et al., 2008). Adaptive filtering for static clutter removal and constant false alarm rate (CFAR) thresholding was employed. The dotted black line represents the actual movement of the human and the solid red line represents the reconstructed track. (© EuMA 2008, Permission of European Microwave Association, *Proc European Radar Conf. Oct 2008* pp. 408–411.)

9.4.4 MIMO, Bistatic, and Multistatic Noise Radar

In 2006, Gray first proposed the application of noise radar to multiple-input multiple-output (MIMO) configurations and further developed it with Fry (Gray, 2006) (Gray and Fry 2007). The MIMO architecture consisted of K transmit antennas and M receive antennas.

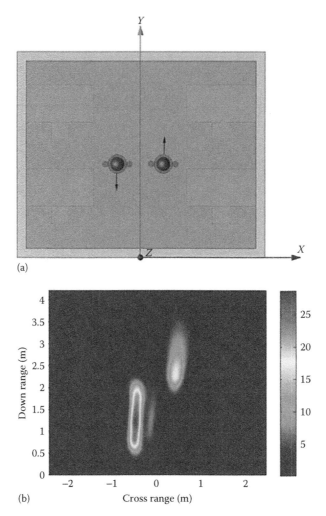

FIGURE 9.14
Tracking of two humans moving continuously, one toward and one away from the Radar. (a) Simulation model and (b) reconstructed tracks of two moving humans. (© 2009 IEEE. Reprinted, with permission, from Wang, H. et al., *IEEE Antennas and Wireless Propagations Letters*, 8, 802–805, 2009.)

When independent noise sources are transmitted from each antenna, the approach became a special case of MIMO radar. Two transmission approaches, termed element space (ES) and beam space (BS), were considered together with both conventional and minimum variance distortionless response (MVDR) beamforming on receive. The ES approach to MIMO noise radar consisted of transmitting independent noise of equal powers from each of the K omnidirectional transmit antennas.

For the BS approach, each independent noise source fed into each antenna either delayed or phase-shifted so as to form a beam illuminating a selected sector of the radar's field of view – effectively coding each sector according to a particular noise source. A total of N beams were formed, with $N < K$. The results revealed that the higher radiated power of the BS results in more power on the scatterers and hence a higher SNR at the receive array resulting in improved target detectability. This higher SNR did not improve resolution for conventional beamforming but did so for the MVDR case. Figure 9.15 shows the architecture of the two approaches and the transmit wavenumber spectrum for each approach.

Chen and Narayanan (2012) considered the estimation of target directions in ultrawideband MIMO noise radar by applying the tapped-delay-line (TDL)-based beamforming

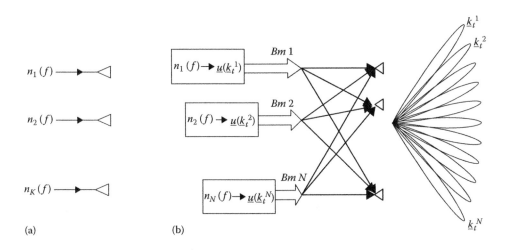

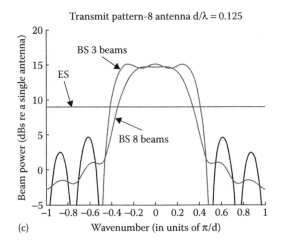

FIGURE 9.15
Performance comparison of element space (ES) and beam space (BS) techniques for MIMO radar. (a) ES architecture, (b) BS architecture, and (c) transmit wavenumber spectra. The ES approach consists of transmitting independent noise from each of the K omnidirectional antennas, while in the BS approach, each independent noise source is fed into each antenna but is either delayed or phase-shifted so as to form a beam illuminating a selected sector of the radar's field of view. The wavenumber spectrum of the ES approach is flat across the whole region of the wavenumber domain and consequently about 75% of the transmit power is wasted as it is turned into nonradiating evanescent power. The BS approach concentrates the transmitted wavenumber spectrum in the region $-\pi/4d$ to $\pi/4d$ and since total transmit power is the same, a much higher power is turned into radiating plane waves. (© 2007 IEEE. Reprinted, with permission, from Gray, D.A. and Fry, R. *IEEE Proc International Waveform Diversity and Design Conf.*, Pisa, Italy, Jun 2007. pp. 344–347.)

technique to concentrate on receiving signals from a certain direction and to suppress interference from others. The conditional generalized likelihood ratio test (CGLRT) was applied for detecting the target, given the target impulse responses estimated using the CLEAN algorithm used to perform a deconvolution on images created in radio astronomy. They extended the application of the iterative CGLRT (ICGLRT) technique, developed for narrowband MIMO radar to ultrawideband noise MIMO radar in order to determine their directions by implementing these two techniques iteratively. Simulation results showed how this approach could successfully extract targets which were originally embedded in other targets reflections and to sequentially improve the accuracy of target direction estimation. The IGLRT does not only iteratively examine target presence for the entire scanned area, but also exploits the information about observed targets to help detect new targets. Figure 9.16 shows the TDL beamformer structure and its performance in detecting multiple targets while suppressing jammer interference.

There are several applications that locate radar receivers in bistatic and multistatic configurations at large separation distances from the transmitter, making it difficult to provide the transmit reference at the receivers. Malanowski and Roszkowski (2011) proposed a scheme to generate and store the reference signal locally at the receiver. Successfully realizing this concept requires appropriate time and frequency synchronization and correction of the local reference signal. They described the concept of the bistatic radar with a locally generated reference signal and proposed methods for synchronization and calibration. The time and frequency synchronization worked by correlating the local copy of the reference signal with the received signal, in search of the direct path signal. The strongest peak indicates the time shift with respect to the direct signal. They obtained frequency synchronization by shifting one of the signals in frequency and correlating them. The frequency shift corresponding to maximum of correlation indicates the required frequency shift. This method is applicable only to cases where there exists a relatively strong direct path leakage. Figure 9.17 shows the architecture and the performance of the approach.

9.4.5 Tomographic Imaging Using Noise Radar

Vela et al. (2012) explored the use of a 4–8-GHz CW noise radar to obtain tomographic images of various target configurations. They obtained images of metallic and dielectric cylinders and blocks in various positions and under both unobscured and concealed conditions. Through rotation of 3D objects, the developed system could obtain 2D image slices of the object under test, with SNR values ranging between 12 and 25 dB.

Shin, Narayanan, and Rangaswamy (2013) developed the theoretical foundations of noise radar tomography using random field cross-correlation generated using the back-propagation algorithm. This approach bypasses the shortcomings of traditional coherent radar techniques which can cause ghosts in the image. In the case of a completely random waveform, the cross-correlation of signals recorded between two points converges to the complete Green's function of the medium, including all reflection, scattering, and propagation modes. Using an array of probes located around the region to be imaged, they showed the detected noise correlation to be a function of the spatial overlap of the EM fields at the probes and the spatial distribution of the dielectric constant and/or conductivity of the sample. After performing simulations over the 8–10-GHz frequency range, they concluded that a single transmission of the noise waveform was not sufficient to generate the tomographic image. However, tomographic images were successfully generated by averaging multiple transmissions of independent and identically distributed (i.i.d.) noise waveforms, which compared well with the image generated using a widely used

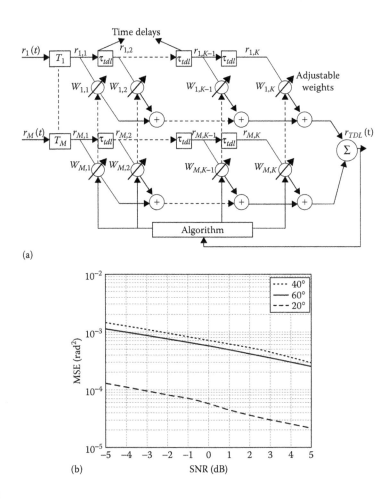

FIGURE 9.16
Performance of the tapped-delay-line (TDL)-based beamforming approach. (a) TDL beamformer architecture and (b) target direction estimation mean square error (MSE) performance. Received signals at each antenna are fed into a tapped-delay line consisting of K taps, each attached to an adjustable weight, and the weighted outputs are summed. Except for the first tap, the time delays for the remaining $K-1$ taps in all TDLs are identical. The smaller lobe width at smaller angles provides better accuracy at smaller angles. In (b), three targets are located at angular directions 20°, 40°, and 60°, and a jamming signal exists at 45° (close to the direction of the 40°-target). This makes the angular accuracy at 40° worse than that at 60°. (© 2012 IEEE. Reprinted, with permission, from Chen, W.J. and Narayanan, R.M., *IEEE Trans. Aerospace and Electronic Systems*, 48(3), 1858–1869, 2012.)

first-derivative Gaussian waveform used as a comparison standard. Calculating the mean square error provided a gauge of the quality of the reconstructed tomographic image using multiple averaged noise waveforms by comparing it with the standard waveform. Shin, Narayanan, and Rangaswamy (2014) further investigated tomographic reconstructions of several symmetric and random arrangements of cylindrical targets. Simulated tomographic images using a noise waveform and a first derivative Gaussian waveform for comparison are shown in Figure 9.18.

Lukin et al. (2013) used a MIMO arrangement in combination with the SAR approach to generate 3D coherent radar images using a 36-GHz noise waveform with a 480-MHz bandwidth. Two linear synthetic apertures used – one for the transmit antenna and the other for the receive antenna. Spatial scanning with these antennas was performed in a manner

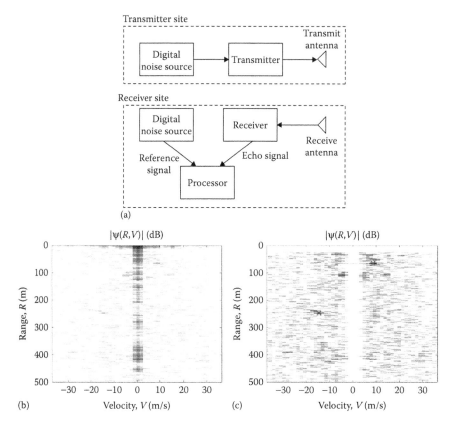

FIGURE 9.17
Performance of bistatic noise radar with locally generated reference signal (Malanowski and Roszkowski, 2011). (a) System block diagram, (b) the cross-ambiguity function between the synchronized reference signal with the received surveillance signal before adaptive filtering, and (c) the cross-ambiguity function between the synchronized reference signal with the received surveillance signal after adaptive filtering. Before adaptive filtering, some reflections at zero Doppler are present and the strongest peak corresponds to the direct path interference. After adaptive filtering, the echoes at zero Doppler (as well as certain Doppler span for close ranges) were removed together with the sidelobes. After this operation, moving targets (cars) can be seen at $R = 80$ m, $V = 10$ m/s and at $R = 250$ m, $V = -15$ m/s. (Permission Mateusz Malinovski © German Institute of Navigation 2011.)

which provided data similar to the ones obtained using a 2D scanner. By varying the delay of the recorded reference noise signal, range focusing was accomplished, which enabled the generation of 2D image (tomographic slice) at specific range bins, as shown in Figure 9.19.

9.4.6 Multifunctional Noise Radar

Surender and Narayanan (2011) proposed a scheme to implement a multifunctional UWB noise radar, which replaced a portion of the transmitted noise spectrum with noise-like orthogonal frequency-division multiplexing (OFDM) signals that could work for communications, telemetry, or signaling. They found the composite waveform mimicked the properties of a pure noise waveform. They evaluated the radar performance and the communications functionalities through the proposed waveform's range resolution, ambiguity function and bit-error-rate (BER) formulations. Such a multifunction radar system achieved surveillance with embedded security-enabled OFDM-based communications, multiuser capability, and

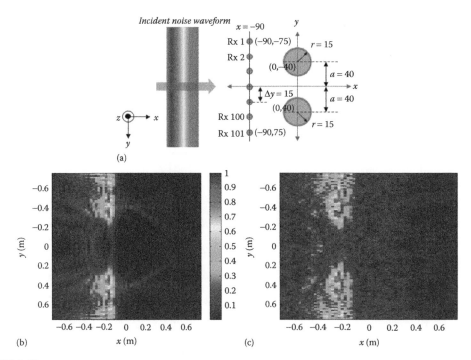

FIGURE 9.18
Simulated X-Band tomographic imaging of two cylinders (Shin, Narayanan, and Rangaswamy, 2014). (a) Imaging geometry, (b) tomographic image using a first-derivative Gaussian pulse of 0.1-ns pulse width, and (c) tomographic image obtained by averaging ten noise waveform realizations over the 8–10-GHz frequency range. (Permission of authors and *International Journal of Microwave Science and Technology*, Hindawi, Open access 2014.)

physical layer security. The system's physical layer design was comprehensively analyzed for the composite UWB noise-OFDM waveform's dual performance, communications reliability, confidentiality, and message integrity protection. Developing a noise-correlation-based frame timing estimation method in the netted radar sensor's receiver, which exploited the cross-correlation properties of band-limited white noise, achieved frame synchronization.

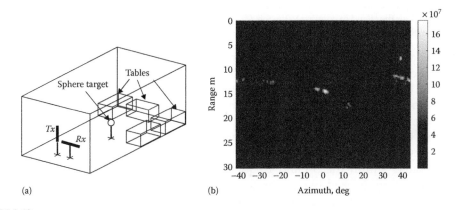

FIGURE 9.19
Tomographic imaging using a 36-GHz noise radar (Lukin et al., 2013). (a) Imaging geometry and (b) horizontal slice of 3D image of the laboratory room with sphere reflector in the middle. (© 2013 IEEE. Reprinted, with permission, from Lukin, K. A. et al., *IEEE Proc. International Conf. on Antenna Theory and Techniques (ICATT)*, Odessa, Ukraine, Sept 2013, pp. 190–192.)

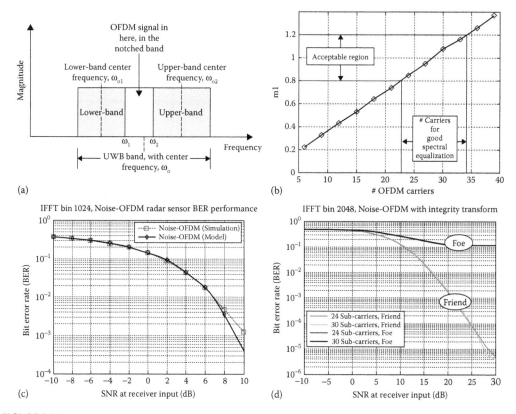

FIGURE 9.20

Multifunctional noise-orthogonal frequency-division multiplexing (OFDM) waveform for combined sensing and communications. (a) Spectrum of transmitted noise-OFDM waveform showing OFDM insertion within the spectral notch in the UWB noise waveform, (b) spectral equalization metric m_1 (defined as the ratio of the OFDM bandwidth to that of the notch) as a function of number of OFDM carriers, (c) BER performance of noise-OFDM waveform with 32 carriers in an additive white Gaussian noise (AWGN) channel, and (d) BER performance of noise-OFDM system incorporating integrity transform and data redundancy, wherein the friend knows the secret key and the foe does not. The ideal value of m_1 in (b) is 1 (indicating that the OFDM signal falls exactly within the notch spectrum) and it is achieved using 28 OFDM carriers, although between 23 and 34 carriers are acceptable. In (d), good BER performance is achieved for the friend in contrast to the foe. (© 2011 IEEE. Reprinted, with permission, from Surender, S.C. and Narayanan, R. M., *IEEE Trans. Aerospace and Electronics Systems*, 47(2), 1380–1400, 2011.)

This method did not require any preamble or transmitter preprocessing in the transmitted data signal. Additionally, critical issues presented underscored the need for a unique medium access control (MAC) algorithm as a duplex multiplexing technique for this *ad hoc* multiradar communications network. Figure 9.20 shows the scheme and its BER performance.

Narayanan, Smith, and Gallagher (2014) described a multifrequency radar system for detecting humans and classifying their activities at both short and long ranges with low-frequency components being shared between both systems. The short-range radar system operated within the S-Band frequency range for through-wall applications at distances up to 3 m. It used two separate waveforms selected via switching between a wide-band noise waveform or a continuous single tone. The long-range radar system operating in the W-Band millimeter-wave frequency range performed at distances of up to about 100 m in free space and up to about 30 m through light foliage. It employed a composite multimodal signal consisting of two waveforms: a wide-band noise waveform and an embedded single tone, which were summed and transmitted simultaneously.

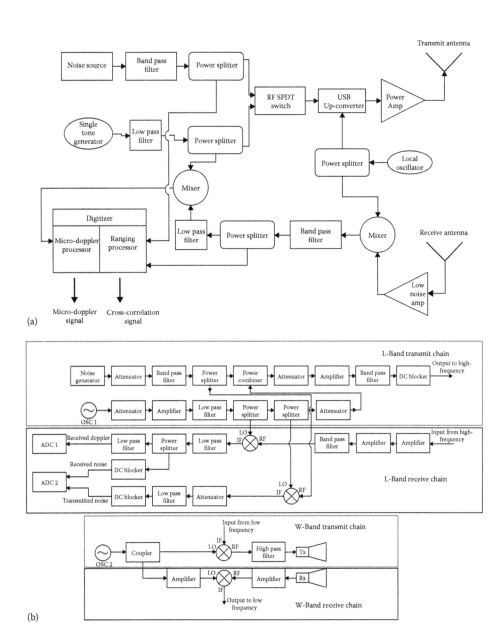

FIGURE 9.21
Multifunctional waveform generation schemes for combined ranging and micro-Doppler measurements (Narayanan, Smith, and Gallagher, 2014). (a) L-Band system block diagram, which used switching to select between the noise and the CW waveforms and (b) W-Band system block diagram which simultaneously transmitted and processed the noise and CW waveforms. (Permission *International Journal of Microwave Science and Technology*, Hindawi, Open access 2014.)

Cross-correlating between the received and transmitted noise signals detected targets with high-range resolution. Doppler analysis on the received single tone signal provided velocity information. Figure 9.21 shows the multifunctional waveform generation schemes for both systems. It implemented background subtraction and distance correction to suppress clutter and to enhance the detectability of targets at far ranges. Employing high-pass filtering eliminated the wind-influenced Doppler signals when detecting targets obscured by foliage. Doppler measurements processing could distinguish between different human movements and gestures using the characteristic micro-Doppler signals. The measurements established the ability of this system to detect and range humans and distinguish between different human movements at different ranges under different obscuring scenarios.

9.5 Advances in Noise Radar Theory and Design

In recent years, several hardware component advances have significantly improved noise radar performance and extended the range of their applications. Specifically, these advances have resulted in real-time operation of noise radar, which was not possible earlier because of the data processing slow speed for operations such as signal generation, sampling, and correlation processing. Full digital implementations of noise radar have made great advances possible.

9.5.1 Quantization Considerations

Xu and Narayanan (2003) studied the impact of different correlation receiving techniques on the imaging performance of UWB random noise radar. They investigated three types of correlation receivers: the ideal analog correlation receiver, the digital–analog correlation receiver, and the fully digital correlation receiver shown in Figure 9.22. The ideal analog correlation receiver faithfully preserved both the target amplitude and the initial phase. The digital–analog correlation receiver is similar to the ideal analog correlation receiver except that it implements the transmit delay using a digital radio frequency memory (DRFM). This approach first digitizes the a replica of the transmitted, the required time delay is realized digitally, and the delayed digital signal is finally converted back to an

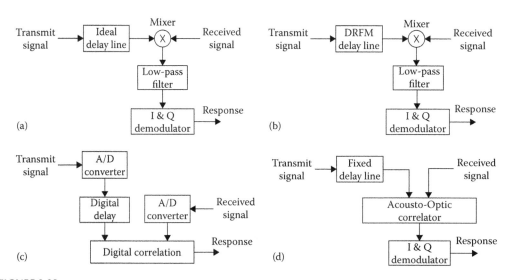

FIGURE 9.22
Four types of correlation schemes for noise correlation radar (Xu and Narayanan, 2003). (a) Ideal analog correlation, (b) digital–analog correlation using a digital radio frequency memory (DRFM) delay line, (c) fully digital correlation, and (d) acousto-optic correlation. (© 2003 IEEE. Reprinted, with permission, from Xu, X. and Narayanan, R.M., *IEEE International Geoscience and Remote Sensing Symposium (IGARSS'03)*, Toulouse, France, 21–25 July 2003., pp. 4525–4527.)

analog signal using a digital-to-analog converter (DAC) and fed to the correlator. They found the digital–analog correlation receiver could preserve both the target amplitude and initial phase information with a processing gain of $\sqrt{2/\pi}$ (~2 dB processing power loss) when using a binary DRFM.

The fully digital correlation receiver digitizes both the received target echo and the replica of the transmit signal before the correlation process starts. Digital signal-processing chips perform the cross-correlation of transmitted and returned signal. They found that amplitude information of the received signals was not preserved in a one-bit by one-bit correlation receiver, and the processing gain was $2/\pi$ (~4 dB processing power loss).

Velocity resolution in radars using traditional Doppler processing degrades as signal bandwidth increases. A technique developed for a random noise radar had limitations due to its lengthy processing time and high memory usage. To mitigate the time and memory requirements, Thorson and Akers (2011) proposed a binary ADC instead of an 8-bit ADC, sufficiently cutting the signal memory in order to enable parallel processing and drastically reduce the overall processing time required to simultaneously estimate a target's range and velocity. It was concluded that a binary ADC achieved the same detection performance when compared with an 8-bit ADC in a single target environment. Figure 9.23 shows the performance of the binary ADC when compared with the 8-bit ADC.

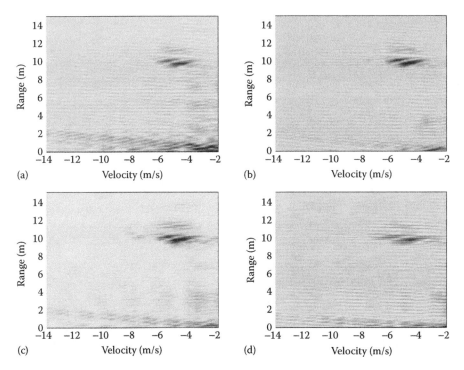

FIGURE 9.23

Range–Doppler imaging using a noise radar operating at 550 MHz with a bandwidth of 400 MHz (Thorson and Akers, 2011). (a) Using an 8-bit ADC, requiring a peak memory of 40 GB and a processing time of 37.5 minutes, (b) using a binary ADC, requiring a peak memory of 21 GB and a processing time of 15.5 minutes, (c) using a binary ADC with multiple processors on a single computer, requiring a peak memory of 42 GB and a processing time of 5.38 minutes, and (d) using a binary ADC with multiple processors on a single computer and speeding up the FFT operation, requiring a peak memory of 35 GB and a processing time of 4.7 minutes. Cases (c) and (d) show a slight degradation in the velocity resolution. (© 2011 IEEE. Reprinted, with permission, from Thorson, T.J. and Akers, G.A., *Proc. IEEE National Aerospace and Electronics Conference (NAECON)*, Dayton, OH, July 2011, pp. 270–275.)

9.5.2 Adaptive Noise Radar

The use of spectrally tailored noise waveforms enhanced the detection of targets buried in dispersive soil by adapting the transmit spectrum to inversely match the soil spectral characteristics (Narayanan, Henning, and Dawood, 1998).

Rigling (2004) developed an adaptive noise radar approach and developed algorithms for adaptive wireless channel identification (i.e., the least mean square (LMS) algorithm) to accomplish noise radar pseudo-pulse compression. This approach allowed a whitened range profile to be estimated with computations of the same order as simple cross-correlation. Theoretical modeling compared the SNR performance of simple cross-correlation, cross-correlation with signal whitening, and channel estimation via the adaptive LMS algorithm. The LMS algorithm provided detection performance equivalent to the whitened cross-correlation with the same computational complexity as simple cross-correlation. The evaluation concluded that the algorithm used in adaptive noise radar processing was nearly identical in implementation to those used in adaptive wireless channel equalization. The performance of the adaptive noise radar is shown in Figure 9.24, which compares it with simple cross-correlation.

Narayanan and McCoy (2013) explored the use of a composite noise waveform consisting of appropriately delayed and summed versions of the original noise waveform for cases

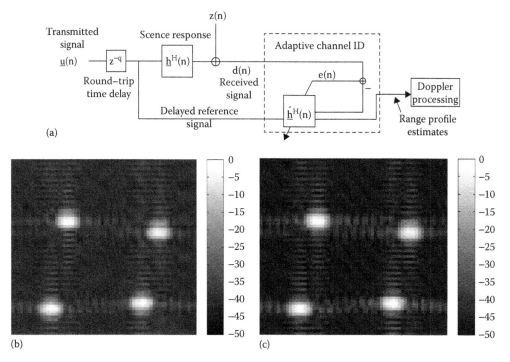

FIGURE 9.24
Performance of adaptive noise radar. (a) Adaptive noise radar receiver architecture that computes range profile estimates through adaptive channel ID processing and least-mean squared (LMS) filtering for noise radar signal compression, (b) mean-squared range–Doppler image of four scatterers computed using simple cross-correlation, resulting in an estimated SNR of 39 dB, and (c) mean-squared range–Doppler image of four scatterers computed using adaptive processing, resulting in an estimated SNR of 44.2 dB. (© 2004 IEEE. Reprinted, with permission, from Rigling, B.D., *Conf. Record of the Thirty-Eighth Asilomar Conference on Signals, Systems and Computers*, Pacific Grove, CA, Nov. 2004, pp. 3–7.)

where the target scattering center had a known *a priori* arrangement. If the arrangement of the scattering centers exactly matched that of the transmit waveform, this enhanced the composite correlation function due to the summation of correlations at the individual delays, thereby resulting in improved target recognition. Targets that did not match the transmit waveform did not show the enhanced correlation peak, and thus could be discarded as false targets. Experimental validations performed for target arrangements with two and three scattering centers using a 1–2-GHz UWB noise radar helped to validate the concept and demonstrate performance. Figure 9.25 shows that the scheme can exploit the knowledge of the target scattering center arrangement to architect an adaptive waveform for target recognition.

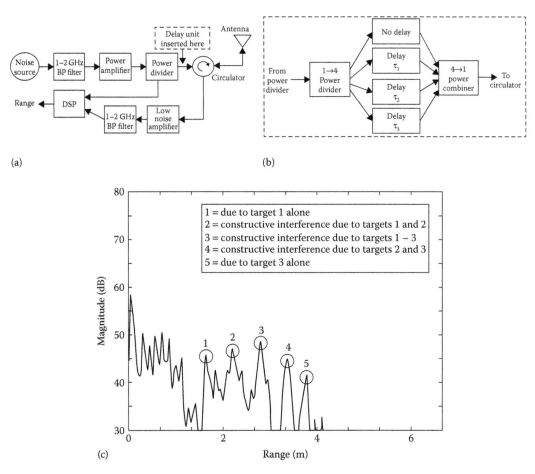

(a)

(b)

(c)

FIGURE 9.25
Delayed and summed adaptive noise radar operating over the 1–2-GHz frequency range. (a) Basic noise radar block diagram, (b) structure of delay unit, and (c) experimental correlation plots of three targets with intertarget separation of 45 cm using three equally spaced delays of 3 ns for the transmit waveform. In (c), Peak 1 is entirely due to the Target 1, and it provides the range to Target 1. Peak 2 is due to constructive interference from Targets 1 and 2, and yields the range to Target 2. Peak 3 is due to constructive interference from all three targets, and provides the range to Target 3. Peak 4 is due to constructive interference from Targets 2 and 3, while Peak 5 is entirely due to Target 3. (© 2013 IEEE. Reprinted, with permission, from Narayanan, R.M. and McCoy, N.S., *Proc. 22nd International Conference on Noise and Fluctuations (ICNF 2013)*, Montpellier, France, Jun 2013.)

9.5.3 Applications of Arbitrary Waveform Generators in Noise Radar

Lukin, Konovalov, and Vyplavin (2010) proposed and analyzed an alternative approach to the evaluation of the cross-correlation function in wideband noise radar, which enabled a significant enhancement in the dynamic range using digital signal-processing techniques. They based their approach on digital generation of both the sounding and the reference signals using fast arbitrary waveform generators (AWGs) with the required delay combined with coherent analog reception of the radar returns. Upconverting both the sounding and the reference signals to the required carrier frequency and cross-correlation between the reference and radar returns was realized at the carrier frequency using an analog mixer and low-pass filter. Since the output signal of the correlator was in the low-frequency band, a comparatively slow ADC could work for sampling. Later, they extended this approach to a stepped frequency radar using narrowband frequency modulation of the sounding wave (Lukin et al., 2011). This enabled the accrual of benefits from both the stepped frequency and noise radar concepts, such as low signal bandwidth at the output of the correlator, high electromagnetic compatibility, and high immunity to interference.

9.5.4 Photonic Applications in Noise Radar

Jiang, Wolfe, and Nguyen (2000) described a novel application of superfluorescent fiber sources (SFSs) with erbium-doped fiber amplifier (EDFA) for random noise radar. The UWB noise coupled with the use of optical fiber delay lines makes SFS ideally suited for random noise radar applications. The optical fiber makes long and multiple-step delay lines of a few kilometers feasible. Its advantages include extremely low-loss, compactness, light-weight, and low-cost. In addition, the dispersion and nonlinearity associated with RF delay lines is avoided. The optical 1.2 THz at 1530-nm wavelength spectrum of the SFS gave a free-space full width at half maximum (FWHM) resolution of 15.2 cm. They proposed a method for optical injection of coherence into the radar system employing EDFA light sources to extract and process the Doppler and polarimetric target responses. Their measurements using a 100-m-long optical fiber validated the system's range resolution and target detection capability.

Kim et al. (2005) described a time-integrating acousto-optic correlator (TIAOC) developed for imaging and target detection using a wideband random noise radar system. This polarization interferometric inline TIAOC used an intensity-modulated laser diode for the random noise reference and a polarization-switching, self-collimating acoustic shear-mode gallium phosphide (GaP) acousto-optic device (AOD) for traveling-wave modulation of the radar returns. The time-integrated correlation output was detected on a 1-D charge-coupled device (CCD) detector array and calibrated and demodulated in real time to produce the complex radar range profile. The complex radar reflectivity was measured in more than 150 radar range bins in parallel on the 3000 pixels of the CCD, significantly improving target acquisition speeds and sensitivities by a factor of 150 over previous serial analog correlator approaches. The polarization interferometric detection of the correlation using the undiffracted light as the reference permitted the use of the full acousto-optic device (AOD) bandwidth as the system bandwidth. Experimental results showed that fully complex random-noise signal correlation and coherent demodulation were possible without an explicit carrier, demonstrating that optically processed random-noise radars did not need a stable local oscillator. The architecture and the performance of the TIAOC are shown in Figure 9.26. The TIAOC was implemented in a 1–2-GHz noise radar system and its performance assessed by Narayanan et al. (2004). Measurements demonstrated the ability of the noise radar using the acousto-optic correlator to provide range profiles of target arrangements, as shown in Figure 9.27.

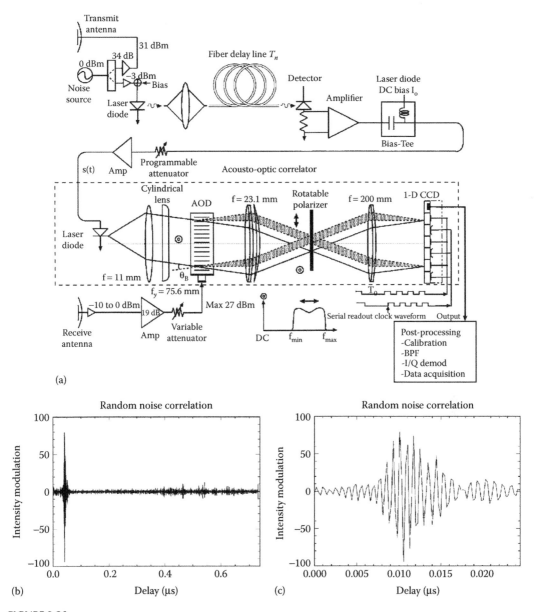

FIGURE 9.26

Implementation and correlation performance of the time-integrating acousto-optic correlator (TIAOC) integrated in a noise radar. (a) System topology employing a fiber-optic delay line for the coarse range delay and an acousto-optic time-integrating correlator to allow parallel range bin processing, (b) cross-correlation after subtraction of calibrated fixed pattern noise, and (c) zoomed-in cross-correlation shown in (b). The replica of the noise signal, delayed in the fiber delay line with delay T_n, modulates the laser diode of the TIAOC. The received antenna signal drives the polarization switching wideband AOD. Cross-correlation is achieved by imaging the AOD aperture onto the CCD. (© 2005 SPIE, with permission from Kim, S. et al., *Optical Engineering*, 40(10), 2005.)

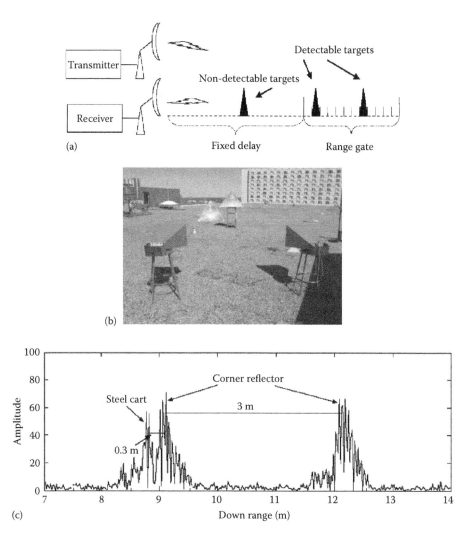

FIGURE 9.27

Performance of a 1–2-GHz noise radar employing a fixed optical delay line and a time-integrating acousto-optic correlator (TIAOC) for achieving fine delays. (a) Function of fixed and variable delay lines, (b) experimental geometry showing the radar pointed toward a corner reflector on the ground and another corner reflector on a small cart, and (c) processed amplitude (arbitrary units) as a function of range. The fixed delay line sets the minimum range, while the variable delay line provides the range profiling function. The main feature of this system is that the range profile was gathered instantaneously as opposed to sequentially switching through the range delays as in a conventional noise radar. The reflection from the corner reflectors is clearly observed, as is the reflection from the front of the steel cart. (© 2004 IEEE. Reprinted, with permission, from Narayanan, R.M. et al., *IEEE Geoscience and Remote Sensing Letters*, 1(3), 166–170, 2004.)

Grodensky, Kravitz, and Zadok (2012) proposed and demonstrated a microwave-photonic ultrawideband noise radar system. The system combined photonic generation of UWB waveforms with fiber-optic distribution. They generated the UWB noise waveform using the amplified spontaneous emission (ASE) associated with either stimulated Brillouin scattering (SBS) in a standard optical fiber or with erbium-doped fiber amplifier (EDFA). The generator produced waveforms of more than 1-GHz bandwidth and arbitrary radio-frequency carriers and distributed the signals over 10-km fiber length to a remote antenna unit. The system combined the antenna remoting capabilities, broad

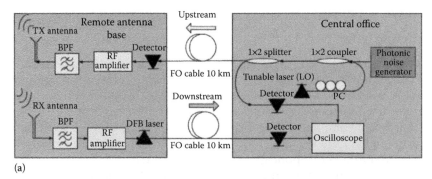

(a)

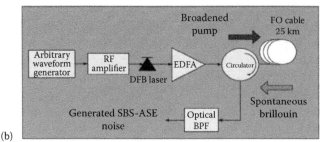

(b)

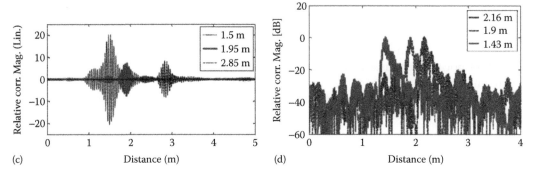

(c)   Distance (m)   (d)   Distance (m)

FIGURE 9.28

Architecture and performance of microwave-photonic ultra-wideband (UWB) noise radar system operating over the 2.5–4-GHz frequency band. (a) Simplified block diagram, (b) setup for stimulated Brillouin scattering-amplified spontaneous emission (SBS-ASE) noise generation, (c) measured correlation between SBS-ASE noise waveform reflected from a metal target and a reference replica for various target distances, and (d) measured correlation between EDFA-ASE noise waveform reflected from a metal target and a reference replica for various target distances. In SBS, a strong pump wave and a weaker counter propagating probe wave optically interfere to generate a traveling longitudinal acoustic wave, which couples these optical waves to each other. Although EDFAs could provide ASE with a near-uniform PSD over a bandwidth of several THz, fiber Bragg gratings (FBGs) were used to restrict the bandwidth to a few GHz in this application. (© 2012 IEEE. Reprinted, with permission, from Grodensky, D. et al., *IEEE Photonics Technology Letters*, 24(10), 839–841, 2012.)

bandwidth, and flexibility of reconfiguration provided by microwave photonics, together with the potentially superior resilience against jamming and interception of noise radars. Photonic noise generation could be scaled toward high frequencies of electrical noise in the millimeter-wave range, provided a high degree of randomness based on a physical process, and readily integrated into a radio-over-fiber (RoF) system. Laboratory experiments demonstrated ranging measurements with 10-cm resolution. Figure 9.28 shows the performance of both approaches when implemented in a 2.5–4-GHz noise radar.

9.6 Conclusion

The half century since Horton's seminal 1959 paper has seen significant advances in the theoretical development, algorithm implementation, and applications of random, pseudo-random, noise, pseudo-noise, and noise-like waveforms in radar systems. Based upon the significant developments in noise radar described above and upon the rapid development in hardware componentry and speed of computing, we believe noise radar has a bright future and is here to stay. The performance of noise radar clearly rivals that of conventional radar in several applications. This coupled with the fact that noise radar has advantages such as immunity from detection, immunity from jamming, immunity from RF interference, spectral parsimony, excellent ambiguity function, and so on, makes it very suitable for both military and commercial applications. It is hoped that this chapter will serve as a starting point to spur further development in the novel applications of noise radar based on advances in signal processing and circuit development.

9.6.1 Summary of the Limits of Noise Radar Technology

The totally random or near-random waveform limits noise radar performance. As a result, each realization is akin to a snowflake and different from another. This requires significant averaging to achieve desirable results and imposes constraints on time and hardware. In addition, for applications at higher frequencies (e.g., in the millimeter-wave regimes), high-speed samplers may not be available for direct digitization, which results in a partly analog implementation. This requires upconverting a low-frequency noise signal prior to transmission and the downconverting the received signal for sampling.

Another noise radar disadvantage is the difficulty of precisely shaping and tailoring the spectrum of the transmitted noise waveform to avoid denied or protected frequency band slices, which can be achieved in traditional stepped frequency waveforms. However, the introduction of high-speed arbitrary waveform generators somewhat alleviates this limitation.

9.6.2 Future Practical Noise Radar Applications

Several applications abound for noise radar and some progress already made in these areas. These include (1) through-wall target detection and imaging, (2) synthetic aperture radar (SAR) imaging, (3) inverse synthetic aperture radar (ISAR) imaging, (4) interferometry and antenna beamforming, (5) high-range resolution target imaging, (6) automobile collision avoidance, (7) covert RF tagging of assets, (8) extension to passive radar, that is radars using transmitters of opportunity, (9) covert radar networks, (10) radar polarimetry, and (11) active radiometry. These future noise radars will have extensive utility in military and commercial applications.

Note: Some of the figures and images in this chapter are published in color in the original references. Readers are suggested to see these in color in the original papers if they wish.

References

Axelsson, S.R.J. 2003. "Noise radar for range/Doppler processing and digital beamforming using low-bit ADC." *IEEE Transactions on Geoscience and Remote Sensing* 41(12): 2703–2720.

Axelsson, S.R.J. 2006. "Ambiguity functions and noise floor suppression in random noise radar." In *Proc. SPIE Conference on SAR Image Analysis, Modeling, and Techniques VIII*, Stockholm, Sweden, September 2006, pp. 636302-1–636302-10.

Bourret, R. 1957. "A proposed technique for the improvement of range determination with noise radar." *Proceedings of the IRE* 45(12): 1744.

Carpentier, M.H. 1968. "Analysis of the principles of radars." Chapter 4 in *Radars: New Concepts.* New York, NY: Gordon and Breach, pp. 121–178.

Carpentier, M.H. 1970. "Using random functions in radar applications." *De Ingenieur* 82(46): E166-E172.

Chadwick, R.B., and Cooper, G.R. 1971. "Measurement of ocean wave heights with the random-signal radar." *IEEE Transactions on Geoscience Electronics* 9(4): 216–221.

Chadwick, R.B., and Cooper, G.R. 1972. "Measurement of distributed targets with the random signal radar." *IEEE Transactions on Aerospace and Electronic Systems* 8(6): 743–750.

Chapursky, V.V., Sablin, V.N., Kalinin, V.I., and Vasilyev, I.A. 2004. "Wideband random noise short range radar with correlation processing for detection of slow moving objects behind the obstacles." In *Proc. Tenth International Conference on Ground Penetrating Radar*, Delft, The Netherlands, June 2004, pp. 199–202.

Chen, W.J., and Narayanan, R.M. 2012. "GLRT plus TDL beamforming for ultrawideband MIMO noise radar." *IEEE Transactions on Aerospace and Electronic Systems* 48(3): 1858–1869.

Craig, S.E., Fishbein, W., and Rittenbach, O.E. 1962. "Continuous-wave radar with high range resolution and unambiguous velocity determination." *IEEE Transactions on Military Electronics* MIL-6(2): 153–161.

Dawood, M., and Narayanan, R.M. 2001. "Receiver operating characteristics for the coherent ultrawideband random noise radar." *IEEE Transactions on Aerospace and Electronic Systems* 37(2): 586–594.

Dawood, M., and Narayanan, R.M. 2003. "Generalised wideband ambiguity function of a coherent ultrawideband random noise radar." *IEE Proceedings on Radar, Sonar, and Navigation* 150(5): 379–386.

Dukat, F. 1969. *A Simplified, CW, Random-Noise Radar System.* M.S.E.E. Thesis, United States Naval Postgraduate School, October 1969.

Forrest, J.R., and Meeson, J.P. 1976. "Solid-state microwave noise radar." *Electronics Letters* 12(15): 365–366.

Forrest, J.R., and Price, D.J. 1978. "Digital correlation for noise radar systems." *Electronics Letters* 14(18): 581–582.

Furgason, E.S., Newhouse, V.L., Bilgutay, N.M., and Cooper, G.R. 1975. "Application of random signal correlation techniques to ultrasonic flaw detection." *Ultrasonics* 13(1): 11–17.

Grant, M.P., Cooper, G.R., and Kamal, A.K. 1963. "A class of noise systems." *Proceedings of the IEEE* 51(7): 1060–1061, July 1963.

Gray, D.A. 2006. "Multi-channel noise radar." In *Proc. 2006 International Radar Symposium (IRS)*, Krakow, Poland, May 2006, doi: 10.1109/IRS.2006.4338086.

Gray, D.A., and Fry, R. 2007. "MIMO noise radar – element and beam space comparisons." In *Proc. International Waveform Diversity and Design Conference*, Pisa, Italy, June 2007, pp. 344–347.

Grodensky, D., Kravitz, D., and Zadok, A. 2012. "Ultra-wideband microwave-photonic noise radar based on optical waveform generation." *IEEE Photonics Technology Letters* 24(10): 839–841.

Gupta, M.-S. 1975. "Applications of electrical noise." *Proceedings of the IEEE* 63(7): 996–1010.

Herman, M.A., and Strohmer, T. 2009. "High-resolution radar via compressed sensing." *IEEE Transactions on Signal Processing* 57(6): 2275–2284.

Hochstadt, H. 1958. "Comments on 'A proposed technique for the improvement of range determination with noise radar." *Proceedings of the IRE* 46(9): 1652.

Horton, B.M. 1959. Noise-modulated distance measuring systems. *Proceedings of the IRE* 47(5): 821–828.

Huelsmeyer, C. 1904. "The telemobiloscope." *Electrical Magazine (London)* 2: 388.

Jiang, H., Zhang, B., Lin, Y., Hong, W., Wu, Y., and Zhan, J. 2010. "Random noise SAR based on compressed sensing."In *Proc. IEEE International Geoscience and Remote Sensing Symposium (IGARSS)*, Honolulu, HI, July 2010, pp. 4624–4627.

Kim, S., Wagner, K., Narayanan, R.M., and Zhou, W. 2005. "Broadband polarization interferometric time-integrating acousto-optic correlator for random noise radar." *Optical Engineering* 40(10): 108202, doi: 10.1117/1.2084807

Krehbiel, P.R., and Brook, M. 1979. "A broad-band noise technique for fast-scanning radar observations of clouds and clutter targets." *IEEE Transactions on Geoscience Electronics* 17(4): 196–204.

Kwon, Y., Narayanan, R.M., and Rangaswamy, M. 2013. "Multi-target detection using total correlation for noise radar systems." *IEEE Transactions on Aerospace and Electronic Systems* 49(2): 1251–1262.

Lukin, K.A., Konovalov, V.M., and Vyplavin, P.L. 2010. "Stepped delay noise radar with high dynamic range." In *Proc. International Radar Symposium (IRS)*, Vilnius, Lithuania, June 2010, pp. 1–3.

Lukin, K.A., Vyplavin, P.L., Kudriashov, V.V., Palamarchuk, V.P., Sushenko, P.G., and Zaets, N.K. 2013. "Radar tomography using noise waveform, antenna with beam synthesis and MIMO principle." In *Proc. International Conference on Antenna Theory and Techniques (ICATT)*, Odessa, Ukraine, September 2013, pp. 190–192.

Lukin, K.A., Zemlyaniy, O.V., Vyplavin, P.L., Lukin, S.K., and Palamarchuk, V.P. 2011. "High resolution and high dynamic range noise radar." In *Proc. Microwaves, Radar and Remote Sensing Symposium (MRRS)*, Kiev, Ukraine, August 2011, pp. 247–250.

Malanowski, M., and Roszkowski, P. 2011. "Bistatic noise radar using locally generated reference signal." In *Proc. International Radar Symposium (IRS)*, Leipzig, Germany, September 2011, pp. 544–549.

Mogyla, A.A. 2012. "Detection of radar signals in conditions of full prior information when using stochastic signals for probing." *Radioelectronics and Communications Systems* 55(7): 299–306.

Mogyla, A.A., Lukin, K.A., and Shyian, Y.A., 2002. "Relay-type noise correlation radar for the measurement of range and vector range rate." *Telecommunications and Radio Engineering* 57(2–3): 175–183.

Narayanan, R.M., and Dawood, M. 2000. "Doppler estimation using a coherent ultra-wideband random noise radar." *IEEE Transactions on Antennas and Propagation* 48(6): 868–878.

Narayanan, R.M., Henning, J.A., and Dawood, M. 1998. "Enhanced detection of objects obscured by dispersive media using tailored random noise waveforms." *Proc. SPIE Conference on Detection and Remediation Technologies for Mines and Minelike Targets III*, Orlando, FL, April 1998, 604–614.

Narayanan, R.M., and McCoy, N.S. 2013. "Delayed and summed adaptive noise waveforms for target matched radar detection." In *Proc. 22nd International Conference on Noise and Fluctuations (ICNF 2013)*, Montpellier, France, June 2013, doi: 10.1109/ICNF.2013.6578958.

Narayanan, R.M., Smith, S., and Gallagher, K.A. 2014. "A multi-frequency radar system for detecting humans and characterizing human activities for short-range through-wall and long-range foliage penetration applications." *International Journal of Microwave Science and Technology* 2014: 958905, doi: 10.1155/2014/958905.

Narayanan, R.M., Xu, Y., Hoffmeyer, P.D., and Curtis, J.O. 1995. "Design and performance of a polarimetric random noise radar for detection of shallow buried targets." In *Proc. SPIE Conference on Detection Technologies for Mines and Minelike Targets*, Orlando, FL, April 1995, pp. 20–30.

Narayanan, R.M., Xu, Y., Hoffmeyer, P.D., and Curtis, J.O. 1998. "Design, performance, and applications of a coherent ultrawideband random noise radar." *Optical Engineering* 37(6): 1855–1869.

Narayanan, R.M., Xu, Y., and Rhoades, D.W. 1994. "Simulation of a polarimetric random noise spread spectrum radar for subsurface probing applications." In *Proc. IEEE International Geoscience and Remote Sensing Symposium (IGARSS'94)*, Pasadena, CA, August 1994, pp. 2494–2498.

Narayanan, R.M., Zhou, W., Wagner, K.H., and Kim, S. 2004. "Acousto-optic correlation processing in random noise radar." *IEEE Geoscience and Remote Sensing Letters* 1(3): 166–170.

Poirier, J.L. 1968. "Quasi-monochromatic scattering and some possible radar applications." *Radio Science (New Series)* 3(9): 881–886.

Rigling, B.D. 2004. "Performance prediction in adaptive noise radar." In *Conference Record of the Thirty-Eighth Asilomar Conference on Signals, Systems and Computers*, Pacific Grove, CA, November 2004, pp. 3–7.

Sachs, J., Aftanas, M., Crabbe, S., Drutarovský, M., Klukas, R., Kocur, D., Nguyen, T.T., Peyerl, P., Rovňáková, J., and Zaikov, E. 2008. "Detection and tracking of moving or trapped people hidden by obstacles using ultra-wideband pseudo-noise radar." In *Proc. European Radar Conference (EuRAD)*, Amsterdam, The Netherlands, October 2008, pp. 408–411.

Shastry, M.C., Narayanan, R.M., and Rangaswamy, M. 2010. "Compressive radar imaging using white stochastic waveforms." In *Proc. 5ᵗʰ International Waveform Diversity and Design Conference*, Niagara Falls, Canada, August 2010, pp. 90–94.

Shastry, M.C., Narayanan, R.M., and Rangaswamy, M. 2015. "Sparsity-based signal processing for noise radar imaging." *IEEE Transactions on Aerospace and Electronic Systems* 51(1): 314–325.

Shin, H.J., Narayanan, R.M., and Rangaswamy, M. 2013. "Tomographic imaging with ultra-wideband noise radar using time-domain data." In *Proc. SPIE Conference on Radar Sensor Technology XVII*, Baltimore, MD, April 2013, pp. 87140R-1–87140R-9.

Shin, H.J., Narayanan, R.M., and Rangaswamy, M. 2014. "Ultrawideband noise radar imaging of impenetrable cylindrical objects using diffraction tomography." *International Journal of Microwave Science and Technology*, 2014: 601659, doi: 10.1155/2014/601659.

Smit, J.A., and Kneefel, W.B.S.M. 1971. "RUDAR – an experimental noise radar system." *De Ingenieur* 83(32): ET99-ET110.

Stephan, R., and Loele, H., 2000. "Theoretical and practical characterization of a broadband random noise radar." In *IEEE MTT-S International Microwave Symposium Digest*, Boston, MA, June 2000, pp. 1555–1558.

Surender, S.C., and Narayanan, R.M. 2011. "UWB Noise-OFDM netted radar: Physical layer design and analysis." *IEEE Transactions on Aerospace and Electronic Systems* 47(2): 1380–1400.

Thorson, T.J., and Akers, G.A. 2011. "Investigating the use of a binary ADC for simultaneous range and velocity processing in a random noise radar." In *Proc. IEEE National Aerospace and Electronics Conference (NAECON)*, Dayton, OH, July 2011, pp. 270–275.

Turin, G.L. 1958. "Comments on 'A proposed technique for the improvement of range determination with noise radar.'" *Proceedings of the IRE* 46(10): 1757–1758.

Vela, R., Narayanan, R.M., Gallagher, K.A., and Rangaswamy, M. 2012. "Noise radar tomography." In *Proc. IEEE Radar Conference*, Atlanta, GA, May 2012, pp. 720–724.

Wang, H., Narayanan, R.M., and Zhou, Z.O. 2009. "Through-wall imaging of moving targets using UWB random noise radar." *IEEE Antennas and Wireless Propagation Letters* 8: 802–805.

Xu, X. and Narayanan, R.M. 2003. "Impact of different correlation receiving techniques on the imaging performance of UWB random noise radar." In *Proc. IEEE International Geoscience and Remote Sensing Symposium (IGARSS'03)*, Toulouse, France, July 2003, pp. 4525–4527.

10

Prototype UWB Radar Object Scanner and Holographic Signal Processing

Lorenzo Capineri, Timothy Bechtel, Pierluigi Falorni,
Masaharu Inagaki, Sergey Ivashov, and Colin Windsor

CONTENTS

10.1 Introduction

Subsurface imaging using high-frequency impulse ground-penetrating radar has long been studied and applied in fields such as geotechnical, environmental, and structural engineering [1]. More recently ultrawideband (UWB) radar technology has stimulated new investigations into aerospace and medical (among others) applications [2]. However, it is a common problem that the detection of shallow targets (depth < 20 cm) is hindered by interference between the transmitted and received pulses from shallow electromagnetic impedance contrasts. Mitigating these interference effects requires complex methods of image reconstruction. The authors developed a different approach to short-range materials penetrating radar using holographic radar, which measures the phase of the reflected signal. Demonstration holographic radars operating at 2 GHz and 4 GHz can produce useful images with high spatial resolution over a variety of subsurface impedance contrasts [3,4,5]. However, both impulse and holographic high-frequency radars are commonly limited to scanning small areas on the order of 1 or 2 m^2 using manual scanning [6]. This time-consuming data acquisition method requires the preparation of a grid of reference lines on the scanned surface to guide the manual operation. Deviations from the guiding lines are common, and it is difficult to maintain accurate location control for larger areas (e.g., several square meters). Moreover, the trend toward using higher frequencies (>4 GHz) in subsurface imaging is sustained by the availability of integrated electronic devices operating in this frequency range with improved power consumption and signal-to-noise ratio, and with high-resolution analog-to-digital converters. The shorter wavelength can improve the spatial resolution of phase-based images, but only if a comparable sample step (a few millimeters) is employed and the uncertainty of manual scanning is also limited to a few millimeters. Free-hand scans are possible indoors or outdoors with advanced optical, pulsed microwave, or ultrasonic positioning systems [7,8,9] having millimeter accuracy, but the acquisition rate is slow compared with manual scanning. In addition, set up time and environmental conditions such as natural or artificial light, electromagnetic interference, or air turbulence can limit their applicability. Finally, the cost of the instruments and their set up time is another limiting factor.

A new approach, proposed recently [10,11,12], is the development of a light, compact, robotic holographic radar scanner capable of imaging large areas with fine spatial sampling. The holographic radar was developed with a compact and lightweight (<400 g) monostatic antenna with integrated electronics swept by a mechanical scanner mounted on a robotic vehicle. The trajectory of the robotic vehicle is programmable from a remote PC terminal in order to cover arbitrary areas while avoiding obstacles. The holographic images have a plan-view resolution that is about a quarter of the signal wavelength in the scanned material (e.g., soil, stone, and wood), and they accurately reproduce the dimensions and shape of shallow objects with a contrasting dielectric constant. A full three-dimensional reconstruction containing depth information can be obtained using multiple frequencies and migration software [13].

In addition to the radar itself, the newly available fast developing robotic technologies for the platform/positioning can simplify the design of the mechanics, the electronics, and the trajectory control. This has motivated research into a prototype robotic scanner that combines the benefits of high-spatial-resolution subsurface radar imaging and the automated robotic scanning of indoor or outdoor surveys.

In this chapter, Section 10.2 reviews the underlying theory of the sampling requirements for subsurface imaging with holographic radar. Section 10.3 presents the applicable holographic signal-processing techniques. Section 10.4 describes the principles of operation for the robotic scanner prototype and the laboratory implementation of a high-accuracy automated scanner. Finally, Section 10.5 illustrates investigations in several fields using this subsurface imaging method and emphasizes the real benefits of this robotic approach to high-frequency GPR scanning relative to the standard manual scanning.

10.2 Holographic Subsurface Radars of the RASCAN Type

10.2.1 Principles of Radar Holography

The holographic subsurface radar design uses the classical principles of radio-positioning. As in all radar, the radiated signal reflects from local inhomogeneities if their dielectric constant ε differs from the surrounding medium dielectric constant. The receiving antenna collects the reflected signal. The receiver amplifies and registers the reflected signal. After appropriate signal processing, the operator sees the results displayed on a computer screen in real time [see, e.g., 14].

Traditionally, impulse radar is the most commonly used type of subsurface radar. In general, these repeatedly transmit a single period of a sine wave signal (or impulse), and record the time-domain return signal (or wiggle trace), which contains the reflected impulses. Currently, almost all subsurface radars in commercial production are of this type. The main advantage of impulse radar is the high-effective penetration depth into the surveyed medium achieved by application of time-varying gain, which amplifies the later weaker signals from deeper reflections. A direct determination of reflector depths may be made by measurement of the reflected signal time-of-flight if the electromagnetic wave velocity in the medium is known or determinable [14,15]. Impulse radars have a significant disadvantage from the reverberation effect, that is multiple reflections of the transmitted pulse between the radar antenna and strong reflectors (such as the metal supporting structure to which thermal insulation or heat protection coatings are attached – see Section 10.5). In this case, multiple reflections (often called ghosts or phantoms) of the transmitted impulse signal obscure the primary reflection of interest [16].

RASCAN-type holographic subsurface radars are free from this reverberation effect because they use a continuous wave signal. RASCAN radars also have a distinct advantage in lateral resolution over impulse radars because of the specific design of the radar antenna which combines the transmitter and receiver antennas into one lightweight and compact apparatus with a small footprint.

Holographic subsurface radars get their name from the process of recording the interference pattern on the surface of the medium between the reference wave and the wave (object wave) reflected from subsurface targets. It is worth noting that for a long time, there was a widespread opinion that, due to strong attenuation in typical media and the inapplicability of time-varying gain to continuous wave radar reflections, holographic subsurface radar was unlikely to find any significant practical application [17]. However, the recent development of holographic subsurface radar of the RASCAN type, their commercial production, and sufficiently broad practical application have shown that for examination of low electrical conductivity media at shallow depths, this type of device has many advantages including real-time, plan-view imaging, and high lateral resolution. The design details of the various types of RASCAN radar and their areas of application are described in [18,19] and [3,5].

The principles of recording microwave holograms using RASCAN radar are analogous to optical holography. Consider a plane monochromatic wave with a constant phase (the reference wave) encountering a point target and scattering. The summation of the incident (or reference) and scattered (or object) waves on a flat screen located at some distance behind the object forms an interference pattern as in Figure 10.1a. If the screen is normal to the propagation direction of the reference wave, the interference pattern forms a Fresnel pattern of concentric rings. In optics, after development of the pattern recorded on a photographic plate, illuminating the pattern by a reference wave will form a virtual image of the object, which seems to float behind the screen as in Figure 10.1b. A similar phenomenon occurs when a RASCAN subsurface holographic radar records a microwave hologram of a point target in a uniform medium [13].

In some cases, microwave holograms recorded by RASCAN are remarkably similar to the optical holograms shown in the classical work of D. Gabor [20]. The essential difference is the much lower spatial density of interference lines on the microwave hologram

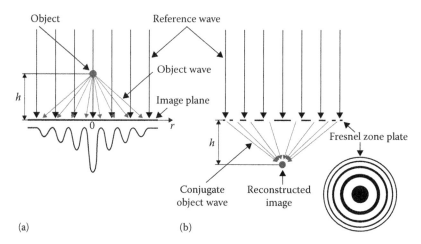

FIGURE 10.1
The holographic radar uses the same principles as an optical hologram. (a) The simplest case of recording of the holographic diffraction pattern for a point target and (b) reconstruction of the target image using the diffraction pattern.

interference patterns [21]. This is because at approximately the same characteristic dimensions of the systems, the radar wavelength is several orders of magnitude greater than wavelengths in the optical waveband.

RASCAN radars have fully implemented this holographic principle. Figure 10.2 shows the subsurface radar RASCAN-5/15000 used in some of the experiments described in this chapter. Radars of RASCAN-5 type use a quadrature signal receiver that allows recording of complex microwave holograms of hidden objects. The radar transducer head containing both a transmitter and a receiver connects via a network cable to a microcontroller unit through a USB link to a computer. The microcontroller unit drives the transmitter and receiver, digitizes the received data, and transmits it to the computer.

Image formation using holographic radar results from scanning a surface using a monostatic antenna. The defined antenna position (x, y) over this surface assumes an ideal planar surface. Computation assumes a reference system (x, y, and z) with the origin on this surface and the perpendicular z-axis representing the depth (Figure 10.3).

FIGURE 10.2
The holographic subsurface radar RASCAN-5/15000 as supplied to customers. The wheel on the radar transducer head measures movement of the radar during scanning.

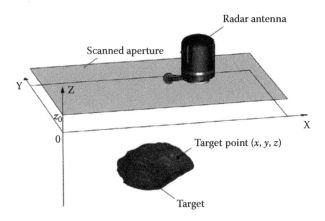

FIGURE 10.3
Geometry of holographic imaging system with a RASCAN head.

Single-frequency holographic image reconstruction used the following set of basic equations:

$$F(k_x, k_y) = \frac{1}{(2\pi)^2} \iint E(x, y) e^{-i(k_x x + k_y y)} dx dy \tag{10.1}$$

$$S(k_x, k_y, z) = F(k_x, k_y) e^{i\sqrt{4(\omega\sqrt{\varepsilon}/c)^2 - k_x^2 - k_y^2} \cdot (z_0 - z)} \tag{10.2}$$

$$E_R(x, y, z) = \iint S(k_x, k_y, z) e^{i(k_x x + k_y y)} dk_x dk_y \tag{10.3}$$

where:
  $F(x, y)$ is the plane wave decomposition of the complex conjugate to hologram $E(x, y, z)$ registered at the interface $z = 0$.
  $S(k_x, k_y, z)$ expresses the propagation of this pattern back to the focusing plane.
  $E_r(x, y, z)$ is the reconstructed image.
  ω is the temporal angular frequency.
  ε is the dielectric permittivity of the medium.
  c is the speed of light.
  k_x and k_y are the spatial frequencies corresponding to the x and y dimensions, respectively.

10.2.2 Spatial and Frequency Sampling Requirements

Generally, the holographic radar assumes a homogeneous scanned medium with a constant propagation velocity v. Implementing the mathematical integral formulation in Equations 10.1, 10.2, and 10.3 requires considering continuous spatial sampling at a discrete grid of points. The imaging process defines the spatial sampling according to the well-known Nyquist criterion. The Nyquist criterion is satisfied if the phase shift from one sample point to the next is less than π radians. For a spatial sample interval of Δx, the worst case will have a phase shift of not more than $2k_x \Delta x$, with similar formulation along the y coordinate. Therefore, the derived spatial sampling criterion for shallow objects is as follows:

$$\Delta x < \frac{\lambda}{4}; \quad \Delta y < \frac{\lambda}{4} \tag{10.4}$$

TABLE 10.1

Holographic Radar Characteristics and Sampling Requirement

Parameter	Rascan-4/2000	Rascan-4/4000	Rascan-4/7000
Frequency range, GHz	1.6–2.0	3.6–4.0	6.4–6.8
Number of operating frequencies		5	
Number of recording signal polarizations		2	
Resolution in the plane of sounding at shallow depths, cm	4	2	1–5
Maximal sounding depth R_{max} (depends on medium permittivity), cm	35	20	15
Spatial sampling $\Delta x < \lambda_c/4$; $\Delta y < \lambda_c/4$ (hp: $v = c/\sqrt{\varepsilon} = c/3 = 10^8 m/s$), cm	<1.3	<0.65	<0.38
Frequency sampling $\Delta f < v/(4R_{max})$, MHZ	71	125	166

where:

$\lambda = 2\pi/k$ is the wavelength

k is the spatial frequency

The frequency sampling requires that $2\Delta k \, R_{max} < \pi$, where R_{max} is the maximum target range, which means $\Delta f < c/(4R_{max})$. Table 10.1 presents the sampling requirements for three different models of RASCAN holographic radar. We can observe that the spatial sampling has to be done in steps less than 13 mm for the lower frequency and 3.8 mm for the highest frequency. It is straightforward that the high-resolution capability can only be achieved if the spatial sampling has a precision much better than these limits. The manual scanning method is intrinsically prone to errors and it is hard to reach these requirements even with skilled operators. Moreover, the acquisition of data from large areas involves a time-consuming process with consequences for the cost of the survey.

10.3 Holographic Radar and Electronic Signal Processing

The detection of a target by holographic radar relies on the complex reflection coefficient of the target and also the distance from the monostatic transmitter-receiver (see Figure 10.1). The measurement of the phase and magnitude (or I and Q components) of the received signals is necessary to allow a complete holographic reconstruction of subsurface targets. The acquisition of both I and Q components has been implemented in a recent version of the holographic radar that can also operate at higher frequency (7 GHz) than previous versions. The block scheme of the hardware is shown in Figure 10.4.

To maintain compatibility with earlier versions, the first version of this system achieved manual scan positioning using a wheel and optical encoder with the scan started by pressing a button on the bottom left of the transducer in Figure 10.2. The present version has enhanced versatility by programming an interface with a PC through a USB port, which allows automated collection of precisely located I/Q data for processing into a full 3D holographic reconstruction.

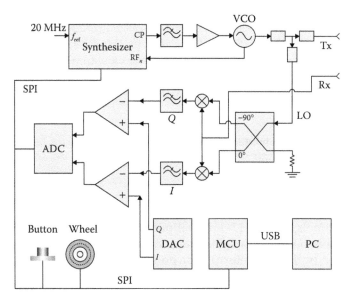

FIGURE 10.4
Block diagram of the electronic system for I/Q component acquisition with holographic radar.

10.4 Practical Scanning Methods and Trade-Offs

10.4.1 Scanner Design Objectives and Constraints

The scanning of large areas (several square meters) is of particular importance for inspecting civil engineering structures such as floors, pavements, and corridors; for minimum metal antipersonnel landmine detection; and shallow imaging of archaeological sites. Even the inspection of smaller areas as in the nondestructive testing (NDT) of engineering materials or of historic artworks or architectural elements requires scanning methods that must ensure adequate speed with high resolution and accuracy. The spatial accuracy is very important in holographic radar imaging because the phase coherence must be preserved in order to reconstruct the full hologram.

In the laboratory, it is possible to design accurate mechanical scanning systems with submillimeter accuracy and millimeter resolution that are adequate for preserving signal coherence. Moreover, the cost of such systems is prohibitive and limits the scanner applicability to areas with typical size of about a square meter. Generally, 2D scanners work best to spatially sample flat surfaces. For 3D surfaces, alternate solutions include mounting the radar head on an articulated robotic arm.

In outdoor environments, the proposed solution is a robotic platform that guarantees sufficient accuracy in the spatial sampling and requires little set up time. In our robotic platform, the antenna moves laterally (across-track) with a mechanical scanner, whereas the entire system moves (along-track) using four wheels with encoders attached to the axes of brush-less DC motors. This robotic configuration is implemented in a proof-of-concept prototype.

10.4.2 Robot Object Scanner (ROS) Design and Realization

Figure 10.5 shows the block diagram of the prototype robotic scanner. The open architecture allows the connection of different electronic units (custom or commercial) interfaced by an RS485 bus. This standard bus is simple to manage and robust for connections with low data rate as in the case of this application. In Figure 10.5, there are three main units as follows:

1. Communication and master control board
2. Movement control system
3. Lateral scanning control system

The Communication and master control board has eight analog input channels, eight analog output channels, and an I/O digital port based on the PIC18F6722 (microchip) microcontroller. This board allows the transmission of commands to other electronic units and reception of data by interrupt or polling methods. It also establishes remote communication using a Bluetooth version 3 protocol. This board acquires the position of the driving wheels connected to DC motors from the Movement Control Board, and the lateral position of the scanner at the front of the robotic platform as shown in Figure 10.5. The board acquires simultaneously the positioning data, the horizontal and cross polarization data from the two holographic radar channels demodulated at low frequency (up to 2 kHz bandwidth), and the tri-axial accelerometer and the scanner height from

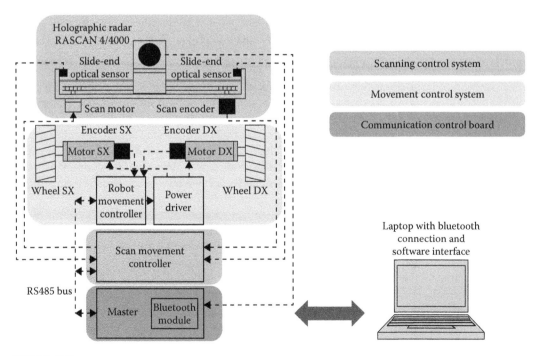

FIGURE 10.5
Block scheme of the robotic platform with interface for holographic radar and other sensors.

FIGURE 10.6
Lateral mechanical movement (along *y*) of the holographic radar head (RASCAN-4/4000) mounted on the robotic scanner. The electronics unit is mounted in a sealed PVC box with connectors for external battery, motors, and sensors.

the ultrasonic channel [22]. The movement electronic unit calculates the assigned trajectory (linear, piecewise linear, or circular) and minimizes the positioning errors with a Lyapunov controller. Similarly, the lateral scanner provides position control of the radar head, covering a swath about 28 cm wide. For each acquired position, all the information is packed into a custom frame structure and sent via the serial protocol to the remote PC (Figure 10.6).

Because the electromagnetic properties of the environment can change in time and space, the robotic system design can host multiple sensors, which allows different data to be spatially correlated. The signal processing of this information can enhance the detectability, positioning, and characterization of shallow targets. Essentially, the robotic platform has been developed in an open manner, with the option of using additional sensors, in addition to the more advanced versions of holographic radar with I/Q output channels, as previously described.

A possible set of single-point sensors useful for mapping environment parameters are as follows:

1. Metal detector for discriminating metallic versus nonmetallic targets imaged by radar.

2. Low-frequency ultrasonic transducers for ranging the air gap between the antenna and soil: this information is necessary to correct the disturbance (phase shift) effects on the holographic radar image.

3. Noncontact sensors for soil temperature: high-sensitivity pyroelectric or silicon sensors can detect temperature gradients within the soil surface.

4. Noncontact electromagnetic sensors to measure soil moisture [23].

5. An optical or infrared camera to provide images that can be overlaid over the radar hologram.

Additional location systems are under development, but not yet integrated in the prototype:

- Optical systems for absolute positioning of the radar head for use in indoors or in an environment in the presence of EMI restrictions. These would supplement the robot's position detected by the encoder.
- High-accuracy GPS Sensors for use in open areas.

The surface relief effects due to the air gap can be compensated by adding a phase term into the reconstruction algorithm of the holographic image. The problem here is the accuracy of the measurement of the air gap with ultrasonic methods because the wavelength in air dictated by the range resolution for the phase compensation (a few mm) becomes comparable with irregularities of the soil surface (stones, holes, etc.). In order to mitigate this problem, an arrangement of two cross hemi-cylindrical sensors operating at 100 kHz with real-time acquisition and a ranging algorithm has been developed [22].

10.5 Examples of Real-Time High-Resolution Holographic Imaging with Robotic and Automatic Scanners

10.5.1 High-Resolution Imaging for Rapid Investigation *in situ* of Structures Made of Dielectric Materials

The initial experiment tested the scanner's ability to obtain a high-resolution image by scanning two metal letters "W and I" with a total expanse of 210 mm and height 110 mm. These two letters were fixed on the bottom of the plain wooden table in Figure 10.6 with 30 mm thickness. The two letters were placed close to the central longitudinal iron bar of the table that has a dimension of 20 mm along y-axis.

The robotic scanner was set to scan at 5 frequencies, 3.6, 3.7, 3.8, 3.9, and 4.0 GHz, over a length of 380 mm along the x direction. Spatial sampling along both the x and y directions was 5 mm. The robotic scanner moved along the x direction, which corresponds to the cross-polar receiver channel, while the direction y corresponds to the parallel receiver. Assuming a relative permittivity of the wooden plank of 5, the corresponding wavelength at 4 GHz is $\lambda_{\text{wood}} = c/f\sqrt{5} = 3 \times 10^8 / (4 \times 10^9 \sqrt{5}) = 33.5$ mm.

Figure 10.7 compares the optical scan of the two metal letters "W I" with the radar image at the frequency of 3.6 GHz. The image was obtained in real time and the total time for the experiment was about 3 minutes.

The interesting result is that the obtained detected image of the shape of the two letters had a resolution comparable with the expected theoretical one (see Table 10.1), namely: $\lambda_{\text{wood}}/4 = 33.5/4 = 8.3$ mm. The spatial coherence of the acquired data allows the reconstruction of the shape with only small distortion. In particular, the straight line of the metal bar is well reconstructed and a slight inclination of the robot trajectory with respect to the bar can be seen. Moreover, the possible subtle variations in dielectric due to the wood grain may have produced the small difference in the image reconstruction of the "W."

Finally, the beam divergence at the depth of 30 mm tends to enlarge the size of the object represented by the holographic image. For a more quantitative evaluation of the resolution, the software interface on the laptop PC provides a cross-hair for the generation of x–y section of the image. We selected the central image of Figure 10.7 because the cross-polar

FIGURE 10.7
Comparative optical and radar scan. (Left) The optical scan of the metal letter glued on a paper sheet and a 20 cm ruler. (Center) Cross-polar channel at 3.6 GHz. (Right) Parallel channel at 3.7 GHz.

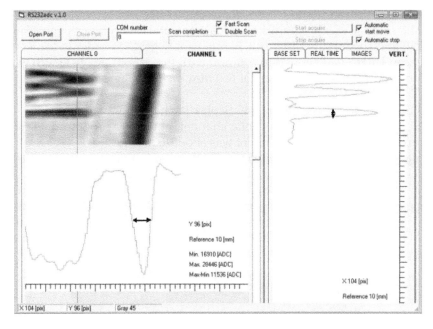

FIGURE 10.8
Evaluation of two targets width: "I" letter on the right and metal bar on the bottom left. The diagram reference is 10 mm.

channel provides a more accurate profile of the metal bar. In Figure 10.8, it is possible to estimate the width of the letter I and the metal bar: the -6 dB width of the metal bar is 32 mm and the width of the "I" is 19 mm against the real width of 21 mm. In both cases, the dimension has been estimated correctly within the theoretical resolution of the 4 GHz holographic radar. Such results are difficult to achieve with manual scanning since the operator would have to follow an ideal line with a deviation less than 8.3 mm. Even, if possible, for an expert operator in laboratory on a small area, it is very difficult to achieve this in the field when investigating area with dimensions in meters.

10.5.2 Outdoor Experiments for Landmine Detection

It is intended that the robotic scanner will find applications to surveys in critical/risky environments, such as the detection of minimum-metal antipersonnel landmines [5,10]. This application requires a high radar probability of detection and low rate of false alarms. The enhanced robotic platform positioning precision will enhance the performance. In such environments, one of the main issues will be the variability of soil characteristics (moisture content, grain size variations, surface roughness, etc.), and signal-processing strategies will play a key role in mitigating detrimental effects.

A demonstration of the developed system searched for inert landmines and clutter objects buried for several months at shallow depth in natural sandy soil in the outdoor test bed at the University of Florence in Italy. After several months, the originally flattened soil surface had roots and grass present at the edges of the area. The irregular surface had undulations up to 3 cm height as shown in Figure 10.9.

The buried objects in this outdoor test bed have already been described in a previous work [24], but in this experiment, their relative positions were found to have changed by natural settling of the sandy soil.

The robotic scanner was programmed for fast operation with two frequencies (3.6 GHz and 3.9 GHz) and spatial sampling on a 10 mm square grid. In order to limit cumulative errors, the scan length was limited to 650 mm with the overall image length of 1300 mm covered in two successive acquisitions of 650 mm each. The width covered by the lateral movement is 250 mm.

The real-time images obtained on the display of the remote laptop PC are shown in Figure 10.10 for the parallel and cross-polar channels at the two programmed frequencies. The total time to complete the experiment was about 18 minutes.

The results obtained and shown in Figure 10.10 indicate that the scanner has adequate spatial resolution to describe the circular shape of the iron disk and enough sensitivity to detect an empty plastic bottle. A circular plastic box with almost the same diameter as the iron disk, and buried close to it, is more difficult to detect because of the much lower dielectric contrast with respect to the sandy soil. Moreover, the surface irregularities at the edges of the iron disk generate additional noise that masks the reflection from the low reflectivity plastic box.

FIGURE 10.9
The robotic scanner searching the test bed of the University of Florence. Inert mines and clutter had been buried for several months before the test.

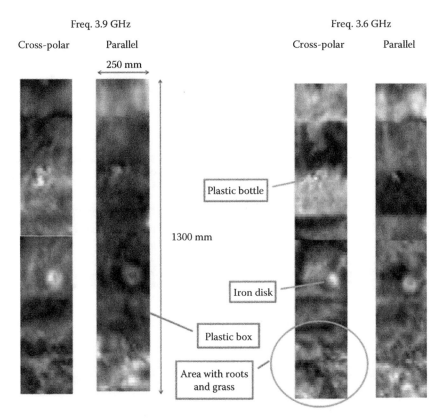

FIGURE 10.10
Results of scanning surrogate mines and clutter in the outdoor test bed.

Apart from possible improvements in the selection of the lower operating frequency of the radar (e.g., 2 GHz) for achieving better penetration and the full reconstruction of the hologram with I/Q signal components, the experiment demonstrated the ease and efficiency of high-resolution imaging using holographic radar even in an outdoor environment.

10.5.3 Nondestructive Test and Evaluation of Thermal Insulation for the Aerospace Industry

10.5.3.1 Objectives in Thermal Insulation Inspection

The Space Shuttle Columbia disaster of February 1, 2003 killed all seven crew members. Figures 10.11 and 10.12 show the launch and recovered debris laid out during the investigation. This and other incidents, which fortunately did not lead to such catastrophic consequences, have aroused interest in the development of new methods for nondestructive testing of insulation and thermal protection coatings on spacecraft and fuel tanks [25,26,27,28,29,30].

In the opinion of NASA investigators, one of the causes of the Columbia disaster was voids in the thermal protection coating on the shuttle's external fuel tank [31,32] as shown in Figures 10.13 and 10.14. The external tank contains liquid oxygen and hydrogen propellants stored at −183 and −253°C, respectively. To reduce fuel vaporization and prevent icing of tank surface that could fragment and damage the shuttle, the tank has an insulating

FIGURE 10.11
Space Shuttle *Columbia* take-off on January 16, 2003.

FIGURE 10.12
Result of *Columbia* disaster on February 1, 2003. Remains of the spaceship collected on the ground were laid-out in a hangar.

polyurethane foam coating [33]. The thickness of the foam is within the range of 25 mm to 50 mm [34]. If the super-cold external tank is not sufficiently insulated from the ambient warm and moist air, then atmospheric water vapor condenses inside the foam voids.

According to this hypothesis, during the launch of Columbia's 28th mission, water that had condensed inside voids rapidly vaporized (boiled) as a result of lowering pressure with increasing altitude following launch [32]. As a result of this explosive boiling, a piece of foam insulation broke off from the external tank and struck the left wing, damaging heat-protective leading edge panel. When Columbia reentered the atmosphere after the mission,

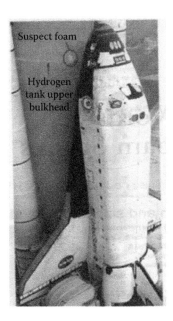

FIGURE 10.13
Space Shuttle Columbia on the launch pad. Yellow design elements are thermal insulation coating of the external cryogenic tank. (From *Aviation Week & Apace Technology*, 27–28, 2003.)

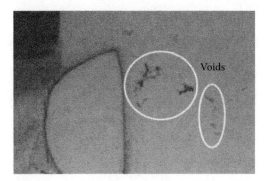

FIGURE 10.14
Cross section of thermal protection coating of shuttle external tank. Voids are visible inside of insulation polyurethane foam. (From *Aviation Week & Space Technology*, 31, 2003.)

this damage allowed plasma (produced ahead of the craft during its flight in stratosphere) to penetrate and destroy the wing structure, causing the spacecraft to break up as shown in Figure 10.12. Most previous shuttle launches had seen similar, but more minor, damage and foam shedding, but the risks were deemed acceptable as shown in Figure 10.15 [35].

It is well known that the tiled thermal protection coatings of return vehicles like the Space Shuttle are exposed to high mechanical, and especially thermal, influence on reentry. In fact, after the first flight of Columbia (April 12, 1981), 16 tiles were lost and 148 tiles were damaged [31]. Similar problems with more serious after-effects arose after the first and only flight of the Russian *Buran* on November 15, 1988. Postflight inspection showed partial destruction to complete loss of thermal shielding tiles as shown in Figure 10.16 [27]. Such damage could lead to a repeat of the *Columbia* disaster in future missions for similar spacecraft.

FIGURE 10.15
A torn off piece of the insulating foam falls along the surface of the external fuel tank. (From Report of Columbia Accident Investigation Board, 2003)

FIGURE 10.16
Destruction of the three tiles placed directly behind 21st section of the wing leading edge of the spacecraft *Buran*. (From *Aviation Week & Space Technology*, 31, 2003.)

Shedding of thermal insulation is connected with impurities and/or insufficient quality control in the bonding of foam or tiles to a space vehicle surface. Gluing tiles is carried out manually, and in these circumstances, it is difficult to maintain the necessary quality control. A variety of control methods are described in detail in [27], mainly involving destructive testing by "tearing off."

Widely applied ultrasonic diagnostic methods for nondestructive testing of different constructions [25] are insufficiently effective for foam insulation diagnostics due to polyurethane's high porosity, which leads to high levels of incoherent acoustic scattering and attenuation [36]. Similar considerations apply to the silicate fiber tiles that shield the outer surface of the Space Shuttle and *Buran*.

Microwave diagnostics using holographic subsurface radars [28,37] could become a good alternative to ultrasonic testing. The basic advantage of microwave diagnostics in comparison with ultrasonic ones is the fundamental difference in physical properties affecting the propagation of electromagnetic versus acoustic waves in heterogeneous media. Electromagnetic waves reflect from inhomogeneities only when their dielectric contrast is sufficient. Thus, electromagnetic waves propagate practically without loss in porous materials such as polyurethane foam insulation wherein the dielectric of air in pores almost matches that of the matrix foam [26]. Moreover, the pore dimensions are much smaller than the length of electromagnetic wave, so the foam can be considered as the continuous medium.

The thermal protection tiles received hydrophobization to prevent moisture penetration. New tiles are checked by immersing them in water for 24 hours followed by weighing [27]. However, this method is not suitable for testing of tiles already installed on the spacecraft, especially for critical postflight inspection. However, in this case, it is possible to use microwave methods. In particular, this example describes the use of holographic radar, which has a high sensitivity to the presence of moisture due to the high relative dielectric constant of water (approximately 80) relative to air or foam (approximately 1) [38].

10.5.3.2 Radar Inspection of a Thermal Protection Coating Sample

The test sample of thermal insulation with artificial flaws used polyurethane foam of 40 mm thickness glued on an aluminum alloy plate of 5 mm thickness. Design of the sample and positions of the flaws are shown in Figure 10.17. The sample form was chosen to mimic the actual construction of spacecraft fuel tank insulation coating.

The sample dimensions were 500 by 400 mm and made in two stages. On the first stage, the central circle, 270 mm in diameter, was sprayed with adhesive, but with three round cuts on the bottom surface of the foam. At the cuts, prime coating and glue are missing (usual total thickness = 200 microns). Instead, the inner surface of the cuts had the prime coating and glue. On the second stage, the rest of the sample was filled. This sample imitated defective gluing on the foam–metal contact.

10.5.3.3 Experimental Results

The experiment used the highest available frequency version of RASCAN-5 holographic radars, with an operating frequency of 15 GHz. The low attenuation factor of polyurethane foam for electromagnetic waves and the need for high resolution to detect small anomalies determined the frequency band choice. The dielectric permittivity of this material is essentially the same as for vacuum, that is about 1. According to data presented in [26], the complex permittivity of the foam sprayed on the external fuel tanks of the Space Shuttle is $\varepsilon = (1.05 - j0.003)$ and the density is only 4% of that for water. This is because the polyurethane foam consists mostly of pores filled by air. Laboratory experiments with RASCAN-5/15000 radar have shown that penetration depth in polyurethane foam is more than 16 cm, which with margin exceeds the Space Shuttle foam insulation thickness [34].

However, it should be noted that the properties of the bulk polyurethane material depend on the fabrication process, and may vary within wide limits, so the density of its industrial samples lay in the range of 48–287 kg/m^3 [36]. Polyurethane foam has the lowest heat conductivity among all modern materials. Depending on its density, polyurethane foam thermal conductivity varies in the range of 0.019–0.033 W/m·K. Due to these

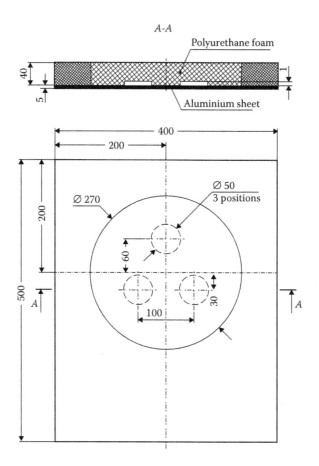

FIGURE 10.17
Sketch of the thermal coating sample showing the polyurethane foam coating and cut out regions mimicking the space craft fuel tank insulation.

characteristics, polyurethane foam is widely used as insulation material in various fields of industry, as well as military and space engineering.

Note that using frequencies around 15 GHz for diagnostics of other common structural materials is hardly possible because of the frequency-dependent attenuation of electromagnetic waves, that is generally a sharp increase in the absorption factor for frequencies above 10 GHz. Therefore, for diagnostics of building structures with concrete, wood, and plaster construction, it is more appropriate to use frequencies in the range of 1.5–7 GHz to obtain good results [38].

The experiments on thermal protection coatings used manual scanning of the sample surface as shown in Figure 10.18 [39]. It required about 4–5 min to scan the sample with dimensions of 40 by 50 cm as shown in Figure 10.17. The robotic scanner described earlier could have performed the inspection faster. The complex microwave hologram (real and imaginary components) and the digital reconstruction (1)–(3) for a signal frequency of 14.6 GHz are shown in Figure 10.19.

The procedure for hologram reconstruction using the algorithm described previously is very fast – requiring less than 1 second. One can readily see on these microwave images the three flaws, and the round sample border created by the two-stage production of the test sample.

FIGURE 10.18
Scanning of thermal insulation coating sample.

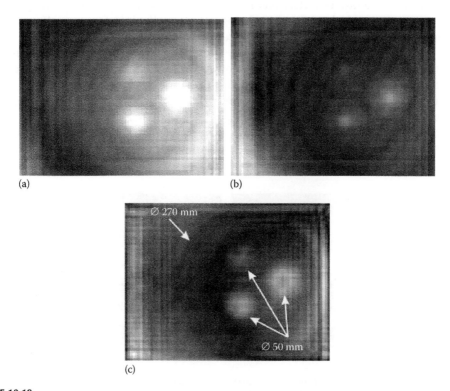

FIGURE 10.19
Result of the experiments in Figure 10.18: (a) real part of the hologram, (b) imaginary part of the hologram, and (c) the reconstructed image.

The vertical and horizontal stripes in Figure 10.19 are the result of the reflection of electromagnetic waves, emitted by the radar in the medium, from the end faces of the sample insulation, primarily from the border of the underlying aluminum plate, that is they are edge effects. Cyclical variation with distance in the reflected wave phase combined with the constant phase of the reference signal forms a striped pattern. This phenomenon is essentially the "zebra effect" that was described and explained in [40].

The examination of heat protection coatings and thermal insulation of space vehicles glued to a metal fuselage is a specialized task for radar system. This is because the metal surface is a perfect mirror for microwaves. The specular reflection ensures certain specificity in the process of hologram reconstruction and requires that this be accounted for when interpreting the results.

An experiment demonstrating the processes occurring in a microwave-transparent material (polyurethane foam) above a perfectly reflective metal surface used the configuration of Figure 10.20. Metal spokes inserted through the side of a sample are labeled on the left of this figure as 1 and 2. Spoke 1 was inserted parallel with the metal surface at a depth of 13 cm (distance above the metal surface of 2.3 cm). The second spoke was inserted in the polyurethane foam at an average depth of 13.5 cm but with a slight dip relative to the surface to demonstrate the above-mentioned "zebra effect" [40].

To determine the true depth of the objects inside the foam, it was necessary to take into account the 3 mm Plexiglas sheet that covers the scanning surface for improving the sliding of the radar head. Another important effect is related to the presence of the metal plate that underlies the polyurethane foam and reflects the microwave as an ideal mirror. The same as in optics, the mirror reflection of spokes must be observed behind the plate.

Upon reconstruction of the microwave holograms from the interference pattern recorded by the radar, the image is focused at a distance specified by the parameter z (1)–(3). The distance, at which the operator-determined "best focusing" for a target occurs, is presumed to be its real depth, $z = z_0$. Results of microwave hologram reconstruction for different depths in the experiments with the spokes are shown in Figure 10.21.

In Figure 10.21, the two top images correspond to the original complex holographic interference pattern recorded in two radar signal components; real and imaginary. Images below correspond to successively deeper hologram reconstructions: where depth $z = 0$, 2.7, 4.7, and 6.7 cm, respectively. Depth 2.7 cm corresponds approximately to the depth of

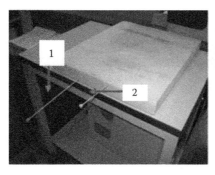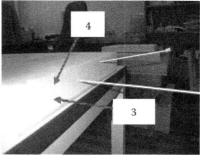

FIGURE 10.20

An experiment demonstrating the processes occurring in a microwave-transparent material (polyurethane foam) above a perfectly reflective metal surface. The radar scanned for metal spokes embedded into the foam covering of the sample: (1) spoke 1, (2) spoke 2, (3) a 5-mm aluminum alloy plate, and (4) a 40-mm-thick polyurethane foam layer.

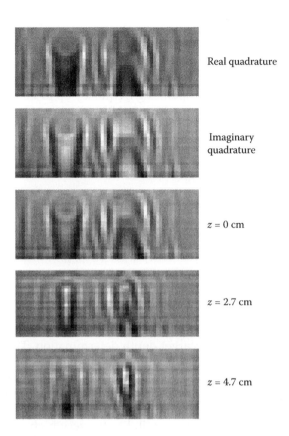

Real quadrature

Imaginary quadrature

$z = 0$ cm

$z = 2.7$ cm

$z = 4.7$ cm

FIGURE 10.21
The results of experiments with spokes inserted into foam as shown in Figure 10.20. The original complex holographic interference pattern (real and imaginary) is at the top, with successively deeper hologram reconstructions below. Depth 2.7 cm corresponds approximately to the depth of the object, 4.7 cm is the position of the underlying metal surface, and 6.7 cm is a virtual image that corresponds to a mirror reflection of the spokes from surface of the metal surface.

the object, 4.7 cm is the position of the metal surface, and 6.7 cm is a virtual image that corresponds to mirror reflection of the spokes from surface of the metal plate.

It is necessary to note that the contrast of the virtual image of spoke 1 at 6.7 cm with respect to the background has changed to the opposite sign in comparison with the real image of the object at a depth of 2.7 cm. This effect is associated with the phase inversion of the electromagnetic wave by 180° upon its reflection from the surface of the metal [41], leading to the polarity reversal in the virtual image contrast pattern. Such complexities indicate needed further investigation to understand the details of microwave holograms recorded in dielectric media located above a metal surface.

Diagnosis of thermal insulation and heat protection coatings glued onto a metal surface is a very specialized task since the metal surface fully reflects the electromagnetic waves. For impulse radars, such targets are characterized by multiple reflections of signal between radar antenna and metal substrate, which obscure the desired target in the recorded impulse radar images. Although holographic radar is free from this shortcoming, the presence of a perfectly specular reflective metal substrate beneath desired targets must be considered when interpreting the results of holographic imaging.

These experiments showed that the proposed diagnostic method for thermal insulation using holographic subsurface radars allows detection of internal defects within the coating. However, in the microwave images, the detected defects are subtle due to the low permittivity contrast between the defects and surrounding polyurethane.

It is possible that by increasing the operating frequency up to 24–25 GHz could enhance the radar sensitivity. This possibility is suggested by experiments with a lower frequency Rascan-5/7000 radar frequencies in range of (6.4–6.8 GHz), which did not detect the defects in the sample used for these experiments. This suggests that increasing the radar operating frequency should enhance both spatial resolution and sensitivity, probably with little loss of penetration due to negligible loss in the low electrical conductivity and low-scattering insulating materials.

10.6 Conclusions

Holographic subsurface radars provide specific advantages and disadvantages relative to impulse radar. Advantages include the possibility for very high-resolution, high-sensitivity plan-view imaging in real time, with the additional possibility for near-real-time full hologram reconstruction and associated target depth determination. The principal disadvantage is the inability to apply time-varying gain to enhance reflections from deeper targets (since the radar is continuous wave). However, this disadvantage is not noticeable for very low electrical conductivity media such as foam, and dry wood or sand (or other earth materials).

In order to take full advantage of the possibility for very high-resolution imaging, the positioning of the radar scanning head requires high precision, achieved with the robotic scanner described above. In addition, it can completely scan large areas otherwise impractical and/or tedious (and expensive) for hand-scanning. Finally, the robotic platform design is compatible with additional sensors to measure independent physical properties of subsurface targets. This will allow better detection rates for targets that may be subtle given one specific property (e.g., permittivity), but more distinct on another (e.g., acoustic velocity), and provide the possibility to more positively identify subsurface radar targets (e.g., discriminating landmines from clutter to lower false alarm rates). The authors' current work is focused on multisensor scanning systems.

Finally, the continuous wave holographic radar provides the unique ability to image targets on or near highly reflective surfaces. An example of this is detection of flaws in the attachment of heat-protective tiles to a metal spacecraft body – a task hardly possible for impulse radars.

References

1. Daniels, D. J. (ed.), *Ground Penetrating Radar 2nd Edition*, Vol 1, IET, London, 2004.
2. Taylor, J.D. (ed.), *Ultrawideband Radar: Applications and Design*, CRC Press, Boca Raton, FL, 2012.
3. Ivashov, S.I., Capineri, L. and Bechtel, T.D., Holographic subsurface radar technology and applications, in Taylor, J.D. (ed.), *UWB Radar. Applications and Design,* CRC Press, Boca Raton, FL, 2012, pp. 421–444.

4. Capineri, L., Falorni, P., Borgioli, G., Bulletti, A., Valentini, S., Ivashov, S., Zhuravlev, A., Razevig, V., Vasiliev, I., Paradiso, M., Windsor, C. and Bechtel, T. Application of the RASCAN Holographic Radar to Cultural Heritage Inspections. *Archaeological Prospection*, Vol 16, 2009, pp. 218–230.

5. Ivashov, S.I., Razevig, V.V., Vasiliev, I.A., Zhuravlev, A.V., Bechtel, T.D. and Capineri, L., Holographic Subsurface Radar of RASCAN Type: Development and Applications, *IEEE Journal of Selected Topics in Applied Earth Observations and Remote Sensing*, Vol 4, No 4, 2011, pp. 763–778.

6. Roberts, R., Corcoran, K. and Schutz, A., Insulated concrete form void detection using ground penetrating radar. *Structural faults and repair conference*. Edinburgh, Scotland, UK. 2010.

7. Pasternak, M., Miluski, W., Czarnecki, W. and Pietrasinski, J., An optoelectronic-inertial system for handheld GPR positioning, *15th International Radar Symposium (IRS)*, 2014, pp. 1–4.

8. Trela, C., Kind, T. and Schubert, M., Positioning accuracy of an automatic scanning system for GPR measurements on concrete structures, *14th International Conference on Ground Penetrating Radar (GPR)*, 2012, pp. 305–309.

9. Falorni, P. and Capineri L., "Optical method for the positioning of measurement points", accepted to the *International Workshop on Advanced Ground Penetrating Radar (IWAGPR) 2015*, Florence, July 7–10, 2015, in print.

10. Arezzini, I., Calzolai, M., Lombardi, L., Capineri, L. and Kansal, Y., Remotely controllable robotic system to detect shallow buried objects with high efficiency by using an holographic 4 GHz radar, *PIERS Proceedings*, 1207–1211, March 27–30, Kuala Lumpur, Malaysia, 2012.

11. Capineri, L., Arezzini, I., Calzolai, M., Windsor, C.G., Inagaki, M., Bechtel, T.D. and Ivashov S. I., High resolution imaging with a holographic radar mounted on a robotic scanner, *PIERS Proceedings*, pp. 1583–1585, August 12–15, Stockholm, 2013.

12. Capineri, L., Razevig, V., Ivashov, S., Zandonai, F., Windsor, C., Inagaki, M. and Bechtel, T. RASCAN holographic radar for detecting and characterizing dinosaur tracks, *7th International Workshop on Advanced Ground Penetrating Radar (IWAGPR)*, 2013, pp. 1–6.

13. Razevig, V., Ivashov, S., Vasiliev, I. and Zhuravlev, A., Comparison of Different Methods for Reconstruction of Microwave Holograms Recorded by the Subsurface Radar, *Proceedings of the 14th International Conference on GPR*, June 4–8, Shanghai, China, 2012, pp. 335–339.

14. Finkelstein M.I., Subsurface radar: principal problems of development and practical use, *Proceedings of the 11th Annual International Geoscience and Remote Sensing Symposium*, Espoo, Finland, June 3–6, 1991, Vol 4, pp. 2145–2147.

15. Chapursky, V.V., Ivashov, S.I., Razevig, V.V., Sheyko, A.P., Vasilyev, I.A., Pomozov, V.V., Semeikin, N.P. and Desmond, D.J., Subsurface radar examination of an airstrip, *Proceedings of the 2002 IEEE Conference on Ultra Wideband Systems and Technologies, UWBST'2002*, May 20–23, 2002, Baltimore, MD, pp. 181–186.

16. Iizuka, K. and Freundorfer, A.P., Detection of nonmetallic buried objects by a step frequency radar, *Proceedings of the IEEE*, Vol 71, No 2, February 1983, pp. 276–279.

17. Junkin, G. and Anderson, A.P., Limitations in microwave holographic synthetic aperture imaging over a lossy half-space, *Radar and Signal Processing, IEEE Proceedings F*, Vol 135, No 4, August 1988, pp. 321–329.

18. Zhuravlev, A.V., Ivashov, S.I., Razevig, V.V., Vasiliev, I.A., Türk A.S. and Kizilay, A., Holographic microwave imaging radar for applications in civil engineering, *Proceedings of the IET International Radar Conference*, April 14–16, 2013, Xian, China. pp. 14–16.

19. Remote Sensing Laboratory, Nondestructive Testing Devices, RASCAN-4/4000 radar (4 GHz), http://www.rslab.ru/english/product/rascan4/result/animation.

20. Gabor. D., A new microscopic principle, *Nature*, Vol. 161, 1948, pp. 777–778.

21. Razevig, V.V., Ivashov, S.I., Vasiliev, I.A., Zhuravlev, A.V., Bechtel, T., and Capineri, L., Advantages and restrictions of holographic subsurface radars. Experimental evaluation, *Proceedings of the XIII International Conference on Ground Penetrating Radar*, Lecce, Italy, June 21–25, 2010, pp. 657–662.

22. Cambini, C., Giuseppi, L., Calzolai, M., Giannelli, P. and Capineri, L., Multichannel airborne ultrasonic ranging system based on the Piccolo C2000 MCU, *Embedded Design in Education and Research Conference (EDERC) 2014*, Milano, Italy, September 11–12, 2014, pp. 80–84.

23. Olmi, R., Priori, S., Capitani, D., Proietti, N., Capineri, L., Falorni, P., Negrotti, R. and Riminesi, C., Innovative techniques for sub-surface investigations, *Materials Evaluation*, Vol 69, No 1, pp. 89–96, 2011.

24. Borgioli, G., Bulletti, A., Calzolai, M. and Capineri, L., Detection of the vibration characteristics of buried objects using a sensorized prodder device. *IEEE Trans. Geoscience and Remote Sensing*, Vol 52, No 6, June 2014, pp. 3440–3452.

25. Capineri, L., Bulletti, A., Calzolai, M., and Francesconi, D., Lamb wave ultrasonic system for active mode damage detection in composite materials, *Chemical Engineering Transactions*, Vol 33, 2013, pp. 577–582.

26. Kharkovsky, S. and Zoughi, R., Microwave and millimeter wave nondestructive testing and evaluation, *IEEE Instrumentation & Measurement Magazine*, April 2007, pp. 26–38.

27. Gofin, M.Ya., *Heat resisting and thermal protecting systems of reusable space ships*. Moscow Aviation Institute, 2003. 672 p. (in Russian).

28. Ivashov, S.I., Vasiliev, I.A., Bechtel, T.D. and Snapp, C., Comparison between impulse and holographic subsurface radar for NDT of space vehicle structural materials, *Progress in Electromagnetics Research Symposium 2007*, Beijing, China, March 26–30, 2007, pp. 1816–1819.

29. Ivashov, S., Razevig, V., Vasiliev, I., Bechtel, T. and Capineri, L., Holographic subsurface radar for diagnostics of cryogenic fuel tank thermal insulation of space vehicles, *NDT & E International*, Vol 69, January 2015, pp. 48–54.

30. Ivashov, S., Razevig,V., Vasiliev, I., Zhuravlev, A., Bechtel, T. and Capineri, L., Non-destructive testing of rocket fuel tank thermal insulation by holographic radar, *Proceedings of the 6th International Symposium on NDT in Aerospace*, Madrid, Spain, November 12–14, 2014.

31. Report of Columbia Accident Investigation Board, 2003.

32. *Aviation Week & Space Technology*. April 7, 2003, p. 31.

33. *Aviation Week & Space Technology*. February 17, 2003, p. 27, 28.

34. *Aviation Week & Space Technology*. October 4, 2004, p. 58.

35. *Aviation Week & Space Technology*. August 1, 2005, cover page.

36. Dombrow, B.A., *Polyurethanes*, Reinhold Publishing Corporation, New York, 1957.

37. Lu, T., Snapp, C., Chao, T.-H., Thakoor, A., Bechtel, T., Ivashov, S. and Vasiliev, I., Evaluation of holographic subsurface radar for NDE of space shuttle thermal protection tiles, *Sensors and Systems for Space Applications. Proceedings of SPIE*, Volume 6555, 2007.

38. Ivashov, S.I., Razevig, V.V., Vasiliev, I.A., Zhuravlev, A.V., Bechtel, T.D., and Capineri, L., Holographic Subsurface Radar of RASCAN Type: Development and Applications, *IEEE Journal of Selected Topics in Earth Observations and Remote Sensing*, Vol 4, No 4, December 2011. pp. 763–778.

39. Gofin, M.Ya., BURAN Orbital Spaceship Airframe Creation: The Heat Protection Structure of the Reusable Orbital Spaceship, http://www.rslab.ru/downloads/scan.avi.

40. Inagaki, M., Windsor, C.G., Bechtel, T.D., Bechtel, E., Ivashov, S.I. and Zhuravlev, A.V., Three-dimensional views of buried objects from holographic radar imaging, *Proceedings of the Progress in Electromagnetics Research Symposium, PIERS 2009*, Moscow, Russia, August 18–21, August 2009, pp. 290–293.

41. Born, M. and Wolf, E., *Principles of Optics: Electromagnetic Theory of Propagation, Interference, and Diffraction of Light*, Cambridge University Press, UK, 1959.

11

Ultrawideband Sense-through-the-Wall Radar Technology

Fauzia Ahmad, Traian Dogaru, and Moeness Amin

CONTENTS

11.1 Introduction

Sense-through-the-Wall (STTW) radar has emerged as a viable technology for providing high-quality imagery of enclosed structures in a variety of important civilian and military applications. STTW radar employs electromagnetic (EM) waves to penetrate through building wall materials. Owing to its "see"-through-walls ability, STTW radar has garnered much attention in the last decade. This has led to advances in both algorithmic and component technologies that address challenges specific to STTW radar and allow proper imaging and reliable image recovery. STTW applications include military surveillance, law enforcement, and search-and-rescue operations.

The large number of papers, special journal issues, conference sessions, and two books dedicated to investigations in this area demonstrate the level of interest in advancing this technology [1,2]. Practical systems are currently being developed in several countries, while fundamental research is being carried out by academic groups and government agencies to implement new concepts and understand the performance offerings and limits of this technology.

This chapter primarily focuses on UWB imaging radar systems and discusses suitable antennas, EM phenomenology, and recently devised signal-processing techniques for sensing through walls.

STTW radar systems are usually designed to perform two main functions: obtain images of the building interior (and, related to this, reconstruct the building layout) and detect and localize targets of interest (typically humans) within buildings. Humans belong to the class of animated objects characterized by motion of the torso and limbs, breathing, and heartbeat. While both aforementioned functions may be integrated in the same sensor, the underlying operating principles are somewhat different. Obtaining high-quality radar images typically requires the transmission of waveforms with large bandwidth, as well as the combination of signals received by multiple channels (implemented as either a physical or a synthetic array; in the latter case, the radar platform is in motion). Detecting moving targets typically relies on moving target indication (MTI) or Doppler radar return processing [3], which have the important advantage of rejecting the stationary clutter in the scene, such as the radar returns from walls and furniture objects. In this operational mode, the radar platform must be stationary. It is important to emphasize though that the two functions are not necessarily mutually exclusive – high spatial resolution and station-ary clutter rejection for moving target detection are both desirable features of an STTW radar system.

The requirement for the radar waves to penetrate obstacles before reaching the target makes the STTW technology closely related to ground-penetrating radar (GPR) [4]. Thus, both types of systems operate optimally at relatively low microwave frequencies (under 4 GHz), due to the reasonably good penetration properties of the electromagnetic (EM) waves in that region of the spectrum. To achieve good range and cross-range resolution, these radar systems often employ ultrawideband (UWB) waveforms and synthetic aper-ture radar (SAR) techniques [5]. Nevertheless, there are specific differences in scattering phenomenology and signal processing between the STTW and GPR sensing modalities. In particular, STTW systems have to deal with a propagation environment characterized by strong, discrete clutter items and significant multipath returns; on the other hand, the targets may be moving (translation or otherwise), which creates an opportunity to sepa-rate them from stationary clutter by MTI or Doppler processing techniques.

Many material-penetrating UWB radar systems employ impulse-like waveforms (either baseband or modulated), in a region of the microwave spectrum between 0.5 and 4 GHz. There are several advantages to using impulse waveforms in a UWB radar:

- First, the data acquisition is very fast – in principle, only one transmitted short impulse (with duration on the order of nanoseconds) is required to create a range profile of the scene. On a related note, the narrow pulse width makes operation possible at very high pulse repetition frequencies (PRFs).

- Second, the design of both the transmitter and receiver of the UWB radar system is relatively simple, although there are some stringent requirements on the radio-frequency (RF) electronic components, as explained in a subsequent paragraph.

- Third, the impulse UWB radar can operate at very low average power levels, which reduces the chance of interfering with other RF electronic equipment – this makes the impulse radar particularly suited to short-range applications.

Another option is to operate a UWB radar with stepped-frequency waveforms. The obvi-ous disadvantage of this type of radar design is the slower data acquisition rate, which could become an issue for real-time systems trying to detect moving targets. On the other hand, stepped-frequency waveforms allow for precise control of the transmit-ted and received spectrum by notching out specific subbands that may interfere with other RF equipment (this control is much more difficult to achieve with impulse radar). Additionally, equalization techniques can be readily applied to compensate for various linear distortions along the transmission, propagation, and receiving chain, since the data are collected in the frequency domain.

Finally, UWB radar systems may use pulse compression techniques, based on linear fre-quency modulation (chirp) waveforms, as well as noise or pseudo-random (phase-modulated or frequency-hopped) sequences. These types of waveforms realize a compromise between the impulse and stepped-frequency modalities, in the sense that they allow for relatively fast data acquisition rates without the need of extremely fast sampling rates in the receiver. Radar systems based on pulse compression, as well as on stepped-frequency waveforms, can typically transmit larger average powers than impulse-based systems, which render them suitable for longer range applications. Waveforms based on noise or pseudo-random sequences have the additional advantage of being difficult to detect by adversaries, and have important application to systems that use multiple-input multiple-output (MIMO) techniques [6].

Examples of existing through-the-wall imaging radar systems based on UWB waveforms include the following:

- The Synchronous Impulse Reconstruction (SIRE) radar, developed at the U.S. Army Research Laboratory (ARL) [7]
- The Through-Wall Synthetic Aperture Radar (TWSAR), developed by Defence Research and Development Canada [8]
- The Standoff MTI, Imaging, and Scanning Radar (SOMISR), developed by Syracuse Research Corporation [9]

Note that all these systems involve relatively large antenna arrays and a significant amount of hardware, and are mounted on vehicles. Some of the smaller, portable UWB radar systems, designed specifically for the detection of moving targets, include the following:

- RadarVision, developed by Time Domain Corporation [10]
- The Xaver family of systems, developed by Camero (Israel) [11]
- The AKELA Standoff Through-Wall Imaging Radar (ASITR), developed by AKELA [12]
- Prism, developed by Cambridge Consultants, UK [13]

The distinction between the two categories is not very clear-cut; some of the imaging systems are also capable of detecting moving targets, while some of the moving target detection systems are equipped with imaging capabilities.

Similar to any UWB RF equipment, an STTW UWB radar system presents the designer with several major challenges. In terms of hardware components, one must pay particular attention to the antennas, the various RF amplifiers, and the analog-to-digital converter (ADC). Section 11.2 of this chapter discusses the radar antenna requirements. The amplifiers must have good linearity in both amplitude and phase and a low noise figure. The amplitude linearity is critical for UWB waveforms (particularly in the transmitter, where larger powers are generated), since any spurious harmonics of certain in-band frequencies may also fall inside the signal band and create distortions. The ADCs of an impulse UWB radar must sample the received signals at very high rates (on the order of 4–6 GSamples/second), which are currently at the upper end of performance achievable by electronic component technology. In fact, most operational impulse UWB radar systems use an equivalent sampling scheme, which allows them to employ ADCs with much lower maximum sampling rates. Another major challenge for UWB radar systems, in general, is the frequency allocation regulatory environment. The optimal frequency band for STTW applications, around 1–3 GHz, also happens to be one of the most crowded regions of the RF spectrum, with users that include mobile telephony, global positioning system (GPS), digital television broadcasting, and air traffic control. One mitigating factor for UWB radar operation in this environment is the very low average transmitted power (typically on the order of mW), which is spread over a wide frequency range.

Given the commonalities between STTW radar and GPR in terms of hardware design, we will not delve much on this aspect, but focus instead on the EM scattering phenomenology and recently devised signal-processing techniques that are specific to the former. Section 11.2 is the only part that deals with a radar hardware component and discusses UWB antennas for STTW applications. In Section 11.3, we describe modeling techniques for UWB imaging radar systems and how they can help us understand the EM phenomenology of these sensors. In Section 11.4, we treat the problem of through-the-wall detection of moving targets with a UWB imaging radar system. Section 11.5 presents multipath exploitation using a UWB imaging radar. Section 11.6 contains conclusions regarding STTW radar systems.

11.2 UWB Antennas for STTW Radar

UWB antenna design is an important topic of modern RF engineering and currently attracts a large amount of interest from many research groups. In this chapter, we do not attempt to provide an overview of all possible techniques and configurations used in UWB radar antenna design, but limit ourselves to examples of the antennas used by existing STTW radar systems. A much more complete account of this topic can be found in Chapter 2 of reference 1. Our discussion starts with a list of requirements that make a UWB antenna suitable for these radar systems.

One important consideration for UWB antennas is a good input impedance match over the entire operational band, in systems that are frequently characterized by a bandwidth-to-center-frequency ratio (in short, bandwidth ratio) on the order of 1:1 (or 100%). Equivalently, one would require a low S_{11} parameter (below −10 dB) or a low voltage standing wave ratio (VSWR), with the antenna terminated by the feed line impedance (typically 50 Ω). This is normally achieved by minimizing the reflections at the radiating end of the antenna, at all frequencies of interest – a large number of techniques developed to address this problem are described in the literature [14] and will not be discussed here. Another key requirement for UWB antennas, particularly for impulse-based radar systems, is phase linearity. This condition is equivalent to preserving the phase origin location, as well as ensuring constant group delay over the entire frequency band. Phase linearity is important for achieving an impulse radiated by the antenna with the desired shape and spectral characteristics; antennas that satisfy this condition are called nondispersive.

Since many of the STTW radar system operate at short ranges (up to 20 m), the antennas usually do not require very large gain; 5–10 dBi would be a typical gain at frequencies between 1 and 4 GHz. Moreover, to insure full coverage of the scene at short ranges, as well as provide large integration angle for SAR imaging systems, wide beam patterns (e.g., 90°–120°) are preferable for these antennas. Notice though that long-range applications, such as vehicle- or air-borne imaging systems, will require large-gain antennas with narrower main beams.

STTW radar applications are very sensitive to returns outside the main antenna beam, which typically do not have to propagate through walls, as compared with the returns from the targets of interest, which are highly attenuated by walls. Consequently, antennas for these applications must display low sidelobes and backlobes. A typical front-to-back ratio requirement for STTW radar is 30 dB [15].

In terms of polarization, most STTW radar antennas are linearly polarized. Phenomenological studies do not show any major differences between vertical-vertical (V-V) and horizontal-horizontal (H-H) polarizations for this application. However, exploiting the cross-polarization return has been suggested as a method for detecting particular targets of interest [16,17] – in that case, a fully polarimetric radar system is required. There are specific antenna design challenges involved in such a system, mostly related to the cross-polarization isolation of the radiating elements; 25 dB of polarization isolation has been suggested for these applications [18]. Some UWB antennas (e.g., those with spiral geometry) radiate waves with circular polarization – this modality may exhibit advantages when the transmitted wave penetrates structures with particular geometries that strongly attenuate one type of linear polarization (e.g., rebar rods or meshes).

Other characteristics of the radar antennas relate to feeding, size, ease of fabrication, and scalability for element arrays. UWB antennas that have a symmetric (balanced) feed design require a matching component (typically a wideband balun [14]) when connected to usual coaxial feed lines; in other cases, an impedance transformer is required for impedance matching (the balun and transformer are frequently built into the same component). Small antenna size is desirable for light, compact, and portable radar units. For larger vehicle-mounted systems, these requirements can be somewhat relaxed. Planar antenna designs are particularly suited for this application because of their compact size and ease of fabrication. Finally, since many STTW radar systems use a physical antenna array, the ability to arrange multiple elements into one- or two-dimensional (2D) arrays is very important in selecting a specific design. This creates limitations to the element size (no larger than one half-wavelength in cross-dimension) and puts additional constraints on the mutual coupling between elements.

Many microwave radar antennas are based on the horn geometry. The transversal electromagnetic (TEM) horn in Figure 11.1(a) is close to an ideal, nondispersive UWB antenna. It performs better than many other antenna types at low frequencies (below 1 GHz), since there is no cutoff frequency for a propagating TEM mode. Sheet resistors are typically mounted close to the radiating end in order to attenuate the surface currents that propagate along the metallic plates, thus reducing the reflections from the end of the structure. The construction is bulky and requires a structural support for the metallic plates (typically made of Styrofoam). A balun is needed to match the symmetric 200 Ω input impedance to a coaxial feed line. This type of antenna is used in the SIRE radar transmitter [19]. Another type of horn antenna is the quad-ridge horn in Figure 11.1(b),

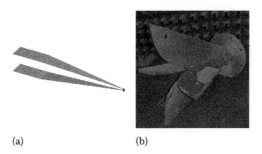

(a)   (b)

FIGURE 11.1
Example UWB MW Horn antennas. (a) 200 Ω TEM horn antenna and (b) quad-ridge horn antenna.

and manufactured by ETS-Lindgren [20]. This is a dual polarization design that comes in different sizes depending on the desired frequency band. It is also a relatively bulky and heavy antenna, which could only be accommodated by a large, vehicle-mounted radar system. This antenna type has frequently been used in radar laboratory setups [21–23].

The Impulse Radiating Antenna (IRA) is typically designed for high power, high gain, short duration microwave impulse transmission. These are characteristically most suitable for a long-range UWB radar system. The 46 cm diameter IRA shown in Figure 11.2(a) is used by the impulse synthetic aperture radar (ImpSAR) developed by Eureka Aerospace [24]. Another type of UWB antenna is the log-periodic dipole array in Figure 11.2(b), which consists of a sequence of dipoles whose length, diameter, and spacing are scaled logarithmically [14]. These antennas can operate over very large bandwidths; however, they are dispersive (i.e., exhibit phase nonlinearity). Additionally, the relatively large size makes them unsuitable for compact designs and arrays.

Among the UWB planar antennas, one of the simplest geometries is the bowtie in Figure 11.3(a), which can be seen as a "fat" dipole with a diameter tapering off toward the feed ends. Although a simple bowtie displays excellent VSWR over a large bandwidth, the radiated pattern has the donut shape characteristic to a dipole, which is unsuitable for STTW applications. In order to obtain a unidirectional pattern, a ground plane must be added, which reduces the bandwidth by at least a factor of two. Additionally, the bowtie antenna is dispersive and requires a matching balun for 50 Ω asymmetric feeds. An alternative planar antenna geometry related to the bowtie is the diamond dipole of Figure 11.3(b), which has reportedly been used in UWB STTW radar applications [25].

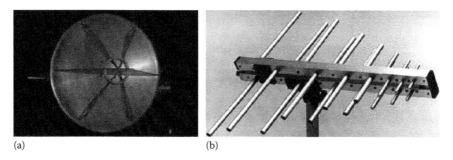

(a)　　　　　　　　　(b)

FIGURE 11.2
UWB impulse radiating antennas. (a) A 46-cm impulse radiation antenna and (b) log-periodic dipole array.

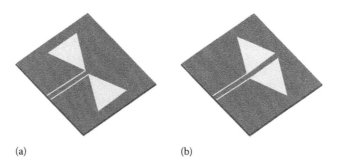

(a)　　　　　　　　　(b)

FIGURE 11.3
Example UWB planar antennas. (a) Bowtie antenna and (b) diamond dipole antenna.

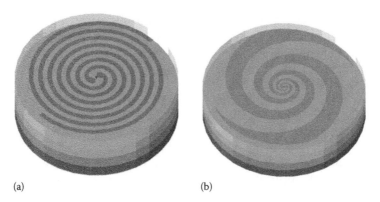

(a)   (b)

FIGURE 11.4
Cavity-backed spiral antennas: (a) Archimedes spiral and (b) logarithmic spiral.

The cavity-backed spiral is another popular UWB antenna design [14]. In Figure 11.4, we show two variants: the Archimedes (or constant width) spiral in Figure 11.4(a) and the logarithmic spiral in Figure 11.4(b). The radiating element is planar; in order to make the pattern unidirectional, the cavity is filled with an absorbing material. The idea of adding an absorber-filled cavity to reduce the backlobes of a UWB antenna is applicable to most other planar geometries. As compared with planar antennas backed by a simple ground plane, this solution typically offers larger bandwidths, but at the same time increases the size. Although it can achieve very large bandwidths, the cavity-backed spiral antenna requires a balun and is dispersive. The polarization is circular – a typical design would involve a right-hand polarized transmitter antenna and a left-hand polarized receiver antenna. A variation in the logarithmic spiral antenna is used by the L3 Communications Handheld STTW radar system [26].

The Vivaldi family of antennas [27,28] make excellent candidates for STTW radar applications. They can typically cover multiple octaves of bandwidth at almost constant gain. The design is relatively simple and the feed can provide direct matching to a 50-Ω coaxial line. They exhibit low sidelobes and backlobes and were reported to provide a clean, non-dispersive response to short UWB impulses [1]. Variants of the Vivaldi antennas include the antipodal of Figure 11.5(a) and the tapered slot of Figure 11.5(b) designs. Since these are end-fire antennas, they can be easily stacked in linear arrays. Examples of STTW systems that include Vivaldi antennas are the SIRE radar receivers [29] and the UWB system built at the University of Tennessee [30].

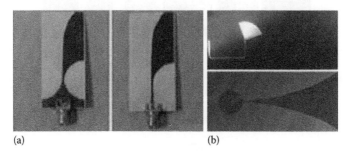

(a)   (b)

FIGURE 11.5
Vivaldi antennas: (a) antipodal Vivaldi and (b) tapered slot antenna. (Reprinted from Yang, Y. et al., Design of compact Vivaldi antenna arrays for UWB see through wall applications, *Prog. Electromagnetic Res.*, vol. 82, pp. 401–418, 2008.)

Another family of UWB antennas uses the slotted microstrip patch geometry. These antennas have very low profiles, simple feed structures and are easily fabricated (especially into arrays of elements) and are, therefore, very suitable for lightweight, portable radar systems. On the downside, they do not provide bandwidths as large as other designs presented in this section (typically, below 50%). Figure 11.6 shows several examples of slotted patch antennas: the U-shape in Figure 11.6(a), the E-shape shown in Figure 11.6(b), and the dual-stacked E-shape patch DSEP of Figure 11.6(c). The last of the three was developed at the Center for Advanced Communications (CAC), Villanova University [31] and operates between 2 and 3 GHz, with a gain of at least 7 dBi.

Any discussion of modern radar antennas would be incomplete without mentioning antenna arrays. In STTW applications, almost all stationary systems use horizontal linear arrays to estimate the target azimuth. When the azimuth is estimated via beamforming [1,21], the elements must be equally spaced at a distance of maximum half a wavelength, in order to avoid grating lobes (aliasing). In this case, the angular resolution is dictated by the total array size. As a numerical example, at 2 GHz, a 1-m-wide receiver array (assuming only one transmitter) offers an angular resolution of about 8.6° (equivalent to 0.75 m cross-range resolution at 5 m downrange), and this array must contain at least 14 elements. It is noted that use of emerging compressive sensing techniques for azimuth estimation can relax these requirements [2]. Examples of linear antenna arrays used in STTW applications include the SIRE radar and University of Tennessee UWB system (both based on Vivaldi elements) and the DSEP dual polarization arrays developed at CAC. When SAR image techniques are employed, a pair of transmit and receive antennas is sufficient to estimate the target location in a 2D plane. However, using a vertical linear array combined with the synthetic aperture along a horizontal track allows the formation of three-dimensional (3D) images. This was demonstrated by researchers at DRDC with the TWSAR system [8]. Stationary systems can also be used to create 3D radar images if they are equipped with 2D antenna arrays. All these are active research topics that will not be further expanded here due to space limitations. For more information, the reader is directed to the references at the end of this chapter.

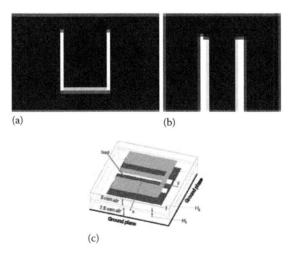

(a)   (b)

(c)

FIGURE 11.6
Slotted microstrip patch antennas: (a) U-shape, (b) E-shape, and (c) dual-stacked E-shape patch (DSEP).

11.3 Modeling of UWB STTW Imaging Radar

Understanding the phenomenology involved in typical sensing scenarios helps the radar engineer not only with the system design, but also with the interpretation of collected data. The results presented in this section were obtained via computer simulations of complex and realistic scenarios. One goal of investigating these computer models is the prediction of the sensor's performance before the actual system is built. This may enable the optimization of certain design parameters and the avoidance of certain mistakes. Another reason is that computer models allow a clear separation of the EM scattering phenomena from the artifacts introduced by the radar hardware and signal processing. Consequently, the modeling results are typically much cleaner than the real radar data and, in that respect, they represent the "best case scenario" for radar performance.

Several computational electromagnetic (CEM) methods have been proposed for modeling UWB radar in general and STTW applications in particular. Algorithms that work in the time domain are a very good choice for these models, since they directly emulate the operation of an impulse-based, UWB radar system. Among these, the most popular is the finite-difference time-domain (FDTD) [32], although other techniques, such as the finite-element time-domain (FETD) [33] and finite-volume time-domain (FVTD), have also been proposed. Many of the simulations presented in this section were obtained with the AFDTD software [34], based on the FDTD algorithm and developed at ARL. Other CEM techniques employed in simulating radar scattering scenarios are based on ray-tracing and other high-frequency methods (physical optics, the geometric theory of diffraction, and their extensions [35]). While much more efficient from a computational standpoint than full-wave methods (such as FDTD), these methods offer only approximate solutions. Note that, although the high-frequency techniques are based in the frequency domain, their speed and efficiency make them reasonable choices for wideband simulations, as required by UWB radar systems.

We present the UWB SAR images of two buildings, obtained by simulations with the AFDTD software. The building layouts are shown in Figure 11.7. The first building in Figure 11.7(a) has only one floor, while the second has two floors as shown in Figure 11.7(b) with the bottom floor identical to the single-floor building. The overall horizontal dimensions are 10 m × 7 m, while the height is 2.2 m for the single-floor building and 4.8 m for the two-floor building. The walls are made of 0.2-m-thick bricks and have glass windows and wooden doors. Interior walls are made of 5-cm-thick sheetrock. The ceiling/floor dividing the two stories is made of plywood (top) and sheetrock (bottom) and includes 2 × 8-inch wooden beams. We also considered a wooden staircase connecting the lower and upper floors. The upper floor ceiling/roof is flat and made of a 7.5-cm-thick concrete slab. The entire building is placed on top of a ground plane with $\varepsilon' = 10$ and $\varepsilon'' = 0.6$. Six standing humans were placed at various positions and azimuth orientations inside the buildings, including four on the ground floor and two on the upper floor. Out of the six humans, two carry AK-47 rifles (one on each floor). The "fit man" model, made of uniform dielectric and described in reference 36, was used throughout these simulations. The rooms of the buildings contain a great deal of detail in terms of furniture, appliances, and bathroom/kitchen fixtures – the reader should consult references 37 and 38 for a complete description, including the dielectric properties of the materials included in the model.

In terms of SAR system geometry, we considered two possible monostatic configurations, one ground-based shown in Figure 11.8(a) and one airborne in Figure 11.8(b). Since the AFDTD software can only analyze far-field scenarios, these models are best suited to emulate the circular spotlight SAR mode [39]. While a straightforward conversion from

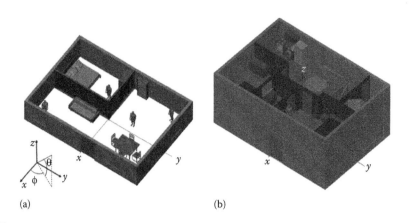

(a)   (b)

FIGURE 11.7
Representation of the building meshes used in the radar imaging simulations in this section: (a) one-floor building and (b) two-floor building (only the upper floor is visible). (Fig 11.7(a): Permission of SPIE, Dogaru, T. et al., Three-dimensional radar imaging of buildings based on computer models, in *Proceedings of SPIE Conference on Radar Sensor Technology XVII*, Baltimore, MD, vol. 8714, pp. 87140L-1-87140L-12, 2013. © SPIE 2013; Fig. 11.7(b): Permission of SPIE, Le, C. et al., Synthetic aperture radar imaging of a two-story building, in *Proceedings of SPIE Conference on Radar Sensor Technology XVI*, Baltimore, MD, vol. 8361, pp. 83610J-1-83610J-9, 2012. © SPIE 2012.)

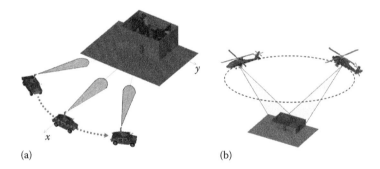

(a)   (b)

FIGURE 11.8
Schematic representation of the SAR imaging geometries considered in this section, showing: (a) ground-based radar system and (b) airborne radar system. (Fig. 11.8(b): Permission of SPIE, Dogaru, T. et al., Three-dimensional radar imaging of buildings based on computer models, in *Proceedings of SPIE Conference on Radar Sensor Technology XVII*, Baltimore, MD, vol. 8714, pp. 87140L-1-87140L-12, 2013. © SPIE 2013.)

spotlight to strip-map SAR modes can be performed for the far-field case, a near-field configuration would require a separate model of the radar system (see references 1 and 38 for examples of similar near-field simulations).

For the AFDTD simulations providing the radar scattering data, the building model was discretized using 5 mm cubic cells in order to keep the numerical dispersion errors under control [32]. In the case of the two-floor building, the total grid size was just over 3 billion cubic cells, which required 180 GB of memory for running the program for one incidence angle and a frequency range between 0.3 GHz to 2.5 GHz. Most 2D images presented here used a 60° aperture in azimuth, with 0.25° increments. For the 3D images, the integration angle in elevation was 40° (from 10° to 50° from the horizontal plane), in 1° increments. For these image parameters, the resolution is on the order of 15 cm in down-range, 25 cm in cross-range, and 35 cm in elevation. From a computational standpoint, there were very large problems that required high-performance computing platforms available at the Defense Supercomputing Resource Centers (DSRC) [40]. A typical set of simulations producing the radar data for one 2D SAR image was run over 128 cores and took about 100,000 CPU hours. The total CPU time required to obtain the 3D building image was on the order of 4 million hours.

With regards to the SAR image-formation technique, we used a frequency-domain version of the back-projection algorithm [5]. While the terminology varies among authors, we call this method the matched filter imaging algorithm [5]. For a far-field monostatic sensing geometry, the complex intensity of an image pixel is computed as

$$I(r) = \sum_{\omega} \sum_{A} W(r_A, \omega) S(r_A, \omega) e^{j \frac{2\omega}{c} |r - r_A|} \tag{11.1}$$

where:

r and r_A represent the position vectors of the image and aperture points, respectively

$\omega = 2\pi f$ with f representing the frequency

$S(r_A, \omega)$ is the scattering matrix element [35] corresponding to the radar polarization (computed by the AFDTD modeling software), and

$W(r_A, \omega)$ is a real-valued window function that operates in both the aperture and frequency dimensions.

Notice that the double sum is computed over all aperture positions and frequencies within the band. The exponential factor $e^{j2\omega/c|r - r_A|}$ is reminiscent of the Green's function (where the $|r - r_A|^2$ factor in the denominator was dropped) and describes the round-trip propagation delay of the signal between radar and target. It can easily be shown [41], by taking an inverse discrete Fourier transform (DFT), that this formulation is equivalent to the delay-and-sum version of the time-domain back-projection algorithm [5]. Additionally, other popular imaging algorithms, such as the polar format algorithm [39] and the time-reversal imaging technique [42], can be recast in a format similar to Equation 11.1.

Note that the characteristics of the excitation UWB impulse do not appear explicitly in Equation 11.1, since $S(r_A, \omega)$ is a frequency-domain quantity, which does not depend on the transmitted waveform. However, we can always account for the latter by setting the frequency dependence of the W window function to match the excitation impulse spectrum. In our case, we applied generic Hanning windows in both frequency and aperture domains, although any other finite-support window function could be used. Importantly, in this section, we do not attempt to compensate the SAR images for the delays in the radar signals caused by propagation through the dielectric walls. Therefore, the targets in these images appear slightly displaced with respect to the ground truth. The two-dimensional (2D) images shown here represent top views in the horizontal plane (for the ground-based case), with pixel magnitudes (in dB) mapped on a color scale. The gray contour overlays represent the ground truth and are not part of the radar images.

The first set of images in Figures 11.9 and 11.10 consider the ground-based system, with the elevation angle $\theta = 0°$, and the radar moving on a trajectory in the horizontal plane. The aperture is centered in a direction perpendicular to one of the exterior walls. This configuration offers the strongest backscatter radar return from the all walls that are perpendicular to the radar line-of-sight (LOS). In Figure 11.9, we show vertical-vertical (V-V) polarization SAR images based on AFDTD data for each individual floor, when the radar aperture is centered along the x axis (on the left side of the page). Figure 11.9(a) shows the SAR image of the lower floor only – in this sparsely furnished interior, the human targets are clearly visible. The "ghost" images projected by three of the humans onto the nearby

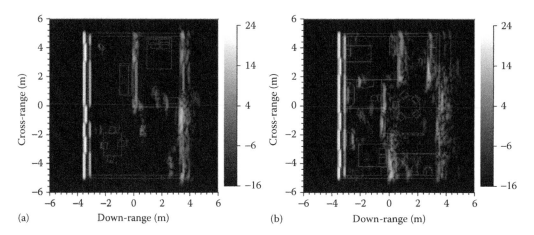

FIGURE 11.9
SAR images for the radar moving in the horizontal plane. Images show each individual floor of the building in Figure 11.7, for θ = 0° and V-V polarization, based on AFDTD data, with the radar aperture centered along the *x* axis: (a) lower floor and (b) upper floor. Note: the staircase was taken out of both floor models. (Permission of SPIE, Le, C. et al., Synthetic aperture radar imaging of a two-story building, in *Proceedings of SPIE Conference on Radar Sensor Technology XVI*, Baltimore, MD, vol. 8361, pp. 83610J-1-83610J-9, 2012. © SPIE 2012.)

side walls are also visible. This phenomenon is caused by multiple scattering between the humans and the surrounding walls. The radar signatures of furniture items appear weaker than that of the humans, with the exception of the dresser and the empty book-shelf directly behind the wooden exterior door. Figure 11.9(b) shows the SAR image of the upper floor by itself. This includes floor/ceiling dividing the two stories and the entire structure placed above it. The increased complexity of the upper floor layout and number of interior objects leads to increased clutter in the SAR image. In such an environment, detecting the human targets becomes problematic, a fact clearly illustrated by the target placed in the kitchen, on the upper floor. In addition to multibounce scattering effects observed in the image, we also noticed other bright objects such as the refrigerator, side wall of the bathroom vanity cabinet, the bathtub, and the shower stall, which further complicates the target detection process.

In Figure 11.10, we show the SAR images of the entire two-story building with the same aperture and bandwidth as in Figure 11.9. The V-V polarization image of the entire building is essentially the pixel-by-pixel sum of the two individual floor SAR images (to which we must add the radar return from the staircase). Notice that the area of the exterior wall of the two-story building is two times the exterior wall area of each individual floor model. Consequently, the magnitude of the radar return from the exterior wall of the two-story model is about four times, or 6 dB higher, as shown in Figure 11.10(a). This causes the interior objects to look less visible than in Figure 11.9 images when the same 40 dB dynamic range is used for representation. Figure 11.10(a), where only the human target contours were overlaid onto the SAR image, illustrates the challenge of detecting stationary human targets in a realistic STTW environment. Another effect worth mentioning in the SAR images of the two-story building is the increased amount of clutter at farther ranges. This is mostly produced by multiple reflections of the radar signals between parallel walls, as well as larger objects and the walls, which create false, delayed replicas of the original objects. We conclude that the information on building interior contained in the raw SAR images has very limited usefulness in regions where the radar signals propagate through more than two walls (one way).

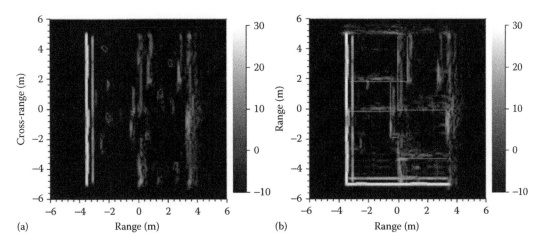

FIGURE 11.10
SAR images of the entire two-story building, for θ = 0° and the radar moving in a horizontal trajectory. Images based on AFDTD data show (a) the aperture along y axis and (b) the combined apertures along x and y axes. (Permission of SPIE, Le, C. et al., Synthetic aperture radar imaging of a two-story building, in *Proceedings of SPIE Conference on Radar Sensor Technology XVI*, Baltimore, MD, vol. 8361, pp. 83610J-1-83610J-9, 2012. © SPIE 2012.)

To obtain a top view of the building layout, we combined the SAR images viewed from two orthogonal sides of the building in Figure 11.10(b). Here, we present the images obtained for V-V polarization, with the overlay of the interior wall contours. The fused image consists of pixel-by-pixel addition of image magnitudes, performed in linear space and then converted back to dB scale. The small gaps showing in the corners formed by pairs of orthogonal interior walls are due to different spatial delays for the orthogonal directions of propagation characteristic to the two SAR images. One interesting feature noticeable in this image is the periodic beam structure that supports the ceiling of the bottom floor. Only a few of these beams (at the nearest ranges) can be distinguished in the images. Because of the orientation of these beams (most of them parallel to the short side of the building), it is apparent that their contribution to the image clutter is more pronounced for the aperture at the bottom of the page than for the aperture at the left of the page.

The difficulty of separating the scene features by height in 2D images prompted us to investigate the possibility of obtaining 3D images of a building by computer models [41]. Similar studies have been performed by other researchers [43,44], while reference 8 reports an experimental radar system for 3D building imaging. The parameters used by the EM simulations and the image formation algorithm were described in a previous paragraph. While performing these steps was rather straightforward, visualizing the 3D images on a 2D medium support (such as the computer screen) required further processing. Our approach involved a background removal procedure prior to visualization, meaning that we only displayed voxels standing out of the background. More specifically, we processed the image through a 3D constant false alarm rate (CFAR) detector [3], which, in essence, compares each voxel in the image with a threshold that depends on the surrounding background level, such that the detection scheme preserves a constant false alarm probability. Once the voxels indicating target detection have been identified (and assuming they are clustered together around the outstanding features in the image), all voxels within a "target" volume (or more exactly, voxel cluster) are assigned a constant magnitude equal to the maximum voxel magnitude within the cluster). At the same time, the background, consisting of voxels rejected by the detector, is assigned an arbitrarily

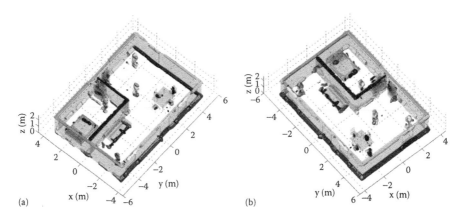

FIGURE 11.11

The 3D building image for the airborne spotlight configuration and V-V polarization, with SNR = 40 dB, as seen from two different aspect angles: (a) $\theta_1 = 60°$, $\phi_1 = -50°$; (b) $\theta_2 = 60°$, $\phi_2 = 50°$. The feature colors correspond to their brightness levels in the raw 3D image.

low magnitude, at the bottom of the image dynamic range. Finally, the visualization is performed by displaying the isosurfaces (2D surfaces of constant magnitude in the 3D space) representing each target within the 3D image volume. While only projections of this 3D image can be rendered on a 2D support, changing the viewing angle can offer a more complete interpretation for the end user.

The 3D images of the single-floor building, obtained by combining radar data collected from two orthogonal sides and seen from two different viewing angles, are shown in Figure 11.11. The obvious features detected in the image are the top and bottom edges of the walls, the humans, the edges of the dresser and sofa, as well as the chairs. Notice that the wall edges (particularly the ones at the bottom) appear with some interruptions. While it is difficult to explain the gap in the bottom edge of the front wall, the gaps in the interior and back walls clearly correspond to shadows (or "ghosts") of the humans or other objects projected onto those walls. The human images also appear fragmented, with two major scattering centers corresponding to the ground plane footprint (multiple scattering due to the ground bounce) and the torso (single scattering). Since the focus of this study was primarily on phenomenology, no attempt was made to separate the human targets from the clutter. Further image analysis (as demonstrated in [45]) would be necessary to classify the image features into different categories.

11.4 Through-the-Wall Moving Target Indication with UWB Radar

As demonstrated in the previous section, the targets of interest in a through-wall sensing scenario (usually humans) are very difficult to detect in static radar images, due to the large amount of clutter introduced by other scatterers in the scene. Several methods based on background subtraction [21], spatial filtering [46], and wall subspace projections [47 and references therein] have been proposed to mitigate the radar clutter, particularly the wall returns. Moreover, the application of emerging compressive sensing techniques has proven successful in clutter mitigation problems [2,48–51]. However, MTI and Doppler processing techniques remain, in many cases, the most effective ways to detect moving targets in heavily cluttered environments. In this section, we describe possible approaches to MTI within the context of UWB radar.

As shown in [52], the conventional Doppler filter bank technique (usually implemented as a DFT in modern radar systems) is not a good approach for target velocity estimation in high-resolution radar. The typical Doppler processing scheme takes a DFT of the slow-time data samples within the same range resolution cell, over a period of time called the coherent processing interval (CPI), with the assumption that the target dwells within the same cell during that time interval. However, when the resolution cell is small (on the order of 10 cm for UWB radar), the target can migrate very rapidly from one resolution cell to another, effectively spanning several cells within a CPI. This leads to a loss of both range and Doppler resolution proportional to the target velocity [52]. That being said, short-time Fourier transform-based Doppler processing can still be useful in understanding the fine structure of the target motion via time–frequency analysis (by means of spectrograms or other time–frequency distributions [53]). (Editor's note: Chapter 4 discusses time–frequency analysis methods.) However, in order to apply these techniques to UWB radar data, the measurement phase must be adjusted (based on the average target velocity estimate) such that, after the correction, the target appears to stay within the same resolution cell during a CPI. Such an approach was demonstrated in [54] and will not be pursued here.

A literature survey of MTI techniques applied to UWB STTW radar indicates that the most common approaches are based on change detection [2,23,55–60]. In the following, we assume that the radar is stationary and receives UWB waveforms over multiple channels, allowing the formation of 2D images of the scene. We call all the radar data involved in the creation of one image a "data frame." The basic idea of the change detection approach is to subtract two consecutive data frames from one another (usually in the image domain): the end result consists of preserving the moving object images, while the stationary clutter is eliminated or at least strongly suppressed. Note that we require the radar platform to be stationary, as all current MTI STTW systems operate. Detecting moving targets from a moving platform is a much more difficult problem, frequently encountered in airborne MTI radar [3]. Several advanced signal-processing methods have been developed to address those sensing scenarios, such as displaced phase center antenna (DPCA) or space-time adaptive processing (STAP); however, to our knowledge, application of these techniques to STTW radar has not been attempted to date.

Two basic change detection approaches are possible. In the first approach, called coherent change detection (CCD), complex image pixel intensities corresponding to two consecutive data frames are subtracted on a pixel-by-pixel basis. The process can be formally described as

$$\Delta_{CCD}\left(\mathbf{r}_q\right) = \left| I\left(\mathbf{r}_q, t + \Delta T\right) - I\left(\mathbf{r}_q, t\right) \right| \tag{11.2}$$

where:
$I\left(\mathbf{r}_q, t\right)$ is the complex value of the pixel index q at time t
$\mathbf{r}_q$ is the position vector of pixel q
ΔT is the time interval between two data frames (in this case, equal to the CPI).

Note that, prior to taking the absolute value of the pixel q, the operation described by Equation 11.2 is linear. Since the beamforming process involved in the radar image formation is linear as well, one can interchange the order of the two processing stages without modifying the final result. In that case, the structure of the receiver resembles the classic delay line canceller

(DLC) [3], where two consecutive range profiles received by the same radar channel are coherently subtracted from one another. The DLC described here is the simplest MTI filter structure; higher order MTI filters, involving more than two pulses and weighting coefficients, can be designed to achieve specific transfer functions that can improve the detection performance [10].

The second approach, called noncoherent change detection (NCD), performs a pixel-by-pixel subtraction of the pixel magnitudes corresponding to two consecutive data frames. The operation is described by the following equation:

$$\Delta_{\text{NCD}}\left(\mathbf{r}_q\right) = \left|I\left(\mathbf{r}_q, t + \Delta T\right)\right|^2 - \left|I\left(\mathbf{r}_q, t\right)\right|^2 \qquad (11.3)$$

Notice that the pixel values in a CCD image are always positive, whereas in an NCD image, they can be either positive or negative. When a target occupies two different pixels for two successive data frames, the subtraction operation creates two separate images of the same target, corresponding to the reference and the current target positions (regardless of the change detection approach). When CCD is applied, these two images cannot be easily separated from one another. However, since the NCD image is bipolar, one can use a zero-thresholding procedure [58] and retain only the positive image pixel, corresponding to the current target position. The thresholding operation is described by

$$\Delta_{\text{TH}}\left(\mathbf{r}_q\right) = \begin{cases} \Delta_{\text{NCD}}\left(\mathbf{r}_q\right) & \text{if } \Delta_{\text{NCD}}\left(\mathbf{r}_q\right) \geq 0 \\ 0 & \text{otherwise} \end{cases}. \qquad (11.4)$$

The two change detection processes are described by the block diagrams in Figure 11.12. It is apparent that if the target does not move outside the current resolution cell (pixel) during the time interval between frames, all pixels in two consecutive images will have the same magnitudes and only CCD, which is sensitive to phase changes, can be applied to detect the moving target. This case corresponds to scenarios where the radar image has low resolution, the target moves slowly, and/or the frame update interval is short. On the other hand, NCD may be readily applied to scenarios with high image resolution, rapidly moving targets, and/or long frame update intervals.

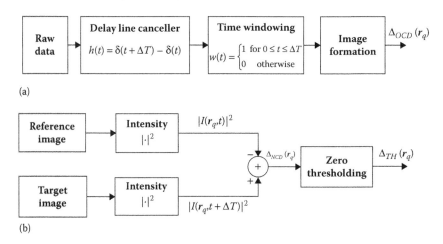

FIGURE 11.12
Block diagrams describing the two change detection approaches: (a) coherent change detection using a delay line canceller and (b) noncoherent change detection. (© 2013 IEEE. Reprinted, with permission, from Amin, M and Ahmad, F., Change detection analysis of humans moving behind walls. *IEEE Trans. Aerosp. Electron. Systems*, 49, 1410–1425, 2013.)

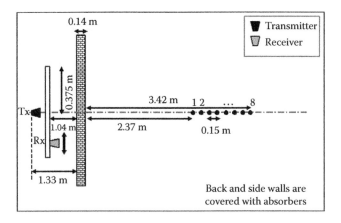

FIGURE 11.13
Layout of the radar experimental scenario used in the change detection study described in [23]. Positions 1,2,..., 8 indicate positions of the human target. (© 2013 IEEE. Reprinted, with permission, from Amin, M and Ahmad, F., Change detection analysis of humans moving behind walls. *IEEE Trans. Aerosp. Electron. Systems,* 49, 1410–1425, 2013.)

An experimental setup was developed at the Radar Imaging Laboratory, Villanova University, to investigate the change detection approaches described in this section in the context of through-the-wall radar imaging [23]. A schematic representation of the scenario is shown in Figure 11.13. The receiving linear array is synthesized by moving the receiving antenna into 11 successive positions along a 0.75-m horizontal aperture. The transmitting antenna is placed in the middle of the array, 0.29 m behind the receivers. The operational bandwidth is from 1.5 to 2.5 GHz. The wall is 0.14-m thick, made of solid concrete blocks and placed 1.05 m away from the receiver array. The target is a human positioned at the points labeled 1 through 8, with ranges between 2.37 m and 3.42 m. The displacement between two adjacent positions is 0.15 m, which roughly matches the downrange resolution of the imaging system.

A set of results obtained in these experiments is presented in Figure 11.14. For both change detection algorithms, we consider target motion between positions 3 and 4 (corresponding to two adjacent pixels) and between positions 3 and 7 (corresponding to a displacement across several pixels). The images in Figure 11.14(a) and (b) were obtained via the CCD algorithm, while Figure 11.14 (c) and (d) show the NCD-processed images. One can observe the double target image obtained by the CCD procedure – this effect is clearly more pronounced in Figure 11.14(b), where we have a large target displacement relative to the radar resolution during the processing interval. The second image is mostly eliminated by zero-thresholding in the NCD images. The large artifacts present in these images can be attributed to the nonlinearity of the modulus function, which magnifies the effect of small errors between successive measurements, leading to imperfect background cancellation for the NCD images. Notice that the CCD images are not artifact-free either. In fact, the CCD method is inherently more sensitive to phase errors than the NCD approach. Although this issue is not evident in the measurements presented here, other STTW scenarios (e.g., those involving cinderblock walls) may be particularly prone to inducing phase errors in the processing.

To summarize the discussion comparing coherent and noncoherent change detection for MTI radar, we suggest the following criterion in deciding which technique is more appropriate for a specific scenario. If the moving target stays within the same resolution cell during a CPI, then CCD is the only available option. (As a side note, this kind

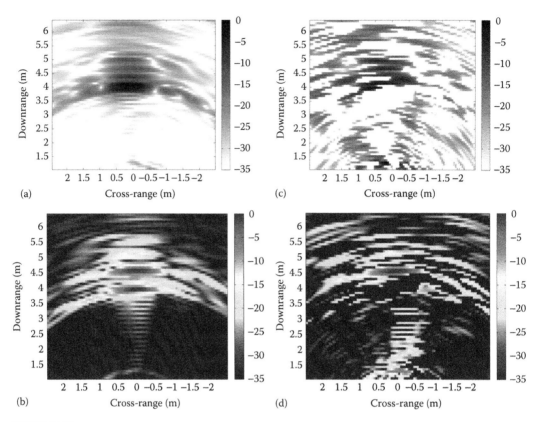

FIGURE 11.14
Change detection images obtained in the experimental setup described in Figure 11.13, showing: (a) CCD image with the target moving between positions 3 and 4, (b) CCD image with the target moving between positions 3 and 7, (c) NCD image with the target moving between positions 3 and 4, and (d) NCD image with the target moving between positions 3 and 7. (© 2013 IEEE. Reprinted, with permission, from Amin, M and Ahmad, F., Change detection analysis of humans moving behind walls. *IEEE Trans. Aerosp. Electron. Systems*, 49, 1410–1425, 2013.)

of scenario always requires a processing scheme sensitive to phase, or a coherent type of processing. Fourier-transform-based Doppler processing, which also works best when the target does not migrate between cells during a CPI, is yet another example of coherent processing technique.) If the target moves over multiple resolution cells during a CPI, then NCD may be the preferable technique, because it allows suppression of the double target image by zero-thresholding. Applying this criterion to UWB STTW MTI radar, we conclude that, for radial motion (along the radar LOS), NCD is the better choice given the fine downrange resolution; however, for radar systems with poor cross-range resolution, CCD seems more appropriate for transversal target motion. As a caveat, one should keep in mind that NCD images typically display a large number of artifacts that need to be addressed by further signal or image processing steps.

One of the obvious artifacts that impact the change detection procedure is the presence of image sidelobes. Since the sidelobes represent "spillage" of the target signature into the neighboring resolution cells, it is apparent that they affect the sensitivity of both coherent and noncoherent change detection schemes [55]. While the analysis of CCD in the presence of sidelobes is rather complicated, in the case of NCD, they clearly reduce the difference between the pixel magnitudes of successive images. Data windowing is often employed to control the sidelobe levels – however, this procedure introduces a penalty in terms of resolution. While this may not constitute a problem in the downrange direction for UWB radar (where the resolution is typically very good), further reduction in the cross-range resolution (which may be fairly poor to start with) may severely compromise the change detection scheme for targets with transversal motion, especially in the noncoherent case. One solution for sidelobe mitigation in STTW MTI radar, dubbed multiple-image noncoherent change detection (MNCD), was described in [55]. In essence, this method considers multiple NCD images obtained over several data frames with respect to the same reference image and performs a pixel-wise minimum amplitude selection operation across this set of images. The effect of this procedure is preserving the main lobes of the target signature, while at the same time suppressing the target sidelobes. It was shown in [55] that the MNCD method can generally outperform the CCD approach, at the cost of increased computational complexity.

Reference 60 reports STTW MTI experiments run by ARL researchers using the SIRE radar in stationary mode. In this study, the signal-processing chain includes, besides the image formation and CCD stages, a CFAR detector, a pixel clustering algorithm, and a tracker in Figure 11.15. The CFAR detector is a standard radar signal-processing tool that allows target detection in variable (spatially nonstationary) background clutter with unknown power. In this case, a 2D version of the cell-average CFAR detector was used, similar to the approach described in [61]. The clustering procedure, based on the k-means algorithm [62], is an intermediate step between the detector and the tracker, meant to reduce the number of points of interest fed into the next processing stage. This technique is particularly useful in high-resolution images, where a target signature can span several pixels. At the outset of this procedure, only the centroids of each cluster are taken into account by the tracking algorithm.

The radar tracker performs several functions: it improves the current estimates of the target state vector based on previous observations made on that target; associates the current radar measurements with the correct targets; and reduces the false alarm rate by eliminating spurious detections that cannot be associated with any valid target tracks [63]. One key component of a tracker is the Kalman filter, which predicts the state vector for the next observation (data frame) based on previous estimates of the same. The other major component of a tracker is the observation-to-track association algorithm. A multiple hypothesis tracker is commonly used to solve this problem in modern radar systems [63]; this technique has been shown to be very effective in STTW scenarios involving multiple targets, while significantly reducing the false alarm rate [64].

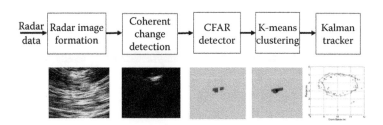

FIGURE 11.15

Signal-processing chain for the STTW MTI study performed at ARL based on SIRE radar experimental data. (From Novak, L. et al., Performance of a high-resolution polarimetric SAR automatic target recognition system, *The Lincoln Laboratory Journal*, 6, 11–24, 1993. With permission.)

11.5 Multipath Exploitation in UWB STTW Imaging Radar

The multipath propagation and scattering phenomenon is of major importance in STTW radars and has attracted significant interest in the research community [1]. In addition to the shortest path between the target and the antenna, the transmitted wave may travel by indirect paths due to secondary reflections arising from interior walls, floor, and ceiling. The energy in such multipath returns may accumulate at locations where no physical targets reside, thus creating "ghosts." The ghosts have an impact on both stationary and dynamic radar sensing scenarios and could lead to high false alarm rates. Even for a simple four-wall room case, the number of possible multipath propagation mechanisms is very large; Figure 11.16 shows examples multipath propagation cases. Typically, only ghosts generated by low-order multiple reflections are strong enough to appear as valid targets at the end of the processing chain.

   Because multipath exists and often observed in STTW radar, imaging techniques must properly describe and address it using accurate analytical models. Broadly, there are two paradigms to deal with indirect propagation, namely, multipath suppression and multipath exploitation. The key idea of multipath suppression is to characterize the returns and mitigate their effects on STTW image formation [23,51,64,65]. Different properties of direct and indirect radar returns can be used to distinguish between the two arrivals and attenuate, if not remove, the indirect returns. While methods proposed in [23,51,65] can achieve ghost suppression without any prior knowledge of the scene, the algorithm in [64] is performed as a post-tracking stage and assumes prior knowledge of the room layout. While the multipath suppression methods are generally straightforward to apply, they do not make use of the energy and target information contained in the multipath returns.

   Multipath exploitation methods, on the other hand, aim at utilizing the multipath for through-the-wall imaging enhancements [66–72]. By properly modeling the indirect propagation paths, whether they are resolvable or not, their energy can be captured and attributed to their respective targets, allowing an increase in target to clutter and noise ratios, and thus culminating in an enhanced image. Further, areas in the shadow region of highly attenuating targets, which cannot be directly illuminated by the radar, can be imaged by utilizing multipath [68]. Although multipath exploitation has potential and tangible benefits, it often requires prior information or is computationally demanding.

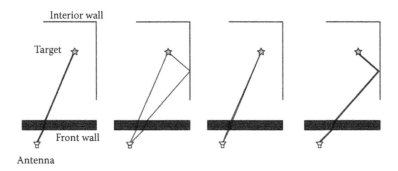

FIGURE 11.16
Various cases for multipath in indoor scenes. From left to right: direct propagation, one secondary reflection at an interior wall, multiple reflections inside front wall, two secondary reflections at an interior wall, and target–target multipath. (© 2014 IEEE. Adapted, with permission, from Leigsnering, M. et al., Multipath exploitation and suppression for SAR imaging of building interiors: An overview of recent advances. *IEEE Signal Processing Magazine*, 31, 110–119, 2014.)

In this section, we describe in detail a multipath exploitation approach that assumes the presence of few targets and employs sparse reconstruction for image recovery. For complete treatment of multipath exploitation based on conventional image reconstruction approaches, the reader is referred to [73,71 and references therein]. In order to establish a signal model under multipath propagation, we employ a Geometric Optics (GO) or ray tracing approach, which uses local plane wave assumption or "ray of light" to model the propagation of the EM wave [74]. We assume perfect knowledge of the building layout and ignore target-to-target interactions. Further, it is assumed that the wall returns have been suppressed and the measured data contain only the target returns. For details on the signal model when the wall returns are present in the measurements, refer to [72].

Considering a monostatic sensing geometry and a maximum of R possible propagation paths for each target–sensor combination, including the direct path, the measurement model can be expressed as

$$\mathbf{y} = \boldsymbol{\Psi}^{(0)}\mathbf{s}^{(0)} + \boldsymbol{\Psi}^{(1)}\mathbf{s}^{(1)} + \cdots + \boldsymbol{\Psi}^{(R-1)}\mathbf{s}^{(R-1)} \tag{11.5}$$

where:

$\mathbf{y}$ is the stacked vector representing the measurements from all aperture points
$\mathbf{s}^{(r)}$ is the vectorized image of the scene corresponding to the rth path
The matrix $\boldsymbol{\Psi}^{(r)}$ is a dictionary of the radar response that embodies the GO propagation model for the rth path.

More specifically, the qth column of $\boldsymbol{\Psi}^{(r)}$ is a concatenation of the received signals at all aperture points for the rth path due to a point target located at pixel q with position vector $\mathbf{r}_q$. As such, the qth column of $\boldsymbol{\Psi}^{(r)}$ depends on both the UWB excitation and the two-way propagation delays between the aperture points and the pixel q.

The various propagation delays can be readily calculated using GO considerations [67]. For illustration, consider the antenna–target geometry shown in Figure 11.17, where the front wall has been ignored for simplicity. Multipath propagation consists of the forward propagation from the antenna to the target along the path P'' and the return from the target via a reflection at the interior wall along the path P'. Assuming specular reflection at the wall interface, we observe from Figure 11.17 that reflecting the return path about the interior wall yields an alternative antenna–target geometry. We obtain a virtual target as shown in Figure 11.17, and the delay associated with path P' is the same as that of the path $\tilde{P}'$ from the virtual target to the antenna. This correspondence simplifies the calculation of the one-way propagation delay associated with path P'. From the position of the virtual target, we can calculate the propagation delay along path P' as follows. Under the assumption of free-space propagation, the delay can be simply calculated as the Euclidean distance from the virtual target to the receiver divided by the propagation speed of the wave. For the specific case of STTW, however, the wave has to pass through the front wall on its way from the virtual target to the receiver. As the front wall parameters are assumed to be known *a priori*, the delay can be readily calculated using Snell's law [67].

Next, using Equation 11.5, a stacked signal model is formed

$$\mathbf{y} = \overline{\boldsymbol{\Psi}}\overline{\mathbf{s}} \tag{11.6}$$

with a combined dictionary

$$\overline{\boldsymbol{\Psi}} = \begin{bmatrix} \boldsymbol{\Psi}^{(0)} & \boldsymbol{\Psi}^{(1)} & \cdots & \boldsymbol{\Psi}^{(R-1)} \end{bmatrix} \tag{11.7}$$

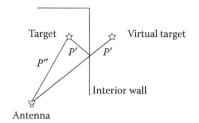

FIGURE 11.17
Multipath propagation via reflection at an interior wall. (© 2014 IEEE. Adapted, with permission, from Leigsnering, M. et al., Multipath exploitation in through-the-wall radar imaging using sparse reconstruction. *IEEE Transactions on Aerospace & Electronic Systems*, 50, pp. 920–939, 2014.)

and stacked image vector

$$\bar{\mathbf{s}} = \left[\left(\mathbf{s}^{(0)} \right)^{T} \quad \left(\mathbf{s}^{(1)} \right)^{T} \quad \cdots \quad \left(\mathbf{s}^{(R-1)} \right)^{T} \right]^{T} \tag{11.8}$$

where the superscript "T" denotes matrix transpose.

Relying on the multipath signal model established in Equation 11.8, we proceed to invert the model by means of sparse reconstruction. We aim at undoing the ghosts, that is inverting the multipath measurement model and achieving an image reconstruction, wherein only the true targets remain.

In practice, any prior knowledge about the exact relationship between the R images $\mathbf{s}^{(0)}, \mathbf{s}^{(1)}, \dots, \mathbf{s}^{(R-1)}$ is either limited or nonexistent. However, we know with certainty that these images describe the same underlying scene. That is, if a certain element in, for example $\mathbf{s}^{(0)}$ has a nonzero value, the corresponding elements in the other images should be also nonzero. This means that corresponding pixels in the image vectors should be grouped, as shown in Figure 11.18, necessitating the application of group sparse reconstruction

$$\hat{\bar{\mathbf{s}}} = \arg\min_{\bar{\mathbf{s}}} \frac{1}{2} \left\| \mathbf{y} - \bar{\mathbf{\Psi}}\bar{\mathbf{s}} \right\|_2^2 + \alpha \left\| \bar{\mathbf{s}} \right\|_{2,1} \tag{11.9}$$

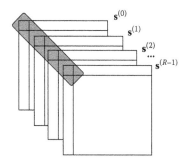

FIGURE 11.18
Group sparse structure of the images corresponding to the various propagation paths. (© 2014 IEEE. Adapted, with permission, from Leigsnering, M. et al., Multipath exploitation and suppression for SAR imaging of building interiors: An overview of recent advances. *IEEE Signal Processing Magazine*, 31, 110–119, 2014.)

where α is the so-called regularization parameter that trades off least square data fitting and sparse solution,

$\| \cdot \|_2$ represents the squared l_2-norm (the sum of the square magnitudes of the elements), and

$$\|\bar{\mathbf{s}}\|_{2,1} := \sum_q \left\| \left[s_q^{(0)}, s_q^{(1)}, \ldots, s_q^{(R-1)} \right]^T \right\|_2 = \sum_q \sqrt{\sum_{r=0}^{R-1} s_q^{(i)} \left(s_q^{(i)} \right)^*} \tag{11.10}$$

is the mixed $l_1 - l_2$ norm term that ensures the group structure in the sparse reconstruction. As defined in Equation 11.10, the mixed $l_1 - l_2$ norm behaves like a l_1 norm (sum of absolute element values) on the vector

$$\left\| \left[s_q^{(0)}, s_q^{(1)}, \ldots, s_q^{(R-1)} \right]^T \right\|_2, \forall q$$

and therefore, induces group sparsity. In other words, each

$$\left\| \left[s_q^{(0)}, s_q^{(1)}, \ldots, s_q^{(R-1)} \right]^T \right\|_2,$$

and equivalently each

$$\left[s_q^{(0)}, s_q^{(1)}, \ldots, s_q^{(R-1)} \right]^T,$$

is encouraged to be set to zero. On the other hand, within the groups, the l_2 norm does not promote sparsity [75]. The convex optimization problem in Equation 11.9 can be solved using SparSA [76], YALL group [77], or other available schemes [78,79]. Optionally, a downsampling of the measurements in Equation 11.6 can be done to reduce the amount of data. However, special care has to be taken to ensure that the coherence (maximum normalized inner product between any two distinct columns) of the product of the sampling matrix and the dictionary is sufficiently small in order to guarantee reliable image reconstruction [80].

Once a solution $\hat{\mathbf{s}}$ is obtained, the images corresponding to the R paths can be noncoherently combined to form an overall image with improved target to clutter and noise ratio, with the elements of the composite image $\hat{\mathbf{s}}_{GS}$ defined as

$$\left[\hat{\mathbf{s}}_{GS} \right]_q = \left\| \left[s_q^{(0)}, s_q^{(1)}, \ldots, s_q^{(R-1)} \right]^T \right\|_2, \forall q. \tag{11.11}$$

The Radar Imaging Lab at Villanova University conducted an experiment in a semicontrolled environment to demonstrate multipath suppression. The STTW radar imaged a single aluminum pipe (61 cm long, 7.6 cm diameter) placed upright on a 1.2-m high foam pedestal at 3.67 m downrange and 0.31 m cross-range, as shown in Figure 11.19. Imaging used a 77-element uniform linear monostatic array with an interelement spacing of 1.9 cm. The origin of the coordinate system is set at the center of the array. The 0.2-m thick concrete front wall was located parallel to the array at 2.44 m downrange. The left sidewall was at a cross-range of −1.83 m, whereas the back wall was at 6.37 m downrange as shown in Figure 11.19. Also, there was a protruding corner on the right at 3.4 m cross-range and 4.57 m downrange. A stepped-frequency signal, consisting of 801 equally spaced

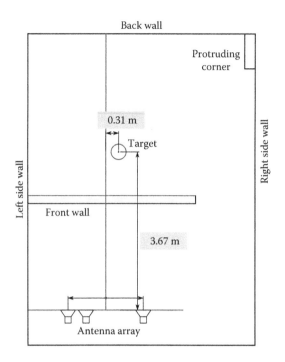

FIGURE 11.19
Layout of the radar experimental scenario used in the sparsity-based multipath exploitation study described in [72]. (© 2014 IEEE. Adapted, with permission, from Leigsnering, M. et al., Multipath exploitation in through-the-wall radar imaging using sparse reconstruction. *IEEE Transactions on Aerospace & Electronic Systems*, 50, pp. 920–939, 2014.)

frequency steps covering the 1–3-GHz band was employed. The left and right side walls were covered with RF-absorbing material, but the protruding right corner and the back wall were left uncovered.

Background subtracted data were considered to focus only on target multipath. Figure 11.20(a) depicts the back-projection image using all available data. Apparently, only the multipath ghosts due to the back wall, and the protruding corner in the back right, are visible. Hence, only these two multipath propagation cases were considered for the group sparse scheme. A downsampling of the measurements was performed and only 25% of the aperture positions and 50% of the frequencies were employed in sparse reconstruction. The corresponding result is shown in Figure 11.20(b). The multipath ghosts have been clearly suppressed.

Recall that the multipath exploitation method described above leverages prior information of the indoor scattering environment to remove ghost targets and improve image quality. In practice, precise prior knowledge of the interior wall locations is usually not available. The wall locations rather have to be estimated from the radar returns using building layout estimation techniques, such as in references [81] and [2], Chapter 2. These estimates are subject to errors that can be on the order of STTW system wavelengths. Multipath exploitation requires accurate knowledge of the room layout in order to deliver high-quality images. In the presence of wall location errors, two mechanisms lead to degradation of the reconstructed image quality. First, the returns from a specific multipath are coherently combined in the measurement model. Since the wall location errors cause the expected multipath delays to deviate from the actual delays, the coherence of the multipath

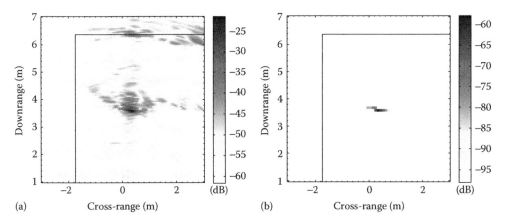

FIGURE 11.20
Multipath exploitation can produce ghost-free images from the experiment shown in Figure 11.19. (a) Back-projection image with full data volume and (b) group sparse reconstruction with 25% of the aperture positions and 50% of the frequencies. (© 2014 IEEE. Adapted, with permission, from Leigsnering, M. et al., Multipath exploitation in through-the-wall radar imaging using sparse reconstruction. *IEEE Transactions on Aerospace & Electronic Systems*, 50, pp. 920–939, 2014.)

returns is lost, resulting in a mismatch between the dictionary and the received signal. Second, wall location errors may lead to a misalignment of the various subimages in the signal model, thereby violating the group sparse structure of $\bar{\mathbf{s}}$. The perfect alignment of the various images is only guaranteed if the multipath delays are equal to the assumed ones.

In order to deal with the aforementioned issues, it is imperative to take wall position uncertainties into account in the sparse reconstruction process. To this end, a parameterization of the wall locations is used in the signal model of Equation 11.6 to accommodate uncertainties in the knowledge of the scattering environment [82]. This leads to a joint minimization over the image vector $\bar{\mathbf{s}}$ and the wall position vector $\mathbf{w}$

$$\min_{\bar{\mathbf{s}},\mathbf{w}} \frac{1}{2}\left\|\mathbf{y} - \bar{\mathbf{\Psi}}(\mathbf{w})\bar{\mathbf{s}}\right\|_2^2 + \alpha\left\|\bar{\mathbf{s}}\right\|_{2,1} \tag{11.12}$$

The image reconstruction in Equation 11.12 is now a nonconvex optimization problem as the dictionary has a nonlinear dependence on the wall locations [82]. This is addressed by expressing (12) as a nested optimization problem

$$\min_{\mathbf{w}} \min_{\bar{\mathbf{s}}} \frac{1}{2}\left\|\mathbf{y} - \bar{\mathbf{\Psi}}(\mathbf{w})\bar{\mathbf{s}}\right\|_2^2 + \alpha\left\|\bar{\mathbf{s}}\right\|_{2,1} \tag{11.13}$$

which consists of a convex part and a nonconvex part.

More specifically, the inner optimization over $\bar{\mathbf{s}}$ is convex, which is exactly the same as the reconstruction problem in Equation 11.9 and can be solved efficiently. The outer minimization is nonconvex; however, the dimension of the solution space is much smaller and, thus, easier to search. The outer nonconvex problem in Equation 11.12 can be solved by general nonlinear optimization methods; examples being Quasi-Newton (QN) methods using finite-difference gradients [83] or heuristic methods, such as particle swarm optimization (PSO) [84].

11.6 Conclusion

In this chapter, we considered the problem of imaging building interiors using UWB radar emissions. We provided an overview of EM scattering phenomenology and recent algorithmic advances in UWB STTW radar. First, we reviewed the antenna requirements for a UWB STTW radar and provided examples of antennas employed in existing STTW radar systems. Second, in an effort to provide an understanding of the phenomenology involved in STTW, we presented computer models of complex and realistic STTW scenarios and provided corresponding imaging results, which represent the "best case scenario" for radar performance. Third, we discussed two methods, namely, coherent and noncoherent change detection, for moving target indication in STTW radar applications. The advantages and disadvantages of each method were highlighted and performance comparison was provided using real data. Finally, a sparsity-based approach to exploit the rich indoor multipath environment for improved target detection was described under both known and uncertain wall locations. This multipath exploitation technique was shown to produce an image of the scene without "ghosting."

Although the progress reported in this chapter is substantial and notable, many challenging scenarios remain unresolved using the current techniques. As such, further research and development efforts are required. However, with continuing advances in hardware and system architectures, a definite increase in opportunities for handling more complex building scenarios is expected.

References

1. Amin, M. (Ed.). 2011. *Through-the-Wall Radar Imaging*, CRC Press, Boca Raton, FL.
2. Amin, M. (Ed.). 2015 *Compressive Sensing for Urban Radar*, CRC Press, Boca Raton, FL.
3. Skolnik, M. 2001. *Introduction to Radar Systems*, McGraw Hill, New York, NY.
4. Jol, H. 2009. *Ground Penetrating Radar Theory and Applications*, Elsevier Science, Oxford, UK.
5. Soumekh, M. 1999. *Synthetic Aperture Radar Signal Processing*, Wiley, New York, NY.
6. Li, J. and Stoica, P. 2009. *MIMO Radar Signal Processing*; Wiley, Hoboken, NJ.
7. Nguyen, L., Ressler, M. and Sichina, J. 2008. "Sensing through the wall imaging using the Army Research Lab ultra-wideband synchronous impulse reconstruction (UWB SIRE) radar," *Proceedings of SPIE*, vol. 6947.
8. Sevigny, P., DiFilippo, D., Laneve, T. and Fournier, J. 2012. "Indoor imagery with a 3-D through-wall synthetic aperture radar," *Proceedings of SPIE*, vol. 8361.
9. Clark, B. 2009. "Trial results of a stand-off sense-through-the-wall radar," *Proceedings of the 55th Annual MSS Tri-Service Radar Symposium*, Boulder, CO.
10. Nag, S. and Barnes, M. 2003. "A moving target detection filter for an ultra-wideband radar," *Proceedings of the IEEE Radar Conference*, pp. 147–153.
11. Camero-Technologies Ltd. Web page. http://www.camero-tech.com/products.php.
12. AKELA, Inc. Web page. http://www.akelainc.com/products/astir.
13. Cambridge Consultants Ltd. Web page. http://www.cambridgeconsultants.com/projects/prism-200-through-wall-radar.
14. Balanis, C. 2002. *Antenna Analysis and Design*, Wiley, New York, NY.
15. Lisuzzo, A. et al., 2011. *Sensing through-the-wall technologies*, NATO SET-100 Final Report, RTO-TR-SET-100.

16. Dogaru, T. and Le, C. 2009. *Through-the-Wall Small Weapon Detection Based on Polarimetric Radar Techniques*; ARL-TR-5041, U.S. Army Research Laboratory, Adelphi, MD.

17. Yemelyanov, K., Engheta, N., Hoorfar, A. and McVay, J. 2009. "Adaptive polarization contrast techniques for through-wall microwave imaging applications," *IEEE Transactions on Geoscience and Remote Sensing*, vol. 47, no. 5, pp. 1362–1374.

18. Dogaru, T. and Le, C. 2012. *Through the wall radar simulations using polarimetric antenna patterns*, ARL Technical Report, Adelphi, MD, ARL-TR-5951.

19. Smith, G. 2012. *Wideband transverse electromagnetic horn antenna design for ultra wideband synthetic aperture radar VHF/UHF radar*, ARL-TR-6279, U.S. Army Research Laboratory, Adelphi, MD, Dec. 2012.

20. ETS Lindgren Web page. http://www.ets-lindgren.com/3164-06.

21. Ahmad, F. and Amin, M. G. 2008. "Multi-location wideband synthetic aperture imaging for urban sensing applications," *Journal of the Franklin Institute*, vol. 345, no. 6, pp. 618–639.

22. Soldovieri, F., Ahmad, F. and Solimene, R. 2011. "Validation of microwave tomographic inverse scattering approach via through-the-wall experiments in semicontrolled conditions," *IEEE Geoscience and Remote Sensing Letters*, vol. 8, no. 1, pp. 123–127.

23. Amin, M. and Ahmad, F. 2013. "Change detection analysis of humans moving behind walls," *IEEE Transactions on Aerospace and Electronic Systems*, vol. 49, no. 3, pp. 1410–1423.

24. Tatoian, J. et al. 2008. "Introduction to polychromatic SAR," *Proceedings of the IEEE Antennas and Propagation Symposium*, San Diego, CA.

25. Schantz, H. and Fullerton, L. 2001. "The diamond dipole: a Gaussian impulse antenna," *Proceedings of the IEEE Antennas and Propagation Symposium*, pp. 100–103.

26. L3 Communications Cyterra Web page. http://www.cyterra.com/products/ranger.htm.

27. Janaswamy, R. and Schaubert, D. 1987. "Analysis of the tapered slot antenna," *IEEE Transactions on Antennas and Propagations*, vol. 35, no. 7, pp. 1058–1065.

28. Langley, J., Hall, P. and Newham, P. 1993. "Novel ultrawide-bandwidth Vivaldi antenna with low crosspolarisation," *Electronics Letters*, vol. 29, pp. 204–2005.

29. Smith, G., Harris, R., Ressler, M. and Stanton, B. 2006 *Wideband Vivaldi notch antenna design for UWB SIRE VHF/UHF radar*, ARL-TR-4409, U.S. Army Research Laboratory, Adelphi, MD.

30. Yang, Y. and Fathy, A. 2009. "Development and implementation of a real-time see-through-wall radar system based on FPGA," *IEEE Transactions on Geoscience and Remote Sensing*, vol. 47, no. 5, pp. 1270–1280.

31. Komanduri, V., Hoorfar, A. and Engheta, N. 2005 "Low-profile array design considerations for through-the-wall microwave imaging applications," *Proceedings of the IEEE Antennas and Propagation Symposium 2005*, Washington DC, pp. 338–341.

32 Taflove, A. and Hagness, S.C. 2000. *Computational Electrodynamics: The Finite-Difference Time-Domain*, Artech House, Norwood, MA.

33. Stowell, M., Fasenfest, B. and White, D. 2008. "Investigation of radar propagation in buildings: a 10-billion element Cartesian-mesh FETD simulation," *IEEE Transactions on Antennas and Propagation*, vol. 56, no. 8, pp. 2241–2250.

34. Dogaru, T. 2010. *AFDTD user's manual*, ARL Technical Report, Adelphi, MD, ARL-TR-5145.

35. Knott, E., Tuley, M. and Shaeffer, J. 1993. *Radar Cross Section*, Artech House, Norwood, MA.

36. Dogaru, T., Nguyen, L. and Le, C. 2007. *Computer Models of the Human Body Signature for Sensing Through the Wall Radar Application*; ARL-TR-4290, U.S. Army Research Laboratory, Adelphi, MD.

37. Dogaru, T. and Le, C. 2010. *Through-the-wall radar simulations for complex room imaging*, ARL Technical Report, Adelphi, MD, ARL-TR-5205.

38. Le, C. and Dogaru, T. 2012. "Synthetic aperture radar imaging of a two-story building," *Proceedings of SPIE*, Baltimore, MD, vol. 8361.

39. Jakowatz, C., Wahl, D., Eichel, P., Ghiglia, D. and Thompson, P. 1996. *Spotlight-Mode Synthetic Aperture Radar: A Signal Processing Approach*, Kluwer Academic Publishers, Norwell, MA.

40. ARL DSRC Web page. http://www.arl.hpc.mil.

41. Dogaru, T., Liao, D. and Le, C. 2012. *Three-dimensional radar imaging of a building*, ARL Technical Report, Adelphi, MD, ARL-TR-6295.

42. Fink, M. 1992 "Time reversal of ultrasonic fields – Part I: Basic principles," *IEEE Transactions on Ultrasonics, Ferroelectrics and Frequency Control*, vol. 39, pp. 555–566.

43. Schechter, R. and Chun, S. 2010. "High resolution 3-D Imaging of objects through walls," *Optical Engineering*, vol. 49.

44. Debes, C., Amin, M. and Zoubir, A. 2009. "Target detection in single and multiple-view through-the-wall radar imaging," *IEEE Transactions on Geoscience and Remote Sensing*, vol. 47, no. 5, pp. 1349–1361.

45. Mostafa, A., Debes, C. and Zoubir, A. 2012. "Segmentation by classification for through-the-wall radar imaging using polarization signatures," *IEEE Transactions on Geoscience and Remote Sensing*, Vol. 50, pp. 3425 – 3439.

46. Yoon, Y. and Amin, M. 2009. "Spatial filtering for wall-clutter mitigation in through-the-wall radar imaging," *IEEE Transactions on Geoscience and Remote Sensing*, Vol. 47, pp. 3192–3208.

47. Tivive, F., Bouzerdoum, A. and Amin, M. 2015. "A subspace projection approach for wall clutter mitigation in through-the-wall radar imaging," *IEEE Transactions on Geoscience and Remote Sensing*, vol. 53, no. 4, pp. 2108–2122.

48. Ahmad, F., Qian, J. and Amin, M. 2015. "Wall clutter mitigation using Discrete Prolate Spheroidal Sequences for sparse reconstruction of indoor stationary scenes," *IEEE Trans. Geosci. Remote Sens.*, vol. 53, no. 3, pp. 1549–1557.

49. Zhu, Z. and Wakin, M. 2015. "Wall clutter mitigation and target detection using discrete prolate spheroidal sequences," *Proceedings of the 3rd Int. Workshop on Compressed Sensing Theory and its Applications to Radar, Sonar and Remote Sensing*, Pisa, Italy.

50. Ahmad, F., Amin, M.G. and Dogaru, T. 2014. "Partially sparse imaging of stationary indoor scenes," *Eurasip Journal on Advances in Signal Processing*, Special Issue on Sparse Sensing in Radar and Sonar Signal Processing, 2014:100.

51. Mansour, H. and Liu, D. 2013. "Blind multi-path elimination by sparse inversion in through-the-wall-imaging," in *Proc. IEEE Int. Workshop on Computational Advances in Multi-Sensor Adaptive Processing*, Saint Martin.

52. Dogaru, T. 2013. *Doppler processing with ultra-wideband impulse radar*, ARL Technical Note, Adelphi, ARL-TN-0529.

53. Cohen, L. 1994. *Time Frequency Analysis*, Prentice Hall, Englewood Cliffs, NJ.

54. Smith, G., Ahmad, F. and Amin, M. 2012. "Micro-Doppler processing for ultra-wideband radar data," *Proceedings of SPIE*, Vol. 8361.

55. Martone, A., Ranney, K. and Le, C. 2014. "Noncoherent approach for through-the-wall moving target indication," *IEEE Transactions on Aerospace and Electronic Systems*, vol. 50, no. 1, pp. 193–206.

56. Ahmad, F. and Amin, M.G. 2012. "Through-the-wall human motion indication using sparsity-driven change detection," *IEEE Transactions on Geoscience and Remote Sensing*, vol. 51, no. 2, pp. 881–890.

57. Maaref, N. et al., 2009. "A study of UWB FM-CW radar for the detection of human beings in motion inside a building," *IEEE Transactions on Geoscience and Remote Sensing*, vol. 47, no. 5, pp. 1297–1300.

58. Soldovieri, F., Solimene, R. and Pierri, R. 2009. "A simple strategy to detect changes in through the wall imaging," *Progress in Electromagnetic Research*, vol. 7, pp. 1–13.

59. Hunt, A. 2009. "Use of a frequency-hopping radar for imaging and motion detection through walls," *IEEE Transaction on Geophysics and Remote Sensing*, vol. 47, no. 5, pp. 1402–1408.

60. Martone, A., Innocenti, R. and Ranney, K. 2009. *Moving Target Indication for Transparent Urban Structures*, ARL-TR-4809, U.S. Army Research Laboratory, Adelphi, MD.

61. Novak, L., Owirka, G. and Netishen, C. 1993. "Performance of a high-resolution polarimetric SAR automatic target recognition system," *The Lincoln Laboratory Journal*, vol. 6, pp. 11–24.

62. Wilpon, J. and Rabiner, L. 1985. "A modified K-means clustering algorithm for use in isolated work recognition," *IEEE Transactions on Acoustics, Speech and Signal Processing*, vol. 33, no. 6, pp. 587–594.

63. Blackman, S., and Popoli, R. 1999. *Design and Analysis of Modern Tracking Systems*, Artech House, Norwood, MA.
64. Dogaru, T. et al., 2014. "Multipath effects and mitigation in sensing through the wall radar," *Proceedings of the 60th Annual MSS Tri-Service Radar Symposium*, Springfield, VA, Jul. 2014.
65. Tan, Q. and Song, Y. 2010. "A new method for multipath interference suppression in through-the-wall UWB radar imaging," *Proceedings of the Int. Conf. on Advanced Computer Control*, vol. 5, Shenyang, China, pp. 535–540.
66. Burkholder, R. 2009. "Electromagnetic models for exploiting multi-path propagation in through-wall radar imaging," in *Proc. Int. Conf. on Electromagnetics in Advanced Applications*, Torino, Italy, pp. 572–575.
67. Setlur, P., Amin, M. and Ahmad, F. 2011. "Multipath model and exploitation in through-the-wall and urban radar sensing," *IEEE Transactions on Geoscience and Remote Sensing*, vol. 49, no. 10, pp. 4021–4034.
68. Kidera, S., Sakamoto, T. and Sato, T. 2011. "Extended imaging algorithm based on aperture synthesis with double scattered waves for UWB radars," *IEEE Transactions on Geoscience and Remote Sensing*, vol. 49, no. 12, pp. 5128–5139.
69. Gennarelli, G. and Soldovieri, F. 2013. "A linear inverse scattering algorithm for radar imaging in multipath environments," *IEEE Geoscience and Remote Sensing Letters*, vol. 10, no. 5, pp. 1085–1089.
70. Gennarelli, G., Catapano, I. and Soldovieri, F. 2013. "RF/microwave imaging of sparse targets in urban areas," *IEEE Antennas Wireless Propagation Letters*, vol. 12, pp. 643–646.
71. Leigsnering, M., Amin, M.G., Ahmad, F. and Zoubir, A.M. 2014. "Multipath Exploitation and Suppression in SAR Imaging of Building Interiors," *IEEE Signal Processing Magazine*, vol. 31, no. 4, pp. 110–119.
72. Leigsnering, M., Amin, M.G., Ahmad, F. and Zoubir, A.M. 2014."Multipath Exploitation in Through-the-Wall Radar Imaging using Sparse Reconstruction," *IEEE Transactions on Aerospace and Electronic Systems*, vol. 50, no. 2, pp. 920–939.
73. Amin, M. and Ahmad, F. 2013. "Through-the-Wall Radar Imaging," in R. Chellappa and S. Theodoridis (Eds.), *Academic Press Library in Signal Processing: Communications and Radar Signal Processing*, vol. 2, Elsevier.
74. Amin, M. and Ahmad, F. 2008. "Wideband synthetic aperture beamforming for through-the-wall imaging [lecture notes]," *IEEE Signal Processing Magazine*, vol. 25, no. 4, pp. 110–113.
75. Bach, F., Jenatton, R., Mairal, J. and Obozinski, G. 2011. "Convex optimization with sparsity-inducing norms," in *Optimization for Machine Learning*, Sra, S., Nowozin, S. and Wright, S. J. Eds., MIT Press, Cambridge, MA.
76. Wright, S., Nowak, R. and Figueiredo, M. 2009. "Sparse reconstruction by separable approximation," *IEEE Transactions on Signal Processing*, vol. 57, no. 7, pp. 2479–2493.
77. Deng, W., Yin, W. and Zhang, Y. 2011. "Group sparse optimization by alternating direction method," Department of Computational and Applied Mathematics, Rice University, Technical Report TR11–06.
78. Baraniuk, R.G., Cevher, V., Duarte, M.F. and Hegde, C. 2010. "Model-based compressive sensing," *IEEE Transactions on Information Theory*, vol. 56, pp. 1982–2001. [Online]. Available: http://arxiv.org/abs/0808.3572
79. Eldar, Y., Kuppinger, P. and Bolcskei, H. 2010. "Block-sparse signals: Uncertainty relations and efficient recovery," *IEEE Transactions on Signal Processing*, vol. 58, no. 6, pp. 3042–3054.
80. Wakin, M.B. 2015. "Compressive Sensing Fundamentals," in *Compressive Sensing for Urban Radar*, Amin, M. (ed.), CRC Press, Boca Raton, FL.
81. Lagunas, E., Amin, M. G., Ahmad, F. and Na´jar, M. 2013. "Determining building interior structures using compressive sensing," *Journal of Electronic Imaging*, vol. 22, no. 2, pp. 021003.
82. Leigsnering, M., Ahmad, F., Amin, M. and Zoubir, A. 2016. "Parametric dictionary learning for sparsity-based TWRI in multipath environments," *IEEE Transactions on Aerospace and Electronic Systems*, vol. 52, no. 2, pp. 532–547.

83. Gill, P.E., Murray, W. and Wright, M.H. 1981. *Practical optimization*. London, UK: Academic Press.
84. Poli, R., Kennedy, J. and Blackwell, T. 2007. "Particle swarm optimization," *Swarm Intelligence*, vol. 1, no. 1, pp. 33–57.
85. Yang, Y. et al., 2008. "Design of compact Vivaldi antenna arrays for UWB see through wall applications," *Prog. Electromagnetic Res.*, vol. 82, pp. 401–418.

12

Wideband Wide Beam Motion Sensing

François Le Chevalier

CONTENTS

12.1 Introduction

Surveillance radars need to accurately determine target positions and movements. This chapter shows how widening the bandwidth of a generic radar improves its Doppler or angular resolution. It develops the required bandwidths for obtaining such gains. It shows the processing algorithms necessary to give access to the required surveillance and target analysis functions.

This chapter uses the results of different studies conducted by N. Petrov, T. Faucon, G. Pinaud, G. Babur, P. Aubry, and Y. He in TU Delft; S. Bidon, F. Deudon, and M. Lasserre in ISAE Toulouse; J.P. Guyvarch, A. Becker, G.E. Michel and G. Desodt in Thales; and O. Rabaste and L. Savy in ONERA.

12.1.1 Surveillance Radar System Requirements

Despite the wide diversity of surveillance radar systems – from long-range air and ground/surface surveillance to short-range or terminal area surveillance or protection, or inside building surveillance – we can identify a few basic and permanent requirements:

1. Detection, tracking, imaging, and classification of all targets on-the-fly – if possible, simultaneously!
2. Detecting difficult targets with slow speeds (a few m/s), low RCS (-20 dBm^2), at low altitude would be a discriminating feature for most air- or ground-based surveillance systems. The requirement applies to long- or short-range defense and security applications.
3. Difficult environments, such as urban or coastal locations and high sea states, have gained importance for defense and security applications. The increasing development of wind farms adds a specific and difficult component to those varied sources of undesired echoes. The fast-growing availability of military and civilian radio-frequency devices presents an omnipresent jamming threat to radar performance.

These long-lasting requirements will lead to designing fast reaction systems than can perform multiple functions quasi-simultaneously, over wide angular coverages, and at short and long ranges.

Obviously, there is no single answer to such wide-ranging requirements. However, radar with its natural ability to search in large volumes or areas in adverse climate conditions is certainly a key component for future surveillance systems. Some limitations still exist, especially regarding the fine analysis capabilities, compared to optronic systems. Often a specific surveillance mission will require integrating multiple and diverse remote sensor systems.

However, we will see in this chapter how wideband systems – and in some cases ultrawideband – will provide increased capabilities to answer the three requirements stated above. Moving target detection and analysis presents the greatest problem in target detection. More precisely, we will demonstrate how wider bandwidths and beamwidths combined with appropriate waveforms and processing techniques are key elements in the design of future surveillance systems. These signal-processing techniques can provide improved angular and velocity discrimination in adverse clutter and jamming conditions.

12.1.2 Radar Spatial Resolution

In standard radar design, it is well known that the frequency bandwidth ΔF of a radar defines its range resolution ($\delta R = c/2\Delta F$), the dimension L (width and height) of the antenna defines its angular resolutions (in azimuth and elevation) as ($\delta\theta = c/2L$), and the coherent time integration T_i defines its velocity (Doppler) resolution ($\delta v = \lambda/2T_i$).

However, these basic principles only apply to standard radar designs with a focused pencil beam, a standard pulse compression waveform, and no ambiguities in range or Doppler. In other words, these principles, loosely based on Cramer–Rao inequalities, only give lower bounds on the achievable resolutions. In actual practice:

- It may prove valuable to use a wider beam for simultaneous coverage of a wide area.

- It is also very often necessary to remove the ambiguities in range or Doppler, which usually require several bursts of narrowband coherent pulses, rather than one long coherent burst.

In other words, the best radar design does not generally correspond to the best achievable resolution in every domain – angle, range, and velocity. The designer must implement design tradeoffs, guided by update rate constraints or ambiguity/eclipses removal.

12.1.3 Range Resolutions with Wideband Systems

Moreover, for wider bandwidths (typically with fractional bandwidths in the order of 10% or higher), the standard reasoning about range–Doppler–angle measurements properties becomes less intuitive. The designer must consider the existence of coupling terms between the different axes. The Doppler effect depends on carrier frequency, as does the angular location measured with a phased array. In this situation, the standard decoupling between the three (or four) basic measurements: range, Doppler, and angle(s) – resulting from analysis in the respective three domains: fast-time, slow-time, and space – is not true anymore. If those couplings do not prevent application of a classical matched filter approach on receive, they still lead to much heavier computations, and also give rise to annoying sidelobes. But on the other hand, these coupling terms provide complementary information, which can help to remove the usual ambiguities, and thus improve the overall radar performance. We will see this later in the chapter.

In this chapter, we analyze these wideband effects in detail and identify generic radar designs, which can benefit from these properties. More precisely, we show how widening the bandwidth of a generic radar may in the end improve its Doppler or angular resolution, and what are the required bandwidths for obtaining such gains, and which processing algorithms are necessary to give access to the required surveillance and target analysis functions.

12.1.4 Availability of Wider Bandwidths

From an operational perspective, trying to improve radar designs by widening their bandwidth also makes sense, because wider bandwidths become available for the different components and subsystems – starting with the receivers, which tend to become digital with high signal throughput, typically 100 MHz per channel. Obviously, extending the dimension of an antenna, or lengthening the coherent integration time, is much more challenging task and directly impacts the operational features including transportability and reactivity. Therefore, playing with the signal's bandwidth looks like a much better solution, even though limitations due to spectrum overcrowding have also to be tackled [1].

12.1.5 Technical Approach to Bandwidth Expansion

In the following sections, we first consider the coupling between fast-time and slow-time, which occurs for wideband signals. This will show the way to take advantage of this coupling for ambiguity removal and target analysis.

We will then analyze the relation between angle and range resolutions for a typical long-range phased array surveillance radar. This relation will demonstrate a clear trade-off between the qualities of these measurements. Analyzing this tradeoff with a class of space–time codes (circulating codes) also clarifies the relations between resolutions and sidelobe levels – which are key drivers of radar performance.

Finally, a careful analysis of a short-range multistatic detection problem will demonstrate that the basic tools required for moving targets detection and classification at long range can also work for short-range detection and target analysis and classification with UWB systems.

All these analyses will essentially deal with moving target surveillance and analysis. This requires canceling clutter echoes – usually at least 50 dBs stronger than the typical target echoes, which puts very strict limits to the admissible sidelobe levels. This necessity prevents adoption of more specific approaches, which exhibit spectacular resolution improvements in nonrealistic environments. More precisely, this chapter will also emphasize the importance of considering the sidelobe level while taking into account realistic dynamic range of the received signals, when trying to improve radar resolutions.

12.2 Diversity: Target Coherence and Diversity Gains

For narrowband radar design purposes, the target is supposed to be an isotropic (in aspect angle) white (in frequency) scatterer, which means received signals can be coherently added on reception. In reality, and for large bandwidths, we may accurately represent the target as distributed isotropic scatterers, characterized by their position $\bar{x}$ relative to a specific point on the target and their complex diffraction coefficient $I(\bar{x})$. This specific nature of the target has consequences for radar system performance since it changes the result of the coherent summation and consequently the accuracy of the measurements and the fluctuation of the signals. The following section briefly summarizes and illustrates the main results.

12.2.1 Target Coherence

When modeling targets, the term *isotropic* means that the diffraction coefficient of the scatterer does not depend on aspect angle. The description *white* means that it does not depend on frequency in the angular sector and bandwidth considered for coherent integration. Though oversimplifying, this assumption is very generally used (e.g., for SAR and inverse SAR imaging) and has been shown [2] to provide excellent quality images, even if some artifacts may be observed (e.g., resonances, moving scatterers on the target).

Within this multiple isotropic and white scatterers model, the scattering coefficient of the target, $H(\vec{k})$, can be written as a function of $\vec{k}$, the wave vector (vector along the incidence angle, modulus $2\pi/\lambda$), as described in Figure 12.1:

$$H(\vec{k}) = \int I(\vec{x}) e^{-2j\vec{k}.\vec{x}} \, d\vec{x}, \tag{12.1}$$

Since this expression is a Fourier transform, it can be inverted to provide the image $I(\vec{x})$ of the target (to within a scalar coefficient, obtained through calibration), based on the available measurements $H(\vec{k})$:

$$I(\vec{x}) = \int H(\vec{k}) e^{+2j\vec{k}.\vec{x}} \, d\vec{k}, \tag{12.2}$$

where the limits of the integral are determined by the measurement system (usually a frequency bandwidth Δf around f_0 and an angular sector $\Delta\theta_0$ around θ_0).

This target model is the basis of holographic measurements [2,3] routinely used for target analysis. It also provides the basic parameters for sampling in the $\vec{k}$ domain and for resolution in the $\vec{x}$ domain, when observing a target with depth Δx and transverse dimension Δy, with an observation bandwidth Δf associated with an angular sector $\Delta\theta$, as described in Figure 12.2. These relations are mere consequences of the Fourier transform relationship between the measurements hologram $H(\vec{k})$ and the target image $I(\vec{x})$.

To show the effects of sampling and resolution, Figure 12.3 shows the image obtained on a real target drone (CT20). It shows a radar observation in the horizontal plane, with an angular sector width of 20° around 40° and a frequency bandwidth of 2 GHz centered at 9 GHz.

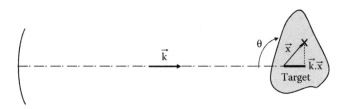

FIGURE 12.1
The target scattering coefficient assumes a collection of isotropic scatterers independent of frequency.

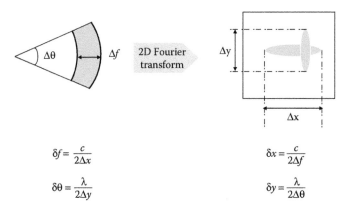

$$\delta f = \frac{c}{2\Delta x} \qquad\qquad \delta x = \frac{c}{2\Delta f}$$

$$\delta\theta = \frac{\lambda}{2\Delta y} \qquad\qquad \delta y = \frac{\lambda}{2\Delta\theta}$$

FIGURE 12.2
Sampling and resolution criteria.

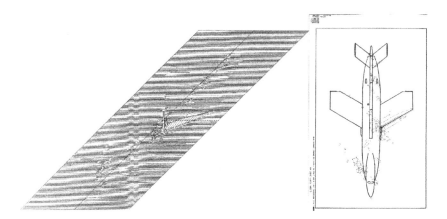

FIGURE 12.3
Example of a real ultrawideband radar target image using a 9-GHz center frequency signal with a 2-GHz bandwidth (Courtesy of ONERA). The high spatial resolution signal sees the CT 20 drone as a collection of multiple scattering centers at different ranges. Note how only certain areas have significant reflections.

These basic analysis tools provide the critical instantaneous bandwidth and the critical antenna array extent, which are essential parameters for target coherence. The critical bandwidth Δf_c is defined as the maximum bandwidth such that the target coefficient remains coherent

$$\Delta f_c = \frac{c}{2\Delta x} \tag{12.3}$$

In other words, if the bandwidth transmitted in the direction of the target is equal to or larger than Δf_c, then the received signals cannot be coherently summed; that is, the target is in fact resolved in range by the signal. For instance, if the maximum dimension of the target is $\Delta x = 30$ m, then $\Delta f_c = 5$ MHz. (Editor's note: See the discussion of radar spatial resolution and bandwidth selection based on target resonances in Chapter 3.)

As a consequence, when the bandwidth transmitted in the direction of the target is equal to or larger than Δf_c, this results in the targets being overresolved in range. This implies implementing some kind of distributed target integration, thus providing a diversity gain for fluctuating targets, as will be seen in the next Section.

Similarly, as shown in Figure 12.4, the critical array extent D_c is the maximum extent of the antenna array such that the target coefficient, observed at range R, remains coherent, so

$$D_c = \delta\theta \cdot R = \frac{\lambda}{2\Delta y} R \tag{12.4}$$

For instance, if $R = 100$ km, $\Delta y = 30$ m, and $\lambda = 3$ cm, we obtain $D_c = 50$ m. Thus, very generally, for monostatic systems, the extent of the antenna array is much smaller than the critical extent, and only bistatic or multistatic systems (also called multisite, or statistical MIMO, in the literature) can be used for resolving the target in angle.

These two critical parameters, maximum bandwidth Δf_c and critical array extent D_c, become the basic parameters in designing multiple-input multiple-output (MIMO) systems, as was also correctly analyzed by Chernyak [4], Dai et al. [5], and Wu et al. [6], where the full-rank observation matrix is equivalent to using an interelement spacing larger than D_c.

FIGURE 12.4
Critical array extent D_c is the maximum extent for antenna size, so the observed target coefficient at some range remains coherent.

12.2.2 Coherent vs Noncoherent Integration

Detection of fluctuating radar targets is limited by the presence of noise and by the fact that the target may provide only very small signals for certain presentation angles or frequencies of illumination (a phenomenon also known as target fading in the literature). In order to mitigate target fading, most radars use frequency agility:

1. They transmit successive bursts at different carrier frequencies.
2. When received, each burst is coherently processed as usual (Doppler filtering in each range cell).
3. The outputs of these coherent summations are noncoherently summed (sum of the modulus, or the squared modulus), before final detection thresholding is applied.

This way improves the signal-to-noise ratio (SNR) through each coherent burst processing. The resulting noncoherent summation allows consideration of observations at different frequencies, involving different target radar cross section (RCS).

For a high required probability of detection, "some" noncoherent integration is preferable, in order to avoid getting trapped in a low RCS zone, especially for highly fluctuating targets (e.g., Swerling 1 targets, in the standard classification of targets fluctuations [7]). "*Some*" means that coherent integration must first be used to get a sufficient signal-to-noise ratio (SNR), which should typically be larger than 0 dB after coherent integration, so that it is not too much degraded by the modulus operation, which, as every nonlinear operation, severely reduces the detection capability if it is done at low SNR. This explains the often used "Golden Rule": first improve SNR through coherent integration and then mitigate the low RCS zones by sending a few bursts with frequency agility from burst to burst. The price to pay for that noncoherent integration (and the associated diversity gain), when the available time-on-target is limited, is a lower Doppler resolution because of shorter coherent bursts.

Turning to wideband radars and using Parseval's theorem, we first observe that the energy in the squared modulus of the impulse response (range profile) is the same as the energy in the squared modulus of the frequency response. So, summing the energy of the impulse response along the length of the target is equivalent, from a detection point of view, to summing the energy of the corresponding frequency response.

Integration along the range profile of the target for a preassumed length of the targets of interest (e.g., 15 m for air targets) is a way to combine coherent integration, used to obtain the range profile with its associated Doppler spectra in each range cell, with noncoherent integration. In other words, for wideband radars, coherent integration time – and the clutter separation that it provides – need not be reduced to take benefit of diversity gain. Thus, summing the N bursts in each range cell of a narrowband agile radar is equivalent to summing the N samples of the range profile of a high-range resolution radar.

However, from a radar design point of view, there is a significant difference with a narrowband radar, because because the Doppler resolution is limited by the duration of each coherent burst. However, with a wideband radar, the Doppler resolution is limited by the overall time on target. This means that for similar global time on target and similar overall bandwidth, the Doppler resolution of the high-range resolution radar is N times higher than for a narrowband radar.

The next question is: how many resolution cells are required on the target, or what is the best resolution for detection? Figure 12.5 gives a generic answer by showing the required SNR per range cell as a function of the number of cells on the target: $N = 6$, or 10 (or equivalently the required SNR per frequency as a function of the number of frequencies used), for Swerling targets [7]. The diversity gain can be defined as the difference in the required SNR per sample, between the coherent summation case and the noncoherent summation case. It clearly appears that the diversity gain is maximum for $N = 6$: between 2.6 and 7 dB, depending on the required probability of detection, for this case of Swerling case 1 and 2. There is still a gain for higher resolutions, but it is much smaller.

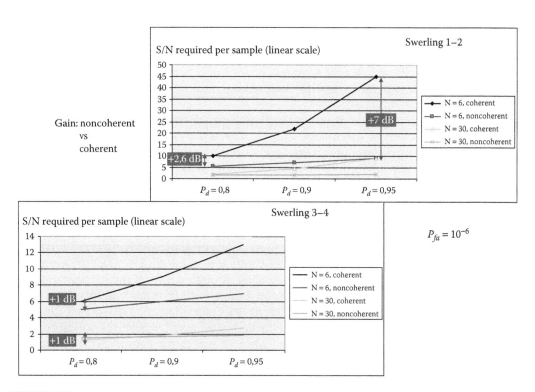

FIGURE 12.5
The effect of spatial resolution cells for fluctuating target detection. The traces show noncoherent versus coherent integration for $P_{fa} = 10^{-6}$.

Similar analyses with Swerling case 3 and 4 targets show that the gain is lower – as expected, since the fluctuations of Swerling 3 targets are smaller than those of Swerling 1 – but still exists, at least for detection probabilities larger than 0.8 (i.e., gain between 2.7 and 1 dB). The diversity gain would then become a loss for very high resolution of Swerling 3 targets (1 dB loss for a target analyzed in 30 cells and a required $P_d = 0.8$).

This diversity analysis leads to the following conclusions:

- Even if the target RCS per range cell is reduced compared with the low-resolution (standard) radars, the summation of the energies of adjacent range cells along the range profile provides a diversity effect that more than overcomes this RCS reduction.
- The more the target fluctuates, the higher the diversity gain.
- Higher requirements in the probability of detection P_d lead to a higher diversity gain.
- The smallest targets should be divided into 5–10 range cells, not more, which means that metric resolutions are sufficient for detecting standard air targets.
- For fluctuating targets, a few dBs can be gained from cutting the targets into pieces.
- This diversity gain can be obtained without incurring the penalty of lower Doppler resolution, as is the case with standard frequency agility in narrowband radars. In other words, with wideband radars, it is possible to benefit simultaneously from high Doppler resolution (e.g., improved slow target detection, target classification) and from the diversity gain.

A very similar reasoning could apply in the angular/spatial domain, as just shown here in the range/frequency domain. For a multistatic system with a few radar sites, some noncoherent integration will nicely complement coherent integration of the signals received by each site. Depending on the exact signature of the target, spatial diversity and frequency diversity could be preferable: frequency diversity when the scatterers are distributed in range, spatial diversity when the scatterers are distributed in angle. A good solution, if possible, consists in combining both, for example with two or three frequencies per site, and two or three transmitting and/or receiving sites. However, it should be emphasized that frequency diversity is very generally an existing feature on most medium-/long-range monostatic radars (because most of them use multibursts operation, for ambiguity/eclipses removal), whereas spatial diversity, requiring multisite implementations, is only applicable for specific situations, such as passive radars as discussed by Cherniakov in [8] and Chernyak in [9].

12.3 Wideband Unambiguous Moving Target Indication

12.3.1 Using Range Migration Information

Having demonstrated the potential of wideband waveforms from a power budget and Doppler resolution point of view, it is now necessary to look in more details to the range–Doppler processing and to the associated range–Doppler ambiguities. An essential limitation for standard narrowband radars using bursts of periodic pulses comes from the well-known pulsed radar range–Doppler ambiguity relation, which states that the ambiguous velocity V_a and the ambiguous range D_a are related by $D_a \times V_a = \lambda \times c/4$. That relation means many ambiguities, either in range or velocity (or both), need consideration. This in turn implies the transmission of successive pulse trains with different repetition frequencies, requiring more time to be spent on target for ambiguity and blind speeds removal (without a corresponding gain in Doppler resolution, since the successive coherent pulse trains are then processed incoherently) [2].

We can find an alternative solution by improving the range resolution (through increasing the instantaneous bandwidth), so that the moving target range variation (range walk or range migration) during the pulse train becomes nonnegligible compared with the range resolution. This is equivalent to stating that the Doppler effect is varying across the whole bandwidth compared with the Doppler resolution and can no longer be considered a mere frequency shift. Such radars may use bursts with a low pulse repetition frequency (no range ambiguities) and wideband pulses, such that the range walk phenomena during the whole burst is significant enough to remove the velocity ambiguity (the range walk being a nonambiguous measurement of the radial velocity). It then becomes possible to detect the target and measure range and velocity with only one long coherent pulse burst.

If M is the number of pulses in the burst, T_r the repetition period, V_a the standard ambiguity velocity [$V_a = \lambda/(2T_r)$], ΔF the instantaneous bandwidth, and δR the range resolution [$\delta R = c/(2\Delta F)$], then the condition for sufficient migration is written:

$$MV_aT_r \gg \delta R \Leftrightarrow \frac{\lambda}{2}M \gg \delta R$$

$$\Leftrightarrow M \gg \frac{F_0}{\Delta F}$$

(12.5)

For example, a burst of $M = 60$ pulses at 1 kHz repetition frequency with 500 MHz bandwidth in X band would be a possible candidate for nonambiguous MTI detection. Alternatively, in S band with 200 MHz bandwidth, a burst of 50 pulses could be used with similar results.

The coherent signal processing of such radars (whose range resolution is in the order of a few wavelengths, typically less than 10) involves, as shown in Figure 12.6, for each velocity hypothesis, a coherent summation of the received echoes (Fourier transform), after range walk compensation:

Xr,t:received signal from r^{th} pulse, at t^{th} time sample
Hypothesis: range t δR, speed v

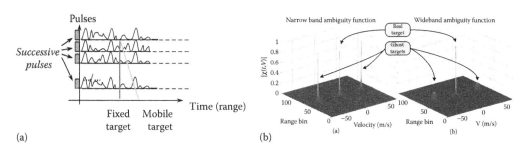

FIGURE 12.6
Wideband signal processing and ambiguity functions. (a) Signal processing for multiple pulses and (b) ambiguity functions. Left: Narrowband, 1/10,000 bandwidth successive signal pulses. Right: Wideband, 1/10 bandwidth. (From Le Chevalier, F., Space-time transmission and coding for airborne radars, *Radar Science and Technology*, 6, 411–421, 2008. With permission.)

$$T_{t\,\delta R,V} = \sum_{r=0}^{M-1} x_{r,\Gamma}\left[t - r\frac{VT_r}{\delta R}\right] e^{-2\pi j\,r\frac{F_0}{F_r}\frac{2V}{c}},$$

With $\Gamma(u)$ = nearest integer from u.

More precisely, the matched filter for wideband MTI simply consists in a coherent summation of the received samples (i.e., a matched filter), for each possible velocity v and delay t $(t = 2R/c)$ of the target. The signal from a migrating target is illustrated in Figure 12.7, first in the range/slow-time domain, then in the subband slow-time domain (after Fourier transform of each column). In this last representation, each line is a sinusoid, with frequency varying from line to line (i.e., from subband to subband), since the Doppler is varying with frequency. Similarly, each column is also a sinusoid, with frequency varying from column to column (i.e., from pulse to pulse), since the range is varying with slow-time.

The literal expression of the matched filtering of the received signal v_{fr} is [2]

$$T(v,t) = \sum_{\substack{f=0,\dots,N-1 \\ r=0,\dots,M-1}} y_{f,r} \exp\left(2\pi j\,\frac{f\,t}{N}\right)\exp\left(-2\pi j\,r\frac{2v}{\lambda_0}T_r\right)\exp\left(-2\pi j\,rf\frac{\delta F}{F_0}\frac{2v}{\lambda_0}T_r\right) \quad (12.6)$$

$$T(v,t) = \sum_{\substack{f=0,\dots,N-1 \\ r=0,\dots,M-1}} y_{f,r} \exp\left(2\pi j\,\frac{f\,t}{N}\right)\exp\left(-2\pi j\,r\frac{2v}{\lambda_0}T_r\left(1+f\frac{\delta F}{F_0}\right)\right) \quad (12.7)$$

In this expression, v_{fr} is the received signal as a function of frequency (subband) f and pulse number r, N is the number of frequencies (subbands), M is the number of pulses in the coherent burst, T_r is the repetition period, δF is the frequency step, F_0 is the central carrier frequency, and $\lambda_0 = c/F_0$ is the corresponding wavelength. The coupling between velocity and range, introduced by the migration effect (or, equivalently, by the fact that the Doppler shift is varying with frequency), is taken into account by the last term in

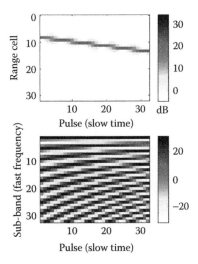

FIGURE 12.7
The signal from a migrating target, in the range/slow-time domain, and in the subband slow-time domain. The real part of the signal is represented, making apparent the 2D-sinusoidal modulation due to the range and velocity of the target.

this expression, which becomes negligible when the migration during the burst, MvT_r, is smaller than the range resolution, $\delta r = c/(2N\delta F)$.

This processing leads to an ambiguity function of Figure 12.6, which does not exhibit the periodic ambiguities in Doppler.

12.3.2 Adaptive Processing

For better rejection of clutter ambiguities, S. Bidon and F. Le Chevalier showed in [10,11] that it is preferable to implement an adaptive processing, such as Iterative Adaptive Approach (IAA), or Bayesian approaches, which optimize the separation between adjacent targets, while still accepting multiple or diffuse targets such as clutter. An illustrative example is shown in Figure 12.8, from [10], showing the range–Doppler image obtained with the Polarimetric Agile Radar in S- and X-band (PARSAX) radar (100 MHz bandwidth at 3 GHz carrier, 500 Hz repetition frequency, 64 pulses): the residuals of clutter ambiguities at ± 25 m·s^{-1} are visible, but the clutter attenuation provided by the migration (migration of only 2 range cells at first velocity ambiguity) and the adaptive processing is larger than 25 dB.

Compared with the standard narrowband radar with 100 m range resolution, these ambiguous clutter lines have thus been reduced by 20 dB (through the improvement in range resolution) and a +25 dB (gain obtained by exploitation of migration): globally, this amounts to 45 dB rejection of ambiguous clutter lines.

Other techniques have been proposed for the detection of such migrating targets [12,13], and this is still an area of research and development. We should expect such wideband surveillance modes to appear in future radar design as possible answers to the requirements outlined at the beginning of this Chapter.

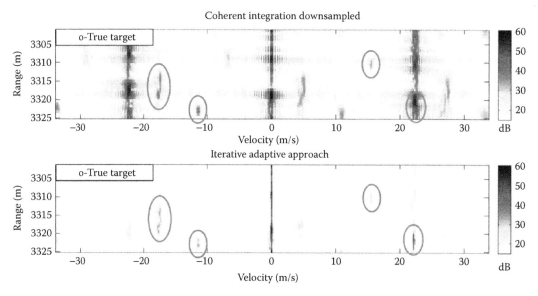

FIGURE 12.8
Experimental results with the Polarimetric Agile Radar in S- and X-band (PARSAX) radar. Upper: Coherent integration (matched filter); lower: Iterative Adaptive Approach. The targets, circled in green, clearly appear after Iterative Adaptive Approach (IAA) processing, even when superposed to ambiguous clutter residuals, and most of the ambiguous sidelobes are appropriately canceled. (From Petrov, N. et al., Wideband spectrum estimators for unambiguous target detection, *Proceedings of 16th International Radar Symposium (IRS)*, 676–681, 2015. With permission.)

For a more realistic application, let us consider a typical example, for surface-air surveillance. Take the case of an S-band radar (3 GHz), with 1 kHz repetition frequency ($D_a = 150$ km, $V_a = 50$ ms^{-1}), with a 200 MHz bandwidth providing a 0.75 m theoretical range resolution, using one burst of 45 pulses. Based on these conditions, it would experience a migration of 3 range cells at 50 m-s^{-1}, and provide the following advantages:

- Nonambiguous detection of moving targets in clutter, with only one burst of high-resolution pulses (possibly obtained with pulse compression or stepped frequency techniques).
- Clutter attenuation (~20 dB, for a metric range resolution), due to the high range and Doppler resolutions, and better than 45 dB at blind speeds.
- Simultaneous detection of fixed and moving targets (SAR + ground moving target indication), with the high range resolution low PRF pulse train appropriate for SAR imaging.
- High-resolution range–Doppler classification on the fly: Figure 12.9 illustrates this possibility with the image of a hovering helicopter, with 50 cm resolution, where the main rotor and the tail rotor are clearly visible at ranges 5 m and 13 m (signals obtained by electromagnetic modeling of a Puma helicopter).
- ECCM properties (spread spectrum signals, requiring specific interception for ELINT or ESM, and specific devices for simulation of the wideband Doppler compression effect).
- Monopulse angular resolution of extended targets (essential for air–ground high target density situations), as shown in Figure 12.10. Since an angular measurement can be made in each range cell, if the SNR allows, then the histogram of angular measurements on an extended target will provide the information that two targets are present and their respective angular locations. Schematically, the angular resolution comes down to the better angular accuracy.

This list of benefits clearly shows the improvements that can be obtained with wideband radar – which naturally can be combined with the space–time waveforms – discussed on the next Section – to bring about longer time on targets with wider bandwidths and thus improve clutter and disturbances rejection.

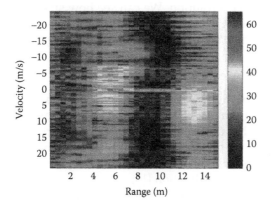

FIGURE 12.9
Helicopter image taken with high-resolution (50 cm) range–Doppler (HRR) signature. The main rotor appears at 5 m and the tail rotor at 13 m. (From Le Chevalier, F., Space-time transmission and coding for airborne radars, *Radar Science and Technology*, 6, 411–421, 2008. With permission.)

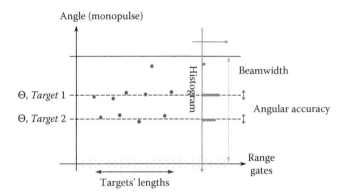

FIGURE 12.10
Wideband angular measurement for 2 extended targets: histograms of the angular measurements made in every range cell provide information about the number of targets inside the main lobe of the antenna.

12.4 Trading Range and Angle Resolutions: Space–Time Coding (Coherent Collocated MIMO)

The following sections show how modern active antenna arrays featuring simultaneous wideband and multiple channels, on transmit and receive, open the way to new beamforming techniques and waveforms. This method uses different signals simultaneously transmitted in different directions, thus jointly coding space and time, and then coherently processed in parallel on receive. Such concepts, first proposed and demonstrated by Drabowitch and Dorey [14,15], should now be considered as mature techniques to be implemented in operational systems. Basically, the main advantages to be gained are a better extraction of targets – especially slow targets – from clutter, multipath, and noise, and a better identification of targets obtained through longer observation times, and possibly wider bandwidths. More specifically, we will analyze the relation between bandwidth, range, and angular resolutions in such space–time coded systems, with special emphasis on the critical – and too often neglected – issue of sidelobes.

This analysis starts with digital beamforming, which for long-range radar is a technique allowing increased Doppler resolution. We can see how space–time coding will improve angular resolution at the cost of widening the bandwidth – or more generally will allow increased degrees of freedom for a more flexible radar design.

12.4.1 Principle of Wide Area Surveillance

Standard digital beamforming helps to obtain wide angular sector instantaneous coverage with a wide beam illumination on transmit by transmitting through one relatively small subarray, or through multiple subarrays with appropriate phase coefficients. In this technique, known as "beam spoiling," the multiple directive beams are simultaneously formed on receive through coherent summations of signals received on different subarrays, in parallel for each aiming direction [16].

If $s_T(t)$ is the signal transmitted over the wide angular sector, from antenna position $\vec{x}(0)$, the signal received by a target in a direction θ_0, at range $c\tau_0/2$, is written:

$$s_R(t, \theta_0) = e^{j\vec{k}(\theta_0)\vec{x}(0)} \cdot s_T\left(t - \frac{\tau_0}{2}\right) \tag{12.8}$$

We assume here that there is no Doppler effect during the duration of one pulse. Any Doppler effect will then be processed, as usual, from pulse to pulse.

The signal received by one antenna element at position $\vec{x}(r)$ is written $s(t, \theta_0)$:

$$s(t, \theta_0) = A_0 \, e^{j\varphi_0} \, e^{j\vec{k}(\theta_0)\,\vec{x}(r)} \, e^{j\vec{k}(\theta_0)\,\vec{x}(0)} \, s_T(t - \tau_0) \qquad (12.9)$$

where $A_0 \, e^{j\varphi_0}$ is the complex reflection coefficient of the target.

Assuming that there are N receiving antennas, at positions $\vec{x}(n), n = 1 \ldots N$, and that these antennas are within the critical angle of the target previously defined (quasi-monostatic situation), the received signals are processed, as usual, through matched filtering for every possible position (τ, θ) of an expected target, thus providing, within an insignificant complex coefficient, the output function $\chi_{\theta_0}(\tau, \theta)$:

$$\chi_{\theta_0}(\tau, \theta) = \sum_{n=1}^{N} e^{j\left(\vec{k}(\theta_0)\,\vec{x}(0) - \vec{k}(\theta)\,\vec{x}(n)\right)} \cdot \int s_T(t)\left(s_T(t+\tau)\right)^* dt \qquad (12.10)$$

This is the result of processing the ambiguity function for one received pulse, giving for one assumed target at position $(\tau_0 = 0, \theta_0)$, the output for every possible position (τ, θ). It clearly appears that, in this standard situation, the angular response and the time response are decoupled. In other words, the output is the product of the angular response (antenna diagram) and the time response (pulse compression). This will not be the case anymore for space–time coded signals, as will be seen in the next paragraph (Figure 12.11).

Real applications generally use weightings to reduce the sidelobes in range and angle. The resulting ambiguity function is then, if $W_s(n)$ is the spatial weighting, along the antenna array, and $W_t(t)$ is the time weighting, along the time replica:

$$\chi_{\theta_0}(\tau, \theta) = \sum_{n=1}^{N} W_s(n) e^{j\left(\vec{k}(\theta_0)\,\vec{x}(0) - \vec{k}(\theta)\,\vec{x}(n)\right)} \cdot \int W_t(t) s_T(t)\left(s_T(t+\tau)\right)^* dt \qquad (12.11)$$

A typical example of such an ambiguity function is shown in Figure 12.11.

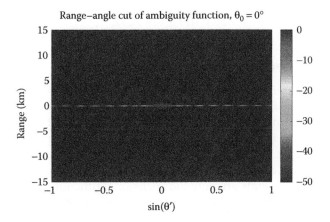

FIGURE 12.11
Transmit–receive ambiguity function of a linear chirp for a linear array of 15 equispaced antennas on receive. The transmitted signal is a chirp with a signal duration $T = 100$ µs and time–bandwidth product $TB = 255$. Processing used a time weighting Hamming (along the chirp replica) and a spatial Taylor 30 dB weighting (along the receiving antenna array).

Digital beamforming generally does not essentially change the power budget, compared with standard focused exploration, since the lower gain on transmit (due to wider illumination) is traded against a longer integration time (made possible by the simultaneous observation of different directions). In fact, the main benefit provided by digital beamforming is an *improved velocity resolution* obtained through this longer integration time, especially useful for target identification purposes, or for detection of slow targets in clutter – key design drivers identified at the beginning of this Chapter.

12.4.2 Space–Time Coding

However, the improved velocity resolution of this wide beam exploration comes at a cost: that is, the nondirective beam on transmit, which induces a poorer rejection of echoes coming from adjacent directions:

- For ground or surface applications, this means that detection of small targets in the presence of strong clutter will become more difficult: the clutter echoes from different directions, which were canceled not only through the Doppler rejection, but also through the angular separation on transmit and on receive, are now less easily rejected.

- For airborne applications, a severe limitation arises from the clutter spreading in Doppler, due to the wider beam on transmit (the clutter spreading is due to the movement of the platform, relative to clutter reflectors): this leads to a poor minimum detectable velocity, and to a poor clutter rejection, since only half the dBs are obtained, compared with focused pencil beam illumination.

- The use of a wide beam on transmit implies that the angular resolution – and accuracy – is only obtained on receive; the angular resolution is thus poorer – approximately by a factor $\sqrt{2}$ – compared with the standard pencil beam solution.

In order to recover this angular separation on transmit (which was basic to standard focused beam techniques), it is necessary to code the transmitted signals (space–time coding), such that the signals transmitted in the different directions be different – and then become separable on receive as shown in Figure 12.12.

12.4.2.1 Principles of Space–Time Coding

The principle of multiple simultaneous transmissions consists of simultaneously transmitting different waveforms in the different directions, thus achieving space–time coding. This technique is also known as colored transmission since the spatial distribution is then colored, as opposed to the white distribution corresponding to the wide beam. In the literature, this technique is also known as collocated coherent multiple-input multiple-output (MIMO). A more general and explicit denomination is space–time coding, meaning that the transmitted signals are now bidimensional functions of time and space.

In Figure 12.13, the coding is supposed to be a succession of M subpulses, coded in phase or frequency, but any type of code can be used [18], such as transmitting different frequencies through the different subarrays (frequency diverse array, [19]). The directivity on transmit, simultaneously in the different directions, is then recovered by signal processing on receive. For signal processing on receive, the transmitted waveforms should – ideally – be "orthogonal" so that they can be separated on each receiving channel.

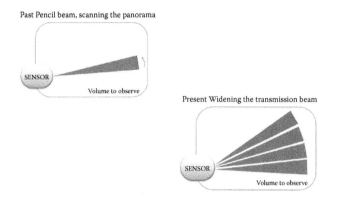

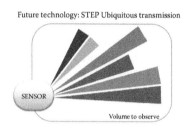

FIGURE 12.12
A history of the radar exploration of space. The past depended on brute force scanning with a pencil beam. Current methods include widening the transmit beam and using multiple receive beams. Future methods will send different signals in different directions, widen bandwidths, and use multiple transmitters and receivers.

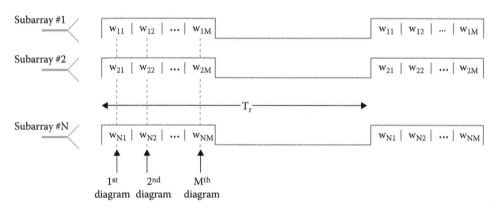

FIGURE 12.13
Colored transmission (space–time coding) uses a series of pulses coded in frequency or phase. Reception signal processing recovers the signal directivity.

The transmitted waveforms are still periodic in time, since that is a necessary condition for an efficient Doppler cancelation of long-range clutter, as demonstrated in [20]. For example, in the mountains, the use of nonperiodic waveforms is generally a very bad solution, since the far-away returns from large structures, such as mountains or coasts, cannot be eliminated if they are not present in all the successive samples collected in the range gate.

Another way of considering such concepts is to describe them as the transmission, during each subpulse m, through successive diagrams, the m-th diagram $D_m(\theta)$ resulting from the illumination law $w_{1m}, w_{2m}, ..., w_{Nm}$ on the array, as illustrated in Figure 12.14 for three concepts: frequency coding (identical diagrams at different carrier frequencies,

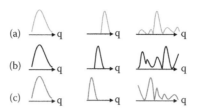

FIGURE 12.14
Successive space–time coding diagrams: (a) Frequency coding, (b) fast angular scanning, and (c) pseudo-random orthogonal diagrams.

also known as frequency diverse array, [19]); fast angular scanning; and pseudo-random orthogonal diagrams.

If the antenna is made of N subarrays arranged on a line Ox, where $D_m(\theta)$ is the successive diagram, and x_n is the position of the subarrays, the analytical expression for a given coding sequence w_{nm} is

$$D_m(\theta) = \left| \frac{1}{N} \sum_{n=0}^{N-1} w_{nm}\, e^{-2i\pi \frac{x_n}{\lambda}\sin\theta} \right| \tag{12.12}$$

The optimum processing then basically consists of the operations described in Figure 12.15, which essentially does a coherent summation of the received samples, for each angle–(Doppler) range hypothesis:

- Transverse filtering separates the signals received from the different transmitters. For instance, if the transmitted signals have different frequencies, this is just a bank of filters separating the different frequencies; the v_n signals correspond to each transmitter n, for all the P receiving channels.

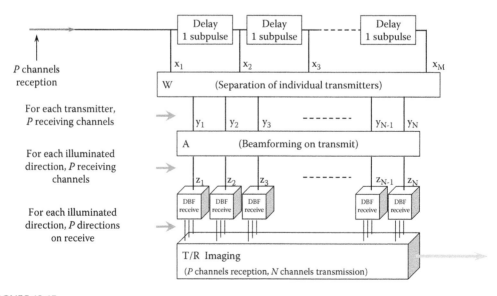

FIGURE 12.15
Block diagram of the optimum reception of colored signals. The process separates the signals from different transmitters.

- Digital beamforming on transmit (basically a Fourier transform), coherently summing the transmitted signals, for each receiving antenna or channel; the z_n signals correspond to each direction n, for all the P receiving channels.
- Digital beamforming on receive for each direction on transmit.

Thus, two beamforming algorithms operate—one on transmit and one on receive—but not all the directions have to be examined because, in general, for each transmitting direction, only one receiving direction should be examined. There is a simplification for fast angular (intrapulse) scanning because this method physically creates the transmitted pencil beams on transmission.

For increased performance in cluttered environments or adverse conditions, digital beamforming will preferably be performed with appropriate adaptive algorithms [2] on transmit and receive.

12.4.2.2 The Transmit and Transmit–Receive Ambiguity Function

Let us first consider a simple case of only one receiving antenna (known as MISO, for multiple input single output, in the literature). The radar shown in Figure 12.16 transmits N coherent signals $s_T^1(t)$, $s_T^2(t)$, ..., $s_T^N(t)$ through a linear array of N identical antennas or subarrays, at positions $\vec{x}(1)$, $\vec{x}(2)$, ..., $\vec{x}(N)$, and only one antenna receives the reflected signal. Because our objective is to analyze the specificities of space–time coding during transmission, we shall examine a MISO system knowing that the gain, angular resolution, and sidelobe level on receive can be decoupled from the transmit part, as will be shown later in Equation 12.17.

The transmitted signal in a given direction θ_0 is the sum of all transmitted signals, with appropriate phase shifts for this direction defined by the wave vector $\vec{k}(\theta_0)$ as defined in Figure 12.16:

$$s_T(t, \theta_0) = \sum_{n=1}^{N} e^{j\vec{k}(\theta_0).\vec{x}(n)} s_T^n(t)$$

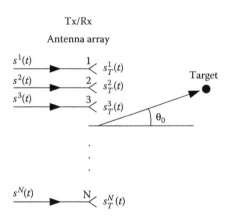

FIGURE 12.16
Space–time coding with a multiple input and single output (MISO) radar. The summation of multiple transmitted signals determines the direction of the beam. A single receiving antenna collects the reflection from the target.

The signal received by a target in this direction θ_0, at range $c\tau_0/2$, is written:

$$s(t,\theta_0) = \sum_{n=1}^{N} e^{j\bar{k}(\theta_0).\bar{x}(n)} s_T^n\left(t - \frac{\tau_0}{2}\right) \tag{12.13}$$

We again assume here *the Doppler effect during the duration of one pulse is negligible* and the Doppler effect will then be processed, as usual, from pulse to pulse. The ambiguity function then depends only on the range (delay) and angle.

The signal received by one antenna element at position $\bar{x}(r)$ is written:

$$s^r(t,\theta_0) = A_0 e^{j\varphi_0}\, e^{j\bar{k}(\theta_0).\bar{x}(r)} \sum_{n=1}^{N} e^{j\bar{k}(\theta_0).\bar{x}(n)}\, s_T^n(t - \tau_0) \tag{12.14}$$

The received signal is processed, as usual, through matched filtering for every possible position θ, τ of an expected target, thus providing, to within an insignificant complex coefficient, the output function χ:

$$\chi_{\theta_0}(\theta,\tau) = \sum_{\substack{n=1 \\ m=1}}^{N} e^{j\left(\bar{k}(\theta_0).\bar{x}(n)-\bar{k}(\theta).\bar{x}(m)\right)} \int s_T^n(t)\left(s_T^m(t+\tau)\right)^* dt \tag{12.15}$$

Thus, the transmit ambiguity function $\left|\chi_{\theta_0}(\theta,\tau)\right|^2$ is a three-dimensional function (five-dimensional for 2D arrays, six-dimensional with Doppler). This gives the delay–angle ambiguity for each aiming direction θ_0.

For a complete analysis, it is now also necessary to take into account the beamforming on receive by the coherent summation on the different receiving channels, as described in Section 12.4.1:

$$\chi_{\theta_0}(\theta,\tau) = \sum_{p=1}^{N} e^{j\left(\bar{k}(\theta_0)-\bar{k}(\theta)\right).\bar{x}(p)} \times \sum_{\substack{n=1 \\ m=1}}^{N} e^{j\left(\bar{k}(\theta_0).\bar{x}(n)-\bar{k}(\theta).\bar{x}(m)\right)} \int s_T^n(t)\left(s_T^m(t+\tau)\right)^* dt \tag{12.16}$$

For simplicity, we will consider in the following a linear array of antennas with $\lambda/2$ spacing between antennas. The "transmit–receive ambiguity function" is then written, with $\sin\theta = u$ and $\sin\theta_0 = u_0$:

$$\chi(u_0,u,\tau) = \left[\sum_p e^{j\pi(u_0-u)p}\right] \times \int \left[\sum_n e^{j\pi u_0 n} s_T^n(t)\right]\left[\sum_m e^{-j\pi u m}\left(s_T^m(t+\tau)\right)^*\right] dt \tag{12.17}$$

For the analysis of ambiguity function, it is useful to consider two bidimensional cuts, expressed as functions of cosines and range variables:

- $\left|\chi(u_0,u,0)\right|^2 = D(u,u_0)$, which is the angular transmit diagram (at the exact range of the target), as a function of the angular aiming position u_0.
- $\left|\chi(u_0,u,c\tau/2)\right|^2$, which is the range–angle ambiguity function, for u_0 aiming direction – indeed, ideally, this range–angle ambiguity function should be analyzed for each possible aiming direction u_0.

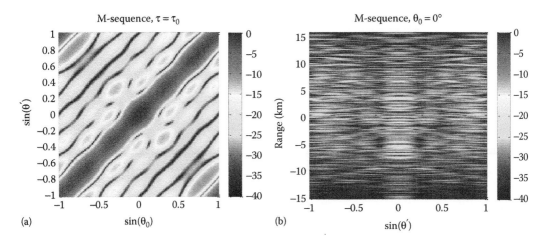

FIGURE 12.17
The two cuts of the transmit ambiguity function of a space–time code made of maximal-length sequences, inside a pulse 100 μs, $BT = 256$, $B = 2.56$ MHz, 8 antenna elements. (a) Generally, the fluctuation along the angle–angle diagonal is unacceptable and (b) the average sidelobes at 25 dB are considered unacceptable.

An example of those two cuts is given in Figure 12.17, for a space–time code made of 8 maximal-length sequences, inside a pulse of 100 μs, with a time–bandwidth $BT = 256$, bandwidth $B = 2.56$ MHz, and 8 antenna elements. From this simple example, two features are of particular interest:

- The fluctuation level along the diagonal of the angle–angle cut. Generally, such fluctuations are not acceptable, since they imply a nonhomogeneous exploration of space. Fluctuations smaller than 1 dB are generally required along this diagonal.
- The average sidelobe level on the range–angle cut. These are around 20–25 dB in this case, which again is generally not acceptable, since the dynamic range between the different targets is much larger. As a general rule, modern radar systems require at least 30 or 40 dB average sidelobe levels.

12.4.2.2.1 Circulating Pulse Analysis

This section will investigate space–time codes leading to acceptable ambiguity functions. We shall start with the circulating pulse case.

The circulating pulse is a simple example, illustrated in Figure 12.18, where a subpulse is successively transmitted through each subarray. If the subarrays are regularly spaced (i.e., uniform linear array) horizontally, this is equivalent to moving the phase center very rapidly through the whole array, thus creating an artificial Doppler (SAR effect) on transmit. For example, if the subpulse is 100 ns long, with 10 subarrays, this produces an artificial Doppler of ±5 MHz – clearly distinct from the standard Doppler effect, which can be measured only as a phase shift from pulse to pulse. Formally, the code w_{kl} is

$$w_{nm} = \delta(n - m)$$

where $M = N$ is the number of moments in the code (number of subpulses), equal to the number of subarrays. This constraint needs not be satisfied for all types of space–time codes.

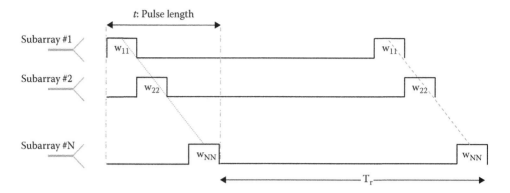

FIGURE 12.18
The circulating pulse transmits a pulse through each antenna of an array to create a wide beam where the angle is coded as frequency.

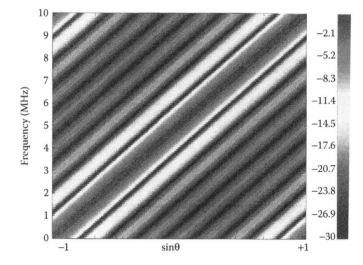

FIGURE 12.19
This example of a circulating pulse shows the angle–frequency coding. The signal is a subpulse, 100 ns long, with 10 subarrays on transmit and only one on receive. Each column (cut along a vertical line) gives the diagram at the corresponding angle (abscissa).

The global effect is equivalent to a frequency coding in azimuth, as is obvious in Figure 12.19. In the figure, each column represents the spectrum of the signal transmitted in each direction, evaluated through a Fourier transform on a duration equal to 1 μs (duration of the global pulse that provides approximately 1 MHz resolution). This coding is similar to Figure 12.14a: the diagram is identical from subpulse to subpulse, but the phase center of the antenna is changed (rather than the frequency, as supposed in Figure 12.14).

This example shows that the standard radar ambiguity is now a range–angle–Doppler ambiguity, since the coding is indeed a space–time coding. The resulting range–angle ambiguity function is shown in Figure 12.20, for the same example: the widening of the peak in angle in adjacent ranges is clearly visible.

The essential limitation of this simple space–time coding is that only one transmitter operates at each instant. It is generally preferable to use all the transmitters simultaneously to maximize the effective radiated power. This condition depends on the precise

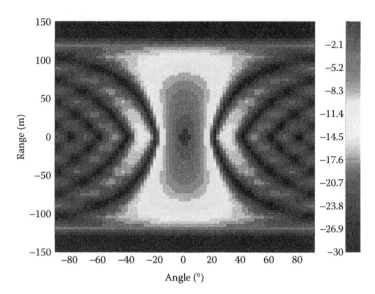

FIGURE 12.20
The circulating pulse range–angle ambiguity for a subpulse 100 ns long with 10 subarrays on transmit and only one on receive.

characteristics of the active elements, such as the maximum tolerable duty factor. The circulating code, presented later in Section 12.4.3, will alleviate this limitation.

Finally, it can be noted that this circulating pulse can also be used for simulating a continuous transmission with pulsed transmitters—a function that might be used for integrating pulsed radar systems in communication networks for specific applications.

12.4.2.2.2 Fast or Intrapulse Scanning

In this mode, the angular diagram is rapidly scanned from subpulse to subpulse, as described in Figure 12.14b [36]. There is a total ambiguity between time (range) and angle, as shown in Figure 12.21, which can be removed, for example, by a symmetrical scanning in the opposite direction, by changing the transmitted frequency from subpulse to subpulse, or more simply by digital beamforming (DBF) on receive or, more generally, by coding the subpulses in phase or frequency (which implies a widened bandwidth). The latter technique is a beam-widening technique on transmit, trading a wider bandwidth against a wider beam.

An important advantage is that, compared with the other techniques, it does not require any digital beamforming on transmit, so the implementation is much less demanding in computer power.

The following section will show the use of Delft codes, which can provide both high resolution in range and angle and adequate sidelobe rejection. These are derived from circulating codes by combination with a fixed spatial code [21], processed with an adequate mismatched filter.

12.4.3 Circulating Codes

A well-known basic example of angular coding uses a dispersive antenna, which effectively sends different frequency beams in the different directions. On reception, the different frequencies are then separated and digital beamforming is performed as usual for each

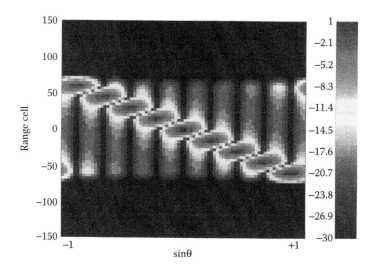

FIGURE 12.21
Intrapulse scanning results in a range–angle ambiguity for the 100 ns-long subpulse, with 10 subarrays on transmit and only one on receive.

transmitted beam/frequency. For example, Air Traffic Control radars use such dispersive antennas for creating stacked beams by frequency coding in elevation.

A space–time coding that produces the same effect in a more flexible way uses a "circulating" signal, which can be any complex waveform (e.g., code or chirp) having good autocorrelation properties and which normally has a large time–bandwidth product. The signal $s(t)$ "circulates" along the array if the n-th channel signal $s^n(t)$ is defined as shown in Figure 12.22.

We can obtain the circulating codes using one single code, shifted in time from one antenna element to the following one:

$$s_T^n(t) = s_0\left(t - \frac{n}{B}\right) \tag{12.18}$$

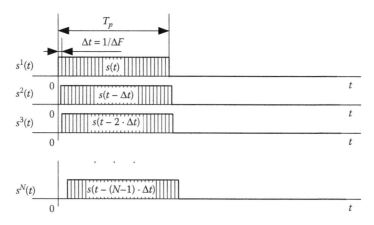

FIGURE 12.22
Space–time circulating signals used for digital beamforming.

The transmit ambiguity function of a circulating code is then written:

$$\chi(u_0, u, \tau) = \sum_{n,m} e^{j\pi u_0 n}\, e^{-j\pi u m}\quad C_0\!\left(\tau + \frac{n-m}{B}\right) \tag{12.19}$$

where $C_0(\tau)$ is the autocorrelation function of the circulating code.

Circulating codes have been analyzed in detail in [22] and shown to provide very clean ambiguity functions, especially when the circulating code is a chirp – at the expense of a loss in range resolution. As an example, consider the case of a circulating chirp, with the following characteristics: an array made of 15 elementary antennas, a pulse duration $T = 100$ µs, bandwidth $B = 2.57$ MHz ($BT = 257$), and a time delay $\Delta t = 1/B = 0.34$ µs between adjacent chirps (Figure 12.23).

These chirps can be considered as an example of multifrequency code – multiple chirps, with carriers shifted from 10 kHz, from antenna to antenna – or as a circulating code – the same chirp, with time origin shifted from 0.4 µs, from antenna to antenna – if the edge effects can be neglected.

In Figures 12.24 and 12.25, the images previously defined (angle–angle cut $\chi(u_0, u, 0)$ of the ambiguity function at $\tau = 0$ and range–angle cut $\chi(0, u, \tau)$ at $\theta_0 = 0$) are shown for the standard wide beam (no space–time coding) and the circulating chirps, respectively.

An overall time Hamming weighting has also been used on receive to improve the chirp filtering. Digital beamforming was implemented on receive, with a Taylor 30 dB weighting along the 13 antennas array. For clearer analysis, two "cuts of these cuts" are also shown: the angular diagram $\chi(0, u, 0)$ for $\theta_0 = 0$ and the range profile $\chi(0, 0, \tau)$.

The main property of this circulating chirp space–time coding is that the ambiguity function is very clean (sidelobe level lower than −50 dB almost everywhere), but the range resolution is degraded by a factor 15, as is obvious by comparing the range profiles on Figure 12.25 (circulating code) and Figure 12.24 (standard wide beam on transmit, no space–time coding). This is due to the special combination of time shifts and frequency spacing where the correlation peaks resulting from the matched filtering of the different chirps are adjacent to each other, thus combining to build an 8 times larger peak. Another interpretation is that only part of the spectrum is radiated in each direction – this is essentially a very dispersive antenna – and so the range resolution is degraded in each direction. As will be seen in the next section, Delft codes have been proposed [21] to alleviate this range resolution degradation.

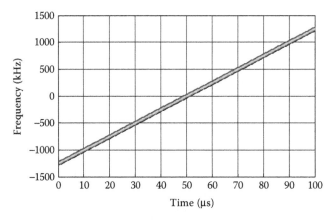

FIGURE 12.23
Circulating chirps coded signal shown in frequency–time representation.

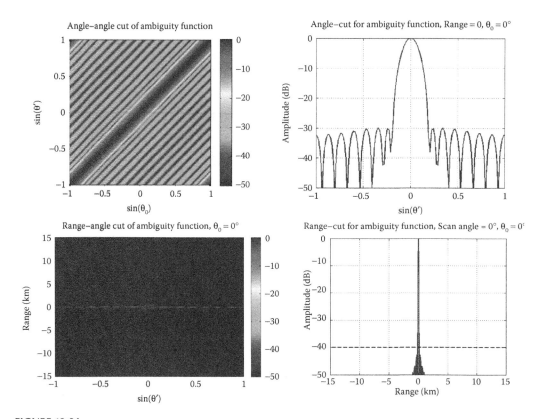

FIGURE 12.24

Transmit–receive ambiguity function for standard wide beam, 15 elementary antennas on receive (weighting Taylor 30 dB), pulse duration $T = 100$ μs, bandwidth $B = 2.57$ MHz, ($BT = 257$); upper left: angle–angle cut; upper right: diagram for $u_0 = \tau = 0$; lower left: range–angle cut; lower right: range profile for $u_0 = u = 0$.

The comparison of this multiple simultaneous transmissions radar with the more standard radar using a wide beam on transmit demonstrates that space–time coding has allowed *trading the angular resolution on transmit against range resolution*. Indeed, the angular diagram is drastically improved (more than 20 dB improvement over most of the domain), but the range profile is degraded by a factor N. In order to recover the full range resolution, one must now increase the bandwidth correspondingly.

In other words, for this space–time coding, increasing the bandwidth on transmit by a factor N is a way to obtain the full angular resolution (and associated rejection of strong echoes) on transmit, while keeping the original range resolution, in comparison with the wide beam standard illumination (no angular resolution on transmit). Since the final angular resolution is the geometric mean of the transmit and receive resolutions (product of the diagrams, for Gaussian beams assumption), the final angular resolution is improved by a factor $\sqrt{2}$.

Coming back to the very basic pencil beam system described on the top of Figure 12.12, we can also compare this basic radar with the wide-beam space–time coded radar just described, with N times bandwidth: it then appears that, globally, we have *improved the Doppler resolution* of the radar and kept all other characteristics (angular resolution and rejection, range resolution and rejection) equal, *by widening the bandwidth* of the same factor. For each pencil beam radar with an active antenna, it is possible to improve the Doppler resolution at the cost of widening the bandwidth if all other things remain the same.

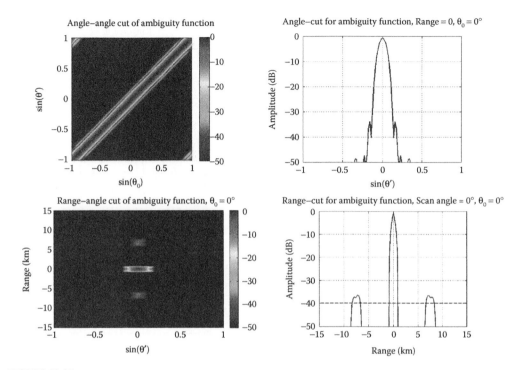

FIGURE 12.25
A circulating chirp using 15 elementary antennas on transmit and receive for a pulse duration $T = 100$ μs, bandwidth $B = 2.57$ MHz, $(BT = 257)$, time delay between adjacent chirps: $\Delta t = 1/B = 0.34$ μs; upper left: angle–angle cut; upper right: diagram for $u_0 = \tau = 0$; lower left: range–angle cut; lower right: range profile for $u_0 = u = 0$.

This very simple relation, though quite unsettling at first for any radar expert, opens ways for future modes dedicated to slow targets detection and analysis.

Moreover, as in the previous analysis of wideband radars, it must be emphasized that, contrary to "high-resolution" processing techniques, these gains in resolution are obtained without any assumption about the number of targets, the levels or spectra of clutter, or their nature (point scatterers vs. distributed echoes, both for targets and clutter). Furthermore, such high-resolution techniques could also be implemented on the space–time coded system, thus improving the performances for more specific applications.

12.4.4 Delft Codes with Mismatched Filtering

12.4.4.1 Delft Codes Ambiguity Function

In order to recover the full-range resolution, it is possible to combine the circulating chirp with a fixed, purely spatial coding. A fixed phase shift c_n is then implemented on each antenna element on transmit:

$$s_T^n(t) = c_n s_0\left(t - \frac{n}{B}\right) \tag{12.20}$$

leading to the following transmit ambiguity function:

$$\chi(u_0, u, \tau) = \sum_{n,m} c_n e^{j\pi u_0 n}\, c_m^* e^{-j\pi u m}\quad C_0\left(\tau + \frac{n-m}{B}\right) \tag{12.21}$$

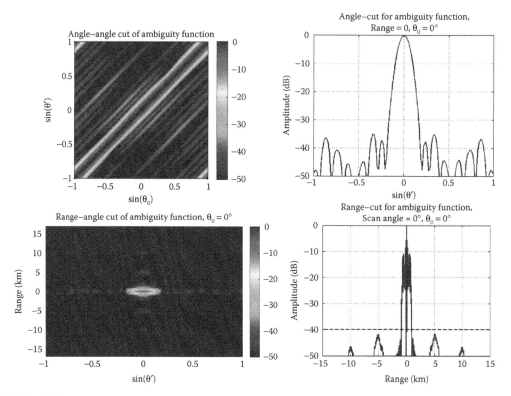

FIGURE 12.26

Delft codes, fixed spatial pseudo-random binary phase code, with time circulating chirp, for 15 elementary antennas on transmit and receive, pulse duration $T = 100$ μs, bandwidth $B = 2.57$ MHz, ($BT = 257$), time delay between adjacent chirps: $\Delta t = 1/B = 0.34$ μs; upper left: angle–angle cut; upper right: diagram for $u_0 = 0$; lower left: range–angle cut; lower right: range profile for $u_0 = u = 0$.

An example of the corresponding transmit ambiguity function is shown in Figure 12.26, for $N = 15$ antennas, and a fixed pseudo-random binary phase shift code along the antenna, with a circulating chirp (assuming again $T = 100$ μs, $BT = 257$). An overall Hamming time weighting has also been used on receive to improve the chirp filtering. As previously shown, digital beamforming was implemented on receive, with a Taylor 30 dB weighting along the 15 antennas array. The cut along the range axis clearly demonstrates the recovering of the range resolution achieved through this combination of circulating time code with a fixed spatial code.

We can formally establish an interesting property of this ambiguity function. Let us assume that the autocorrelation of the chirp is "perfect," that is, a Dirac function: $C_0(\tau) = \delta(\tau)$. Then, the transmit ambiguity function is written:

$$\chi\left(u_0, u, \tau\right) \cong \sum_{n,m} c_n\, c_m^*\, e^{2\pi j\left(n\frac{u_0}{2} - m\frac{u}{2}\right)} \delta\left(\tau + \frac{n-m}{B}\right)$$

$$\chi\left(u_0, u, \tau\right) \cong \sum_n c_n\, c_{n-B\tau}^*\, e^{2\pi j n\left(\frac{u_0 - u}{2}\right)} e^{2\pi j B\tau\frac{u}{2}}$$

(12.22)

where the last phase term can be omitted, since only the modulus of the ambiguity function is of interest.

$$\chi\left(u_0, u, \tau\right) \text{ proportional to } \sum_n c_n\, c_{n-B\tau}^*\, e^{2\pi j n\left(\frac{u_0-u}{2}\right)} \tag{12.23}$$

The transmit ambiguity function is then null for $B\tau > N$: the transmit ambiguity function is approximately zero outside of a corridor centered on the axis $\tau = 0$ (this corridor is the same which was apparent on the ambiguity function of the circulating code in Figure 12.25). Moreover, *inside this corridor, the ambiguity function is simply the standard ambiguity function* (range–Doppler) of the phase code c_n, the Doppler axis being replaced here by the u axis.

Finally, it appears that the range–angle transmit ambiguity function of Delft codes is the convolution (along τ) of the autocorrelation of the circulating code (chirp) with the standard ambiguity function of the fixed spatial code c_n, here expressed as a function of range and angle (rather than range and Doppler as usual).

We will use this property to design a mismatched filter for Delft codes based on the optimal mismatched filters of standard time codes. Such a mismatched filter is necessary, since the sidelobe level obtained with the matched filter will generally not be considered as acceptable.

12.4.4.2 Mismatched Filtering of Delft Codes

For defining a mismatched filter for Delft codes, a new mismatched phase code d_m is used for the replica, with a length M larger than N (typically $3N$), but *keeping the same circulating chirp in the replica*. This allows optimization of the "corridor" part of the range ambiguity function, which depends essentially on the spatial code c_n, but does not affect the clean pattern of the ambiguity function outside of this corridor (which essentially results from the chirp autocorrelation), and it will not lengthen the overall replica, whose length is that of the chirp. The "mismatched ambiguity function" is then:

$$\chi\left(u_0, u, \tau\right) = \sum_{n=1}^{N}\sum_{m=1}^{M} c_n\, e^{j\pi u_0 n}\; d_m^*\, e^{-j\pi u m}\; C_0\!\left(\tau + \frac{n-m}{B}\right) \tag{12.24}$$

The computation of d_m is conveniently done with the original technique described in [10], based on quadratically constrained quadratic programming. This method, based on a reformulation of the optimization problem as a convex quadratically constrained quadratic program (QCQP), insures that any obtained solution is a global solution of the problem. Furthermore, the quadratically constrained quadratic program (QCQP) optimization technique can apply the sidelobe constraint not only at 0 Doppler (corresponding to 0 angle in our case), but over a wider Doppler (or angular) domain – a detailed presentation of this technique is given in [23].

12.4.5 Examples of Equispaced Antennas

Some results are presented below, for a linear array of 15 equispaced antennas, with pseudo-random binary phase spatial codes, chirp circulating code $T = 100\ \mu s$, $BT = 257$. An overall time weighting Hamming has also been used on receive, for improving the chirp filtering. As previously, digital beamforming was implemented on receive, with a Taylor 30 dB weighting along the 15 antennas array.

For Delft codes processed with mismatched filtering in Figure 12.27, a significant improvement is obtained in the range profile, while still keeping the full range and angle

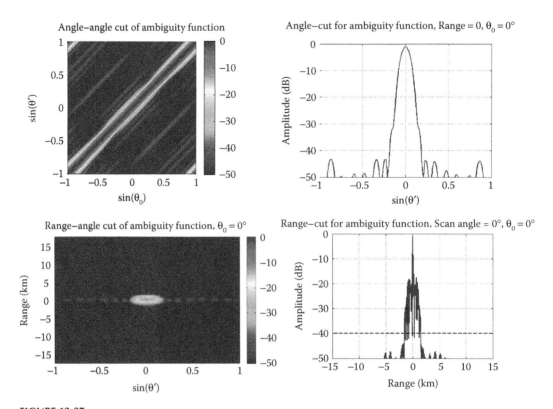

FIGURE 12.27
Delft code, mismatched filtering: angle–angle cut, cut of this figure for $u_0 = 0$, range–angle cut for $u_0 = 0$, and range profile for $u_0 = u = 0$.

resolution, and the very low sidelobes elsewhere (with a slight rise in angular sidelobes, at −45 dB, as seen on the angle–angle cut).

This improvement in the range profile is obtained with a loss essentially due to the time Hamming windowing, lower than 1.5 dB, and with no degradation in range and angle resolutions.

Figure 12.28 presents examples of transmit–receive ambiguity functions and shows that different space–time circulating codes allow different tradeoffs in the range–angle domain, with sidelobes localized around the interfering target, or in angle only, or in range only, and so on.

12.4.6 Space–Time Coding as a Diversity Technique Conclusions

As a whole, as shown in Figure 12.29, these colored transmission techniques can provide a desired angular instantaneous coverage (wide angular sector). The designer must trade this angular coverage against a larger instantaneous frequency bandwidth. Alternatively, starting from the standard wide digitally formed beam, space–time coding will provide angular separation on transmit (hence better clutter rejection), obtained at the cost of a larger instantaneous frequency bandwidth. These interpretations fit exactly with the situation described above with circulating codes. The use of Delft codes allows further improvement, providing high-range resolution while still keeping a very clean ambiguity function over most of the range–Doppler domain.

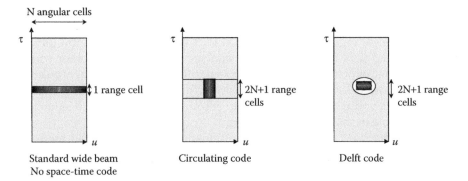

FIGURE 12.28
Example space–time diversity transmit–receive ambiguity functions for a standard wide beam, circulating code, and a Delft code.

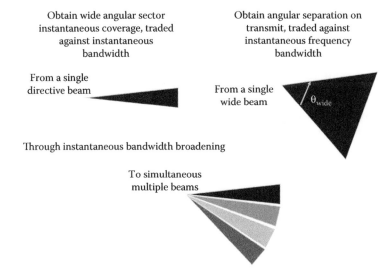

FIGURE 12.29
The colored transmission tradeoffs for providing desired angular coverage.

A provocative interpretation, as we have seen, is that space–time coding can improve the Doppler resolution as previously mentioned; this is the main benefit to be gained from widening the instantaneous coverage. The improvement comes at the expense of an increased bandwidth, which is necessary to get multiple simultaneous transmissions, all other things equal. In this case, improving the Doppler resolution by increasing the bandwidths is just a way to increase the instantaneous angular coverage and, hence, the time on target, and finally, the Doppler resolution.

More generally, this overview of circulating and Delft space–time codes shows that they should be considered as a supplementary tool for diversity. Just as most radars send several successive bursts with different frequency carriers (for taking advantage of target fluctuations, as described in Section 12.2.2), it now becomes possible to also use different space–time codes, so that the interfering targets or clutter do not affect detection similarly on every burst.

The analysis done here on the basis of circulating codes clearly outlines the possibility of such tradeoffs. Obviously, this does not mean that circulating codes are the only solution for space–time coding. A few space–time codes with acceptable ambiguity functions have been published [25]. Future work will analyze and implement other codings with ambiguity functions, which offer specific properties for each operational requirement [26]. Our purpose here was essentially to demonstrate this space–time coding diversity with clear-cut properties and tradeoffs, taking into account realistic levels of sidelobes, which play an essential role in the analysis and design.

Such coding techniques could have applications for air-to-air combat mode antennas where they can provide the necessary instantaneous wide coverage while still maintaining a high visibility in clutter. Their strong resistance to jamming must also be emphasized, since any repeating jammer will give its position by repeating the received code, thus making it easy for the radar system to identify sidelobe jamming as such and cancel the corresponding false plots.

These colored transmission techniques are also the solution to the well-known beams rendezvous problem for bistatic systems. Allowing a wide beam on transmit without incurring the widening of the main beam clutter spectrum (since transmission directivity is recovered on receive), they provide the well-known benefits of bistatic systems (namely, an improved detection in clutter through decreased clutter ambiguities, covertness, and ECCM).

For airborne surveillance radars, they also provide the solution to a classic dilemma of how to increase the Doppler resolution needed for slow-target detection and target classification, without widening the clutter spectrum as with a standard wide beam on transmit.

12.5 The Range–Doppler Surface for Moving Targets Analysis

12.5.1 Range–Doppler Signatures

Radar observations generally consist of a range–Doppler analysis of the scene. In many surveillance situations, the targets are not single scatterers, localized in range and Doppler, but rather straddle a certain number of Doppler cells:

- Airborne targets such as aircrafts, helicopters, and drones have moving parts (compressors, turbines, rotors, propellers), which produce specific Doppler signatures.
- Vehicles and tanks include moving or rotating parts contributing to an extended spectral signature.
- Human targets also have specific Doppler signatures, due to the relative movements of the different parts of the bodies: arms, legs, and so on.

These Doppler signatures of each target have been dubbed "micro-Doppler," and used for classification purposes for decades. For instance, Doppler ground surveillance radars, such as the Thales *Radar d'Acquisition et de Surveillance Terrestre* (RASIT), have long since been fitted with special devices for transferring this Doppler signature in the acoustic domain, so that the operator can listen to the modulations of the signal and recognize the types of targets. Automatic versions of such equipment have also been implemented for aircraft or helicopter classification with surface or airborne radars.

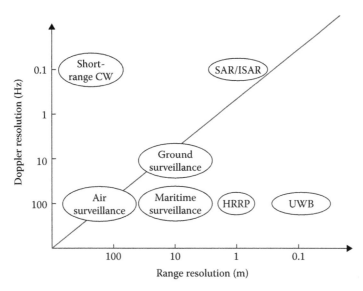

FIGURE 12.30
Range versus Doppler resolutions for typical radar systems.

In many situations nowadays, the target is also extended in range, because the range resolution is higher than the dimension of the target: the target signature then becomes a range–Doppler signature, which should be analyzed as such (and not only as range profile – a.k.a. High-Range Resolution Profile, HRRP – as is often the case). In this section, we will analyze such signatures, and show how they can be described as a range–Doppler surface for further classification purposes.

As illustrated in Figure 12.30, there is a wide variety of such situations, with different range and Doppler resolutions.

Acknowledging the wide variety of operational situations and radar characteristics, we will base the analysis on a kind of extreme situation: human target analysis with UWB radar (essentially for indoor operation): this can indeed be considered as a difficult situation, because the relative radar bandwidth is large – typically larger than 100%, for obtaining a good range resolution (typically 15 cm) at low carrier frequencies (typically between 0.5 and 3 GHz, for preserving the possibility of through-the-wall detection) – and the signature itself is complex and intricate, extending both in range and Doppler. The extension of that approach to other situations will be briefly considered in the end of this section.

12.5.2 Human Signatures

Human target analysis is acknowledged to be useful for a wide range of security and safety applications, such as through-wall detection and ground surveillance. The analysis has usually been conducted in the time–frequency domain (micro-Doppler). Time–frequency transforms, such as the short-time Fourier transform, are used to analyze target Doppler signatures in slow time. Since then, studies have investigated micro-Doppler-based target feature analysis [28]. The high-range resolution profile (HRRP) of human targets has also been studied [29]. Micro-Doppler profile and HRRP, which are both generated by micro-motions, have their shortcomings because they only contain information from either the time–frequency or the time–range domain. Micro-Doppler analysis neglects range information, while HRRP analysis neglects Doppler information. Therefore, in order to analyze the target signature more comprehensively, a new representation called range–Doppler surface (RDS) has been proposed. As an alternative to the micro-Doppler profile and HRRP, RDS is a radar target representation extracted from a three-dimensional data cube – the range–Doppler video sequence [30].

As such, the RDS will be shown to combine all the important information contained in both HRRP and micro-Doppler signatures. In the following, the exact definition will be given, and illustrated by experimental results on human targets with different activities. More detailed analysis can be found in [27], where comparisons between models and measurements are also examined.

12.5.3 The Range–Doppler Surface Concept

The range–Doppler surface can be considered as a byproduct of the range–Doppler video sequence, which is the sequence of range–Doppler images. These range–Doppler images are obtained by standard matched filtering of the received signals, taking into account the migration effect due to the wideband characteristic: the signal is affected by migration of the range of a moving target during a burst of periodic pulses – or, expressed in other words, the Doppler effect is proportional to the carrier frequency. Figure 12.31 shows how the radar return from fixed and mobile targets may vary with each successive pulse.

The coherent signal processing of such radars (whose range resolution is in the order of a few wavelengths, typically less than 10) involves, as shown in Figure 12.31), for each velocity hypothesis, a coherent summation of the received echoes (Fourier transform), after range walk compensation:

$X_{r,t}$: received signal from p^{th} pulse, at t^{th} time sample

Hypothesis: range $t\,\delta R$, speed V

$$T_{t\,\delta R,V} = \sum_{p=0}^{N-1} x_{r,\Gamma\left[t-p\frac{VT_r}{\delta R}\right]}\, e^{-2\pi j p \frac{F_0}{F_r}\frac{2V}{c}},$$

with $\Gamma(u)$ = nearest integer from u

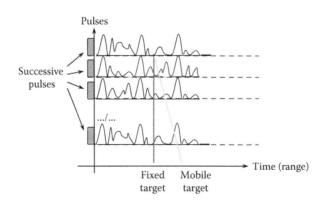

FIGURE 12.31
The radar return from fixed and mobile targets may vary with each successive pulse. Wideband coherent signal processing can eliminate the associated problems and provide successive range–Doppler images.

More precisely, the matched filter for wideband MTI simply consists in a coherent summation of the received samples, for each possible speed v and delay t ($t = 2R/c$) of the target. The literal expression has been shown in Section 12.3.1 to be written:

$$T(v,t) = \sum_{\substack{f=0,\dots,N-1 \\ r=0,\dots,M-1}} y_{f,r} \exp\left(2\pi j \frac{f\,t}{N}\right) \exp\left(-2\pi j\, r \frac{2v}{\lambda_0} T_r\right) \exp\left(-2\pi j\, rf \frac{\delta F}{F_0} \frac{2v}{\lambda_0} T_r\right)$$

In this expression, y_{fr} is the received signal as a function of frequency (subband) f and pulse number r, N is the number of frequencies (subbands), M is the number of pulses in the coherent burst, T_r is the repetition period, δf is the frequency step, F_0 is the central carrier frequency, and $\lambda_0 = c/F_0$ is the corresponding wavelength. The coupling between speed and range, introduced by the migration effect (or, equivalently, by the fact that the Doppler shift is varying with frequency), is taken into account by the last term in this expression. The output $T(v,t)$ is the delay-Doppler image obtained from the M pulses 0 to $M-1$.

Defined as a sequence of range–Doppler images, the range–Doppler video sequence [27] was proposed to describe the slow-time evolution of target range–Doppler signatures. One pulse repetition interval can be used as a typical interval between two frames. Five out of all the frames are shown in Figure 12.32, and the human range–Doppler shapes between consecutive frames are also visible.

For implementing this range–Doppler processing (matched filter) in UWB radar, the Keystone-transform-based range migration compensation approach has been proposed [31]. The goal of the Keystone-transform-based approach is to keep the full Doppler resolution of UWB radar while eliminating the migration phenomenon. Keystone transform was originally proposed for target migration compensation in synthetic aperture radar image processing. It has also been used in wideband radar to align range profiles and increase the coherent integration gain for detecting dim/high-speed targets. We will not elaborate on Keystone transform here – the interested reader may refer to [31] for more details.

In Figure 12.33, the range–velocity images of one simulated human gait are obtained by the Keystone-transform-based approach.

The operational frequency band of the simulated UWB signal is 3.1–5.3 GHz. The pulse repetition frequency is 400 Hz, which leads to a maximum unambiguous velocity of 7.14 m/s. After Keystone-transform-based range migration compensation, the echoes of these fast-moving body segments (e.g., hand or foot) are well focused and clearly visible.

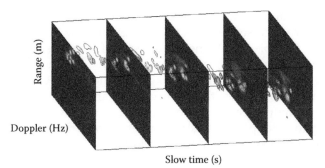

FIGURE 12.32
Construction of a range–Doppler video sequence to show the slow time evolution of the range–Doppler signature.

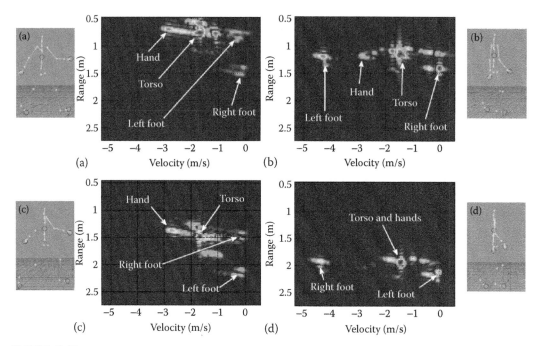

FIGURE 12.33
Range–velocity images of one simulated human gait: (a) double support, (b) right stance, (c) double support, and (d) left stance.

Note that the arm responses are difficult to separate from the thorax response, since they heavily overlap with each other.

For modeling the human target, five dominant markers were selected (i.e., torso, left hand, right hand, left foot, right foot as shown in Figure 12.33), and an equal RCS is assumed for the reflection from all the markers for simplicity. The movements of the scatterers are provided by the motion capture system designed by Carnegie Mellon University. The human range profiles are then obtained by coherently summing the echoes from different parts of the human body.

Before constructing the RDS, it is necessary to detect the target in the range–Doppler domain, since detection allows the extraction of targets and elimination of false alarms. The cell-average constant false alarm rate (CA-CFAR) procedure [32] is a classical approach of detecting a target in noise and clutter. Detection is performed employing a two-dimensional CA-CFAR procedure in the range–Doppler domain. For each range–Doppler image in the range–Doppler video sequence, a sliding 2D window is applied to scan this RD image pixel by pixel. One pixel is claimed as detected if its intensity exceeds an estimated threshold. Figure 12.34 shows a typical 2D window. The cell under test covers the target reflections. The reference cells estimate background noise for computing the detection threshold. The guard cells separate the cell under test and reference cells as a barrier. The sizes of these cells strongly affect the performance of the CFAR detection, and thus should be tuned according to radar parameters and target characteristics (e.g., signal bandwidth, maximum unambiguous Doppler, and target velocity).

In Figure 12.35, the detected scatterers of a simulated human target are shown in a three-dimensional (3D) volume, where the intensities of different scatterers are represented by various colors. Note that the simulated radar system uses the same parameters as used in generating Figure 12.33. Finally, the RDS of Figure 12.36 is constructed by creating a surface

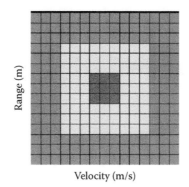

Velocity (m/s)

FIGURE 12.34
2D CA-CFAR window [32] (inner square: cell under test; white ring: guard cells; outer ring: reference cells).

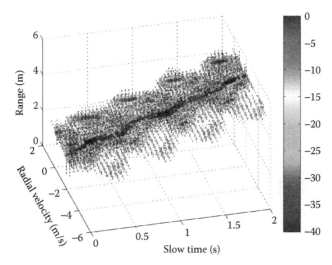

FIGURE 12.35
The range–Doppler surface of a simulated walking human target.

that has the same intensity value within the 3D range–Doppler–time volume (i.e., range–Doppler video sequence) in Figure 12.35. The isosurface plots are similar to contour plots, in that they both indicate where values are equal. The MATLAB® function isosurface is applied to extract the isosurface from the volume using a user-defined isosurface threshold.

The isosurface connects points that have the specified value in much the same way contour lines connect points of equal elevation. Note that the difference in the surface color in Figure 12.36 is not due to different intensities, but due to the lighting effect used to illustrate the 3D object in MATLAB®. Selecting a reasonable threshold is important in this procedure, because this affects the final output significantly. Although currently the threshold is set manually, automatic approaches to construct the volume surface have to be designed.

As previously mentioned, target analysis is frequently performed in the time–range domain (High-Range Resolution Profiles) or in the time–frequency (micro-Doppler) domain. As mentioned above, HRRP neglects Doppler information, while micro-Doppler neglects range information. Furthermore, micro-Doppler is of limited use in multitarget situations, since the Doppler spectra of different targets may overlap. The RDS shows the target surface in the 3D range–Doppler–time space. All the important information of

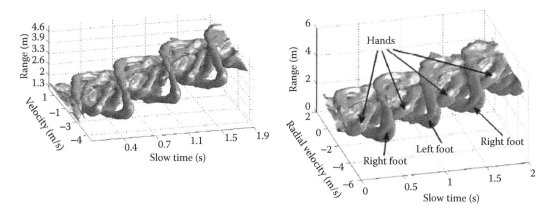

FIGURE 12.36
Range–Doppler surface of a simulated walking human target, showing the contributions of different parts of the body (From Širman, J. D., Computer Simulation of Aerial Target Radar Scattering, Recognition, Detection, & Tracking, 2002. With permission.)

the target, which might be contained in HRRP and micro-Doppler, is included in RDS. Figure 12.37 (a) and (c) show that the body segments are nonseparable in both HRRP and micro-Doppler image because of the severe overlap of the echoes.

However, the RDS in Figure 12.36 and its 2D projections in Figure 12.37 (c) and (d) provide a new way to show the overall view of the target information. The scatterer overlap

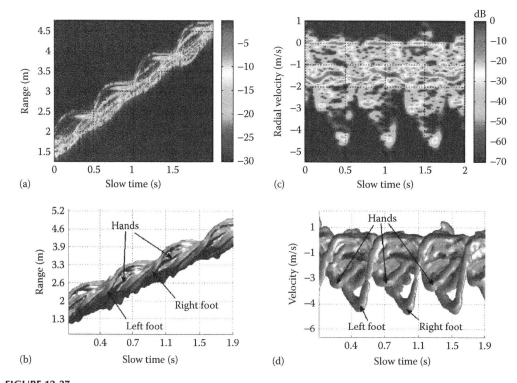

FIGURE 12.37
Different RDS projections: (a) range profiles, (b) micro-Doppler, (c) range–time projection of RDS, and (d) Doppler–time projection of RDS.

that exists in HRRP and micro-Doppler is mitigated in RDS. The feet and hands can be separated directly in RDS without involving additional processing that is normally demanded in HRRP and micro-Doppler images.

The RDS of two common human activities, running and jumping, has also been investigated, and some results are given in [27], demonstrating the possibility to discriminate between these activities. Real experimental results are also described in this same reference, showing also the ability to analyze multiple targets with this same Range Doppler Surface.

12.5.4 Extension to Different Radar Situations

To show other range–Doppler surface constructions, we shall use a simulated moving helicopter (SNR = 25 dB) [34]. The range profiles and time–frequency images of the different parts of the helicopter are presented. The range–Doppler surface of the helicopter is also illustrated in Figure 12.38.

1. Radar parameters:

Waveform	Center Frequency	Bandwidth	PRF	Sampling Rate	CPI
pulse	10 GHz	1 GHz	3000 Hz	2 GHz	60 PRI = 0.02 s

2. Target: helicopter hovering in the air with fixed blades' rotation speed.

Helicopter	Main Rotor	Tail Rotor	Length	Width
AH-64 (Apache)	4 blades, each 7.46-m long, 6 cycle/s rotation rate	4 blades, each 1.5-m long, 25 cycle/s rotation rate	15.3 m	5.4 m

This new concept, called range–Doppler surface (RDS), was proposed to analyze human targets using ultrawideband radar. The RDS is extracted from range–Doppler video sequence. The effectiveness of using the RDS was demonstrated by simulated and experimental data. As an alternative to micro-Doppler and high-resolution range profiles, RDS preserves all the significant range–Doppler–time information.

12.6 Conclusions

This chapter explored different techniques taking advantage of widened instantaneous bandwidths, for removing ambiguities, reducing the angular sidelobes, or analyzing target signatures. The different techniques [35] briefly presented may be combined for simultaneous optimization of space exploration and target analysis and detection. For example, multifrequency transmission through different subarrays, or circulating chirp coding, with a total instantaneous bandwidth between 200 MHz and 500 MHz, or interleaved multifrequency pulse trains, will allow beamforming on transmit and receive and high resolution in range and Doppler, thus providing better detection, location, and classification of multiple surface and low-flying targets, within only one burst of coherent radar pulses.

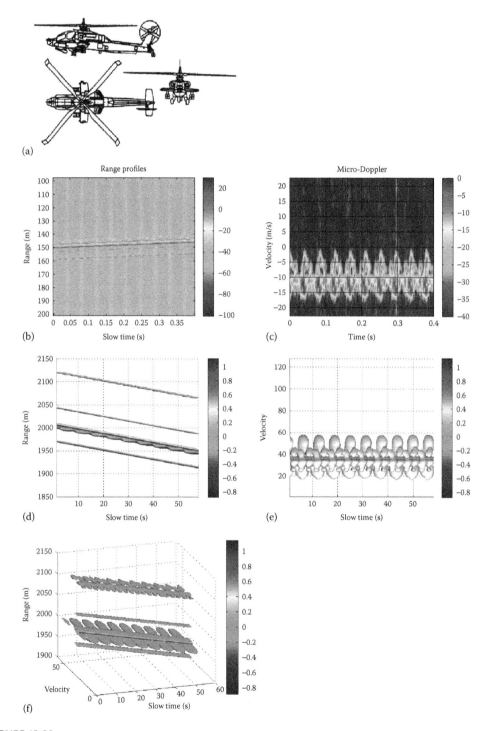

FIGURE 12.38
Different RDS projections of a helicopter signature: (a) Views of the helicopter, (b) range profiles, (c) micro-Doppler, (d) range–time projection of RDS, (e) Doppler–time projection of RDS, and (f) range–Doppler surface.

Basically, the main advantages to be gained are a better extraction of targets (especially slow targets) from clutter, multipath, and noises (e.g., repeater jammers), and a better identification of targets, obtained through longer observation times and wider bandwidths.

More generally, it may be interesting to note that while the second half of the 20th century has seen major developments in radar waveform design and time–Doppler signal processing, this early decade of the 21st century is now focused on antenna developments around phased array design and space–time processing on transmit and receive.

Cost reduction of active electronic scanning arrays and wideband-integrated front ends will enable generalization of these technologies for the more demanding applications, and intelligent radar management will of course be required to take full advantage of the bandwidth and agility available on surveillance radar systems.

References

1. Struzak, R., Tjelta, T. and Borrego, J., 2015. On Radio-Frequency Spectrum Management, *The Radio Science Bulletin* No. 354 (September 2015), pp. 10–34.
2. Le Chevalier, F., 2002. *Principles or Radar and Sonar Signal Processing*. Artech House, Norwood, MA.
3. Pouit, C., 1978. Imagerie Radar à Grande Bande Passante. *Proceedings of the International Colloquium on Radar*, Paris, France.
4. Chernyak, V., 2007. About the 'New' Concept of Statistical MIMO Radar, *Proceedings of the Third International Waveform Diversity & Design Conference*. Pisa, Italy, June.
5. Dai, X.-Z., Xu, J. and Peng, Y.-N., 2006. High Resolution Frequency MIMO Radar, *Proceedings of the CIE Radar Conference*. Shanghai.
6. Wu, Y., Tang, J. and Peng, Y.-N., 2006. Analysis on Rank of Channel Matrix for Mono-static MIMO Radar System, *Proceedings of the CIE Radar Conference*. Shanghai.
7. Richards, M. A., Scheer, J. A. and Holm, W. A., 2010. *Principles of modern radar: basic principles*. SciTech Publishing, The IET, Johnson City.
8. Cherniakov, M. (ed.) 2007. *Bistatic Radar: Principles and Practice*. Wiley, Chichester, UK.
9. Chernyak, V., 1998. Fundamentals of Multisite Radar Systems. CRC Press.
10 Petrov, N. and Le Chevalier, F. 2015. Wideband spectrum estimators for unambiguous target detection. *Proceedings of 16th International Radar Symposium (IRS)*. Dresden, pp. 676–681.
11. Deudon, F., Bidon, S., Besson O., Tourneret, J.-Y. and Le Chevalier, F., 2010. Modified Capon and APES for Spectral Estimation of Range Migrating Targets in Wideband Radar. *Proceedings of IEEE National Radar Conference*, June.
12. Bidon, S., Tourneret, J.-Y., Savy, L. and Le Chevalier, F., 2014. Bayesian sparse estimation of migrating targets for wideband radar. *IEEE Trans. Aerosp. Electron. Syst.*, vol. 50, no. 2, pp. 871–886, April.
13. Deudon, F., Le Chevalier, F., Bidon, S., Besson O. and Savy, L., 2010. A Migrating Target Indicator for Wideband Radar, *Proceedings of the Sensor Array and Multichannel signal Processing Workshop*. Israel, October.
14. Drabowitch, S. and Aubry, C., 1969. Pattern Compression by Space-Time Binary Coding of an Array Antenna, *Proceedings of the AGARD CP 66*. Advanced Radar Systems.
15. Dorey, J., Blanchard, Y., Christophe, F. and Garnier, G., 1978. Le Projet RIAS, Une Approche Nouvelle du Radar de Surveillance Aérienne, *L'Onde Electrique*, vol. 64, no. 4.
16. Kinsey, R., 1997. Phased Array Beam Spoiling Technique, *IEEE Antennas and Propagation Society International Symposium*. Vol. 2, pp. 698–701.
17. San Antonio, G. and Fuhrmann, D.R., 2007. MIMO Radar Ambiguity Functions, *IEEE Journal of Selected Topics in Signal Processing*, Vol. 1, no. 1, pp. 167–177.

18. Levanon, N. and Mozeson, E., 2004. *Radar Signals.* John Wiley & Sons, Interscience Division, New York, NY.

19. Antonik, P., Wicks, M. C., Griffiths, H. D., et al. 2006. Frequency diverse array radars, *Proc. IEEE Radar Conf. Dig.*, Verona, NY, USA, April. pp. 215–217.

20. Le Chevalier F., 2013. Space-time coding for active antenna systems, Ch.11. *"Principles of Modern Radar, Vol II: Advanced Techniques"*, J. Sheer and W. Melvin, (eds). Scitech Publishing, The IET.

21. Babur, G., Aubry, P. and Le Chevalier, F., 2013. Research Disclosure: Delft Codes: Space-Time Circulating Codes Combined With Pure Spatial Coding For High Purity Active Antenna Radar Systems; *Research Disclosure N° 589037*, May.

22. Babur, G., Aubry, P. and Le Chevalier, F., 2013. Space-time radar waveforms: circulating codes, *Journal of Electrical and Computer Engineering*, Vol. 2013, Article ID 809691, 8 pages; Hindawi Publishing Corp.

23. Faucon, T., Pinaud, G. and Le Chevalier, F., 2015. Mismatched filtering for space-time circulating codes, *Proceedings of IET International Radar Conference*, Hangzhou, PR China, October.

24. Le Chevalier, F., 2015. *Wideband wide beam motion sensing*, Tutorial at the IET International Radar Conference, Hangzhou, PR China, October.

25. Rabaste, O., Savy, L., Cattenoz, M. and Guyvarch, J.-P., 2013. Signal Waveforms and Range/Angle Coupling in Coherent Colocated MIMO Radar, *Proc. IEEE International Radar Conference*, Adelaide, Australia.

26. Tan, U., Rabaste, O., Adnet, C., Arlery, F. and Ovarlez, J.-P., 2016. Comparison of Optimization Methods for Solving the Low Autocorrelation Sidelobes Problem, submitted to the *IEEE Radarconf 2016*. Philadelphia, USA, May.

27. He, Y., Molchanov, P., Sakamoto, T. Aubry, P., Le Chevalier, F. and Yarovoy, A., 2015. Range-Doppler surface: a tool to analyse human target in ultra-wideband radar, *IET Radar, Sonar & Navigation*, Vol. 9, no. 9, December, pp. 1240–1250.

28. Chen, V. C., 2011. *The micro-Doppler effect in radar.* Artech House, Norwood, MA.

29. Fogle, O. R., 2011. Human micro-range/micro-Doppler signature extraction, association, and statistical characterization for high resolution radar, PhD thesis. Wright State University.

30. He, Y., Aubry, P., Le Chevalier, F. and Yarovoy, A. G., 2014. Self-similarity matrix based slow-time feature extraction for human target in high-resolution radar, *Int. Journal of Microwave and Wireless Technologies*, Vol. 6, no. 3–4, pp. 423–434.

31. He, Y., Aubry, P., Le Chevalier, F. and Yarovoy, A. G., 2014. Keystone transform based range-Doppler processing for human target in UWB radar, *Proc. IEEE Radar Conf.*, Cincinnati, OH, USA, pp. 1347–1352.

32. Oppenheim, A. V., Schafer, R. W. , Buck, J. R., et al., 1999. *Discrete-time signal processing*, Prentice Hall, Upper Saddle River, NJ.

33. He, Y., 2014. 'Human Target Tracking in Multistatic Ultra-Wideband Radar'. PhD thesis, Delft University of Technology.

34. Širman, J. D., 2002. Computer Simulation of Aerial Target Radar Scattering, Recognition, Detection, & Tracking. Artech House Publishers, Norwood, MA.

35. Le Chevalier, F., 1999. Future Concepts for Electromagnetic Detection, *IEEE Aerospace and Electronic Systems Magazine*, Vol. 14, no. 10, October.

36. Le Chevalier, F., 2008. Space-Time Transmission and Coding for Airborne Radars, *Radar Science and Technology*, Vol. 6, no. 6, pp. 411–421, December.

Index

Note: Page numbers followed by f and t refer to figures and tables, respectively.

Printed and bound by CPI Group (UK) Ltd, Croydon, CR0 4YY

01/11/2024

01782601-0010